Methods in Molecular Biology™

Series Editor
John M. Walker
School of Life Sciences
University of Hertfordshire
Hatfield, Hertfordshire, AL10 9AB, UK

For other titles published in this series, go to
www.springer.com/series/7651

METHODS IN MOLECULAR BIOLOGY™

Toll-Like Receptors

Methods and Protocols

Edited by

Claire E. McCoy and Luke A.J. O'Neill

School of Biochemistry and Immunology, Trinity College Dublin, Dublin 2, Ireland

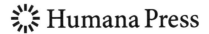

Humana Press

Editors
Claire E. McCoy
School of Biochemistry and Immunology
Trinity College Dublin,
Dublin 2, Ireland

Luke A.J. O'Neill
School of Biochemistry and Immunology
Trinity College Dublin,
Dublin 2, Ireland

ISBN: 978-1-934115-72-5
e-ISBN: 978-1-59745-541-1
ISSN: 1064-3745
e-ISSN: 1940-6029
DOI: 10.1007/978-1-59745-541-1

Library of Congress Control Number: 2008943238

Printed on acid-free paper

springer.com

Preface

The discovery of Toll-like receptors (TLRs) initiated a renaissance for the field of innate immunity, leading to the discovery of multiple receptors and signalling pathways important for host defence. TLRs are expressed highly on cells of the immune system and play a crucial role in initiating an effective immune response that protects the host against invading pathogens. Ten TLRs have been identified to date in humans. They recognised so-called pathogen-associated molecular patterns (PAMPs) found within micro-organisms such as bacteria, viruses, and fungi *(1)*. Association of PAMPs with a specific TLR results in receptor dimerisation and activation of intra-cellular signalling cascades, leading to the expression of cytokines, chemokines, and interferons required to activate effector mechanisms both innate and adaptive, which lead to the elimination of the invading pathogen *(2)*.

Although they play an invaluable role in this fight against infection, it has become apparent that either under- or overactive TLRs can lead to the pathogenesis of disease. Overstimulation of TLR pathways can lead to the excessive production of cytokines and interferons that are associated with chronic inflammatory diseases such as rheumatoid arthritis and inflammatory bowel disease as well as diseases such as asthma and sepsis; while underactive TLRs can lead to a greater susceptibility to viruses and bacteria *(3)*. This may be due to the recent discovery of polymorphisms. For example, polymorphisms within TLR2, TLR4, TLR5, and receptor–adaptor molecules such as Mal have been shown to alter protein behaviour and thereby increase susceptibility of human populations to diseases such as leprosy, atherosclerosis, Legionnaires' and malaria, respectively *(4)*. TLRs can also respond to products from damaged cells and other endogenous ligands resulting in host activation of TLRs and initiation of autoimmune diseases such as systemic lupus erythematosus.

The importance of TLRs in the immune response as well as their implication in a range of diseases has led TLRs to become the focal point of many research laboratories. This book aims to condense the various techniques that have been used to study TLRs, their downstream signalling pathways, and their role in the pathogenesis of disease. There are four parts within this book. Part I focuses on receptors and describes how to determine the expression of each TLR within a cell type or tissue of choice, along with explaining the types of experimental ligands used to activate each TLR. This part also provides information on bioinformatic tools that can be used to identify TLRs and TLR-like proteins. FRET analysis used to study receptor–receptor interaction is also discussed. Part II focuses on how to measure downstream signalling events upon TLR activation, namely protein–protein interactions, phosphorylation and ubiquitination of target proteins, apoptosis, and negative regulation. Readouts such as measurement of cytokine secretion and reporter gene assays are also discussed in depth.

Part III describes how genetic techniques and microarray analysis can be applied to TLR research. The generation of knockout mice using ENU mutagenesis is described as is the use of small interfering RNA used to analyse signalling pathways. Methods to identify polymorphisms within TLRs are also described. Finally, Part IV describes techniques used to measure TLR expression and TLR activation in different disease models such as sepsis, rheumatoid arthritis, inflammatory bowel disease, systemic lupus erythematosus, and atherosclerosis.

TLR protocols will therefore be an extremely useful manual to a wide range of biologists and medical researchers, who are currently studying TLRs or aim to study TLRs in the future.

Claire E. McCoy
Luke A. J. O'Neill

References

1. Akira, S. (2003). Mammalian Toll-like receptors. *Curr Opin Immunol* **15**, 5–11.
2. Medzhitov, R. (2001). Toll-like receptors and innate immunity. *Nat Rev Immunol* **1**, 135–45.
3. O'Neill, L. A. (2003). Therapeutic targeting of Toll-like receptors for inflammatory and infectious diseases. *Curr Opin Pharmacol* **3**, 396–403.
4. Carpenter, S. & O'Neill, L. A. (2007). How important are Toll-like receptors for antimicrobial responses? *Cell Microbiol* **9**, 1891–901.

Contents

Contributors

MARIA T. ABREU • *Division of Gastroenterology, University of Miami Miller School of Medicine, Miami, USA*

MOSHE ARDITI • *Division of Pediatric Infectious Diseases and the Atherosclerosis Research Center, Burns and Allen Research Institute, University of California, Los Angeles, USA*

ANA M. AVALOS • *Department of Microbiology, Boston University School of Medicine, Boston, USA*

JESSICA K. BELL • *Department of Biochemistry and Molecular Biology, Virginia Commonwealth University, Richmond, Virginia, USA*

NADIA BEN LARBI • *Molecular & Cellular Therapeutics, Royal College of Surgeons in Ireland, Dublin, Ireland*

RUDI BEYAERT • *Department of Molecular Biomedical Research, Ghent University, Ghent, Belgium*

HARRY BJÖRKBACKA • *Department of Clinical Sciences, Malmö University Hospital, Lund University, Sweden*

ANDREW G. BOWIE • *School of Biochemistry and Immunology, Trinity College Dublin, Dublin, Ireland*

KIVA BRENNAN • *Molecular & Cellular Therapeutics, Royal College of Surgeons in Ireland, Dublin, Ireland*

FABIA BRENTANO • *Department of Rheumatology, University Hospital Zürich, Zurich, Switzerland*

DAVID J BUTTLE • *School of Medicine and Biomedical Sciences, University of Sheffield, Royal Hallamshire Hospital, Sheffield, United Kingdom*

ISABELLE CARPENTIER • *Department of Molecular Biomedical Research, Ghent University, Ghent, Belgium*

MICHAEL CARTY • *School of Biochemistry and Immunology, Trinity College Dublin, Dublin, Ireland*

ROBERT L. COFFMAN • *Dynavax Technologies, Berkeley, California, USA*

BEATRICE COORNAERT • *Department of Molecular Biomedical Research, Ghent University, Ghent, Belgium*

JÉROME DELLACASAGRANDE • *OPSONA Therapeutics Ltd, Institute of Molecular Medicine, Trinity Centre for Health Sciences, St. James' Hospital, Dublin, Ireland*

STEFAN DREHER • *Institut fuer Medizinische Mikrobiologie, Immunologie und Hygiene, Technische Universitaet Muenchen, Munich, Germany*

PADRAIC G. FALLON • *Institute of Molecular Medicine, Trinity Centre for Health Sciences, St. James's Hospital, Dublin, Ireland.*

RUTH FERSTL • *Institut fuer Medizinische Mikrobiologie, Immunologie und Hygiene, Technische Universitaet Muenchen, Munich, Germany*

SYLVIA FICHTE • *Institut fuer Medizinische Mikrobiologie, Immunologie und Hygiene, Technische Universitaet Muenchen, Munich, Germany*
MASAYUKI FUKATA • *Division of Gastroenterology, University of Miami Miller School of Medicine, Miami, USA*
MONIQUE GANGLOFF • *Department of Biochemistry, University of Cambridge, Cambridge, UK*
NICHOLAS J. GAY • *Department of Biochemistry, University of Cambridge, Cambridge, UK*
STEFFEN GAY • *Department of Rheumatology, University Hospital Zürich, Zurich, Switzerland*
PEARL GRAY • *Division of Pediatric Infectious Diseases and the Atherosclerosis Research Center, Burns and Allen Research Institute, University of California, Los Angeles, USA.*
EDITH M. HESSEL • *Dynavax Technologies, Berkeley, California, USA*
KASPER HOEBE • *The Scripps Research Institute, Department of Genetics, La Jolla, California, USA*
GABOR HORVATH • *University of Massachusetts Medical School, Worcester, Massachusetts, USA*
CAROLINE A. JEFFERIES • *Molecular & Cellular Therapeutics, Royal College of Surgeons in Ireland, Dublin, Ireland*
SINÉAD KEATING • *School of Biochemistry and Immunology, Trinity College Dublin, Dublin, Ireland*
CHIEA-CHUEN KHOR • *Singapore Institute for Clinical Sciences, Section for Genetic Medicine, Brenner Centre for Molecular Medicine, Singapore*
CARSTEN J. KIRSCHNING • *Institut fuer Medizinische Mikrobiologie, Immunologie und Hygiene, Technische Universitaet Muenchen, Munich, Germany*
DIEGO KYBURZ • *Department of Rheumatology, University Hospital Zürich, Zurich, Switzerland*
EICKE LATZ • *University of Massachusetts Medical School, Worcester, Massachusetts, USA*
DELPHINE LAVENS • *Department of Medical Protein and Research Department of Biochemistry, Faculty of Medicine and Health Sciences, Ghent University, Ghent, Belgium*
IRMA LEMMENS • *Department of Medical Protein and Research Department of Biochemistry, Faculty of Medicine and Health Sciences, Ghent University, Ghent, Belgium*
JOSHUA N. LEONARD • *Experimental Immunology branch, National Cancer Institute, Bethesda, Maryland, USA*
NIAMH E. MANGAN • *Institute of Molecular Medicine, Trinity Centre for Health Sciences, St. James's Hospital, Dublin, Ireland.*
ANN MARSHAK-ROTHSTEIN • *Department of Microbiology, Boston University School of Medicine, Boston, USA*
CLAIRE E. MCCOY • *Department of Biochemistry and Immunology, Trinity College Dublin, Ireland.*
ANDREI E. MEDVEDEV • *Department of Microbiology and Immunology, School of Medicine, University of Maryland, Baltimore, USA*

SINEAD MIGGIN • *Institute of Immunology, National University of Ireland Maynooth, Maynooth, Co. Kildare, Ireland*
TOM P. MONIE • *Department of Biochemistry, University of Cambridge, Cambridge, UK*
LUKE A.J. O'NEILL • *Department of Biochemistry and Immunology, Trinity College Dublin, Ireland.*
LISA C. PARKER • *School of Medicine and Biomedical Sciences, University of Sheffield, Royal Hallamshire Hospital, Sheffield, United Kingdom*
WENJI PIAO • *Department of Microbiology and Immunology, School of Medicine, University of Maryland, Baltimore, USA*
LYNNE R. PRINCE • *School of Medicine and Biomedical Sciences, University of Sheffield, Royal Hallamshire Hospital, Sheffield, United Kingdom*
IAN SABROE • *School of Medicine and Biomedical Sciences, University of Sheffield, Royal Hallamshire Hospital, Sheffield, United Kingdom*
DAVID M. SEGAL • *Experimental Immunology branch, National Cancer Institute, Bethesda, Maryland, USA*
JAKUB SIEDNIENKO • *Institute of Immunology, National University of Ireland Maynooth, Maynooth, Co. Kildare, Ireland*
PAUL SIMS • *Dynavax Technologies, Berkeley, California, USA*
PHILIP SMITH • *Institute of Molecular Medicine, Trinity Centre for Health Sciences, St. James's Hospital, Dublin, Ireland.*
ROSALINDA SORRENTINO • *Division of Pediatric Infectious Diseases and the Atherosclerosis Research Center, Burns and Allen Research Institute, University of California, Los Angeles, USA*
STEPHAN SPILLER • *Institut fuer Medizinische Mikrobiologie, Immunologie und Hygiene, Technische Universitaet Muenchen, Munich, Germany*
JULIANNE STACK • *School of Biochemistry and Immunology, Trinity College Dublin, Dublin, Ireland*
XIAO TAO • *Laboratory of Molecular Neurobiology and Biophysics, Rockefeller University, New York, USA*
JAN TAVERNIER • *Department of M edical Protein and Research Department of Biochemistry, Faculty of Medicine and Health Sciences, Ghent University, Ghent, Belgium*
LIANG TONG • *Department of Biological Sciences, Columbia University, New York, USA*
MELISSA B. UCCELLINI • *Department of Microbiology, Boston University School of Medicine, Boston, USA*
PETER ULRICHTS • *Department of Medical Protein and Research Department of Biochemistry, Faculty of Medicine and Health Sciences, Ghent University, Ghent, Belgium*
RYAN UNGARO • *Department of Gastroenterology, Mount Sinai School of Medicine, Madison Avenue, New York, USA*
KELLY VERSHELST • *Department of Molecular Biomedical Research, Ghent University, Ghent, Belgium*

LYNN VERSTREPEN • *Department of Molecular Biomedical Research, Ghent University, Ghent, Belgium*
GREGORY A. VIGLIANTI. • *Department of Microbiology, Boston University School of Medicine, Boston, USA*
SCOTT YOUNG • *Leica Microsystems CMS, Mannheim, Germany*

Part I

Methods to Detect and Analyze Toll-Like Receptors

Chapter 1

Expression Analysis of the Toll-Like Receptors in Human Peripheral Blood Mononuclear Cells

Jakub Siednienko and Sinead M. Miggin

Summary

Toll-like receptors (TLRs) are key regulators of the innate and adaptive immune response to bacterial, viral, and fungal pathogens. To date, 10 human TLRs and 13 mouse TLRs have been identified and they exhibit tissue-specific mRNA/protein expression patterns. Thus, it is essential that the TLR expression profile of model cell lines be delineated prior to experimentation in order to establish whether the requisite TLRs are expressed in the cell line/type of interest. This may be quickly achieved by employing a reverse transcription-polymerase chain reaction (RT-PCR) approach whereby total RNA isolated from the cell type of interest is used as a template for RT-PCR analysis of TLR expression using TLR1–TLR10 specific oligonucleotides. Herein, total RNA was isolated from human peripheral blood mononuclear cells (PBMCs) and its integrity was confirmed by formaldehyde–formamide RNA gel electrophoresis. Thereafter, total RNA was used as a template for RT-PCR analysis using oligonucleotides specific for the amplification of TLR1–10. We have shown that PBMCs express mRNA encoding TLR1–10. These findings suggest that PBMCs may represent a useful TLR-responsive model cell line for examining TLR1–10 signalling events.

Key words: Toll-like receptor, Expression analysis, RNA isolation, RT-PCR, PBMC.

1. Introduction

Toll-like receptors (TLRs) are key regulators of the innate and adaptive immune response and are activated by specific pathogen-associated molecular patterns (PAMPs). Activation of TLRs initiates a signalling cascade through the adapter molecules for induction of expression of various pro-inflammatory cytokines, e.g. IL-1β and IL-18 through activation of nuclear-

Claire E. McCoy and Luke A.J. O'Neill (eds.), *Methods in Molecular Biology, Toll-Like Receptors, vol. 517*
DOI: 10.1007/978-1-59745-541-1_1

factor-κB (NF-κB), ultimately leading to the eradication of microbes *(1, 2)*.

To date, 10 human TLRs and 13 mouse TLRs have been identified. It has been shown that TLRs are activated by specific PAMPs and that the ability of specific TLRs to heterodimerise adds further to the diverse range of pathogens that may be recognised. For example, lipopolysaccharides (LPS) from Gram-negative bacteria are recognised by TLR4. Flagellin is recognised by TLR5. TLR2 in combination with either TLR1 or TLR6 recognises triacyl lipopeptides and diacyl lipopeptides, respectively. Mouse TLR11 recognises uropathogenic bacteria. In contrast to the plasma-membrane-localisation of the aforementioned TLRs, the so-called anti-viral TLRs 3, 7, 8, and 9 are endosomally localised, membrane-bound receptors. TLR3 recognises double-stranded (ds) RNA that is produced from many viruses during replication. TLR7 recognises synthetic imidazoquinoline-like molecules, guanosine analogues such as loxoribine and single-stranded (ss) RNA derived from viruses such as human immunodeficiency virus. TLR8 mediates the recognition of imidazoquinolines and ssRNA. TLR9 recognises bacterial and viral CpG DNA motifs. After recognition of PAMPs, a cascade of intracellular signalling events are activated that culminate in the induction of pro-inflammatory cytokines such as tumor necrosis factor (TNF)α, IL-6, IL-1β, and IL-12. In addition, anti-viral type-I interferons (IFNβ and multiple IFNα) are induced by TLR3, 4, 7, 8, and 9. Moreover, TLRs induce dendritic cell (DC) maturation, essential for the induction of pathogen-specific adaptive immune responses *(3)*.

With regard to the study of TLR signalling, much of our knowledge comes from either over-expression studies using HEK 293 cells or from cells derived from TLR-deficient mice. With the emergence of new and divergent TLR signalling pathways in many novel cell types, it is vital that the expression of the relevant TLRs is established in the chosen model cell line at the onset of any experiment. This is evidenced by the fact that certain cell types are deficient in specific TLRs, for example U373 human astrocytoma cells and HEK 293 cells are deficient in TLR2 and thus stimulation with the TLR2 ligand Pam_3Cys will not activate NF-κB (Miggin, personal observation; *[4]*). A reverse transcription-polymerase chain reaction (RT-PCR) approach using total RNA isolated from the model cell type allows the analysis of TLR expression at the transcriptional level. Firstly, formaldehyde–formamide RNA gel electrophoresis ascertains the integrity of the freshly isolated RNA. Once the RNA integrity has been established, RT-PCR analysis using TLR-specific oligonucleotides allows for the specific amplification of TLR1–TLR10.

2. Materials

2.1. Cellular Material

Blood samples were obtained from healthy donors. Peripheral blood mononuclear cells (PBMCs) were prepared by standard Lymphoprep™ (ready-made solution for the isolation of a pure lymphocyte solution; Nycomed, Oslo, Norway) density gradient centrifugation. Ethical approval was obtained from the Ethics Committee of the National University of Ireland, Maynooth, Ireland.

2.2. Total RNA Preparation

1. Tri® Reagent (Sigma). Store at 4°C (*see* **Note 1**).
2. Chloroform (99% purity, GPR). Store at room temperature (*see* **Note 2**).
3. Isopropanol (99.5% purity, GPR). Store at room temperature (*see* **Note 2**).
4. Nuclease-free water. Store at room temperature.
5. Ethanol (75%) (prepare by mixing 7.5 ml of molecular biology grade absolute ethanol with 2.5 ml nuclease-free water in a sterile tube). Store at –20°C (*see* **Note 1**).

2.3. RNA Formaldehyde–Formamide Gel Electrophoresis

1. Agarose. Store at room temperature.
2. Ethidium bromide (10 mg/ml stock in water). Store in an area dedicated to DNA/RNA gel electrophoresis at room temperature (*see* **Note 3**).
3. DEPC (diethyl pyrocarbonate). Store at 4°C.
4. DEPC-treated water. (Add DEPC to give 0.1% solution in water. Mix vigorously then leave at room temperature overnight. Autoclave.)
5. A 10x MOPS. (0.2 M MOPS, 50 mM sodium acetate, 10 mM EDTA, pH 7.0; add two drops of DEPC to 1 L of 10x MOPS, mix vigorously the leave stand overnight at room temperature. Autoclave. Dilute to 1x for running gel and dissolving agarose matrix using DEPC-treated water.) Store at room temperature.
6. DNA loading buffer (0.025 g xylene cyanol, 0.025 g bromophenol blue, 1 mM EDTA, 10 ml glycerol in a total volume of 20 ml; use exclusively for RNA). Store at room temperature.
7. Formaldehyde (37%, GPR; use exclusively for RNA). Store and use in fume hood.
8. Formamide (deionised GPR; for RNA use only). Store at room temperature.
9. DNA/RNA gel visualisation system (e.g. Eagle Eye® II Imaging System).

2.4. First Strand cDNA Synthesis

1. Random hexamers (Promega (20 μg; 500 μg/ml)). Store at –20°C (*see* **Note 4**).
2. dNTPs (10 mM). Store at –20°C.
3. RNasin® (Promega, 2,500 U; 40 U/μl). Store at –20°C (*see* **Note 4**).
4. MMLV-RT with 5x buffer (Promega, 200 U/μl). Store at –20°C (*see* **Note 4**).
5. Nuclease-free water. Store at room temperature.
6. Thin wall tubes (0.2 ml). Autoclave and dry prior to use.

2.5. PCR

1. dNTPs (10 mM). Dilute to 2.5 mM working stock with nuclease-free water. Store at –20°C.
2. GoFlexi DNA polymerase with 5x Green GoTaq Flexibuffer and 25 mM $MgCl_2$ (Promega). Store at –20°C (*see* **Note 4**).
3. Thin wall tubes (0.2 ml). Autoclave and dry prior to use.
4. Nuclease-free water. Store at room temperature.
5. Oligonucleotide primers (unmodified; **Table 1** ; make 100 μM master stock and 10 μM working stock using nuclease-free water). Store at –20°C.
6. Fifty per cent glycerol (50% w/w glycerol in nuclease-free water; autoclave before use). Store at room temperature.
7. Thermocycler with heated lid.

2.6. Agarose Gel Electrophoresis

1. DNA molecular weight marker (0.5 μg/μl).
2. Agarose.
3. Ethidium bromide (100 mg/ml). Working solution is prepared by dilution of 100 μl stock ethidium bromide with 1 L of water. Store in an area dedicated to RNA/DNA gel electrophoresis (*see* **Note 3**).
4. A 50x TAE (242 g Tris-base, 57.1 ml glacial acetic acid, 100 ml 0.5 M EDTA, pH 8.0 in a total volume of 1 L; dilute to 1x using distilled water and use for running gel and dissolving agarose matrix).
5. DNA Loading buffer (0.025 g xylene cyanol, 0.025 g bromophenol blue, 1.25 ml 10% SDS, 12.5 ml glycerol in a total volume of 20 ml).
6. TE buffer (10 mM Tris-HCl, pH 8.0, 1 mM EDTA pH 8.0).
7. DNA gel visualisation system (e.g. Eagle Eye® II Imaging System).

Table 1
Oligonucleotides used for the amplification of human TLR1-10 and HPRT housekeeping gene

Name	Primer sequence	T_m (°C)	Product size (bp)
TLR1Forward	ACTAATAGGGGTACCAGGC	57	787
TLR1Reverse	GCTACAGTCTTGACTGACAC	57	
TLR2Forward	GCCCATTGCTCTTTCACTGC	59	743
TLR2Reverse	GCAGTTCCAAACATTCCACG	57	
TLR3Forward	ATATGCGCTTTAATCCCT	50	792
TLR3Reverse	AATGTACAGAGTTTTTGGATC	52	
TLR4Forward	CAAAATCCCCGACAACCTCC	59	416
TLR4Reverse	TGCTGGAAAGGTCCAAGTGC	59	
TLR5Forward	TCTCCACAGTCACCAAACCAG	59	784
TLR5Reverse	CAGCATCAGAGAGACCACAG	59	
TLR6Forward	CCTTAGAAGAACTCCAAAGA	53	497
TLR6Reverse	TTAAGATTTCACATCATTG	46	
TLR7Forward	ATTCCATTTTGGAAGAAGAC	51	565
TLR7Reverse	GCCCAGGTAGAGTATTTCTATG	58	
TLR8Forward	GCTGCTGCAAGTTACGGAATG	59	842
TLR8Reverse	TGGCAGTTTGGGTGGCACGTG	63	
TLR9Forward	TACTTCCTCTATTCTCTGAGC	55	416
TLR9Reverse	TTCCACTTGAGGTTGAGATG	55	
TLR10Forward	ATCCTGACTTACCTCAACAACG	58	499
TLR10Reverse	ACCAAGTTACACTCTTCAGTC	56	
HPRTForward	AGTGATGATGAACCAGGTTA	53	558
HPRTReverse	ATTATAGTCAAGGGCATATC	51	

3. Methods

The aim of these methods is to successfully isolate RNA from human cells and determine by RT-PCR analysis whether TLR1–10 are present at the transcriptional level. This approach is employed in preference to TLR detection at the translational level by western blot analysis using TLR-specific antibodies. This is because the endogenous TLR protein may be expressed at levels below the detection limit of the TLR-specific antibody and thus will be extremely difficult to detect by western blot analysis. Here, we have chosen to use human peripheral blood mononuclear cells as they are routinely employed by researchers wishing to study the mechanisms involved in regulating TLR signalling cascades. When isolating RNA from cells, the researcher must be acutely aware of the dangers associated with the introduction of unwanted nucleases into the cellular material, particularly the RNA sample. Thus, to confirm that the RNA integrity has been maintained, formaldehyde–formamide RNA gel electrophoresis is performed. This step is not essential for researchers familiar with the isolation of nuclease-free RNA.

Regarding the first strand cDNA reaction, a negative control reaction is performed in the absence of reverse transcriptase (i.e. –RT) as first strand cDNA should not be generated in the absence of reverse transcriptase; thus the subsequent PCR with –RT template should not yield an amplification product. Should a product be seen in the –RT negative control, it suggests possible contamination of a reagent/RNA, e.g. a vector encoding the same TLR that you are trying to amplify. This is potentially serious as false positives are thus created.

With regard to the PCR step, intron-spanning gene-specific primers are designed. Whereas introns are present in genomic DNA, they are not present in mRNA and so their presence/absence may be used to distinguish between genomic DNA and mRNA, respectively. The RT-PCR amplification product generated from the mRNA will be the correct size as indicated in **Table 1** since the mRNA lacks an intronic region. In contrast, the genomic DNA will yield an incorrect amplification product that is larger than predicted due to the presence of an intronic region. In addition to RT-PCR amplification of the TLRs, a positive control for the RT-PCR is employed, namely the housekeeping gene hypoxanthine phosphoribosyltransferase 1 (HPRT). This will ascertain whether the RT-PCR procedure itself was successful.

3.1. Cellular Material

Human PBMCs are prepared from whole blood by density gradient centrifugation.

3.2. Total RNA Preparation

1. Take a pellet of 1x 10^7 PBMCs (*see* **Note 5**) and place on ice.
2. Add 1 ml of Tri® Reagent to the cell pellet and pipette up and down 5–10 times. Allow to stand at room temperature for 5 min.
3. Add 0.2 ml of chloroform to the sample and shake vigorously for 15 s. Allow the samples to stand at room temperature for 15 min.

4. Centrifuge for 15 min at max speed (14,000 × *g*) in a bench-top centrifuge at 4°C.
5. Transfer the upper clear aqueous phase to a clean sterile 1.5 ml tube (*see* **Note 6**).
6. Add 0.5 ml of isopropanol, invert 4–5 times and allow to stand at room temperature for 10 min. An RNA precipitate may be visible in the solution at this stage.
7. Centrifuge at 12,000 × *g* for 10 min at 4°C. The RNA pellet will form a fine white pellet/precipitate on the side/bottom of the tube.
8. Remove the supernatant and add 1 ml of ice-cold 75% ethanol. Invert/shake 5–10 times (*see* **Note 7**).
9. Centrifuge at 12,000 × *g* for 5 min at 4°C. Again, the RNA pellet will form a fine white pellet/precipitate on the side/bottom of the tube (*see* **Note 7**).
10. Remove the supernatant and allow the pellet to dry at room temperature. This is achieved by covering the open tube with parafilm followed by piercing of the parafilm with a fine needle.
11. Resuspend the RNA pellet in 50–70 μl of nuclease-free water (*see* **Note 8**). The RNA may be stored at –20°C for up to 1 week or at –80°C for a number of weeks.
12. Spectrophotometrically, determine the A_{260}. Typically, take 5 μl of RNA and add 995 μl of nuclease-free water. Mix and read A_{260} using a 1 ml quartz cuvette with nuclease-free water serving as a blank where 1 A_{260} unit = 40 μg/ml in 1 cm cuvette. The $A_{260/280}$ ratio should be <1.8.

3.3. RNA Formaldehyde–Formamide Gel Electrophoresis

1. Set up a dedicated area for the running, staining, and destaining of agarose gels such that any possible ethidium bromide contamination will be contained (*see* **Note 3**).
2. Prepare a 1% agarose gel by melting in a microwave 0.7 g agarose with approximately 70 ml of 1x MOPS ensuring that all the agarose has dissolved, allowing to cool till hand hot followed by the addition of 3.57 ml of formaldehyde to the agarose gel matrix and pouring into the mini-gel casting tray and comb insertion. Allow to cool and set (*see* **Note 9**).
3. Carefully remove the comb and the aluminium sealing gates. Place the casting tray containing the gel into the base unit. Pour 1x MOPS into the reservoir such that the gel is completely submerged (~700 ml of 1x MOPS).
4. Prepare the RNA by mixing 10 μg of total RNA with 1 μl of 10x MOPS, 3.5 μl of formaldehyde, 10 μl of formamide, and 2 μl ethidium bromide (1 mg/ml) to a total volume of 21 μl. Heat to 65°C for 15 min followed by cooling on ice for 5 min.

Add 2 µl of DNA loading buffer, mix, and centrifuge at 14,000 × *g* for 5 s. Load the entire RNA sample in the well.

5. Complete the assembly of the gel unit and connect to a power supply. The gel can be run at 80 V (constant), 250 mA, 20 W for approximately 1 h or until the dye front runs to within 5 cm of the end. As MOPS is a poor buffer, it may be advisable to mix the buffer in the gel box after 30 min using a large syringe or small beaker.
6. Remove the gel and visualise the RNA by using a system such as an Eagle-Eye camera system. Place the gel directly on the surface of the transilluminator, turn on the UV source and close the doors to the box. Do not look at the UV light as it will harm your eyes. Make all the necessary adjustments (brightness/contrast) to optimise your view of the gel on the monitor and then save/print the image. **Figure 1A** illustrates PBMC RNA following formaldehyde–formamide RNA electrophoresis. RNA of suitable quality for RT-PCR is shown in lane 1 and unsuitable degraded RNA is shown is lane 2.
7. Dispose of the RNA gel in a dedicated ethidium bromide waste disposal unit.

3.4. First Strand cDNA Synthesis

1. Allow the random hexamers, RNA, 5x RT buffer and dNTP to thaw on ice. When thawed, mix and centrifuge at 14,000 × *g* for 5 s.
2. On ice, prepare two 0.2 ml thin wall tubes with each containing 1 µl of random hexamers, 2 µg of total RNA, and nuclease-free water to a total volume of 15 µl. Label one tube as +RT and the other as –RT. Mix gently and centrifuge at 14,000 × *g* for 5 s.

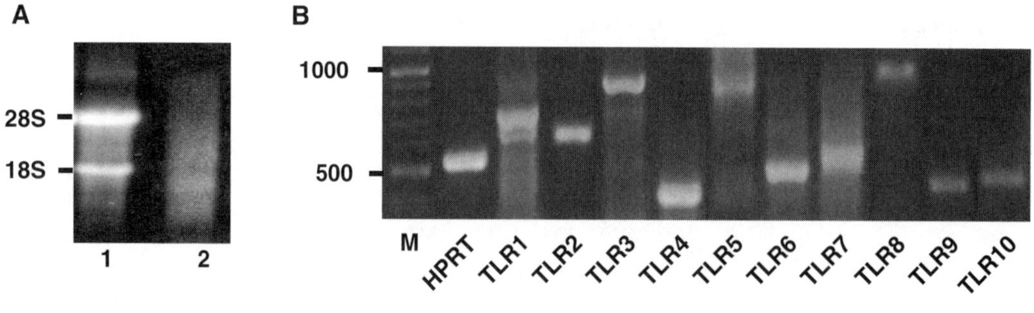

Fig. 1. Analysis of RNA integrity and TLR1–10 expression in peripheral blood mononuclear cells (PBMCs). (**A**) Total RNA was isolated from PBMCs and subject to formaldehyde–formamide RNA gel electrophoresis. Lane 1 illustrates intact RNA that is suitable for reverse transcription-polymerase chain reaction (RT-PCR) as evidenced by the presence of the 18S and 28S ribosomal subunits. Lane 2 illustrates degraded RNA as shown by an absence of 18S and 28S subunits. (**B**) Agarose gel electrophoresis of RT-PCR products (7 µl/lane) derived from PBMC first strand cDNA template. HPRT was amplified using the oligonucleotides HPRTForward versus HPRTReverse; Toll-like receptors (TLRs) were amplified using TLRForward versus TLRReverse oligonucleotide combinations.

3. Heat at 70°C for 5 min followed by cooling on ice for 5 min. Centrifuge at 14,000 × *g* for 5 s.
4. Add the other components in the following order 5 µl of 5x RT buffer, 2 µl of 5 mM dNTP, 1 µl RNasin, and 2 µl MMLV Reverse transcriptase to the +RT tube. Mix the solution gently followed by a 14,000 × *g* for 5 s (*see* **Note 10**). To the negative control –RT tube, add all the components but replace the MMLV reverse transcriptase with 2 µl of nuclease-free water. The –RT control will establish whether any of the first strand cDNA reagents are contaminated with genomic/other DNA.
5. Incubate the tubes at 37°C for 60 min followed by heating to 80°C for 5 min. The first strand cDNA may be stored at –20°C for up to 1 week.

3.5. PCR

1. Set up the following program in the thermocycler as follows (95°C × 30 s, T_m –5°C × 30 s, 72°C × 1 min) × 35 cycles (*see* **Note 11**).
2. Allow the dNTP, $MgCl_2$, 5x Green GoTaq Flexibuffer, first strand cDNA and oligonucleotide primers to thaw on ice. When thawed, mix and centrifuge at 14,000 × *g* for 5 s.
3. On ice, assemble the PCR reactions in 0.2 ml thin walled tubes in the following order by adding 2 µl of 2.5 mM dNTP, 2 µl of 25 mM $MgCl_2$, 5 µl of 5x Green GoTaq Flexibuffer, 7.5 µl of 50% glycerol, 0.25 µl of nuclease-free water, 2.5 µl of 10 µM TLR/HPRT forward primer, and 2.5 µl of 10 µM TLR/HPRT reverse primer. Next add 0.25 µl Flexi Taq (*see* **Notes 4** and **10**) and 3 µl of either +RT or –RT 1° cDNA. To control for possible contamination of the PCR reagents, negative control PCRs for each primer pair must be prepared whereby the first strand cDNA template is replaced with nuclease-free water.
4. Place the tubes on the thermocycler block and initiate the PCR program.
5. The PCR reactions may be stored at 4°C for a few days or at –20°C for weeks–months.

3.6. Agarose Gel Electrophoresis

1. Set up a dedicated area for the running, staining, and destaining of agarose gels such that any possible ethidium bromide contamination will be contained.
2. Prepare a 1% agarose gel by melting in a microwave 1 g agarose with approximately 100 ml of 1x TAE ensuring that all the agarose has dissolved, allowing to cool till hand hot followed by pouring into the mini-gel casting tray and comb insertion. Allow to cool and set (*see* **Note 9**).

3. Carefully remove the comb and the aluminium sealing gates. Place the casting tray containing the gel into the base unit. Pour 1x TAE into the reservoir such that the gel is completely submerged (~700 ml of 1x TAE).
4. Prepare the DNA marker by mixing 2 μl of the DNA molecular weight marker with 2 μl of loading dye and 6 μl 1x TE buffer in a 1.5 ml tube. Load 10 μl of your PCR sample/marker per well (*see* **Note 12**).
5. Complete the assembly of the gel unit and connect to a power supply. The gel can be run at 100 V (constant), 250 mA, 20 W for approximately 1 h or until the dye front runs to within 3 cm of the end.
6. Remove the gel and place in a tray containing the ethidium bromide for 5 min. Thereafter, transfer the gel to a tray containing tap water and allow to destain for 10 min.
7. Visualise the band by using a UV system such as an Eagle-Eye camera system. Place the gel directly on the surface of the transilluminator, turn on the UV source and close the doors to the box. Do not look at the UV light as it will harm your eyes. Make all the necessary adjustments (brightness/contrast) to optimise your view of the gel on the monitor and then save/print the image. The actual size of the PCR product may be compared to the predicted size by cross-referencing against the DNA molecular weight marker as shown in **Fig. 1B**.
8. Dispose of the DNA gel in a dedicated ethidium bromide waste disposal unit.

4. Notes

1. Tri® Reagent is an improved version of the single-step total RNA isolation by acid guanidinium thiocyanate-phenol-chloroform extraction developed by Chomczynski and Sacchi *(5)*. The researcher must note that the total RNA isolation reagent may be obtained from an alternative supplier and will work equally, though slight supplier-specific differences in the RNA isolation procedure may exist, so the instructions supplied with the reagent must be consulted. Care should be taken to avoid the introduction of nucleases into your RNA and first strand cDNA samples. Wear gloves at all times and change your gloves frequently. Do not reuse gloves. Use nuclease-free reagents as indicated. Close reagent containers and tip boxes immediately after use. Use sterile tips that have

been racked while wearing gloves. Unless otherwise stated, all plasticware is sterile and as such, is assumed to be nuclease-free.

2. The chloroform must not contain any preservatives or isoamyl alcohol. Regarding the chloroform and isopropanol (also known as 2-propanol), we purchase a small volume from the supplier and use it exclusively for RNA work. We find that you do not need to purchase molecular biology grade chloroform or isopropanol.
3. Ethidium bromide is mutagenic; therefore appropriate care must be taken when handling. Wear gloves at all times. Dispose of gloves in a dedicated ethidium bromide solid waste container. A dedicated area in the lab must be maintained exclusively for ethidium bromide work. Dedicated trays must be used exclusively for staining and destaining of agarose gels. The working ethidium bromide solution may be stored at room temperature and reused several times or until the solution is devoid of orange colour. Researchers should be aware that Sybr® green is being used as a less mutagenic alterative to ethidium bromide for the staining of DNA/RNA gels.
4. Routinely, we use reagents supplied by Promega Corporation and the reaction conditions described herein have been optimised for these reagents. It must be noted that alternative reagent sources/suppliers may be readily used, though the optimal reaction conditions may vary slightly. Therefore, the researcher must follow the instructions supplied with the reagent.
5. If you are using an adherent mammalian cell line, approximately 4×10^6 cells/ml Tri® Reagent may be used.
6. During this step, care should be taken to avoid pipetting up the red organic phase which contains the protein and the white interphase containing the DNA.
7. The RNA pellet is loosely attached to the 1.5 ml tube; therefore care must be taken not to dislodge the RNA pellet when you are aspirating off the supernatant, especially if the RNA pellet is particularly small.
8. If the RNA pellet does not dissolve in 50–70 μl of nuclease-free water, you may need to add a further 50 μl of water to allow dissolution to occur. Also, the RNA solution may be allowed to dissolve by placing on ice for 15–30 min with occasional mixing. We find that freezing followed by thawing on ice may also help.
9. Some researchers may prefer to incorporate the ethidium bromide directly to the gel just prior to gel casting. This is achieved by melting the agarose as usual, then adding 2 μl of

ethidium bromide solution (stock at 10 mg/ml) into the gel mix followed by pouring. We find that the images obtained from gels stained postelectrophoresis are of better quality for publication.

10. All enzymes (RNasin, reverse transcriptase, and Taq DNA polymerase) are kept at –20°C until required. When needed, the enzymes are taken from –20°C and placed on ice. The required volume is removed and the enzymes are quickly returned to –20°C.
11. The annealing temperature (T_m – 5°C) is oligonucleotide-pair specific and is defined as the melting temperature (T_m) minus 5°C. For example, to amplify TLR2, you would use the oligonucleotides called TLR2Forward and TLR2Reverse (with a T_m of 59°C and 57°C, respectively (as defined by MWG)). The oligonucleotide with the lower T_m is the one used to determine the overall annealing temperature for the PCR. Therefore, to amplify TLR2, an annealing temperature of 52°C (i.e. 57°C – 5°C) would be used.
12. The 5x Green GoTaq Flexibuffer has the advantage that the PCR samples may be loaded directly into the well without preparation. If you use the colourless-Flexi buffer in your PCR (or use DNA polymerase from another supplier), be aware that the samples must be prepared prior to loading on the gel by mixing 8 μl of PCR reaction with 2 μl of loading dye and 2 μl of TE buffer unless otherwise indicated.

Acknowledgement

This work was supported by the Health Research Board.

References

1. O'Neill, L. A., Fitzgerald, K. A., and Bowie, A. G. (2003) The Toll-IL-1 receptor adaptor family grows to five members. *Trends Immunol* 24, 286–90.
2. Miggin, S. M. and O'Neill, L. A. (2006) New insights into the regulation of TLR signaling. *J Leukoc Biol* 80, 220–6.
3. O'Neill, L. A. (2006) How Toll-like receptors signal: what we know and what we don't know. *Curr Opin Immunol* 18, 3–9.
4. Kurt-Jones, E. A., Sandor, F., Ortiz, Y., Bowen, G. N., Counter, S. L., Wang, T. C., and Finberg, R. W. (2004) Use of murine embryonic fibroblasts to define Toll-like receptor activation and specificity. *J Endotoxin Res* 10, 419–24.
5. Chomczynski, P. and Sacchi, N. (1987) Single-step method of RNA isolation by acid guanidinium thiocyanate-phenol-chloroform extraction. *Anal Biochem* 162, 156–9.

Chapter 2

Ligands, Cell-Based Models, and Readouts Required for Toll-Like Receptor Action

Jérôme Dellacasagrande

Summary

This chapter details the tools that are available to study Toll-like receptor (TLR) biology in vitro. This includes ligands, host cells, and readouts. The use of modified TLRs to circumvent some technical problems is also discussed.

Key words: Toll-like receptors, Cell-based assays, Reporter gene assays, TLR ligands.

1. Ligands

Toll-like receptors (TLRs) are activated by microorganisms such as bacteria, viruses or fungi, and by endogenous ligands. Most of these ligands are by nature complex and undefined. From a technical point of view, defined, specific ligands are needed. This section focusses on commercially available ligands purified for the purpose of TLR study and on the most recent TLR ligands described in the literature. Two suppliers that specialize in TLR-related products are Alexis Biochemicals (Lausen, Switzerland distributed by Axxora) and Invivogen (San Diego, CA) and most of the TLR ligands discussed in this chapter are commercially available from them (*see* **Table 1**). In addition, several pharmaceutical companies are focussing on the development of new TLRs ligands (mainly TLR7/8/9). However, their ligands are usually not commercialized.

Claire E. McCoy and Luke A.J. O'Neill (eds.), *Methods in Molecular Biology, Toll-Like Receptors, vol. 517*
DOI: 10.1007/978-1-59745-541-1_2

Table 1
Prototypic TLR ligands and working concentration

	Ligand	Working concentration
TLR1/2	Pam_3CSK_4	1–10 ng/ml
TLR2	Zymosan	10–50 µg/ml
TLR3	PolyIC	10–50 µg/ml
TLR4	Purified *E. coli* LPS	10–100 ng/ml
TLR5	*S. typhimurium* flagellin	1 µg/ml
TLR6/2	Pam_2CSK_4	1–10 ng/ml
TLR7	Gardiquimod	0.1–1 µg/ml
TLR8	CL075 (3M-002)	0.1 µg/ml
TLR9	ODN 2006 (human)	1 µM
	ODN 1826 (mouse)	
TLR11 (mouse only)	*T. gondii* profilin	1–10 ng/ml

The potency of TLR ligands relies in their ability to induce homo- or hetero-dimerization and/or conformational change of receptor chains *(1)*. Even if it seems simple to activate a given TLR with a specific ligand, it is important to keep in mind that direct binding of a TLR ligand to its receptor has only been experimentally demonstrated for TLR9 *(2)* and TLR1/2 *(3)*.

1.1. Tlr1

TLR1 forms functional heterodimers with TLR2. TLR1/2 heterodimers are receptors for triacyl lipopeptides found in bacteria and mycobacteria *(4)*. An artificial model (*see* **Subheading 4**) suggests that signalling through TLR1 homodimers would trigger a weak signal characterized by the activation of the TNF promoter *(5)*. The ligand of choice for TLR1/2 is the synthetic molecule Pam_3CSK_4 and it is active when used at 10 ng/ml.

1.2. Tlr2

It is difficult to demonstrate ligand specificity for TLR2 as it forms heterodimers with TLR1, TLR6, and possibly TLR10 to recognize a variety of microorganisms *(6)*. Zymosan from yeast cell wall, lipoteichoic acid *(7)*, lipoarabinomannan from bacteria and mycobacteria *(8)*, lipoproteins from mycoplasma, or Gram-negative bacteria *(4)* can all activate TLR2 in the absence of TLR1 or TLR6. In vitro, in the absence of TLR1 and TLR6, TLR2 can be activated with high-mobility group box (HMGB) 1 *(9)*, heat-killed *Listeria monocytogenes (10)*, Pam_3CSK_4 *(4)*, or *Staphylococcus aureus* peptidoglycan *(11)*. With regard to peptidoglycan, it has been shown later that this ligand might be recognized by the cytosolic receptor NOD1 *(12)*.

1.3. Tlr3

TLR3 is an intracellular TLR localized in endosomes; it binds dsRNA of viral origin. In addition, two synthetic TLR3 ligands mimicking dsRNA, polyriboinosinic–polyribocytidylic acid (polyIC) and polyadenylic–polyuridylic acid (polyAU), have been described *(13)*. PolyIC fragments ranging from 15 bp to more than 5,000 bp in size can activate TLR3 but the smaller fragments are less potent than the large fragments. More recently, polyinosinic acid (polyI) has been shown to activate TLR3 in mouse B cells, macrophages, and bone marrow derived dendritic cells *(14)*.

TLR3 ligands are active when added to the medium at 10–50 μg/ml. Complexation with a lipid-based transfection reagent results in a lower effective concentration but such a delivery system might also activate cytoplasmic receptors such as MDA-5 independently of TLR3 *(15)*.

1.4. Tlr4

TLR4 was the first TLR identified in mammals. It is one of the most studied TLRs *(16)*. First considered as the receptor for lipolysaccharide (LPS) from Gram-negative bacteria, it has been later shown that TLR4 requires MD2, LPS binding protein (LBP), and CD14 as coreceptors to function. LPS consists of a polysaccharide moiety and the active component lipid A. Lipid A is composed of a glucosamine disaccharide linked to fatty acids *(17)*. The potency to activate TLR4 depends on the number of fatty acids. Lipid A from pathogenic bacteria (e.g., *E. coli*, *Salmonella* species) contain six fatty acids whereas lipid A containing four or five fatty acids are found in less pathogenic bacteria (mutated *E. coli*, *Rhodobacter sphreroides*, Porphyromonas *gingivalis*). The latter LPS are considered as antagonists because they can inhibit the activation of TLR4 induced by hexaacylated LPS *(18)*.

Agonistic LPS are divided in two categories based on the morphology of bacteria colonies: smooth or rough (S-LPS or R-LPS, respectively). S-LPS contains O-polysaccharide chains which are absent from R-LPS. Wild-type Gram-negative bacteria synthesize S-LPS which needs CD14 to signal through TLR4. Signalling by S-LPS through TLR4/CD14 activates both arms of the TLR4 pathway (i.e., MyD88-dependent and TRAM/TRIF-dependent). R-LPS can signal in the absence of CD14 but it does not activate the TRIF/TRAM pathway *(19)*. This property might be useful for in vitro experiments in the absence of TLR4 coreceptors but such activation of TLR4 is incomplete. Interestingly, it has been recently demonstrated that monophospholipid A, a candidate vaccine adjuvant, stimulates only the TRIF/TRAM arm of the TLR4 signalling *(20)*. This ligand is a promising tool to study the MyD88-independent (i.e., interferon regulatory factors (IRFs)-dependent and NF-κB-independent) response induced by TLR4 activation.

Basic purification protocols lead to LPS containing associated lipoproteins and thereby active on TLR2 and TLR4. Phenol re-extraction of LPS is needed to retain only a TLR4 activity *(21)*. Purified LPS from different bacterial strains are commercially available and are active at 10 ng/ml.

Surprisingly, murine (but not human) TLR4 recognizes the antitumoral agent taxol *(22)*. Proteins of viral origin (e.g., respiratory syncytial virus, mouse mammary tumor virus) can also activate TLR4 *(6)*. Recently, heme has been shown to activate TLR4 in a CD14-dependent and MD2-independent manner *(23)*.

TLR4 does also recognize endogenous ligands *(24)*; these include hyaluronate, fibrinogen, and HMGB1. These substances are generally released after tissue injury and might trigger a danger signal through TLR4 *(25)*. Activation of TLR4 by heat shock proteins is controversial as some authors have shown that this could be due to endotoxin contamination *(26, 27)*.

1.5. Tlr5

TLR5 recognizes monomeric flagellin, a constituent protein of bacterial flagella *(28, 29)*. Purified flagellin from *Salmonella typhimurium* or *Bacillus subtilis* is commercially available (Invivogen, Alexis). Invivogen also proposes a recombinant *S. typhimurium* flagellin produced in mammalian cells, which is devoid of TLR2/4 activity of bacterial origin. Flagellins are active at 0.1–1 μg/ml.

1.6. Tlr6

TLR6 associates with TLR2 to recognize diacyl lipopeptides such as MALP2 and FSL1 *(30)*. Pam_2CSK_4 is a synthetic ligand for TLR2/6; it activates TLR6-expressing cells when used at 1–10 ng/ml. The level of activation is increased by the presence of CD14 and CD36 *(31)*. None of the available TLR ligands seem to activate TLR6 in the absence of TLR2. Nevertheless, artificial activation of a TLR6-specific signalling cascade showed that signalling through TLR6 homodimers is theoretically possible *(5)*.

1.7. Tlr7

TLR7 is expressed in endosomes; it is a receptor for ssRNA from viruses, especially U or GU-rich oligoribonucleotides such as RNA40 from the U5 region of HIV-1 RNA *(32–34)*. Imidazoquinolines, which are synthetic compounds with antiviral activities, including R848 (or resiquimod), imiquimod, and gardiquimod (specific for TLR7 when used at less than 1 μg/ml), have also been identified as TLR7 ligands *(35)*. In addition, TLR7 is specifically activated by guanine nucleotides analogues, e.g. 1 mM loxoribin *(36)*. 3M Pharmaceuticals has developed TLR7-specific ligands, e.g. 3M-001 which is active at 3 μM, but these ligands are not commercially available *(37)*.

Notably, while looking for siRNA directed against human TLR9, Hornung et al. have discovered that siRNA can specifically activate TLR7 *(38)*. This activity was attributed to a nine-base motif within a 19 *mer* sequence.

Immune complexes (ICX) formed of RNA-specific antibodies and self-RNA can activate TLR7; this participates in the development of autoimmune diseases *(24)*. As a result, small nuclear ribonucleoproteins purified from ICX can be used as TLR7 ligands *(39)*.

1.8. Tlr8

TLR8 shares high sequence homology with TLR7 and can recognize most of the TLR7 ligands *(34)*. Technically, it is possible to discriminate between TLR7 and TLR8 activity by using different concentrations of a given ligand. 3M-002 from 3M Pharmaceuticals (sold as CL-075 by Invivogen) can specifically activate human TLR8 when used at 0.1 μg/ml (0.4 μM) whereas higher concentrations of this ligand are required to activate TLR7. Initially, TLR8 was considered to be active only in human and not in mouse but a combination of 10 μM 3M-002 or 3M-003 with 1–3 μM polyT oligodeoxyribonucleotide (ODN) can activate mouse TLR8 *(40)*.

1.9. Tlr9

TLR9 is an intracellular TLR involved in the recognition of DNA from bacterial and viral origin but also self-DNA in ICX *(41)*. Non-self-DNA is detected by the presence of unmethylated CpG motifs. Extensive research has been done to characterize immunostimulatory CpG DNA motifs *(42)*. Two main classes have been described. Class A CpG are active on plasmacytoid dendritic cells (pDC). These reagents contain polyG motifs and a palindromic sequence on a mixed phopsphodiester/phosphorothioate backbone and can multimerize to form large structures. Class B CpG contain one or more CpG and no polyG motifs on a phosphorothioate backbone; they activate B cells. Class C CpG share class A and class B characteristics and properties. Despite the endosomal location of TLR9, CpG ODNs are usually added to the culture medium to activate $TLR9^{+}$ cells. The mechanism of activation by CpG has thus been challenging to understand. Tian et al. *(43)* showed in a recent article that class A CpG interact with HMGB1 and the resulting complex is efficiently internalized by a mechanism involving the receptor for advanced glycation end-products (RAGE) *(43)*. This RAGE-dependent mechanism is thought to facilitate the delivery of class A CpG to TLR9.

In vitro, the optimal concentration for immunostimulatory CpG ODN is 1 μM. Higher concentrations are usually less effective. It should be noted that some CpG ODNs show species specificities (human vs. mouse). The specificity of the activation by CpG ODN can be confirmed by the use of control CpG ODN which contain the same sequence as immunostimulatory CpG

ODN but in which CpG dinucleotides have been replaced by GpC dinucleotides. Purified endotoxin-free *E. coli* DNA at 50 μg/ml can also be used as a TLR9 ligand and, recently, natural CpG sequences from Gram-negative bacteria were characterized and isolated for the first time *(44)*.

Beside immunostimulatory ODNs acting through TLR9, inhibitory ODNs have been identified. The latter bind TLR9, but fail to induce the switch to an active conformation of TLR9. Inhibitory ODNs block signalling by competing with immunostimulatory ODNs for the binding site on TLR9 *(45)*. Two types of inhibitory ODNs have been identified: repeated TTAGGG motifs found in telomeres or ODN containing either unmethylated GC or methylated CG *(46)*.

Immune complexes present in serum from SLE patients activate TLR9 by a mechanism involving the cell surface receptor FcγR (CD32) *(47)*. In vitro, patients' serum can activate TLR9 expressing cells.

Finally, hemozoin from *Plasmodium falciparum (48)* has been shown to be a TLR9 ligand.

1.10. Tlr10

TLR10 is expressed on B cells; it is closely related to TLR1 and TLR6 and may interact with TLR2 *(49)*. The ligand for TLR10 is still unknown. Using chimeric CD4TLR10 (*see* **Fig. 1B** and **Subheading 4**), Hasan et al. showed that promoters activated by signalling through TLR10 include CXCL5, IL4, NF-κB and to a lesser extent IL2, TNF, and AP1 *(5)*.

1.11. Tlr11

TLR11 is only functional in mice. It was first described as a receptor for uropathogenic bacteria *(50)* and more recent studies have shown that it is activated by a protein resembling *Toxoplasma gondii* profilin *(51)*.

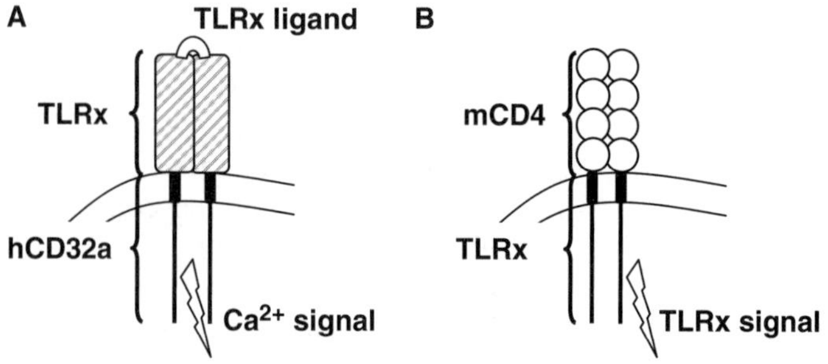

Fig. 1. Engineered Toll-like receptors (TLRs).

2. Host Cells

When it comes to choose a relevant host cell to study TLR biology in vitro, several options are possible. Firstly, cell lines lacking TLR expression can be transfected with one or two TLRs with or without accessory molecules. The second option is to select cell lines naturally expressing TLR(s). Primary cells are another option as they might be more physiologically relevant (i.e., presence of normal expression level of TLRs and presence of cross-talks between signalling pathways) but they are technically more difficult to handle than cell lines.

2.1. Cell Lines Lacking TLR Expression

Cells devoid of endogenous active TLR are a valuable tool to study individual TLRs. The most commonly used cell lines are of epithelial origin; they express all the signalling components needed downstream TLR activation. Stable or transient transfection of one or two TLRs with or without accessory molecules renders them responsive to the corresponding ligand(s). This has been demonstrated in the following cell lines: HEK293 *(52, 53)*, Hela *(54)*, COS7 *(55)*, or CHO *(56)*. It should be noted that HEK293 cells, unlike the variant HEK293T, might have endogenous TLR3 and TLR5 activities *(57)*.

Cells lacking endogenous TLR activity have been used to study TLR signalling, interactions between TLRs *(58)*, requirement for accessory molecules *(31)*, intracellular trafficking *(52)*, or localization. The transfection efficiency in these epithelial cells is very high and the experiments described above can be easily done in transiently transfected cells. Alternatively, stable cell lines expressing TLR, accessory molecules, and even a reporter gene can be generated.

HEK293 cells stably expressing functional human or mouse TLR are commercially available from Invivogen. Two of these cell lines (HEK293 hTLR2/hCD14 and HEK293 hTLR4/hMD2/hCD14) are also available with a stably expressed NF-κB reporter gene which can be used to study TLR activity (*see* **Subheading 3.4**).

The one concern about the use of cells overexpressing a TLR is that overexpression could result in mislocalization, i.e. an intracellular TLR might be found on the cell surface when overexpressed *(59)*. For that reason, cell lines expressing normal levels of TLRs are often the preferred model.

2.2. TLR Expressing Cell Lines

Cell lines expressing TLRs endogenously may also be used to study TLR function. However, they may require transfection of accessory molecules or coreceptors to increase the responsiveness of the cell (see below). It is important to profile cell lines for TLR and accessory molecules expression before using them in an assay. This can be done by RT-PCR and confirmed by flow cytometry or Western blot.

The following cell lines have been used in numerous TLR studies: RAW264.7 (mouse macrophage), THP1 (human acute monocytic leukemia), and U937 (human promonocytic leukemia) belong to the monocyte/macrophage lineage. U373 are epithelial human glioblastoma astrocytoma cells and A549 are human alveolar basal epithelial. Two cell lines having characteristics similar to human pDC have been described: GEN2.2 *(60)* and CAL-1 *(61)*. These cell lines could be a good alternative to the use of primary pDC which are present in very low numbers among PBMC. CD14 expression is needed to render U373 cells responsive to LPS *(62)* or to increase the responsiveness of THP1 cells to several TLR2/4/5 ligands (J. Dellacasagrande, 2006). Differentiation induced by phorbol myristate acetate (PMA) or LPS also increases the responsiveness of THP1 cells to TLR ligands *(63, 64)*. LPS responsiveness of A549 cells through TLR4 is controversial *(65)*.

Table 2 shows the TLR profile of some of these cell lines. In addition, human myeloma cell lines have been shown to express a variety of TLR genes *(66)*.

There are several TLR expressing cells stably transfected with a reporter gene available on the market which can be used to measure TLR activity: THP1-Blue™ derived from THP1 cells (Invivogen, San Diego, CA), Princess® Nina derived from A549 cells (CCS Cell Culture Service, Hamburg, Germany), CellSensor™ (Invitrogen, Carlsbad, CA), and NF-κB-Luc reporter cell lines (Panomics, Fremont, CA) derived from different cell types. All these cells except CellSensor™ cells express a NF-κB-dependent reporter gene.

2.3. Primary Cells

Ideally, cell line studies should be complemented by studies in primary cells. The main problem with the use of primary cells is

Table 2
TLR responsiveness of selected cell lines (no TLR10 ligand has been identified yet; TLR11 is only functional in mouse cells). See text for references

	TLR1/2	TLR2	TLR3	TLR4	TLR5	TLR6/2	TLR7	TLR8	TLR9
RAW264.7	+	+	+	+	±	+	+	+	+
THP1	+	+	−	+	±	+	−	+	+
U373-CD14	−	−	+	+	−	−	−	−	±
GEN2.2	?	?	?	?	?	+	+	−	+
A549	+	+	+	±	+	+	−	−	+

?-not known

Table 3
Functional TLRs in human immune cells

	TLR1/2	TLR2	TLR3	TLR4	TLR5	TLR6/2	TLR7	TLR8	TLR9	TLR10
B cells *(49, 83)*	+	+	–	–	–	+	+	–	+	+
$CD4^+$ T cells *(84)*	+	–	–	–	+	–	+	+	–	?
$CD8^+$ T cells *(85)*	–	–	+	–	–	–	–	–	–	?
Monocytes *(86)*	+	+	–	+	+	+	–	+	–	?
NK cells *(87)*	–	–	+	–	–	–	+	–	+	?
pDC *(49, 88)*	–	–	–	–	–	–	+	–	+	+

?-not known

to choose the right cell type as there are discrepancies in the literature regarding the expression and functionality of TLRs.

It was initially thought that TLRs are restricted to cells of the immune system. However, TLRs have been found in other cell types such as epithelial cells (TLR2, 3, 4, 5, 9 *(67)*), fibroblasts, and even cells from the nervous system *(68)*. Hornung et al. *(69)* and Applequist et al. *(70)* have measured the expression of TLR1–10 mRNA in different types of human or murine immune cells, but the presence of mRNA for a given TLR does not mean that this TLR is active. **Table 3** summarizes the panel of functional TLRs expressed by human immune cells. Expression levels are altered by proinflammatory signals or type I IFN treatment *(63, 71)*. In practice, the most commonly used primary cells are non-sorted PBMC, monocytes, or B cells. Other cell types which are very valuable for the study of TLR biology (e.g., pDC) are more difficult to obtain in sufficient quantity.

Interestingly, mouse cells display a TLR profile similar to the corresponding human cells. One important difference is the LPS responsiveness of murine but not human naive B cells through TLR4 *(72)*. Among primary cells, murine embryonic fibroblasts (MEFs) isolated from mouse embryos are a very good model to study TLRs as they express all known TLRs *(73)*. Ex vivo, MEFs can be grown for over ten passages, their proliferation rate then declines.

3. Readouts

Signalling through the different TLRs involves mainly two families of transcription factors (TF): NF-κB via a MyD88-dependent pathway (all TLRs except TLR3) and the IRFs via a MyD88-independent TRIF-dependent pathway (TLR3) or a MyD88-dependent pathway (TLR7/8/9). NF-κB and IRFs induce the expression of different sets of cytokines, chemokines, and interferons *(74)*.

In order to determine TLR activity in each of the above cell models (*see* **Subheading 2**), the following readouts can be used: measurement of cytokines, chemokines, and/or interferons secretion by ELISA (*see* **Subheading 3.1** and Chapters 19 and 20), detection of activation markers by Western blot (*see* Chapters 10 and 13), RT-PCR or flow cytometry (*see* Chapter 13), phosphorylation of signalling proteins (*see* **Subheading 3.2** and Chapters 10 and 11), and nuclear translocation of transcription factors and/or reporter gene assays (*see* **Subheading 3.4** and Chapters 9, 10, and 14).

3.1. Protein Secretion

In epithelial cells transfected with a TLR (*see* **Subheading 2.1**), stimulation of the recombinant expressed TLR usually triggers the secretion of NF-κB-dependent IL8 which can be measured by ELISA.

In cell lines expressing multiple TLRs (*see* **Subheading 2.2**), the simplest method to assay for the presence of functional TLR is to measure the production of proinflammatory cytokines and/or chemokines after TLR stimulation using specific ligands. In addition, all the readouts described in the following sections are applicable. Alternatively, cells can be transfected with a reporter gene as described in **Subheading 3.4**. Finally, upon TLR stimulation, MEFs (*see* **Subheading 2.3**) produce IL6, MCP-1, and RANTES.

ELISA results can be sufficient to distinguish between two signalling pathways, e.g. TRIF vs MyD88 after TLR4 activation. TRIF activation will result in the production of G-CSF, CXCL10, CCL2, and CCL5 while MyD88 activation induces the secretion of IFNγ, IL1β, IL6, and CCL3 *(20)*. Multiple cytokines and chemokines can be detected in a single small volume sample using multiplex techniques. Cytometric Bead Array (CBA) (BD Biosciences, San Jose, CA) and FlowCytomix (Bender MedSystems, Burlingame, CA) are based on the use of beads coated with an antibody recognizing a specific analyte. For these systems, the acquisition of results requires the use of a flow cytometer. xMAP technology (Luminex, Austin, TX) also uses beads but in 96-well plates format, a compatible microplate reader is needed to process the samples. The multi-array technology by Meso Scale Discovery (MSD, Gaithersburg, MD) uses electroluminescence detection. Wells of dedicated microplates (96- or 384-well) are coated with specific antibodies directed against cytokines, chemokines, or signalling proteins allowing the detection of up to 10 different targets. Secondary antibodies are labelled with the SULFO-TAG™ reagent which emits light upon electrochemical stimulation. Specific readers (SECTOR) are required to process MSD microplates.

3.2. Phosphorylation

The phosphorylation of signalling components, such as the MAPK p38 or the NF-κB subunit p65, occurs usually within minutes after TLR stimulation. There are several techniques to monitor protein phosphorylation, all of them rely on the availability of phospho-specific antibodies. Flow cytometry is done on permeabilized cells whereas Western blots, ELISA-based assays (e.g., TransAM™ from Active Motif (Carlsbad, CA) and CASE™ from SuperArray Bioscience (Frederick, MD)), and AlphaScreen® (Perkin Elmer, Waltham, MA) are done using cell lysates. AlphaScreen® is interesting as it is designed to be used in high throughput screening. This technique is a bead based luminescent assay used to measure interactions between molecules. In the case of protein phosphorylation, two sets of beads containing different dyes are needed: one set of beads is coated with an antibody against the non-phosphorylated form of the target protein and another set of beads is coated with an antibody against the phosphorylated form of the protein. Beads are mixed with cell lysates and, upon excitation, light is emitted if both sets of beads are in close proximity, i.e. when the phosphorylated protein is present in the lysates. Antibody-free FRET-based assays are currently under development *(75)*.

3.3. Translocation of Transcription Factors

Another way to monitor TLR activation is to look at the change in localization of TF activated after ligand binding, e.g. NF-κB, IRFs. This can be assayed using high content screening (HCS) or high content analysis (HCA) assays. The TF are detected using immunochemistry techniques before and after stimulation and the intensity of the signal in the nucleus vs. cytoplasm is measured. Alternatively, cells can be transfected to express a fluorescent fusion protein with the TF (e.g., GFP-TF) and the change in localization can be observed in live cells *(76)*. HCS/HCA techniques require using adherent cells having a low nucleus/cytoplasm ratio.

3.4. Reporter Genes

Reporter gene assays are also used to monitor TLR activity. The most widely used promoter is derived from the ELAM-1 gene promoter modified to contain additional NF-κB motifs *(77)*, an artificial promoter containing 5 NF-κB binding sites is another option. The reporter gene itself can code for a secreted protein (e.g., secreted embryonic alkaline phosphatase (SEAP) *(78)* or secreted luciferase *(79)*) or an intracytoplasmic protein like firefly luciferase. Cells can be transiently or stably transfected with the reporter gene construct and a construct coding for a TLR. TLR activation induces NF-κB-dependent production of the reporter gene. SEAP is secreted in the supernatants of cells and can be detected using colorimetric (e.g., Quantiblue™ from Invivogen) or chemiluminescent (Phosphalight™ from Applied Biosystems, Foster City, CA) techniques. With a secreted reporter protein, it is possible to measure TLR activation over the time simply by sampling the supernatants. It should be noted that fetal calf serum used in culture media contains phosphatase activity which can efficiently be eliminated by heat inactivation (65°C for 10 min). SEAP remains active at this temperature. Secreted luciferase systems are available from Clontech, Palo Alto, CA (Ready-To-Glow™) and from New England Biolabs, Ipswich, MA (*Gaussia* Luciferase). Unlike techniques based on the detection of a secreted protein, "classical" luciferase measurement does not allow to do time course experiments as it requires lysing the cells. Luciferase detection products are available from several suppliers. In reporter gene assays, IRF-dependent and NF-κB-independent promoters are needed to cover the full range of signalling pathways activated by the TLRs. IRF-specific reporter gene assays can be designed based on specific binding sites described for certain cytokines, chemokines, or interferon promoters *(5, 80, 81)*.

4. Engineered TLRs

The study of TLR biology is sometimes rendered difficult by the lack of a known ligand (e.g., for TLR10) or the accessibility of the receptor to its ligand(s) (e.g., TLR3/7/8/9). To overcome these technical problems, two techniques based on the use of engineered TLRs have been described.

In the first technique, the extracellular part of a TLR is fused with the transmembrane (TM) and intracellular (IC) parts of hCD32a *(13)* (*see* **Fig. 1A**). The resulting chimeric protein can be expressed in HEK293 cells where it localizes on the cell surface. The engagement of this TLR by its ligand(s) results in a Ca^{2+} response/influx mediated by CD32a. The two main advantages of this technique are *(1)* intracellular TLRs become accessible to their ligands added exogenously in the culture medium thereby facilitating the identification of new ligands and *(2)* one single readout (Ca^{2+}) can be used to monitor the activity of any TLR. Ca^{2+} influx is a very fast response that has been widely used as a readout in other domains (e.g., 7-TM receptors) and lot of screening tools are available *(82)*.

Using this technique, de Bouteiller et al. *(13)* have demonstrated the need for an acidic pH (~5.7) for TLR3 activity and the minimal requirements for TLR3 ligands (*see* **Subheading 1.3**). Nevertheless, when these chimeras are used to study intracellular TLRs, one should keep in mind that, in normal cells, a ligand identified with this technique will need access to the TLR. In addition, the amount of ligand needed to induce Ca^{2+} mobilization was about 50 times less than the amount needed to induce cytokine secretion by TLR3 expressing cells.

The second technique involves another fusion protein using the extracellular part of mouse CD4 fused with the TM and IC part of a TLR (*see* **Fig. 1B**) *(5, 11, 16)*. These chimeric proteins (CD4TLRx) are expressed on the cell surface when transfected in HEK293T cells. The association of extracellular CD4 domains triggers the signalling cascade corresponding to the TM + IC of the TLR. Hence, it is not needed to know the ligand for this TLR to study the signalling. This technique can also be used to select the best readout for a given TLR. For example, beside repeated κB binding sites which are the most used promoter for reporter gene assays, the promoters for the chemokines IL8 and CXCL5 were shown to be strongly activated by several TLRs. Focussing on TLR10, which has no identified ligand yet, Hasan et al. *(5)* showed that CD4TLR10 specifically activated IL4, CXCL5, and NF-κB promoters after 48 h stimulation; this could be the basis to design an assay to identify TLR10 ligands. These results stress the importance of looking at more than one readout (usually NF-κB) when one studies TLR activity.

A last example of modified TLRs is the dominant negative (DN) forms. These mutants can be obtained by inserting a point mutation in the TIR domain preventing the binding of adaptor proteins or by deleting the intracellular TIR domain. Such mutated TLRs can bind their ligands but they cannot trigger the signalling cascade. Hence, overexpression of one mutated TLR specifically inhibits the activity of the corresponding TLR by competition *(8, 11)*. This technique can be used to show the involvement of each individual TLR in cells expressing several TLRs.

With new ligands discovered regularly, novel accessory molecules identified and maybe more TLRs to be discovered, there is a need for robust cell-based assays to monitor TLR activity. While "empty shells" such as HEK293 transformed to express one TLR are still a very convenient tool, the future will probably see the rise of HCS and HTS assays in primary cells as the preferred techniques.

Acknowledgements

I thank William McCormack and Claudia Wietek for their critical reading of the manuscript.

References

1. Gay, N. J. & Gangloff, M. (2007). Structure and function of toll receptors and their ligands. *Annu Rev Biochem* 76, 141–65.
2. Rutz, M., Metzger, J., Gellert, T., Luppa, P., Lipford, G. B., Wagner, H. & Bauer, S. (2004). Toll-like receptor 9 binds single-stranded CpG-DNA in a sequence- and pH-dependent manner. *Eur J Immunol* 34, 2541–50.
3. Jin, M. S., Kim, S. E., Heo, J. Y., Lee, M. E., Kim, H. M., Paik, S. G., Lee, H. & Lee, J. O. (2007). Crystal structure of the TLR1-TLR2 heterodimer induced by binding of a tri-acylated lipopeptide. *Cell* 130, 1071–82.
4. Takeuchi, O., Sato, S., Horiuchi, T., Hoshino, K., Takeda, K., Dong, Z., Modlin, R. L. & Akira, S. (2002). Cutting edge: role of Toll-like receptor 1 in mediating immune response to microbial lipoproteins. *J Immunol* 169, 10–4.
5. Hasan, U. A., Dollet, S. & Vlach, J. (2004). Differential induction of gene promoter constructs by constitutively active human TLRs. *Biochem Biophys Res Commun* 321, 124–31.
6. Akira, S., Uematsu, S. & Takeuchi, O. (2006). Pathogen recognition and innate immunity. *Cell* 124, 783–801.
7. Schwandner, R., Dziarski, R., Wesche, H., Rothe, M. & Kirschning, C. J. (1999). Peptidoglycan- and lipoteichoic acid-induced cell activation is mediated by toll-like receptor 2. *J Biol Chem* 274, 17406–9.
8. Sandor, F., Latz, E., Re, F., Mandell, L., Repik, G., Golenbock, D. T., Espevik, T., Kurt-Jones, E. A. & Finberg, R. W. (2003). Importance of extra- and intracellular domains of TLR1 and TLR2 in NFkappa B signaling. *J Cell Biol* 162, 1099–110.
9. Park, J. S., Gamboni-Robertson, F., He, Q., Svetkauskaite, D., Kim, J. Y., Strassheim, D., Sohn, J. W., Yamada, S., Maruyama, I., Banerjee, A., Ishizaka, A. & Abraham, E. (2006). High mobility group box 1 protein interacts with multiple Toll-like receptors. *Am J Physiol Cell Physiol* 290, C917–24.
10. Flo, T. H., Halaas, O., Lien, E., Ryan, L., Teti, G., Golenbock, D. T., Sundan, A. & Espevik, T. (2000). Human toll-like receptor 2 mediates monocyte activation by *Listeria monocytogenes*, but not by group B streptococci or lipopolysaccharide. *J Immunol* 164, 2064–9.
11. Ozinsky, A., Underhill, D. M., Fontenot, J. D., Hajjar, A. M., Smith, K. D., Wilson, C. B., Schroeder, L. & Aderem, A. (2000). The repertoire for pattern recognition of pathogens by the innate immune system is defined by cooperation between toll-like receptors. *Proc Natl Acad Sci U S A* 97, 13766–71.
12. Girardin, S. E., Boneca, I. G., Carneiro, L. A., Antignac, A., Jehanno, M., Viala, J., Tedin, K., Taha, M. K., Labigne, A., Zahringer, U., Coyle, A. J., DiStefano, P. S., Bertin, J., Sansonetti, P. J. & Philpott, D. J. (2003). Nod1 detects a unique muropeptide from gram-negative bacterial peptidoglycan. *Science* 300, 1584–7.
13. de Bouteiller, O., Merck, E., Hasan, U. A., Hubac, S., Benguigui, B., Trinchieri, G., Bates, E. E. & Caux, C. (2005). Recognition of double-stranded RNA by human toll-like receptor 3 and downstream receptor signaling requires multimerization and an acidic pH. *J Biol Chem* 280, 38133–45.
14. Marshall-Clarke, S., Downes, J. E., Haga, I. R., Bowie, A. G., Borrow, P., Pennock, J. E., Grencis, R. K. & Rothwell, P. (2007). Polyinosinic acid is a ligand for toll-like receptor 3. *J Biol Chem* 282, 24759–66.
15. Kato, H., Takeuchi, O., Sato, S., Yoneyama, M., Yamamoto, M., Matsui, K., Uematsu, S., Jung, A., Kawai, T., Ishii, K. J., Yamaguchi, O., Otsu, K., Tsujimura, T., Koh, C. S., Reis e Sousa, C., Matsuura, Y., Fujita, T. & Akira, S. (2006). Differential roles of MDA5 and RIG-I helicases in the recognition of RNA viruses. *Nature* 441, 101–5.
16. Medzhitov, R., Preston-Hurlburt, P. & Janeway, C. A., Jr. (1997). A human homologue of the Drosophila Toll protein signals activation of adaptive immunity. *Nature* 388, 394–7.
17. Caroff, M. & Karibian, D. (2003). Structure of bacterial lipopolysaccharides. *Carbohydr Res* 338, 2431–47.
18. Coats, S. R., Reife, R. A., Bainbridge, B. W., Pham, T. T. & Darveau, R. P. (2003). *Porphyromonas gingivalis* lipopolysaccharide antagonizes *Escherichia coli* lipopolysaccharide at toll-like receptor 4 in human endothelial cells. *Infect Immun* 71, 6799–807.
19. Huber, M., Kalis, C., Keck, S., Jiang, Z., Georgel, P., Du, X., Shamel, L., Sovath, S., Mudd, S., Beutler, B., Galanos, C. & Freudenberg, M. A. (2006). R-form LPS, the master key to the activation ofTLR4/MD-2-positive cells. *Eur J Immunol* 36, 701–11.
20. Mata-Haro, V., Cekic, C., Martin, M., Chilton, P. M., Casella, C. R. & Mitchell, T. C. (2007). The vaccine adjuvant monophosphoryl lipid A as a TRIF-biased agonist of TLR4. *Science* 316, 1628–32.
21. Hirschfeld, M., Ma, Y., Weis, J. H., Vogel, S. N. & Weis, J. J. (2000). Cutting edge: repurification of lipopolysaccharide eliminates signaling through both human and murine toll-like receptor 2. *J Immunol* 165, 618–22.

22. Kawasaki, K., Akashi, S., Shimazu, R., Yoshida, T., Miyake, K. & Nishijima, M. (2000). Mouse toll-like receptor 4.MD-2 complex mediates lipopolysaccharide-mimetic signal transduction by Taxol. *J Biol Chem* 275, 2251–4.
23. Figueiredo, R. T., Fernandez, P. L., Mourao-Sa, D. S., Porto, B. N., Dutra, F. F., Alves, L. S., Oliveira, M. F., Oliveira, P. L., Graca-Souza, A. V. & Bozza, M. T. (2007). Characterization of heme as activator of toll-like receptor 4. *J Biol Chem* 282, 20221–9.
24. Marshak-Rothstein, A. (2006). Toll-like receptors in systemic autoimmune disease. *Nat Rev Immunol* 6, 823–35.
25. Matzinger, P. (2002). The danger model: a renewed sense of self. *Science* 296, 301–5.
26. Gao, B. & Tsan, M. F. (2003). Recombinant human heat shock protein 60 does not induce the release of tumor necrosis factor alpha from murine macrophages. *J Biol Chem* 278, 22523–9.
27. Ohashi, K., Burkart, V., Flohe, S. & Kolb, H. (2000). Cutting edge: heat shock protein 60 is a putative endogenous ligand of the toll-like receptor-4 complex. *J Immunol* 164, 558–61.
28. Hayashi, F., Smith, K. D., Ozinsky, A., Hawn, T. R., Yi, E. C., Goodlett, D. R., Eng, J. K., Akira, S., Underhill, D. M. & Aderem, A. (2001). The innate immune response to bacterial flagellin is mediated by Toll-like receptor 5. *Nature* 410, 1099–103.
29. Smith, K. D., Andersen-Nissen, E., Hayashi, F., Strobe, K., Bergman, M. A., Barrett, S. L., Cookson, B. T. & Aderem, A. (2003). Toll-like receptor 5 recognizes a conserved site on flagellin required for protofilament formation and bacterial motility. *Nat Immunol* 4, 1247–53.
30. Okusawa, T., Fujita, M., Nakamura, J., Into, T., Yasuda, M., Yoshimura, A., Hara, Y., Hasebe, A., Golenbock, D. T., Morita, M., Kuroki, Y., Ogawa, T. & Shibata, K. (2004). Relationship between structures and biological activities of mycoplasmal diacylated lipopeptides and their recognition by toll-like receptors 2 and 6. *Infect Immun* 72, 1657–65.
31. Triantafilou, M., Gamper, F. G., Haston, R. M., Mouratis, M. A., Morath, S., Hartung, T. & Triantafilou, K. (2006). Membrane sorting of toll-like receptor (TLR)-2/6 and TLR2/1 heterodimers at the cell surface determines heterotypic associations with CD36 and intracellular targeting. *J Biol Chem* 281, 31002–11.
32. Diebold, S. S., Kaisho, T., Hemmi, H., Akira, S. & Reis e Sousa, C. (2004). Innate antiviral responses by means of TLR7-mediated recognition of single-stranded RNA. *Science* 303, 1529–31.
33. Lund, J. M., Alexopoulou, L., Sato, A., Karow, M., Adams, N. C., Gale, N. W., Iwasaki, A. & Flavell, R. A. (2004). Recognition of single-stranded RNA viruses by Toll-like receptor 7. *Proc Natl Acad Sci U S A* 101, 5598–603.
34. Heil, F., Hemmi, H., Hochrein, H., Ampenberger, F., Kirschning, C., Akira, S., Lipford, G., Wagner, H. & Bauer, S. (2004). Species-specific recognition of single-stranded RNA via toll-like receptor 7 and 8. *Science* 303, 1526–9.
35. Hemmi, H., Kaisho, T., Takeuchi, O., Sato, S., Sanjo, H., Hoshino, K., Horiuchi, T., Tomizawa, H., Takeda, K. & Akira, S. (2002). Small anti-viral compounds activate immune cells via the TLR7 MyD88-dependent signaling pathway. *Nat Immunol* 3, 196–200.
36. Heil, F., Ahmad-Nejad, P., Hemmi, H., Hochrein, H., Ampenberger, F., Gellert, T., Dietrich, H., Lipford, G., Takeda, K., Akira, S., Wagner, H. & Bauer, S. (2003). The Toll-like receptor 7 (TLR7)-specific stimulus loxoribine uncovers a strong relationship within the TLR7, 8 and 9 subfamily. *Eur J Immunol* 33, 2987–97.
37. Gorden, K. B., Gorski, K. S., Gibson, S. J., Kedl, R. M., Kieper, W. C., Qiu, X., Tomai, M. A., Alkan, S. S. & Vasilakos, J. P. (2005). Synthetic TLR agonists reveal functional differences between human TLR7 and TLR8. *J Immunol* 174, 1259–68.
38. Hornung, V., Guenthner-Biller, M., Bourquin, C., Ablasser, A., Schlee, M., Uematsu, S., Noronha, A., Manoharan, M., Akira, S., de Fougerolles, A., Endres, S. & Hartmann, G. (2005). Sequence-specific potent induction of IFN-alpha by short interfering RNA in plasmacytoid dendritic cells through TLR7. *Nat Med* 11, 263–70.
39. Savarese, E., Chae, O. W., Trowitzsch, S., Weber, G., Kastner, B., Akira, S., Wagner, H., Schmid, R. M., Bauer, S. & Krug, A. (2006). U1 small nuclear ribonucleoprotein immune complexes induce type I interferon in plasmacytoid dendritic cells through TLR7. *Blood* 107, 3229–34.
40. Gorden, K. K., Qiu, X. X., Binsfeld, C. C., Vasilakos, J. P. & Alkan, S. S. (2006). Cutting edge: activation of murine TLR8 by a combination of imidazoquinoline immune response modifiers and polyT oligodeoxynucleotides. *J Immunol* 177, 6584–7.
41. Barton, G. M., Kagan, J. C. & Medzhitov, R. (2006). Intracellular localization of Toll-like receptor 9 prevents recognition of self DNA but facilitates access to viral DNA. *Nat Immunol* 7, 49–56.

42. Krieg, A. M. (2006). Therapeutic potential of Toll-like receptor 9 activation. *Nat Rev Drug Discov* 5, 471–84.
43. Tian, J., Avalos, A. M., Mao, S. Y., Chen, B., Senthil, K., Wu, H., Parroche, P., Drabic, S., Golenbock, D., Sirois, C., Hua, J., An, L. L., Audoly, L., La Rosa, G., Bierhaus, A., Naworth, P., Marshak-Rothstein, A., Crow, M. K., Fitzgerald, K. A., Latz, E., Kiener, P. A. & Coyle, A. J. (2007). Toll-like receptor 9-dependent activation by DNA-containing immune complexes is mediated by HMGB1 and RAGE. *Nat Immunol* 8, 487–96.
44. Magnusson, M., Tobes, R., Sancho, J. & Pareja, E. (2007). Cutting edge: natural DNA repetitive extragenic sequences from Gram-negative pathogens strongly stimulate TLR9. *J Immunol* 179, 31–5.
45. Latz, E., Verma, A., Visintin, A., Gong, M., Sirois, C. M., Klein, D. C., Monks, B. G., McKnight, C. J., Lamphier, M. S., Duprex, W. P., Espevik, T. & Golenbock, D. T. (2007). Ligand-induced conformational changes allosterically activate Toll-like receptor 9. *Nat Immunol* 8, 772–9.
46. Lenert, P. (2005). Inhibitory oligodeoxynucleotides – therapeutic promise for systemic autoimmune diseases. *Clin Exp Immunol* 140, 1–10.
47. Boule, M. W., Broughton, C., Mackay, F., Akira, S., Marshak-Rothstein, A. & Rifkin, I. R. (2004). Toll-like receptor 9-dependent and -independent dendritic cell activation by chromatin-immunoglobulin G complexes. *J Exp Med* 199, 1631–40.
48. Coban, C., Ishii, K. J., Kawai, T., Hemmi, H., Sato, S., Uematsu, S., Yamamoto, M., Takeuchi, O., Itagaki, S., Kumar, N., Horii, T. & Akira, S. (2005). Toll-like receptor 9 mediates innate immune activation by the malaria pigment hemozoin. *J Exp Med* 201, 19–25.
49. Hasan, U., Chaffois, C., Gaillard, C., Saulnier, V., Merck, E., Tancredi, S., Guiet, C., Briere, F., Vlach, J., Lebecque, S., Trinchieri, G. & Bates, E. E. (2005). Human TLR10 is a functional receptor, expressed by B cells and plasmacytoid dendritic cells, which activates gene transcription through MyD88. *J Immunol* 174, 2942–50.
50. Zhang, D., Zhang, G., Hayden, M. S., Greenblatt, M. B., Bussey, C., Flavell, R. A. & Ghosh, S. (2004). A toll-like receptor that prevents infection by uropathogenic bacteria. *Science* 303, 1522–6.
51. Yarovinsky, F., Zhang, D., Andersen, J. F., Bannenberg, G. L., Serhan, C. N., Hayden, M. S., Hieny, S., Sutterwala, F. S., Flavell, R. A., Ghosh, S. & Sher, A. (2005). TLR11 activation of dendritic cells by a protozoan profilin-like protein. *Science* 308, 1626–9.
52. Latz, E., Visintin, A., Lien, E., Fitzgerald, K. A., Monks, B. G., Kurt-Jones, E. A., Golenbock, D. T. & Espevik, T. (2002). Lipopolysaccharide rapidly traffics to and from the Golgi apparatus with the toll-like receptor 4-MD-2-CD14 complex in a process that is distinct from the initiation of signal transduction. *J Biol Chem* 277, 47834–43.
53. Yang, R. B., Mark, M. R., Gray, A., Huang, A., Xie, M. H., Zhang, M., Goddard, A., Wood, W. I., Gurney, A. L. & Godowski, P. J. (1998). Toll-like receptor-2 mediates lipopolysaccharide-induced cellular signalling. *Nature* 395, 284–8.
54. Pridmore, A. C., Wyllie, D. H., Abdillahi, F., Steeghs, L., van der Ley, P., Dower, S. K. & Read, R. C. (2001). A lipopolysaccharide-deficient mutant of Neisseria meningitidis elicits attenuated cytokine release by human macrophages and signals via toll-like receptor (TLR) 2 but not via TLR4/MD2. *J Infect Dis* 183, 89–96.
55. Matsuguchi, T., Takagi, K., Musikacharoen, T. & Yoshikai, Y. (2000). Gene expressions of lipopolysaccharide receptors, toll-like receptors 2 and 4, are differently regulated in mouse T lymphocytes. *Blood* 95, 1378–85.
56. Yoshimura, A., Lien, E., Ingalls, R. R., Tuomanen, E., Dziarski, R. & Golenbock, D. (1999). Cutting edge: recognition of Gram-positive bacterial cell wall components by the innate immune system occurs via Toll-like receptor 2. *J Immunol* 163, 1–5.
57. Hasan, U. A., Trinchieri, G. & Vlach, J. (2005). Toll-like receptor signaling stimulates cell cycle entry and progression in fibroblasts. *J Biol Chem* 280, 20620–7.
58. Wang, J., Shao, Y., Bennett, T. A., Shankar, R. A., Wightman, P. D. & Reddy, L. G. (2006). The functional effects of physical interactions among Toll-like receptors 7, 8, and 9. *J Biol Chem* 281, 37427–34.
59. Takeshita, F., Leifer, C. A., Gursel, I., Ishii, K. J., Takeshita, S., Gursel, M. & Klinman, D. M. (2001). Cutting edge: Role of Toll-like receptor 9 in CpG DNA-induced activation of human cells. *J Immunol* 167, 3555–8.
60. Chaperot, L., Blum, A., Manches, O., Lui, G., Angel, J., Molens, J. P. & Plumas, J. (2006). Virus or TLR agonists induce TRAIL-mediated cytotoxic activity of plasmacytoid dendritic cells. *J Immunol* 176, 248–55.
61. Maeda, T., Murata, K., Fukushima, T., Sugahara, K., Tsuruda, K., Anami, M., Onimaru, Y., Tsukasaki, K., Tomonaga, M., Moriuchi, R., Hasegawa, H., Yamada, Y. & Kamihira, S.

(2005). A novel plasmacytoid dendritic cell line, CAL-1, established from a patient with blastic natural killer cell lymphoma. *Int J Hematol* 81, 148–54.

62. Tapping, R. I., Orr, S. L., Lawson, E. M., Soldau, K. & Tobias, P. S. (1999). Membrane-anchored forms of lipopolysaccharide (LPS)-binding protein do not mediate cellular responses to LPS independently of CD14. *J Immunol* 162, 5483–9.
63. Zarember, K. A. & Godowski, P. J. (2002). Tissue expression of human Toll-like receptors and differential regulation of Toll-like receptor mRNAs in leukocytes in response to microbes, their products, and cytokines. *J Immunol* 168, 554–61.
64. Park, E. K., Jung, H. S., Yang, H. I., Yoo, M. C., Kim, C. & Kim, K. S. (2007). Optimized THP-1 differentiation is required for the detection of responses to weak stimuli. *Inflamm Res* 56, 45–50.
65. Hippenstiel, S., Opitz, B., Schmeck, B. & Suttorp, N. (2006). Lung epithelium as a sentinel and effector system in pneumonia – molecular mechanisms of pathogen recognition and signal transduction. *Respir Res* 7, 97.
66. Jego, G., Bataille, R., Geffroy-Luseau, A., Descamps, G. & Pellat-Deceunynck, C. (2006). Pathogen-associated molecular patterns are growth and survival factors for human myeloma cells through Toll-like receptors. *Leukemia* 20, 1130–7.
67. Sha, Q., Truong-Tran, A. Q., Plitt, J. R., Beck, L. A. & Schleimer, R. P. (2004). Activation of airway epithelial cells by toll-like receptor agonists. *Am J Respir Cell Mol Biol* 31, 358–64.
68. Ma, Y., Li, J., Chiu, I., Wang, Y., Sloane, J. A., Lu, J., Kosaras, B., Sidman, R. L., Volpe, J. J. & Vartanian, T. (2006). Toll-like receptor 8 functions as a negative regulator of neurite outgrowth and inducer of neuronal apoptosis. *J Cell Biol* 175, 209–15.
69. Hornung, V., Rothenfusser, S., Britsch, S., Krug, A., Jahrsdorfer, B., Giese, T., Endres, S. & Hartmann, G. (2002). Quantitative expression of toll-like receptor 1-10 mRNA in cellular subsets of human peripheral blood mononuclear cells and sensitivity to CpG oligodeoxynucleotides. *J Immunol* 168, 4531–7.
70. Applequist, S. E., Wallin, R. P. & Ljunggren, H. G. (2002). Variable expression of Toll-like receptor in murine innate and adaptive immune cell lines. *Int Immunol* 14, 1065–74.
71. Siren, J., Pirhonen, J., Julkunen, I. & Matikainen, S. (2005). IFN-alpha regulates TLR-dependent gene expression of IFN-alpha, IFN-beta, IL-28, and IL-29. *J Immunol* 174, 1932–7.
72. Hoshino, K., Takeuchi, O., Kawai, T., Sanjo, H., Ogawa, T., Takeda, Y., Takeda, K. & Akira, S. (1999). Cutting edge: Toll-like receptor 4 (TLR4)-deficient mice are hyporesponsive to lipopolysaccharide: evidence for TLR4 as the Lps gene product. *J Immunol* 162, 3749–52.
73. Kurt-Jones, E. A., Sandor, F., Ortiz, Y., Bowen, G. N., Counter, S. L., Wang, T. C. & Finberg, R. W. (2004). Use of murine embryonic fibroblasts to define Toll-like receptor activation and specificity. *J Endotoxin Res* 10, 419–24.
74. O'Neill, L. A. & Bowie, A. G. (2007). The family of five: TIR-domain-containing adaptors in Toll-like receptor signalling. *Nat Rev Immunol* 7, 353–64.
75. Rothman, D. M., Shults, M. D. & Imperiali, B. (2005). Chemical approaches for investigating phosphorylation in signal transduction networks. *Trends Cell Biol* 15, 502–10.
76. Pranada, A. L., Metz, S., Herrmann, A., Heinrich, P. C. & Muller-Newen, G. (2004). Real time analysis of STAT3 nucleocytoplasmic shuttling. *J Biol Chem* 279, 15114–23.
77. Schindler, U. & Baichwal, V. R. (1994). Three NF-kappa B binding sites in the human E-selectin gene required for maximal tumor necrosis factor alpha-induced expression. *Mol Cell Biol* 14, 5820–31.
78. Berger, J., Hauber, J., Hauber, R., Geiger, R. & Cullen, B. R. (1988). Secreted placental alkaline phosphatase: a powerful new quantitative indicator of gene expression in eukaryotic cells. *Gene* 66, 1–10.
79. Markova, S. V., Golz, S., Frank, L. A., Kalthof, B. & Vysotski, E. S. (2004). Cloning and expression of cDNA for a luciferase from the marine copepod Metridia longa. A novel secreted bioluminescent reporter enzyme. *J Biol Chem* 279, 3212–7.
80. Ehrhardt, C., Kardinal, C., Wurzer, W. J., Wolff, T., von Eichel-Streiber, C., Pleschka, S., Planz, O. & Ludwig, S. (2004). Rac1 and PAK1 are upstream of IKK-epsilon and TBK-1 in the viral activation of interferon regulatory factor-3. *FEBS Lett* 567, 230–8.
81. Civas, A., Genin, P., Morin, P., Lin, R. & Hiscott, J. (2006). Promoter organization of the interferon-A genes differentially affects virus-induced expression and responsiveness to TBK1 and IKKepsilon. *J Biol Chem* 281, 4856–66.
82. Monteith, G. R. & Bird, G. S. (2005). Techniques: high-throughput measurement of intracellular Ca(2+) – back to basics. *Trends Pharmacol Sci* 26, 218–23.

83. Mansson, A., Adner, M., Hockerfelt, U. & Cardell, L. O. (2006). A distinct Toll-like receptor repertoire in human tonsillar B cells, directly activated by PamCSK, R-837 and CpG-2006 stimulation. *Immunology* 118, 539–48.

84. Caron, G., Duluc, D., Fremaux, I., Jeannin, P., David, C., Gascan, H. & Delneste, Y. (2005). Direct stimulation of human T cells via TLR5 and TLR7/8: flagellin and R-848 up-regulate proliferation and IFN-gamma production by memory CD4+ T cells. *J Immunol* 175, 1551–7.

85. Tabiasco, J., Devevre, E., Rufer, N., Salaun, B., Cerottini, J. C., Speiser, D. & Romero, P. (2006). Human effector CD8+ T lymphocytes express TLR3 as a functional coreceptor. *J Immunol* 177, 8708–13.

86. Kadowaki, N., Ho, S., Antonenko, S., Malefyt, R. W., Kastelein, R. A., Bazan, F. & Liu, Y. J. (2001). Subsets of human dendritic cell precursors express different toll-like receptors and respond to different microbial antigens. *J Exp Med* 194, 863–9.

87. Hart, O. M., Athie-Morales, V., O'Connor, G. M. & Gardiner, C. M. (2005). TLR7/8-mediated activation of human NK cells results in accessory cell-dependent IFN-gamma production. *J Immunol* 175, 1636–42.

88. Ito, T., Wang, Y. H. & Liu, Y. J. (2005). Plasmacytoid dendritic cell precursors/type I interferon-producing cells sense viral infection by Toll-like receptor (TLR) 7 and TLR9. *Springer Semin Immunopathol* 26, 221–9.

Chapter 3

Toll-Like Receptor Interactions Imaged by FRET Microscopy and GFP Fragment Reconstitution

Gabor Horvath, Scott Young, and Eicke Latz

Summary

Protein–protein interactions regulate biological networks. The most proximal events that initiate signal transduction frequently are receptor dimerization or conformational changes in receptor complexes. Toll-like receptors (TLRs) are transmembrane receptors that are activated by a number of exogenous and endogenous ligands. Most TLRs can respond to multiple ligands and the different TLRs recognize structurally diverse molecules ranging from proteins, sugars, lipids, and nucleic acids. TLRs can be expressed on the plasma membrane or in endosomal compartments, and ligand recognition thus proceeds in different microenvironments. Not surprisingly, distinctive mechanisms of TLR receptor activation have evolved. A detailed understanding of the mechanisms of TLR activation is important for the development of novel synthetic TLR activators or pharmacological inhibitors of TLRs. Confocal laser scanning microscopy (LSM) combined with green fluorescent protein (GFP) technology allows the direct visualization of TLR expression in living cells. Fluorescence resonance energy transfer (FRET) measurements between two differentially tagged proteins permit the study of TLR interaction and distances between receptors in the range of molecular interactions can be measured and visualized. Additionally, FRET measurements combined with confocal microscopy provide detailed information about molecular interactions in different subcellular localizations. Bimolecular complementation using split fluorescent proteins (BiFC) represents an additional valuable method to study mechanisms of receptor activation in living cells. These techniques permit the dynamic visualization of early signaling events in living cells and can be utilized in pharmacological or genetic screens.

Key words: Confocal microscopy, Laser scanning microscopy (LSM), Fluorescence resonance energy transfer (FRET), Fluorescence lifetime imaging microscopy (FLIM), Toll-like receptor (TLR), GFP fragment reconstitution, Protein fragment reconstitution (PCA), Bimolecular fluorescence complementation (BiFC).

Claire E. McCoy and Luke A.J. O'Neill (eds.), *Methods in Molecular Biology, Toll-Like Receptors, vol. 517*
DOI: 10.1007/978-1-59745-541-1_3

1. Introduction

Investigations of protein–protein interactions are important for the understanding of higher organization levels of molecular complexes, their structure–function relationships, and the regulation of signal transduction processes. Many techniques are available to measure and quantify protein–protein interactions in vitro, for example circular dichroism, isothermal titration calorimetry, surface plasmon resonance, nuclear magnetic resonance spectroscopy, or gel retardation assays. However, some techniques require large amount of proteins and protein interactions are assessed outside the more complex context of a living cell. Biophysical approaches combined with light microscopy permit the study of protein–protein interactions in a nondestructive manner in living cells. The lateral resolution of light microscopy is a function of the numerical aperture (NA) of the objective lens and is influenced by the index of refraction of the medium and the wavelength of light. Regular light microscopy and confocal laser scanning microscopy (LSM) therefore have a practical lateral resolution limit of about 200 nm. Most molecular interactions occur in the range of a few nanometers, and thus, receptor–receptor interactions or conformational changes within receptor complexes cannot be directly visualized by light microscopy due to the limitations given by the resolution of light. However, if light microscopy is combined with techniques such as fluorescence resonance energy transfer (FRET) or bimolecular fluorescence complementation (BiFC), molecular interactions can be studied dynamically in living cells. Furthermore, if combined with confocal imaging, information about the subcellular localization of protein interactions or conformational changes can be obtained.

We will discuss the different methods of microscopic detection of FRET based on sensitized emission intensity, photobleaching, and fluorescence lifetime measurements. Furthermore, we will describe the use of BiFC as an additional method of detecting protein–protein interactions in living cells (**Fig.1**).

1.1. Theory of FRET

FRET is a process by which a fluorescent donor molecule in an excited electronic state transfers its energy to a neighboring acceptor molecule via a nonradiative dipole–dipole interaction *(1, 2)*. The term "fluorescence resonance energy transfer" can be misleading, because the energy is not actually transferred by fluorescence. The acceptor fluorophore molecule can be a dark quencher or another fluorescent molecule (in this article, we only explore the latter). The prerequisites for this process are that (1) there is a substantial overlap between the emission spectrum of the donor and the excitation spectrum of the acceptor fluorophore (gray area in **Fig.2A**), (2) the spatial separation of the two

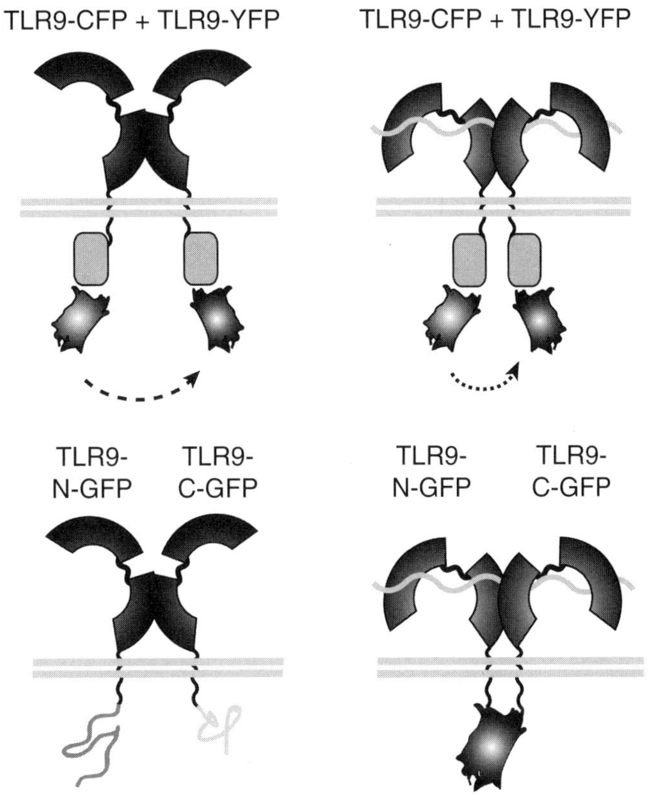

Fig. 1. Model of Toll-like receptor (TLR) association and ligand-induced changes in TLR9 conformation reported by CFP–YFP fluorescence resonance energy transfer (FRET) (**a**) or green fluorescent protein (GFP) fragment fluorescence complementation (**b**).

molecules is between 1 and 10 nm, and (3) the dipole moments of the molecules are correctly aligned.

The most immediate effect of FRET is a decrease in fluorescence lifetime of the donor fluorophore, as the FRET process competes for the available excited donor states with fluorescent and thermal relaxation. This results in the decrease of donor fluorescent quantum efficiency and fluorescent intensity and the simultaneous increase of acceptor fluorescent intensity (as excited at the donor excitation wavelength). These two phenomena are referred to as "donor quenching" and "sensitized emission" (arrows in **Fig.2A**).

The extent of the FRET process can be quantified by FRET efficiency (E), which is a direct measure of the fraction of photon energy absorbed by the donor that is transferred to an acceptor. Transfer efficiency can be calculated by using the rate constants of donor relaxation processes:

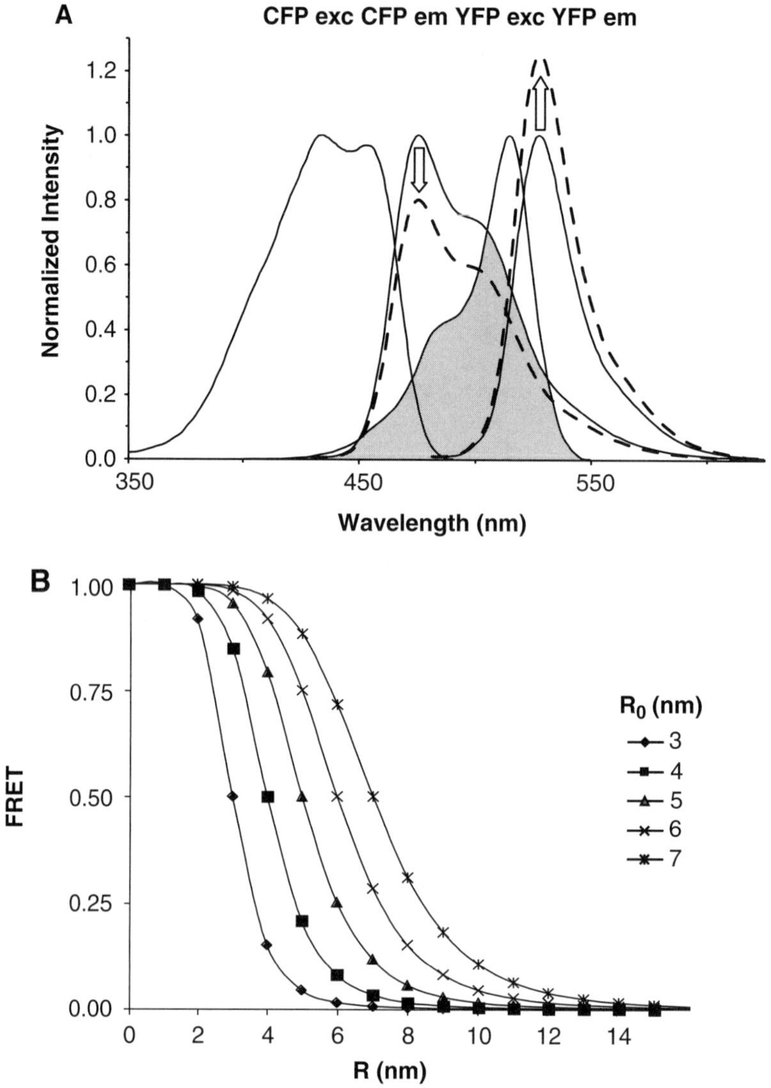

Fig. 2. (**A**) Excitation and emission spectra of CFP and YFP. Shaded area indicates the overlap integral. The down and up arrows represent donor quenching and sensitized emission, respectively. (**B**) Distance dependence of energy transfer efficiency with varying critical Förster distances (R_0). Note that the relationship is almost linear between 20% and 80% transfer efficiency.

$$E = \frac{k_T}{(k_T + k_F + k_0)}, \tag{1}$$

where k_T is the transfer rate constant, k_F is the fluorescent rate constant, and k_0 is the sum of any other relaxation rate constants of donor fluorophore. Since the transfer rate can also be expressed as a function of the separation of the two fluorophores (*see* **Note 1**), one can yield an equation where the transfer efficiency is determined by the distance (R) between the donor and acceptor fluorescent dyes:

$$E = \frac{R_0^6}{R_0^6 + R^6} = \frac{1}{1 + (R / R_0)^6}, \qquad (2)$$

where R_0 is the so-called Förster critical distance corresponding to 50% transfer efficiency for a given donor–acceptor pair (for CFP–YFP, this is 4.9 nm) (*see* **Note 2**). Accordingly, the biggest advantage of any FRET method is the strong dependency of transfer efficiency on the distance between the two fluorescent dyes (**Fig.2b**). The FRET efficiency depends on the donor-to-acceptor separation distance with an inverse sixth power law. Subtle changes in distances between FRET fluorophores at the level of molecular interaction or binding are reflected by changes in FRET efficiencies.

1.2. Experimental Approaches for Measuring FRET

1.2.1. Acceptor Photobleaching

The measurement of fluorescence intensity is one of the easiest and most available ways to characterize protein interactions via the FRET process, and it is adapted to fluorescent plate readers, spectrofluorometers, flow cytometers, wide-field microscopes, and confocal laser scanning microscopes.

One measurable characteristic of energy transfer is the decrease in donor fluorescence intensity (i.e., donor quenching). Thus, the analysis of donor quenching can be used to calculate the transfer efficiency E. **Equation 3** indicates that only two fluorescent intensities are required. The fluorescence intensity of the donor is compared in the presence (I_D^{DA}) or absence (I_D^D) of an acceptor:

$$E = 1 - \frac{I_D^{DA}}{I_D^D}. \qquad (3)$$

One possible way to measure fluorescence dequenching is to bleach the acceptor fluorophore by applying intense laser light at the wavelength of acceptor absorbance to photo-physically destroy the acceptor fluorophore. FRET efficiency is calculated using **Eq. 3** by comparing the donor fluorescence before and after acceptor bleaching. This technique is referred to "donor dequenching after acceptor photobleaching."

1.2.2. Sensitized Emission FRET

Another popular method to study FRET in living samples is to quantify the sensitized emission of the acceptor fluorophore by excitation of the donor fluorophore (termed "sensitized emission FRET"). In this approach, the fluorescence intensity of the acceptor as excited at the donor wavelength in the presence

(I_A^{DA}) or absence (I_A^A) of a donor is compared. The transfer efficiency is

$$E = \frac{\varepsilon_A c_A}{\varepsilon_D c_D}\left(\frac{I_A^{DA}}{I_A^A} - 1\right), \tag{4}$$

where ε_D and ε_A are the extinction coefficients of the donor and acceptor fluorophores as excited at the donor wavelength, and c_D and c_A are the concentrations of the donor and acceptor fluorophores, respectively. The excitation coefficients and concentrations are practically scaling the calculated transfer efficiency values to an absolute scale, which is only dependent on the Förster critical distance. The advantage of this calculation is that the sensitized emission intensity is compared to very low fluorescence intensity, as the acceptor usually cannot be efficiently excited at the donor excitation wavelength. However, considerable spectral bleed-through from the donor can represent a major technical hurdle. Thus, the sensitized emission intensity should be corrected for both donor bleed-through and acceptor cross-excitation. For this reason, sensitized emission FRET measurements always require the presence of appropriate controls (donor and acceptor fluorophore alone).

1.2.3. Donor Photobleaching

Another approach to measure FRET efficiency is the photophysical destruction of the donor fluorophore, which will efficiently decrease the possibility of energy transfer. The complete photo-destruction of the donor molecule would not lead to a feasible method, because it would not leave any measurable fluorescence intensity which can be compared to the initial intensity. Instead, the donor fluorophore is destroyed sequentially by strong excitation light with concomitant imaging of the decreasing fluorescence emission. The photobleaching decay rate of the donor fluorophore is calculated by fitting a double-exponential function to the series of intensities on a pixel-by-pixel basis. The transfer efficiency is obtained by comparing the average photobleaching time constants of a sample with donor only (τ'_D) to a sample with both donor and acceptor present (τ'_{DA}):

$$E = 1 - \frac{\tau'_D}{\tau'_{DA}}, \tag{5}$$

where $\tau'_{DA} > \tau'_D$, because energy transfer introduced an additional pathway for relaxation beside fluorescence emission and nonradiative photo-destruction *(3)*. A disadvantage of donor photobleaching is that it requires two separate measurements to calculate transfer efficiency that can result in statistical artifacts. Additionally, the data analysis requires sophisticated software that is not readily available *(4)*, and long acquisition times (usually 5–15 min, depending on the photostability of donor dye) do not allow for live cell imaging.

On the other hand, if the acceptor molecule is sufficiently photo-labile, its photo-destruction would result in the recovery of donor intensity; thus a simple donor quenching calculation will yield transfer efficiency in the same sample:

$$E = 1 - \frac{I_{\text{Dpre}}}{I_{\text{Dpost}}}, \tag{6}$$

where I_{Dpre} and I_{Dpost} are the donor intensities in pre- and postbleaching conditions. Since donor photobleaching is not frequently used for FRET efficiency analysis, we do not describe this method further in **Subheading 3**.

1.2.4. FRET Analysis by Fluorescence Lifetime Imaging Microscopy (FLIM)

Each of the methods to detect FRET described above is based on measuring fluorescence intensities and changes thereof as a consequence of FRET. These methods make use of the fact that fluorophores display characteristic emission spectra that can be used to separate different fluorophores and that FRET leads to intensity changes in both donor and acceptor fluorophores. Another physical process that can be measured with excited fluorophores is the fluorescence lifetime. Each fluorophore exhibits a unique fluorescence lifetime that can be used to separate fluorophores or to probe for the existence of FRET. The fluorescence lifetime (τ) is the average time during which a fluorescent molecule remains in an excited state before returning to the ground state. FLIM combines the measurement of fluorescence lifetimes with microscopic imaging techniques such as confocal imaging or other imaging modalities. In confocal-based FLIM, for example, fluorescence lifetime decay characteristics of a fluorescent sample are acquired at each position of a confocal scan, i.e., at each pixel of the image. The data can be represented as images, where fluorescence lifetime data are color-coded, and the amount of signal (i.e., the fluorescence intensity) is coded by contrast intensity.

The fluorescence lifetime of a fluorophore is independent of probe concentration, excitation intensity, and photobleaching. In addition, the fluorescence lifetimes are not only different for distinct fluorophores but also depend on the molecular environment of the fluorophore molecules. Since the lifetime does not depend on the concentration of the fluorophore, fluorescence lifetime measurements can directly probe changes of, for example, ion concentrations or oxygen saturation. During the process of FRET, donor fluorophores that transfer energy to acceptor fluorophores show decreased fluorescence lifetimes. This phenomenon can be exploited for sensitive and accurate FRET measurements.

A major advantage of FLIM-based FRET measurements is that only the donor fluorescence decay needs to be measured as the donor lifetime changes upon energy transfer. Thus, spectral bleed-through correction for acceptor fluorescence is not required. By measuring the donor lifetime in the presence and the absence of acceptor one can accurately assess the FRET efficiency)

$$E = 1 - \frac{\tau_{\mathrm{DA}}}{\tau_{\mathrm{D}}}, \quad (7)$$

where τ_{DA} is the donor lifetime in the presence of acceptor (energy transfer situation) and τ_{D} is the donor lifetime in the absence of acceptor.

FLIM can be implemented in wide-field, confocal, and multiphoton excitation microscopes. Point scanning methods are advantageous as subcellular resolution can be obtained, and the lifetime measurements do not reflect average lifetimes of entire excited volumes. Instrumental methods for measuring fluorescence lifetimes can broadly be divided into two categories: frequency domain and time domain *(5, 6)*. Either method can be used in one-photon or two-photon FRET-FLIM. In the frequency domain mode, the fluorescence lifetimes can be determined by a phase-modulated method. The intensity of a laser continuous wave source can be modulated at high frequency, resulting in modulation of the fluorescence. Since the fluorophore in the excited state has a specific lifetime, the fluorescence will be delayed with respect to the excitation signal. Thus, the lifetime can be determined from the phase shift.

Here, we describe the measurement of fluorescence lifetimes using the time domain method of data acquisition. Fluorescence lifetimes can be determined in the time domain by using a pulsed laser source. Time-correlated single photon counting (TCSPC) is typically employed in time domain FLIM measurements. In TCSPC, the laser pulses excite a population of fluorophores and the timing of single-photon emissions is recorded yielding a probability distribution for the emission of single photons. The time-resolved fluorescence decays exponentially and thus, during lifetime analysis, fluorescence decay curves recorded by TCSPC are fitted to (multi) exponential decays. The time response of the instrument components is taken into account by an iterative deconvolution technique.

1.3. Bimolecular Fluorescence Complementation

Protein fragment complementation is another tool to investigate protein interactions in living cells. In general, two matching halves of a reporter protein (e.g., ribonuclease, b-galactosidase, or green fluorescent protein (GFP)) are identified by circular permutation analysis of the protein. The complementing protein fragments expressed separately are not fluorescent unless they are brought into close proximity when fused to the interacting proteins. GFP, as a reporter protein, is of the highest interest, as the two complementary halves of GFP will result in de novo GFP fluorescence upon the interaction of the investigated proteins.

Simple fluorescence intensity analysis can be obtained by various techniques, such as epi-fluorescence or confocal microscopy. This technique of BiFC has been extensively used for mapping interactions between leucin-zippers *(7, 8)* and transcription factors *(9)*. It also provides an advantage compared to FRET measurements, namely the lack of background fluorescence in the absence of protein interaction and the analysis does not require extensive mathematical calculations. A disadvantage of BiFC is that it cannot be used to visualize protein interaction dynamics. Fluorescent proteins only fluoresce after correct folding and maturation of the fluorophore, which takes longer than the process of protein dimerization or conformational changes.

Practically any protein can be fused with the fluorescent protein fragments, but it has to be experimentally determined which termini of the proteins are suitable for successful complementation of the fluorescent protein. Additionally, appropriate controls are necessary in order to control for false-positive interactions. With the evolution of GFP exhibiting different fluorescent properties, it has become possible to use this technique for detecting and even quantifying interactions between multiple components of a signaling network. With the complementary fragments of the GFP derivatives – BFP, CFP, Cerulean, GFP, YFP, Venus, and Citrine – a large amount of possible reconstituted fluorescent proteins are available and the fluorescence emission signature of the reconstituted molecule can indicate which fluorescence protein halves have reconstituted *(10)*. Recent advances in BiFC engineering have led to improved constructs that report protein–protein interactions under physiological conditions *(8)*.

2. Materials

2.1. Instrumentation

In principle, FRET can be analyzed with any instrument that is able to read fluorescence. For example, it is possible to perform FRET measurements with epi-fluorescence microscopes, confocal laser scanning microscopes, spectrofluorometers, flow cytometers, fluorescence plate readers, fluorescence gel documentation systems, or instruments capable of analyzing fluorescence lifetimes. A requirement for intensity-based FRET analysis is a sufficiently sensitive fluorescence acquisition which allows reliable measurements of donor, sensitized emission and acceptor fluorescence. Complete fluorescence separation of donor from acceptor or donor from sensitized emission is not a requirement as the

FRET efficiency can be obtained by using appropriate mathematical algorithms (described above). The analysis of FRET by confocal microscopy is particularly useful for living biological samples, as the spatial resolution of a confocal microscope allows relating the FRET signal to subcellular structures and to fluorescent ligands. A flow cytometry-based FRET analysis does not permit subcellular FRET resolution, but has the advantage that entire cell populations can be analyzed for the existence of or change in FRET *(11, 12)*. A combination of confocal FRET analysis with flow cytometric FRET analysis would be an ideal scenario for comprehensive analysis of receptor interactions.

There are many commercial confocal microscope systems available which allow the analysis of FRET. We will describe the analysis of FRET as performed on a Leica AOBS confocal microscope. The Leica TCS SP5 AOBS confocal microscopes have an optical configuration that does not make use of any filters or dichroic mirrors. As a result of this completely filter-free optical setup, these instruments have less light loss in the optical path than conventional confocal microscopes. In addition, the acousto-optical devices and the spectral detection system, which are described below, allow for a flexible setup of optical paths. A typical optical path for the Leica microscope is shown in **Fig. 3**: (1) Multiparameter

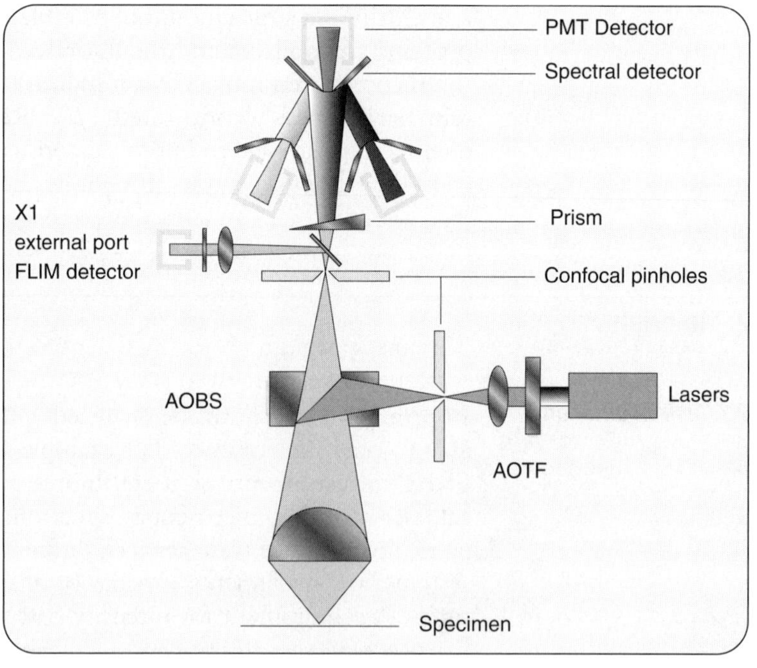

Fig. 3. The principle layout of a Leica confocal microscope is shown. The acousto-optical beam splitter (AOBS) replaces conventionally used dichroic mirrors and the spectral detection system allows a filter-free capture of emission spectra.

analysis requires the selection of several laser lines. To select the laser lines or to attenuate intensity for balanced illumination of different fluorophore densities, acousto-optical tunable filters (AOTF) are used. (2) The fluorescence emission from the specimen is separated by an acousto-optical beam splitter (AOBS), which replaces the conventionally used dichroic mirrors. (3) The spectral detector uses a prism to break up the light. The appropriate wavelength range is selected by a series of spectrophotometer modules. These modules carry reflective mirrors at the edges of the spectrophotometer slits which reflect the wavelengths above and below the captured fluorescence range to other detectors having similar setups. Up to five detectors are implemented in the Leica spectral detection confocal microscopes. The spectral detector replaces the arrays of secondary beam splitters and barrier filters found in conventional confocal microscopes. The nonlinear optical elements (AOTF, AOBS) and the spectral detection system allow for precise wavelength selection in any range at very high light transmission. The complete freedom of selection of laser lines (due to the AOBS) and range of emission capture (due to the spectral detection) is very beneficial for the setup of optimal conditions for FRET experiments.

For FLIM measurements in the time domain on a confocal microscope, a pulsed laser source (e.g., Picoquant, PDL800-B) and a fast detector (e.g., PMC-100-0, Becker & Hickl, Berlin, Germany) along with a time-correlated single photon counting (TCSPC) module (e.g., SPC-830, Becker & Hickl) are necessary. The pulsed excitation can be obtained with a multiphoton laser or a pulsed single-photon laser. For the Leica SP2 AOBS confocal system, a time domain lifetime attachment system (D-FLIM) can be obtained. The Leica TCS SP5 confocal microscopes can be equipped with up to two internal FLIM detectors and with pulsed laser sources. The implementation of TCSCP capability in a laser scanning confocal microscope allows the measurement and analysis of fluorescence lifetimes at each pixel of the confocal scan.

2.2. Data Processing

The image acquisition for FRET measurements can be performed with the Leica confocal software LAS AF. The software provides an easy-to-use interface for different methods of FRET analysis (e.g., acceptor photobleaching FRET or sensitized emission FRET). For lifetime measurements, data analysis can be performed using the SPCImage software provided by Becker & Hickl.

The use of the LAS AF for FRET analysis is not required, as there are many image analysis software packages that can analyze images for the existence of FRET. An excellent freeware program for image analysis is ImageJ, which was developed at the NIH (http://rsb.info.nih.gov/ij/index.html). ImageJ operates with various data formats, can perform a multitude of image manipulations, and it is programmable via a macro interface or using Java.

Table 1
Photo-physical parameters of frequently used fluorescent proteins and fluorophores[a]

Fluorophore names	Extinction coefficient ε (1/(M cm))	Fluorescent quantum yield	Excitation maximum (nm)	Emission maximum (nm)	Laser lines (nm)
ECFP	33,900	0.40	435	475	405,458
EGFP	55,000	0.60	489	508	458,488
EYFP	84,000	0.61	514	527	488,514
mRFP	44,000	0.25	584	607	543,561
FITC	81,000	0.85	495	520	488
Alexa 488	71,000	0.94	495	519	488
R-PE	1,960,000	0.84	498,565	575	488,514,543
Cy3	150,000	0.15	552	570	532,543
Alexa 546	104,000	0.96	556	573	532,543
Cy5	250,000	0.28	643	667	633
Alexa 647	239,000	ND	650	665	633
APC	240,000	0.68	650	660	633

[a]All of these values were obtained from the collective fluorophore database available at George McNamara's web site: http://home.earthlink.net/~pubspectra/.

2.3. Fluorophores

The most commonly used fluorophores applicable for FRET analysis are listed in **Table 1**. Some widely used FRET dye-pairs according to this table are CFP–YFP, YFP (GFP)–mRFP, R-PE–APC, Alexa 488–Alexa 546, Alexa 546–Alexa 647, and Cy3–Cy5. The fluorescence protein pairs CFP–YFP or GFP–mCherry are mostly used for live cell imaging of FRET.

2.4. Reagents

1. DNA transfection reagents: Fugene (Roche, Indianapolis, IN) or Genejuice (EMD Biosciences, Gibbstown, NJ) for transient transfection of cells.
2. Thirty-five millimeter glass-bottom tissue culture dishes (MatTek, Ashland, MA) or lab on a slide imaging slide (ibid.).

3. Methods

3.1. Constructs and Cell Lines

Fluorescent proteins can be fused to either terminus of the protein and the functionality of the fusion partner should be tested empirically. TLRs are type 1 transmembrane receptors with an

N-terminal leader sequence. In our experience, C-terminal fusions of fluorescent proteins with TLRs did not disrupt signaling of the published TLR-FP fusions *(13–30)*. It is advantageous to deliver TLR-FP fusion proteins via viral vectors, such as lentiviruses or retroviruses, as these methods permit dosage of gene integration and thus limit effects of gene overexpression. The most reliable and reproducible FRET measurements can be performed in stably transfected or transduced cells. Transient transfection can act as a cell stressor, which could negatively influence the FRET analysis. For FRET measurements, single color controls (donor and acceptor fluorophores) are necessary for sensitized emission FRET analysis. It is advantageous to generate single color control cells of the fusions used for the FRET experiments, as this would ensure that expression levels of controls are similar to that of the double color sample.

3.2. Data Acquisition and Analysis

3.2.1. Acceptor Photobleaching FRET

As outlined in **Subheading 1**, FRET acceptor photobleaching involves measuring the donor dequenching due to the loss of acceptor fluorescence after acceptor photobleaching. In the acceptor photobleaching method, the change in intensity of the donor fluorophore needs to be monitored as it becomes "dequenched" by the ablation of the acceptor; thus only two channels are detected:

- D = donor, for assessing transfer efficiency
- A = acceptor, for performing photobleaching and determining the efficiency of photobleaching

The actual transfer efficiency is calculated from comparing the donor intensity before (D_{pre}) and after (D_{post}) photobleaching:

$$\mathrm{FRET_{eff}} = \frac{D_{post} - D_{pre}}{D_{post}} \quad \text{for all } \mathrm{D}_{post} > D_{pre}. \tag{8}$$

In theory, all FRET fluorophore pairs can be used for this method. However, in practice, it is better to choose an acceptor fluorophore that is easily photobleached to minimize bleaching times and phototoxic effects of intense laser illumination. Examples of these pairs are Cy3–Cy5, CFP–YFP, GFP–Cy3, and fluorescein–rhodamine.

The Leica confocal microscope software has a built-in wizard that guides the user through an acceptor photobleaching experiment. In the first step, the experimental conditions are defined. Beam path and PMT settings are defined for the donor and the acceptor fluorophore. PMT and laser power should be adjusted so that the PMT is not saturated. In the next step, a region of interest (ROI) is selected for the acceptor photobleaching procedure. Here, the laser illuminating the acceptor fluorophore should be adjusted to 100% power level to ensure efficient photobleaching.

In the bleach configuration window, one can choose to bleach a certain number of scans or one can utilize a "regulator" which ascertains that the bleaching is performed up to a set intensity of the acceptor fluorescence. The experiment is then started and the instrument executes the following steps: (1) acquisition of the prebleach images, (2) photobleaching of the selected ROI, and (3) acquisition of the postbleach images. After the experiment has been executed, the software calculates the FRET efficiency according to **Eq. 8** and displays the pre- and postbleach images.

3.2.2. Sensitized Emission FRET

Sensitized emission FRET is a nondestructive method of FRET analysis and is most frequently used for live cell experiments. Any sensitized emission FRET experiment requires at least three samples: a sample containing donor only, a sample containing acceptor only, and a sample containing both fluorophores. The samples containing single fluorophores are used for calculating spectral overlap and cross-excitation from different laser lights at different detectors. In addition, single color controls are generally useful to determine whether the FRET calculation was accurate and did not lead to false-positive FRET assignments. In addition to these required fluorescent samples, it is also advantageous to prepare a positive FRET control, such as a fusion between donor and acceptor fluorescent proteins. The Leica Confocal Software contains a FRET analysis wizard. The user follows several steps that are briefly outlined below.

In **step 1**, the experimental conditions are set for the acquisition of the donor, FRET (sensitized emission), and acceptor channels. These three images are required for the calculation of the correction factors. Two scans will be performed in a line-by-line fashion. The first scan should be set up with a laser exciting the donor (e.g., 405 or 458 nm for CFP) and two detectors activated for the emission acquisition of donor fluorescence and FRET fluorescence. Adjust the PMT gains so that pixels are not saturated. For the second scan, a laser that excites the acceptor fluorophore is selected (e.g., 514 nm for YFP) and the emission of the acceptor is captured with the same detector that was used in the first scan for the FRET channel. The PMT settings for this channel should not be changed at all. Instead, adjust the image brightness by changing the laser power until the image setting is satisfactory (*see* **Note 3**). Define the zoom level and number of averages for best imaging conditions and proceed to the next step.

In **step 2**, the actual images are acquired. The software is set up for the acquisition of three samples: FRET, donor, and acceptor (A, B, and C in the FRET algorithm). It is necessary to have single positive cells for donor and acceptor fluorophores in the same or independent dishes (*see* **Note 4**). In **step 3**, the calibration is performed. Ensure that you have the appropriate image set active and draw an ROI around a representative cell that expresses either donor or acceptor fluorophore alone. If the background between the samples differs or the background fluorescence is above zero, perform a background correction by selecting ROIs in regions of the sample that are not covered by cells. After the ROIs have been correctly assigned to acceptor and donor fluorescence and the background has been sampled, press button "Next" to obtain the correction factors. The calculated calibration factors will be applied to the current FRET sequence and to all subsequent images that are acquired with unchanged settings. It is possible to save the correction factors and reuse them. However, the precondition for reusing saved settings is that the measurements are performed with exactly the same imaging conditions.

The theoretical background of FRET analysis is given in **Subheading 1**. There are many different ways to calculate FRET efficiency using sensitized emission FRET. The equations differ mostly in the use of correction factors. The Leica FCS software for the SP5 generation instruments gives the user the choice to select a method for the FRET efficiency calculation (*see* **Note 5**).

The FRET output is provided as an image in which the FRET efficiencies found at each pixel of the image are false-colored using a look-up table that codes changes in FRET efficiency as changes in color. It is also possible to analyze particular regions of interest in the image for the amount of FRET efficiency.

3.2.3. FLIM-FRET

There are many different microscopy systems available that can analyze fluorescence lifetimes of samples. We only describe the general principle of image acquisition and analysis. The acquisition of FLIM data for FRET analysis is very simple. The first step is to utilize regular intensity-based imaging to find the sample ROI for the fluorescence lifetime analysis. This area is then scanned by, for example, a pulsed laser light source. For CFP–YFP FRET experiments, a pulsed diode laser emitting at 405 nm can be used at a frequency of 20 MHz. The pulsed laser illuminates the sample over a number of image frames until enough photons have been acquired for the analysis of lifetimes. The acquisition time depends on the amount of fluorescence in the sample and on the required analysis method (see below). The photons detected by the FLIM detector are processed in the TCSPC module, which is controlled by the FLIM SPCM software. The data recorded by the FLIM system are multidimensional. There are two-dimensional arrays of pixel (i.e., in *XY* direction) and each pixel contains the information of fluorescence lifetime, which is the counted photon number in a series of time channels. The SPCM software is able to display the data in various display modes. The number of photons per pixel is typically displayed as shades of gray in an intensity image, which allows an overview on photon counts over the entire image.

The next step of FLIM analysis is the decay data analysis, which is also performed by the SPCM software. To obtain fluorescence lifetimes decay curves, the photon decay information at each recorded pixel must be fitted with an appropriate mathematical model. Since the measurement system is not infinitely fast, the fitting algorithms have to take the so-called "instrument response function" into account. The instrument response function is a pulse shape that is recorded for an infinitely short fluorescence lifetime. The fitting procedure convolutes the model decay function with the instrument response function and relates the results with the photon numbers in the subsequent time channels in each individual pixel. The algorithm varies the model parameters until the best fit between the convoluted model function and the measured data is obtained. This procedure is termed iterative deconvolution. During the analysis procedure, the user has influence on a variety of parameters including the selection of an appropriate fitting model. Typical models used for lifetime analysis are single exponential or a sum of exponential terms. The selection depends on the sample and one can be guided by the goodness of fit of a particular model that was used.

Biological samples often contain fluorescence components of several fluorescent dyes. For example, when CFP–YFP pair is used, one would expect at least two components: fluorescent decay time of CFP and of YFP, and if FRET exists, another decay time of CFP engaged in energy transfer is observed. The decay curves are then multiexponential. If more exponential terms are necessary for the curve fitting procedure, calculation times will become longer and more photon counts are necessary in each pixel in order to achieve an accurate curve fitting. It is advisable to restrict the amount of dyes that are sampled by TCSCP by use of barrier filters that can be placed in front of the FLIM detector (or in the Leica TCS SP5 model by use of the spectral detection system). For example, one could sample CFP-derived photons only by collecting light up to a wavelength of 495 nm. This would ensure that the recorded fluorescence lifetimes are not contaminated with fluorescence decay data of YFP. Thus, a double-exponential decay model can be applied to analyze the lifetime components of CFP, and the CFP molecule that is engaged in energy transfer. The fitting procedure delivers lifetimes and amplitude coefficients for the individual exponential components. For example, a double-exponential decay function is described by

$$[\mathrm{N}(t) = a_1 \cdot \mathrm{e}^{-t/\tau_1} + a_2 \cdot \mathrm{e}^{-t/\tau_2} \left(a_1 + a_2 = 1\right). \qquad (9)$$

If this model is used for CFP–YFP FRET under conditions where YFP lifetimes are not recorded, the CFP lifetimes, τ_1 and τ_2, and the amplitudes, a_1 and a_2, are obtained. A slow lifetime component (CFP, non-FRET) can be separated from a fast lifetime component (CFP, FRET). These data can then be utilized to calculate FRET efficiencies at each pixel of the image. Lifetimes can be reported in a variety of ways. For example, mean lifetimes can be reported as distribution plots of lifetimes observed in the images and the color-coded lifetime images represent the number of photons per pixel as brightness and the fluorescence lifetime as color.

3.2.4. Bimolecular Fluorescence Complementation

The fluorescence analysis of fluorescence complementation is not different from the analysis of fluorescence obtained from the full-length (nonsplit) fluorescent proteins. In the absence of fluorescence complementation, no fluorescence signal is obtained and a positive signal is visible if complementation between the two fluorescent protein peptides occurs. Proximity-assisted folding of fluorescent protein fragment is distance dependent. However, the minimal distance between two fluorescent protein fragments that is needed for effective complementation is not well established. The distance between the two fluorescent proteins is influenced by a number of factors. Firstly, the formation of the receptor complex influences the position of the proteins termini that carry the fluorescent protein fragments and the overall dimension of an active receptor complex also influences the distance between the fragments. Additionally, the linker length between the fluorescent peptides and the fusion partner can influence the complementation of the peptides.

For efficient fluorescent protein complementation to occur, we utilized TLR receptors lacking the Toll/interleukin-1 receptor (TIR) domain. The lengths of the linkers need to be optimized empirically for each construct. The linker can be too short and thus could restrict efficient protein folding. Conversely, it is also possible to choose linkers that are too long leading to reconstitution of the fluorescent protein in the absence of receptor activation.

For receptors that form homodimers, it is important to ensure that the receptor version carrying the N- or C-terminal portion of the fluorescent protein fragments isexpressed at similar ratios. If one construct is expressed at relatively higher levels, homodimerization of this construct is favored. This would lead to a receptor dimer that brings together either only N- or C-terminal fragments of the fluorescent protein, which does not allow fluorescent protein maturation and fluorescence.

4. Notes

1. According to Förster's theory *(1)*, the rate constant of the FRET interaction can be described by the equation

$$k_T = (k_F + k_0) \cdot \left(\frac{R_0}{R}\right)^6, \tag{10}$$

which combined with **Eq. 1** gives the distance dependence of transfer efficiency (**Eq. 2**). The Förster critical distance, R_0 (in nanometers), is calculated from the photo-physical parameters of the interacting fluorophores:

$$[R_0 = 978 \cdot (Q_D \cdot \kappa^2 \cdot n^{-4} \cdot J_{DA})^{1/6}, \tag{11}$$

where Q_D is the fluorescence quantum efficiency of the donor, n is index of refraction of the conveying medium (assumed to be 1.4 for cells in aqueous media), κ^2 is the orientation factor (which has a value between 0 and 4, and is assumed to be 2/3 for dynamic averaging *[31]*), and J_{DA} is the overlap integral. The overlap integral (in the unit of M^{-1} cm^3) can be determined from the emission spectrum of the donor and the excitation spectrum of the acceptor dye normalized to unity:

$$J_{DA} = \frac{\int_0^\infty F_D(\lambda) \cdot \varepsilon_A(\lambda) \cdot \lambda^4 d\lambda}{\int_0^\infty F_D(\lambda) d\lambda} \tag{12}$$

2. Seeing **Eq. 2**, it is very tempting to calculate the distance of two proteins from transfer efficiencies; however, extra care should be taken as how to interpret those values. If the protein-labeling scheme involves antibodies, one has to note that the size of a Fab fragment is about 5.0 nm and the fluorophores can be unevenly distributed along the protein. If fluorescent proteins were used as protein tags, then the closest separation of two fluorescent proteins is approximately 5.0 nm, as the fluorophore is situated inside the protein. This implies that using the fluorescent protein FRET pair, the FRET transfer efficiency has a practical limit of around 50% transfer efficiency. Another factor that should be considered is the way the FRET experiment and the transfer efficiency evaluation were carried out. It is possible to obtain transfer efficiency values as an average for a population of cells based on the mean fluorescence intensities of the population or the mean of transfer efficiencies of individual cells, or the mean for a single cell, or the mean of several hundreds of proteins located in one pixel. All these procedures result in mean transfer efficiencies; however, the averaging is fundamentally different, and accordingly the average molecular distances should have different interpretations.

3. Depending on the expression level of the protein, it may be required to open the pinhole to above one airy unit in order to keep the detector gain at levels below 600 V, which is required for appropriate image quality.

   For CFP–YFP FRET the following settings could be used: Scan 1: CFP excitation 405 nm (if available) or alternatively 458 nm; CFP emission: 462–485 nm; FRET emission: 518–580 nm. Scan 2: YFP excitation 514 nm. The images are acquired in a line-by-line fashion and it is important to keep the emission acquisition settings for acceptor exactly the same as for the FRET channel. In addition, it is vital to keep the PMT settings of the acceptor emission exactly the same as for the FRET channel (set in scan 1). In practice, only three things in the setup for the second scan need change: Firstly, reduce the laser light of the donor laser to 0%. Secondly, deactivate the CFP channel, leaving the YFP channel (i.e., the FRET channel in scan 1) at 518–580 nm. Thirdly, adjust the laser power of the 514 nm acceptor laser so that the acceptor cell emission is within the dynamic range of the PMT (no saturation of pixels) and do not change the PMT setting of the acceptor channel at all. It is vital to keep all measurements under identical conditions as the calculation of bleed-through and cross-excitation is based on these settings. If your controls do not match the actual imaging conditions, i.e., one of the control is brighter than the FRET sample and saturates the channel under the conditions where the FRET sample is optimally illuminated, one should try to use cells that express lower amounts of protein (comparable to the FRET sample) for the calibration. One should not readjust the settings of the PMT between control and sample. All measurements performed under differing conditions are invalid and should be discarded!

4. It is not mandatory to have three independent culture dishes with the different samples and to acquire three images. In fact, it is beneficial to have all three cell populations in the same culture dish. It is then possible to only acquire one image having all three cell populations in one field. In the following step, regions of interests are defined for donor and acceptor fluorescence and this can be done from one image containing donor and acceptor single positive cells or from independent images (which can then be background corrected).

5. The Leica SP2 generation FCS software utilizes the following equations for the FRET efficiency calculation. As outlined above, three channels have to be acquired for the FRET calculation:

   – A = donor emission by excitation of the donor

   – B = FRET emission by excitation of the donor

   – C = acceptor emission by excitation of the acceptor

As mentioned earlier, in a sample containing both fluorophores, these channels do not represent pure donor or acceptor signals, but contain some spectral spillovers from the other dyes, and thus have to be corrected for bleed-through and cross-excitation. These correction factors are obtained from samples containing single fluorophore:

- $a = A/C$ (acceptor-only sample)
- $b = B/A$ (donor-only sample)
- $c = B/C$ (acceptor-only sample)

The transfer efficiency is calculated as follows on a pixel-by-pixel basis:

$$FRET = B - b \cdot A - (c - a \cdot b)C, \quad (13)$$

$$FRET_{eff} = FRET / C. \quad (14)$$

The Leica SP5 instrument software allows the user to choose different methods for FRET efficiency calculation.

6. If fluorescent proteins are used for FRET analysis by the acceptor photobleaching method, one should remember that accurate measurements of FRET can only be obtained from fixed cells. In living cells, proteins diffuse through the cells with a diffusion constant that depends on the subcellular compartment in which the protein is situated. The time between the acceptor photobleaching and the acquisition of the postbleach image allows diffusion of unquenched acceptor fluorophores into the region that is already bleached. It is possible to bleach an entire living cell and compare the donor fluorescence before and after photobleaching. However, depending on cell size and fluorophore expression levels, photobleaching can take longer than the bleaching of smaller regions of interest.

References

1. Forster, T. (1948) Zwischenmolekulare Energiewanderung und Fluoreszenz. *Ann Phys* 2, 55–75.
2. Stryer, L. (1978) Fluorescence energy transfer as a spectroscopic ruler. *Annu Rev Biochem* 47, 819–46.
3. Jovin, T. M. and Arndt-Jovin, D. J. (1989) Luminescence digital imaging microscopy. *Annu Rev Biophys Biophys Chem* 18, 271–308.
4. Szentesi, G., Vereb, G., Horvath, G., Bodnar, A., Fabian, A., Matko, J., Gaspar, R., Damjanovich, S., Matyus, L., and Jenei, A. (2005) Computer program for analyzing donor photobleaching FRET image series. *Cytometry A* 67, 119–28.
5. Gratton, E., Limkeman, M., Lakowicz, J. R., Maliwal, B. P., Cherek, H., and Laczko, G. (1984) Resolution of mixtures of fluorophores using variable-frequency phase and modulation data. *Biophys J* 46, 479–86.
6. Suhling, K., Siegel, J., Phillips, D., French, P. M., Leveque-Fort, S., Webb, S. E., and Davis, D. M. (2002) Imaging the environment of green fluorescent protein. *Biophys J* 83, 3589–95.
7. Hu, C. D., Chinenov, Y., and Kerppola, T. K. (2002) Visualization of interactions among

bZIP and Rel family proteins in living cells using bimolecular fluorescence complementation. *Mol Cell* 9, 789–98.

8. Shyu, Y. J., Liu, H., Deng, X., and Hu, C. D. (2006) Identification of new fluorescent protein fragments for bimolecular fluorescence complementation analysis under physiological conditions. *Biotechniques* 40, 61–6.
9. Grinberg, A. V., Hu, C. D., and Kerppola, T. K. (2004) Visualization of Myc/Max/Mad family dimers and the competition for dimerization in living cells. *Mol Cell Biol* 24, 4294–308.
10. Hu, C. D., and Kerppola, T. K. (2003) Simultaneous visualization of multiple protein interactions in living cells using multicolor fluorescence complementation analysis. *Nat Biotechnol* 21, 539–45.
11. Horvath, G., Petras, M., Szentesi, G., Fabian, A., Park, J. W., Vereb, G., and Szollosi, J. (2005) Selecting the right fluorophores and flow cytometer for fluorescence resonance energy transfer measurements. *Cytometry A* 65, 148–57.
12. Szentesi, G., Horvath, G., Bori, I., Vamosi, G., Szollosi, J., Gaspar, R., Damjanovich, S., Jenei, A., and Matyus, L. (2004) Computer program for determining fluorescence resonance energy transfer efficiency from flow cytometric data on a cell-by-cell basis. *Comput Methods Programs Biomed* 75, 201–11.
13. Latz, E., Visintin, A., Lien, E., Fitzgerald, K. A., Espevik, T., and Golenbock, D. T. (2003) The LPS receptor generates inflammatory signals from the cell surface. *J Endotoxin Res* 9, 375–80.
14. Espevik, T., Latz, E., Lien, E., Monks, B., and Golenbock, D. T. (2003) Cell distributions and functions of Toll-like receptor 4 studied by fluorescent gene constructs. *Scand J Infect Dis* 35, 660–4.
15. Fitzgerald, K. A., Rowe, D. C., Barnes, B. J., Caffrey, D. R., Visintin, A., Latz, E., Monks, B., Pitha, P. M., and Golenbock, D. T. (2003) LPS-TLR4 signaling to IRF-3/7 and NF-kappaB involves the toll adapters TRAM and TRIF. *J Exp Med* 198, 1043–55.
16. Flo, T. H., Ryan, L., Latz, E., Takeuchi, O., Monks, B. G., Lien, E., Halaas, O., Akira, S., Skjak-Braek, G., Golenbock, D. T., and Espevik, T. (2002) Involvement of toll-like receptor (TLR) 2 and TLR4 in cell activation by mannuronic acid polymers. *J Biol Chem* 277, 35489–95.
17. Massari, P., Henneke, P., Ho, Y., Latz, E., Golenbock, D. T., and Wetzler, L. M. (2002) Cutting edge: immune stimulation by neisserial porins is toll-like receptor 2 and MyD88 dependent. *J Immunol* 168, 1533–7.
18. Parroche, P., Lauw, F. N., Goutagny, N., Latz, E., Monks, B. G., Visintin, A., Halmen, K. A., Lamphier, M., Olivier, M., Bartholomeu, D. C., Gazzinelli, R. T., and Golenbock, D. T. (2007) Malaria hemozoin is immunologically inert but radically enhances innate responses by presenting malaria DNA to Toll-like receptor 9. *Proc Natl Acad Sci U S A* 104, 1919–24.
19. Rowe, D. C., McGettrick, A. F., Latz, E., Monks, B. G., Gay, N. J., Yamamoto, M., Akira, S., O'Neill, L. A., Fitzgerald, K. A., and Golenbock, D. T. (2006) The myristoylation of TRIF-related adaptor molecule is essential for Toll-like receptor 4 signal transduction. *Proc Natl Acad Sci U S A* 103, 6299–304.
20. Sandor, F., Latz, E., Re, F., Mandell, L., Repik, G., Golenbock, D. T., Espevik, T., Kurt-Jones, E. A., and Finberg, R. W. (2003) Importance of extra- and intracellular domains of TLR1 and TLR2 in NFkappa B signaling. *J Cell Biol* 162, 1099–110.
21. Compton, T., Kurt-Jones, E. A., Boehme, K. W., Belko, J., Latz, E., Golenbock, D. T., and Finberg, R. W. (2003) Human cytomegalovirus activates inflammatory cytokine responses via CD14 and Toll-like receptor 2. *J Virol* 77, 4588–96.
22. Fitzgerald, K. A., McWhirter, S. M., Faia, K. L., Rowe, D. C., Latz, E., Golenbock, D. T., Coyle, A. J., Liao, S. M., and Maniatis, T. (2003) IKKepsilon and TBK1 are essential components of the IRF3 signaling pathway. *Nat Immunol* 4, 491–6.
23. Husebye, H., Halaas, O., Stenmark, H., Tunheim, G., Sandanger, O., Bogen, B., Brech, A., Latz, E., and Espevik, T. (2006) Endocytic pathways regulate Toll-like receptor 4 signaling and link innate and adaptive immunity. *Embo J* 25, 683–92.
24. Latz, E., Franko, J., Golenbock, D. T., and Schreiber, J. R. (2004) Haemophilus influenzae type b-outer membrane protein complex glycoconjugate vaccine induces cytokine production by engaging human toll-like receptor 2 (TLR2) and requires the presence of TLR2 for optimal immunogenicity. *J Immunol* 172, 2431–8.
25. Latz, E., Verma, A., Visintin, A., Gong, M., Sirois, C. M., Klein, D. C., Monks, B. G., McKnight, C. J., Lamphier, M. S., Duprex, W. P., Espevik, T., and Golenbock, D. T. (2007) Ligand-induced conformational changes allosterically activate Toll-like receptor 9. *Nat Immunol* 8, 772–9.

26. Latz, E., Visintin, A., Espevik, T., and Golenbock, D. T. (2004) Mechanisms of TLR9 activation. *J Endotoxin Res* 10, 406–12.
27. Sau, K., Mambula, S. S., Latz, E., Henneke, P., Golenbock, D. T., and Levitz, S. M. (2003) The antifungal drug amphotericin B promotes inflammatory cytokine release by a Toll-like receptor- and CD14-dependent mechanism. *J Biol Chem* 278, 37561–8.
28. van der Kleij, D., Latz, E., Brouwers, J. F., Kruize, Y. C., Schmitz, M., Kurt-Jones, E. A., Espevik, T., de Jong, E. C., Kapsenberg, M. L., Golenbock, D. T., Tielens, A. G., and Yazdanbakhsh, M. (2002) A novel host-parasite lipid cross-talk. Schistosomal lyso-phosphatidylserine activates toll-like receptor 2 and affects immune polarization. *J Biol Chem* 277, 48122–9.
29. Visintin, A., Halmen, K. A., Latz, E., Monks, B. G., and Golenbock, D. T. (2005) Pharmacological inhibition of endotoxin responses is achieved by targeting the TLR4 coreceptor, MD-2. *J Immunol* 175, 6465–72.
30. Visintin, A., Latz, E., Monks, B. G., Espevik, T., and Golenbock, D. T. (2003) Lysines 128 and 132 enable lipopolysaccharide binding to MD-2, leading to Toll-like receptor-4 aggregation and signal transduction. *J Biol Chem* 278, 48313–20.
31. van der Meer, B. W. (2002) Kappa-squared: from nuisance to new sense. *J Biotechnol* 82, 181–96.

Chapter 4

Predicting Toll-Like Receptor Structures and Characterizing Ligand Binding

Joshua N. Leonard, Jessica K. Bell, and David M. Segal

Summary

Toll-like receptor (TLR) ligand-binding domains comprise 18–25 tandem copies of a 24-residue motif known as the leucine-rich repeat (LRR). Unlike other LRR proteins, TLRs contain significant numbers of non-consensus LRR sequences, which makes their identification by computer domain search programs problematic. Here, we provide methods for identifying non-consensus LRRs. Using the location of these LRRs, hypothetical models are constructed based on the known molecular structures of homologous LRR proteins. However, when a hypothetical model for TLR3 is compared with the molecular structure solved by x-ray crystallography, the solenoid curvature, planarity, and conformations of the LRR insertions are incorrectly predicted. These differences illustrate how non-consensus LRR motifs influence TLR structure. Since the determination of molecular structures by crystallography requires substantial amounts of protein, we describe methods for producing milligram amounts of TLR3 extracellular domain (ECD) protein. The recombinant TLR3-ECD previously used to solve the molecular structure of TLR3-ECD has also been used to study the binding of TLR3-ECD to its ligand, double-stranded RNA (dsRNA). In the last section, we describe the preparation of defined TLR3 ligands and present methods for characterizing their interaction with TLR3-ECD.

Key words: Toll-like receptor, TLR, Homology-based modelling, Receptor specificity, Double-stranded RNA, dsRNA.

1. Introduction

Pathogen recognition by Toll-like receptors (TLRs) is essential for initiating immune responses, but the molecular mechanism by which receptor:ligand interactions trigger this response remains poorly understood *(1)*. TLR extracellular domains (ECDs) recognize and bind ligands through tandem copies of a motif known

Claire E. McCoy and Luke A.J. O'Neill (eds.), *Methods in Molecular Biology, Toll-Like Receptors, vol. 517*
DOI: 10.1007/978-1-59745-541-1_4

as the leucine-rich repeat (LRR) *(2)*. The molecular structures of several LRR-containing proteins are known, making it possible to propose hypothetical structural models of TLR-ECDs from their amino acid sequences *(3)*. These models may fail to accurately predict features such as solenoid curvature, planarity, and the conformations of non-consensus LRRs. Such limitations indicate that true TLR molecular structures can be determined only by high resolution x-ray crystallography, which requires several milligrams of recombinant protein. In addition, by using recombinant protein for solution-phase ligand-binding studies, the stochiometry, affinity, and binding kinetics of the receptor:ligand complex can be determined. These parameters are essential for the production of receptor:ligand co-crystals, thus completing the atomic-scale picture of ligand recognition. Recently, we and another group solved the molecular structure of the TLR3-ECD *(4, 5)*, and here we describe some of the procedures we have used to propose hypothetical structural models, generate TLR3-ECD protein, and measure its interaction with its ligand, double-stranded RNA (dsRNA). Many of these procedures can likely be adapted to characterize other TLRs.

2. Materials

2.1. Modelling TLR-ECD Structures

2.1.1. Locating Leucine-Rich Repeats

1. TLR amino acid sequence.
2. SMART (Simple Modular Architecture Research Tool) web site (http://smart.embl-heidelberg.de/).

2.1.2. Using LRRs to Predict Structures of TLRs

1. TLR amino acid sequence with LRRs identified.
2. Amino acid sequence with individual LRRs identified and structural coordinates for homologous proteins to be used as model scaffold such as the Nogo receptor (PDB 1OZN *[6]*).
3. Molecular software program O from Uppsala Software Factory (http://xray.bmc.uu.se/usf).

2.2. TLR3-ECD Production and Purification

1. High Five™ cells (Invitrogen, Carlsbad, CA).
2. Express Five® SFM medium (Invitrogen) supplemented with 20 mM l-glutamine, 10 μg/mL gentamicin, or Hy-SFX medium (HyClone, Logan, UT) supplemented with 10 μg/mL gentamicin media.
3. Complete EDTA-free protease inhibitor cocktail tablets (Roche Applied Science, Indianapolis, IN).
4. Immobilized metal affinity chromatography (IMAC) resin (NiNTA, Qiagen, Valencia, CA).

5. Binding buffer (for IMAC): phosphate-buffered saline (PBS), pH 7.4 supplemented with 350 mM NaCl, 5% glycerol, and 1 mM β-mercaptoethanol (BME).
6. Elution buffer (for IMAC): binding buffer + 500 mM imidazole.
7. Anti-FLAG® M2 agarose and FLAG® peptide (Sigma, St. Louis, MO).
8. Binding/wash buffer (for Anti-FLAG®): tris-buffered saline (TBS), pH 7.4.
9. Elution buffer (for Anti-FLAG®): TBS + 100 μg/mL FLAG® peptide.
10. Centriprep YM-30 concentrator (Millipore, Billerica, MA).
11. Akta™ system with Superdex 200 (2.6 × 100 cm) size exclusion chromatography (SEC) column (GE Healthcare, Piscataway, NJ).
12. SEC buffer: 20 mM buffer (dependent upon application), 150 mM NaCl, 5% glycerol, 1 mM BME.

2.3. Ligand Binding to TLR3-ECD

2.3.1. dsRNA Synthesis

1. dsRNA template vector: pGEM-T Easy (Promega, Madison, WI). Store at −20°C.
2. NdeI restriction enzyme + 10X buffer (New England Biolabs (NEB), Ipswich, MA). Store at −20°C.
3. QIAquick PCR purification kit (Qiagen).
4. dsRNA synthesis kit: T7 Ribomax (Promega), includes Ribomax Express buffer (2X), T7 Express enzyme mix, RQ1 DNAse, RNAse A, nuclease-free water, 3.2 M NaOAc. Store enzyme and buffer components at −20°C (store RNAse A at room temperature), and minimize freeze-thawing the buffer by aliquotting after the first thaw.
5. PIPES-buffered saline (PiBS): 20 mM PIPES (Sigma), 150 mM NaCl, pH 6.0 (unless otherwise indicated). Store at room temperature.
6. Akta™ liquid chromatography system with Superdex 200 10/300 GL column (GE Healthcare).
7. Absolute ethanol.

2.3.2. Ligand Biotinylation

1. Antarctic phosphatase + 10X buffer (NEB). Store at −20°C.
2. Adenosine 5′-(gamma-thio) triphosphate (ATP-γS, Sigma). Dissolve stock in water to 10 mM, and store aliquots at −20°C.
3. Polynucleotide kinase (PNK) + 10X buffer (NEB). Store at −20°C.
4. No-weigh maleimide-PEO$_2$-biotin (Pierce, Rockford, IL). Store at 4°C, and dissolve in water immediately prior to use.
5. Saturated phenol.

2.3.3. TLR3 Binding ELISAs

1. Reacti-Bind Streptavidin HBC strip-wells (Pierce). Store at 4°C.
2. PiBST: PiBS + 0.1% Tween-20 (Sigma). Store at room temperature.
3. PolyI:C: to prepare stocks, dissolve a mixture of polyinosinic and polycytidylic acids (GE Healthcare) in PBS to 2 mg/mL, heat to 70°C for 10 min, then cool slowly to room temperature to anneal strands. Store aliquots at –20°C.
4. Anti-FLAG® M2-HRP monoclonal antibody (mAb) (Sigma).
5. Bovine serum albumin (BSA) (Sigma).
6. HRP substrate reagent (R&D systems, Minneapolis, MN).
7. H_2SO_4 (1 M).

3. Methods

3.1. Modelling TLR-ECD Structures

3.1.1. Locating Leucine-Rich Repeats

TLR-ECDs consist of 18–25 consecutive LRR motifs. The consensus LRR found in these receptors contains 24 amino acid residues (**Fig. 1A**). However, TLRs typically contain several non-consensus LRRs, which are not recognized by domain search programs such as protein families (Pfam) *(7)*. Therefore, LRRs that deviate from the consensus sequence must be located manually. Such an approach was used to locate the LRRs in the ten human TLRs *(3)*, which, in the case of TLR3, was shown to be valid at the structural level *(4, 5)*.

1. The TLR amino acid sequence is copied into the SMART web site (http://smart.embl-heidelberg.de/) and searched for Pfam domains and signal sequences. This will locate the ECD and the consensus LRRs, as shown for TLR9 in **Fig. 1B**. Note also the intervening, undefined sequences between the consensus LRRs.
2. Deviations from the consensus LRR sequence cause search programs to miss LRRs. To find these missed LRRs, the regions between the consensus LRRs should be examined for L-x-x-L-x-L-x-x-N- sequences. Note that leucine residues can be replaced by other hydrophobic amino acids such as I, V, M, and F, and that N is frequently replaced by C, S, and T. The L-x-x-L-x-L-x-x-N- motif is present in the N-terminal portions of most LRRs. By contrast, the C-terminal portions are more variable in both length and sequence.

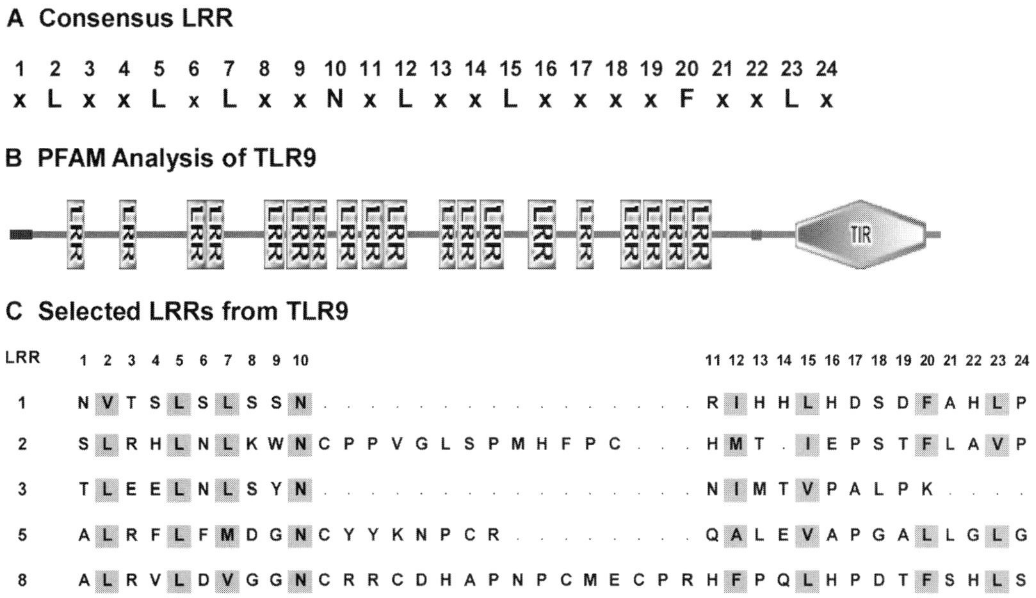

Fig. 1. Location of leucine-rich repeats (LRR) in Toll-like receptors (TLRs). (**A**) The consensus 24-residue sequence found in TLRs and some other LRR proteins such as the Nogo receptor and CD42b. (**B**) Domain organization of TLR9 as determined by protein families (Pfam) computer analysis. TIR refers to the cytoplasmic (Toll/IL-1R/Resistance) signalling domain, and the LRRs are the consensus LRRs located by the search algorithm. A signal sequence is located at the N-terminus, and a transmembrane segment separates the extracellular domain (ECD) from the cytoplasmic domain. (**C**) Representative LRRs from TLR9. LRR1 is a consensus LRR, while LRRs 2, 5, and 8 contain insertions of 13, 8, and 16 residues, respectively, following the consensus asparagine residue at position 10. By contrast, LRR3 is four residues shorter than the consensus.

3. To locate the C-terminal ends of LRRs, look for the next L-x-x-L-x-L-x-x-N- motif. As a guide, the C-terminal portions of LRRs typically contain F-x-x-L or L-x-x-L sequences one or two residues before the beginning of the next LRR. LRRs are usually 24 residues long, but this can vary considerably. For example, **Fig. 1C** shows several non-consensus LRRs from TLR9, which were missed by domain search algorithms but were identified by their N-terminal L-x-x-L-x-L-x-x-N- sequence and the next LRR.

3.1.2. Using LRRs to Predict Structures of TLRs

It is instructive to create a hypothetical model for the TLR-ECDs based upon the known molecular structures of other LRR proteins and the number of LRRs within TLR-ECDs *(3, 8)*. The Nogo receptor contains LRR sequences that conform to the TLR consensus motif shown in **Fig. 1A**. In addition, the nine contiguous LRRs in the Nogo crystal structure provide a large building block for constructing models of TLRs, which contain 18–25 LRRs. Below is an example of how a model structure for the TLR3-ECD was constructed:

1. The N-terminal portion of the TLR3-ECD model is composed of Nogo receptor residues 26–248 (LRR-NT–LRR9) (**Fig. 2A**, segment 1). Coordinates of LRRs 2–9 of the Nogo receptor are extracted from its Protein Data Bank (PDB, http://www.rcsb.org/pdb/home/home.do) file to be used as a "building unit." The next section of the TLR3-ECD model is composed of one building unit (Nogo LRRs 2–9). To align the first two sections of the model, LRRs 2–4 of the building unit are overlayed with LRRs 7–9 of the N-terminal unit (**Fig. 2A**, segment 2) using a molecular software package such as the program O.
2. This procedure is repeated by aligning LRRs 2–4 of a second building unit with LRRs 7–9 of the first building unit (**Fig. 2A**, segment 3). To complete the C-terminal portion of the TLR3-ECD model, Nogo receptor residues 56–309 (LRR2-LRR-CT) are overlayed with the second building unit. In this case, the first three LRRs of the Nogo receptor (residues 56–309) are aligned with LRRs 6–9 of building unit two (**Fig. 2A**, segment 4).

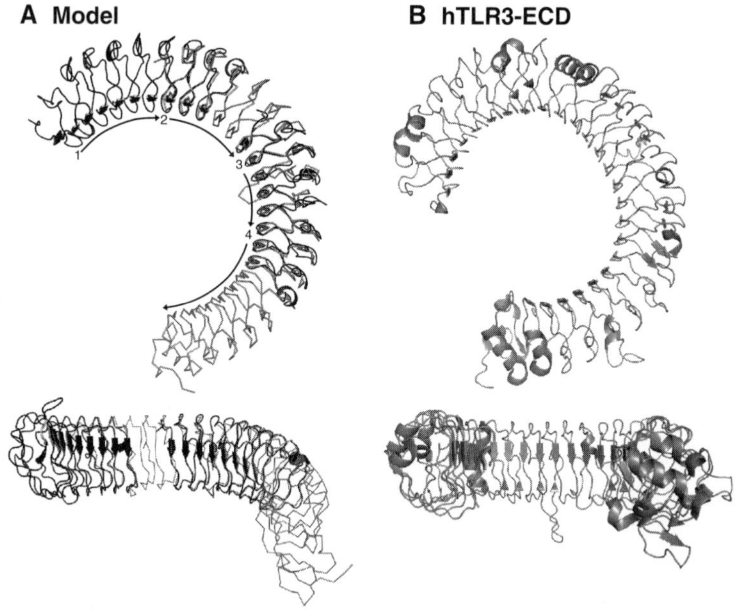

Fig. 2. Cartoon representations of the TLR3-extracellular domain (ECD) structure. Upper figures show lateral views of the Toll-like receptor (TLR) "horseshoe," and lower figures look into the open end of the horseshoe. (**A**) A model based upon the Nogo receptor structure. Segment 1, the N-terminal region, consisting of residues 26–248 of the Nogo receptor; segment 2, building unit 1; segment 3, building unit 2; segment 4, C-terminal unit (residues 56–309 of the Nogo receptor). (**B**) The TLR3-ECD crystallographic structure. Note that the model correctly predicts the overall shape of the TLR3-ECD but fails to predict the planarity of the TLR structure (*bottom views*). Moreover, the actual structure exhibits greater curvature than expected from the model.

3. LRRs that were duplicated in the alignment process (N-terminal unit LRRs 7–9, building unit 1, LRRs 7–9, building unit 2, LRRs 6–9) are removed.

A comparison of this hypothetical TLR3-ECD model with the known crystal structure is shown in **Fig. 2A, B**. Although the model grossly reflects the correct "horseshoe" shape of the TLR3-ECD molecule, the curvature and planarity of the horseshoe are not predicted correctly. Since the model was constructed using consensus LRRs, this comparison suggests that the curvature and planarity of the actual TLR3 structure derive from non-consensus LRRs.

3.2. TLR3-ECD Production and Purification

To determine the molecular structures of TLR-ECDs and to determine how they interact with pathogens at the molecular level, milligram amounts of pure protein are required. In our experience, a baculovirus secretion system has proven to be the best method for producing large amounts of correctly folded protein. However, there is great variation in yield between the different TLRs, and the conditions for expression need to be optimized for each TLR. Here, we describe the method used for expression and purification of TLR3-ECD protein, which is engineered to contain both FLAG® and His_6 affinity tags fused to its C-terminus.

1. High Five™ cells (2×10^6 cells/mL) are infected with TLR3-ECD baculovirus at a multiplicity of infection of 3 at 27°C. Four hours after infection, the temperature is lowered to 21°C (*see* **Note 1**).
2. Forty-eight hours postinfection, cells are removed by centrifugation (12,000 × *g*, 20 min).
3. The supernatant is concentrated by ultrafiltration across a 10 kDa molecular weight cut-off membrane and diafiltered into IMAC binding buffer using a SartoJet diaphragm pump (Sartorius, Goettingen, Germany). A complete EDTA-free protease inhibitor tablet is added at this step.
4. Supernatant is applied to a Ni^{2+}-charged IMAC column that has been pre-equilibrated with IMAC binding buffer, and the column is washed with the same buffer until the $A_{280\ nm}$ approaches the baseline. A gradient of 0–500 mM imidazole in IMAC binding buffer is applied, and fractions containing TLR3-ECD, as identified by SDS-PAGE and Western blotting with Anti-FLAG® HRP, are pooled.
5. Pooled TLR3-ECD IMAC fractions are applied to a 10 mL Anti-FLAG®M2 agarose column pre-equilibrated with TBS (*see* **Note 2**). The column is washed with 20 column volumes of TBS, and the TLR3-ECD is eluted with 5 column volumes of TBS containing 100 μg/mL FLAG®peptide.

Protein containing fractions are pooled and concentrated using a Centriprep YM-30 concentrator.

6. The concentrated TLR3-ECD is eluted on a 2.6 × 100 cm Superdex 200 column pre-equilibrated with SEC buffer. Fractions in the major peak, as detected by absorbance at 280 nM, are pooled, concentrated, and stored at 4°C (*see* **Note 3**).

3.3. Ligand Binding to TLR3-ECD

To test whether TLR3 binds its dsRNA ligand saturably and specifically, we developed an ELISA in which the protein binds to an immobilized ligand and is then detected using antibodies to the C-terminal tags. A similar approach should be applicable for characterizing other TLRs.

3.3.1. dsRNA Synthesis

1. To synthesize dsRNA enzymatically, a dsDNA template is first created by PCR. Since TLR3 recognizes dsRNA molecules of any sequence, any PCR template may be used, but the length of the PCR product determines the length of the final dsRNA molecules. The PCR product is ligated into the pGEM-T Easy vector, which contains a T7 RNA polymerase promoter upstream and a number of restriction sites downstream of the insertion site (*see* **Note 4**). Since the PCR product can insert in either direction, DNA sequencing is used to isolate one clone each having forward (sense) and reverse (antisense) orientations.
2. Plasmid DNA is isolated and linearized by digesting 50 μg DNA with 10 μL NdeI and 20 μL 10X buffer in 200 μL total overnight at 37°C (*see* **Note 5**). Linearized templates are purified using a QIAquick PCR purification kit.
3. Sense and antisense ssRNAs are synthesized simultaneously by mixing 40 μL Ribomax Express T7 2X buffer, 6 μg each of linearized sense and antisense templates, 8 μL T7 Express enzyme mix, and nuclease-free water to bring the total volume to 80 μL. The mix is incubated 2 h at 37°C (*see* **Note 6**).
4. RNA strands are annealed by heating to 70°C for 10 min in a heat block, turning off the heat, and allowing the block and RNA to cool slowly to room temperature (~1 h).
5. Digestion of DNA template and ssRNA (both un-annealed ssRNA and unpaired overhangs due to non-complementary pGEM-derived sequences that are present in ssRNAs) is accomplished by adding 4 μL each of RNAse A (freshly diluted in water to 20 ng/mL) and RQ1 DNAse, then incubating at 37°C for 30 min. Note that when using pGEM-T-based templates, four bases of self-complementary vector-derived sequence on each end of the PCR template (GATT-template-AATC) will survive RNAse A digestion.

6. To remove unwanted by-products of synthesis, the dsRNA is purified by gel filtration using a Superdex 200 10/300 GL column. The column is equilibrated in PiBS, loaded with up to 100 μL of the synthesis reaction, and dsRNA is eluted in PiBS following the manufacturer's instructions (*see* **Note 7**). If multiple peaks are observed, each peak is pooled separately and compared with a dsRNA ladder using gel electrophoresis to determine which peak corresponds to the desired size of dsRNA.
7. Isolated dsRNA is precipitated by adding 2.5 volumes of ethanol and 0.1 volume of 3.2 M NaOAc. After incubation at −80°C for at least 2 h, the sample is pelleted in a microfuge at 4°C at maximum speed for 20 min. Liquid is removed, and pellets are washed with 500 μL of 70% ethanol in water and dried for 15 min. dsRNA is resuspended in PiBS or water and quantified by absorbance at 260 nm (1 OD = 45 μg/mL dsRNA, when pathlength is 1 cm).

3.3.2. Ligand Biotinylation

1. In the first step, phosphate groups are removed from the ends of dsRNA. 125 μL dsRNA in water (~100 μg) is mixed with 75 μL water, 25 μL 10X Antarctic phosphatase buffer, and 25 μL Antarctic phosphatase, and the mix is incubated at 37°C for 40 min. Enzyme is inactivated by heating to 65°C for 10 min.
2. Thiophosphate groups are transferred to the 5′ ends by adding 125 μL water, 50 μL PNK buffer, 25 μL PNK, and 50 μL of ATP-γS (10 mM) to the phosphatase-treated dsRNA, followed by incubation at 37°C for 30 min.
3. Fifty microlitres of water are added to each of two maleimide-PEO_2-biotin microtubes (2 mg/tube) *immediately* prior to use. After the maleimide-PEO_2-biotin is dissolved, it is added to the thiophosphate derivatized dsRNA, and the reaction mixture is incubated at 65°C for 30 min, which covalently conjugates PEO_2-biotin to the ends of the dsRNA.
4. Proteins are removed by extraction with 600 μL saturated phenol, and the aqueous phase is precipitated as in **Subheading 3.3.1**, **step** 7.
5. Biotinylated ligands are re-purified by gel filtration and concentrated as in **Subheading 3.3.1**, **steps 6** and 7.

3.3.3. TLR3 Binding ELISAs

The following ELISAs are used to characterize the binding of TLR3-ECD to dsRNA and related ligands (*see* **Note 8**):

Direct Binding ELISA

1. Reacti-Bind Streptavidin HBC strip-wells are washed three times with 300 μL PiBST and are incubated with biotinylated dsRNA (1 μg/mL in PiBS, 100 μL/well) overnight at 4°C in a humidified box (*see* **Note 9**).

2. The biotinylated dsRNA solution is discarded, wells are washed as above, and a dilution series of TLR3-ECD in PiBS (100 μL/well) is added and incubated at room temperature for 2 h. Each condition should be performed in duplicate or triplicate, and one set containing zero TLR3-ECD should be included as a blank.
3. Solution is discarded, wells are washed as above, and Anti-FLAG® HRP mAb (diluted 1:8,000 in PiBS + 0.1% BSA) is added (100 μL/well) and incubated at room temperature for 1 h.
4. HRP substrate reagents A and B are mixed 1:1 at room temperature immediately prior to use. mAb solution is discarded, wells are washed as above, and mixed HRP substrate is added (100 μL/well). The reaction is allowed to progress for about 10 min or until any well begins to turn dark blue, at which point all wells are halted by adding 1 M H_2SO_4 (50 μL/well).
5. Absorbance at 450 nm is read with a 96-well plate reader, and the absorbances of the blank wells are averaged and subtracted as background. **Figure 3A** illustrates that the binding of TLR3-ECD to immobilized dsRNA is saturable.

Inhibition ELISA

1. Conversely, an inhibition ELISA is used to probe for binding specificity; this is accomplished by slightly modifying the direct binding ELISA protocol. A dilution series of non-biotinylated dsRNA or polyI:C (50 μL/well) is added to wells containing immobilized dsRNA (after the washes as given in the subheading Direct Binding ELISA, **step 2**), followed by 50 μL/well of a subsaturating concentration of TLR3-ECD (e.g., 0.2 μg/mL). Plates are incubated at room temperature for 2 h.

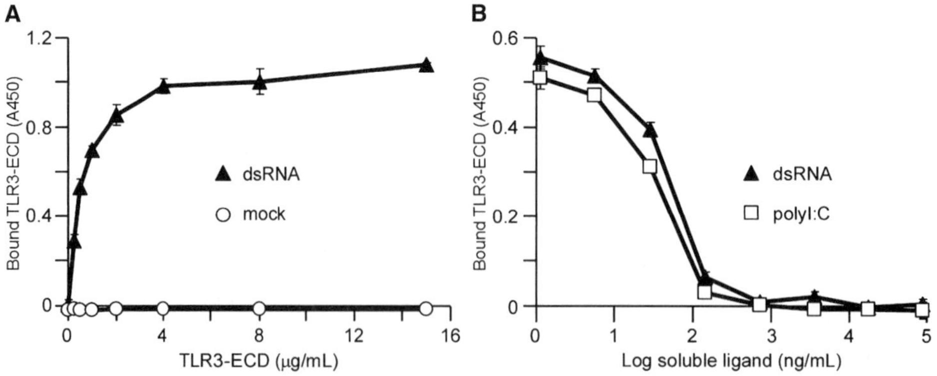

Fig. 3. Binding of Toll-like receptor 3 (TLR3)-extracellular domain (ECD) to immobilized double-stranded RNA (dsRNA). (**A**) Direct binding of TLR3-ECD to immobilized dsRNA as measured by ELISA. Binding reaches a plateau at high TLR3-ECD concentrations, and no binding is observed in the absence of immobilized dsRNA, indicating that binding is saturable and specific. (**B**) Inhibition of TLR3-ECD (0.2 μg/mL) binding to immobilized dsRNA. Soluble dsRNA and polyI:C compete with immobilized dsRNA for TLR3-ECD. The direct binding and inhibition ELISAs were run at pH 6.0 and 5.0, respectively. Both the soluble and immobilized dsRNA used in both ELISAs were 139 bp long.

2. Samples are washed, labeled with Anti-FLAG® HRP, and assayed as in the subheading Direct Binding ELISA, **steps 3–5**. As seen in **Fig. 3B**, both ligands compete for TLR3-ECD, demonstrating that binding is specific.

4. Notes

1. Baculovirus was prepared using Gateway entry and bacmid vectors from Invitrogen. The TLR3-ECD construct contained an N-terminal baculovirus gp67 secretion signal, and at the C-terminal end, a tobacco etch virus (tev) protease cleavage site, followed by FLAG® and His_6 epitope tags. Expression was driven by a polyhedron promoter. For production, 10 L cultures of High Five cells were used. This can be done in-house or commercially (e.g., Kemp Biotechnologies, Frederick, MD).
2. If significant amounts of the protein fail to adhere to the resin, the flow-through fraction can be reapplied to the column. We have also observed that more efficient binding occurs at lower flow rates.
3. TLR3-ECD should be stored at 2–3.5 mg/mL, pH 5.5, and at 4°C for maximum stability. Under these conditions, human TLR3-ECD is stable for at least 6 months. The protein tends to aggregate at higher concentrations.
4. Other dsDNA templates may also be used for RNA synthesis. For example, a T7 promoter sequence can be incorporated at the 5′ end of a PCR primer, as described in the T7 Ribomax manual (Promega). In this case, an ssRNA is generated directly from the PCR product, and the complementary ssRNA is generated from a separate PCR product in which the T7 promoter is instead incorporated into the reverse primer. For short dsRNAs, a single DNA oligonucleotide encoding both the T7 promoter and the entire template sequence can be synthesized chemically (be sure to incorporate a G transcription start site immediately following the promoter). Since T7 polymerase works more efficiently from a dsDNA template, the coding strand is annealed to its complementary oligonucleotide. Separate dsDNA templates are required to make sense and antisense ssRNA oligonucleotides, which are then annealed as in **Subheading 3.3.1, step 4**. RNAse A cleaves ssRNA at the 3′ side of C and U residues, and this activity is compatible with the described use of the pGEM-T Easy vector. If the template used results in an RNA molecule with an unpaired G residue adjacent to the 5′ side of the double-stranded region, RNAse T1 may be used to cleave ssRNA at the 3′ side of the G residue.

5. The pGEM-T Easy vector contains restriction sites other than NdeI that may be used to linearize plasmid templates, but digests that produce 3′ overhangs should be avoided, since they can lead to undesired synthesis by-products.
6. For templates shorter than ~200 bp, incubation for up to 6 h has been found to increase yields significantly.
7. Gel filtration may be run in buffers other than PiBS, but phosphate-containing buffers should be avoided, since they promote bacterial contamination. Other methods for removing unincorporated nucleotides, such as MicroSpin G-25 columns (GE Healthcare), do not efficiently separate by-products and unincorporated nucleotides from the desired dsRNA; this problem becomes more apparent when synthesizing dsRNA smaller than ~200 bp. Note that unincorporated nucleotides cannot be detected using ethidium bromide in agarose gels, but they absorb at 260 nm, so failure to remove these will lead to errors in calculating dsRNA concentration.
8. If a Biacore surface plasmon resonance instrument (GE Healthcare) is available, direct binding of TLR3-ECD to immobilized ligand can be measured quantitatively and in real time using a similar strategy.
9. A mock series lacking immobilized ligand should also be included, especially during initial characterizations. This will test whether the receptor binds non-specifically to the plate (*see* **Fig. 3A**). In cases where non-specific binding is observed, it should be subtracted from the test samples. Non-specific binding might also be reduced by including 5 mg/mL BSA while coating overnight with biotinylated ligand (*see* Direct Binding ELISA, **step 1**).

Acknowledgements

This work was supported by the Intramural Research Program of the NIH (NCI and NIDDK) and by a Trans-NIH/FDA Intramural Biodefense Award from NIAID.

References

1. Gay, N.J. and Gangloff, M. (2007) Structure and function of toll receptors and their ligands. *Annu. Rev. Biochem.* 76, 141–165.
2. Kobe, B. and Kajava, A.V. (2001) The leucine-rich repeat as a protein recognition motif. *Curr. Opin. Struct. Biol.* 11, 725–732.
3. Bell, J.K., Mullen, G.E.D., Leifer, C.A., Mazzoni, A., Davies, D.R., and Segal, D.M. (2003) Leucine-rich repeats and pathogen recognition in Toll-like receptors. *Trends Immunol.* 24, 528–533.
4. Bell, J.K., Botos, I., Hall, P.R., Askins, J., Shiloach, J., Segal, D.M., and Davies, D.R. (2005) The molecular structure of the Toll-like receptor 3 ligand-binding domain. *Proc. Natl. Acad. Sci. U.S.A.* 102, 10976–10980.

Color Plates

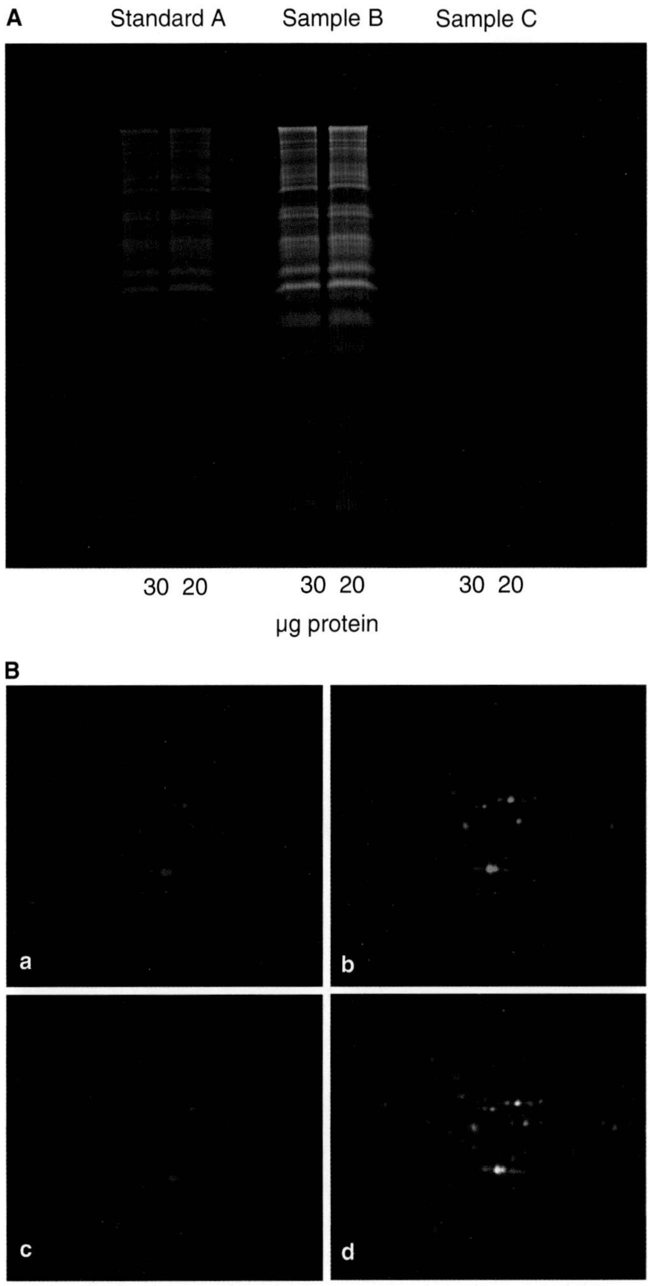

Fig. 3. (**A**) Testing the efficiency labelling for new protein lysates. (*For complete caption refer page 116*).

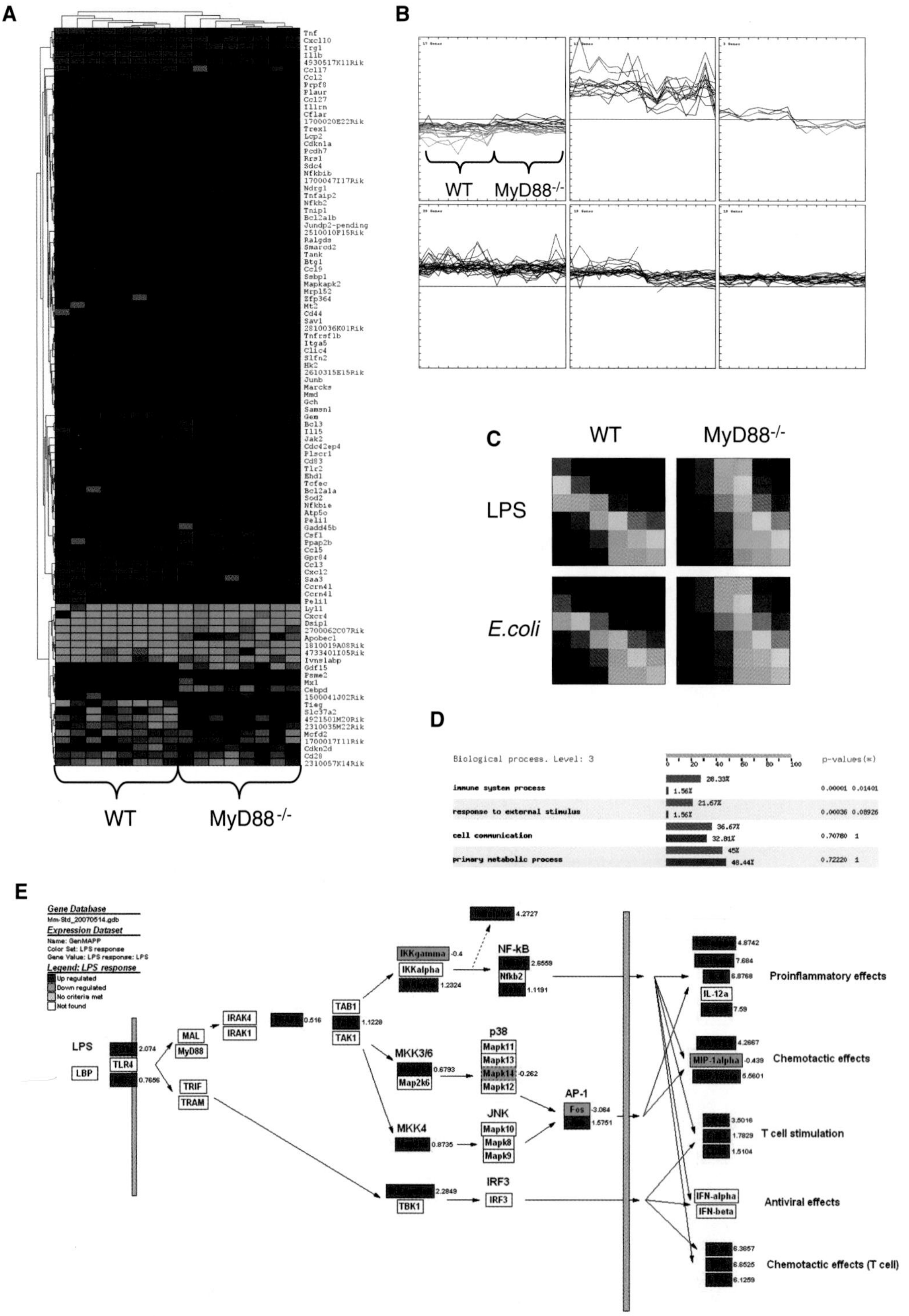

Fig. 3. Data analysis and visualization of the top 100 most significantly up- or down-regulated genes upon LPS stimulation of bone marrow-derived macrophages from C57Bl/6 and MyD88 deficient mice (data from Bjorkbacka et al. 2004 *(9)*. (**A**) Hierarchical clustering (average linkage) using Multi Experiment Viewer (MeV). Both genes and arrays are clustered showing that the arrays are nicely partitioned into MyD88 deficient and C57Bl/6 arrays. Red genes are up-regulated and

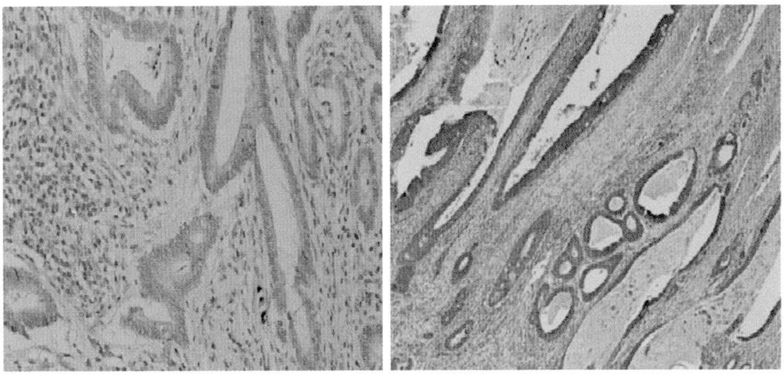

Fig. 2. Comparison between streptavidin–horseradish peroxidase method (*left*) with tyramide amplification method (*right*) in a colon cancer specimen. Notice the marked difference in intensity of staining.

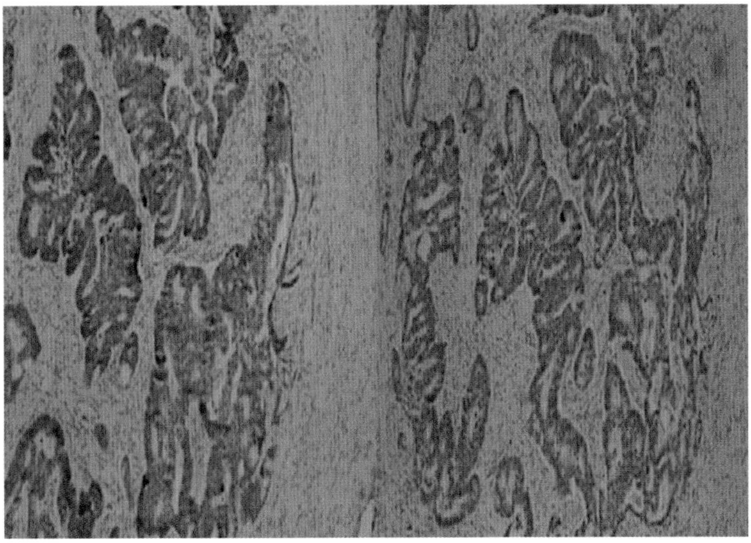

Fig. 3. Comparison between slides omitting primary antibody without avidin and biotin blockers (*left*) and with avidin and biotin blockers (*right*).

green genes down-regulated. (**B**) *K*-means clustering into six clusters using MeV. Both clusters with MyD88-dependent and MyD88-independent genes can be identified. (**C**) Gene Expression Dynamics Inspector (GEDI) mosaic map visualization. Up-regulated genes are colored red and down-regulated blue. Each square in the mosaic consists of a cluster of genes that can be view by clicking that square. The software gives a nice global overview of the gene expression pattern. For instance, the top right corners seem to contain clusters with MyD88-independent up-regulated genes. Also, the LPS and *E. coli* induced gene expression profile is largely similar. (**D**) FatiGO analysis to find significantly over- or underrepresented gene ontology terms within the top 100 genes compared to a random sample of 100 genes also present on the microarray. The gene ontology term "immune system process" is significantly overrepresented among the top 100 genes ($p \sim 0.014$, adjusted for multiple testing), while for instance the term "primary metabolic process" is not. (**E**) GenMAPP visualization of expression data in the Toll-like receptor (TLR) signaling pathway (adapted from Kyoto Encyclopedia of Genes and Genomes (KEGG), http://www.genome.jp/kegg/). As can be seen, LPS treatment of C57Bl/6 macrophages regulates several genes in the TLR signaling pathway.

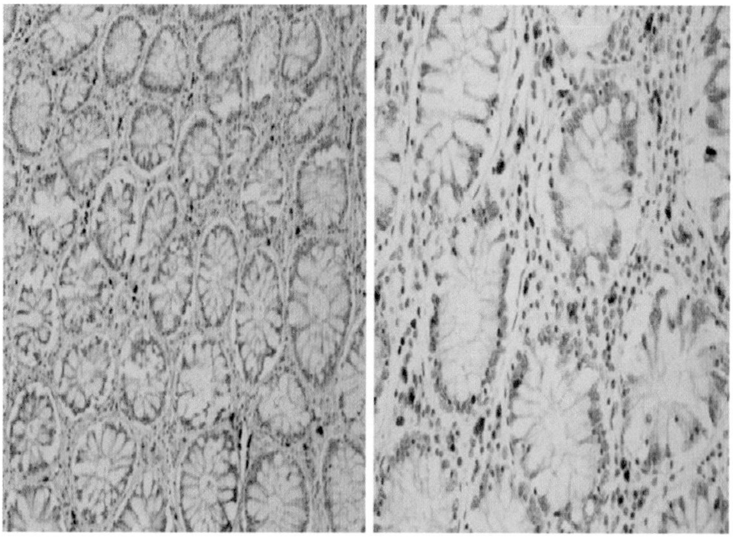

Fig. 5. Comparison of full IgG secondary antibody (*left*) with an $F(ab')_2$ secondary antibody (*right*) with reduction in background staining.

5. Choe, J., Kelker, M.S., and Wilson, I.A. (2005) Crystal structure of human Toll-like receptor 3 (TLR3) ectodomain. *Science* 309, 581–585.
6. He, X.L., Bazan, J.F., McDermott, G., Park, J.B., Wang, K., Tessier-Lavigne, M., He, Z., and Garcia, K.C. (2003) Structure of the Nogo receptor ectodomain: a recognition module implicated in myelin inhibition. *Neuron* 38, 177–185.
7. Finn, R.D., Mistry, J., Schuster-Böckler, B., Griffiths-Jones, S., Hollich, V., Lassmann, T., Moxon, S., Marshall, M., Khanna, A., Durbin, R., Eddy, S.R., Sonnhammer, E.L., and Bateman, A. (2006) Pfam: clans, web tools and services. *Nucleic Acids Res* 34, D247–D251.
8. Rallabhandi, P., Bell, J., Boukhvalova, M.S., Medvedev, A., Lorenz, E., Arditi, M. Hemming, V.G., Blanco, J.C., Segal, D.M., and Vogel, S.N. (2006) Analysis of TLR4 polymorphic variants: new insights into TLR4/MD-2/CD14 stoichiometry, structure, and signaling. *J. Immunol.* 177, 322–332.

Chapter 5

Bioinformatic Analysis of Toll-Like Receptor Sequences and Structures

Tom P. Monie, Nicholas J. Gay, and Monique Gangloff

Summary

Continual advancements in computing power and sophistication, coupled with rapid increases in protein sequence and structural information, have made bioinformatic tools an invaluable resource for the molecular and structural biologists. With the degree of sequence information continuing to expand at an almost exponential rate, it is essential that scientists today have a basic understanding of how to utilise, manipulate, and analyse this information for the benefit of their own experiments. In the context of Toll-interleukin-1 receptor (TIR) domain containing proteins, we describe here a series of the more common and user-friendly bioinformatic tools available as internet-based resources. These will enable the identification and alignment of protein sequences, the identification of functional motifs, the characterisation of protein secondary structure, the identification of protein structural folds and distantly homologous proteins, and the validation of the structural geometry of modelled protein structures.

Key words: Toll-like receptor, TLR, Toll-interleukin-1 receptor (TIR) domain, Bioinformatics, Sequence alignment, Sequence comparison, Homology, Structure validation, FUGUE.

1. Introduction

Toll-like receptors (TLRs) are type I transmembrane receptors. They are constituted of a leucine-rich repeat ligand-binding domain, a single membrane spanning helix, and a signalling Toll-interleukin-1 receptor (TIR) domain *(1, 2)*. TLRs recognise a diverse range of microbial ligands. Following ligand binding, the TLRs undergo conformational change enabling the initiation of signal transduction *(3)*. The TIR domains possess a conserved

Claire E. McCoy and Luke A.J. O'Neill (eds.), *Methods in Molecular Biology, Toll-Like Receptors, vol. 517*
DOI: 10.1007/978-1-59745-541-1_5

$\alpha\beta$ structural organisation essential for signal transduction *(4)*. Indeed, parologs of individual TLR TIRs show particularly high levels of amino acid conservation.

In this chapter we describe the use of the classic bioinformatic tools BLAST (Basic Local Alignment Search Tool) *(5, 6)* and ClustalW *(7)*, for the identification and alignment of TLR TIR paralogs. We also address the identification of structurally homologous proteins and the annotation of a protein's three-dimensional environment through the use of the programs FUGUE *(8)* and JOY *(9)*. Moreover, we describe the use of available resources for the identification of functional motifs within proteins and the validation of the stereochemistry of protein structures. These techniques are highlighted with examples from TIR containing proteins.

These tools provide an important set of resources that, when used either individually or in conjunction with one another, can greatly assist with multiple aspects of the study of TLRs. For example, they enable important functional and structural observations to be made about specific proteins. Additionally, they can aid the design of expression constructs for structural and biochemical studies and assist in the design of rational mutagenesis for functional work.

2. Materials

2.1. Identifying TLR Orthologs

1. Human TLR4 amino acid sequence (accession number O00206).
2. Multiple TLR4 ortholog sequences (*see* **Note 1**).

2.2. Finding a Homologue of Known Three-Dimensional Structure

1. Human TLR4 amino acid sequence (accession number O00206). Select the region from residue 674 to 839.
2. Key to formatted JOY alignments (**Table 1**).

2.3. Three-Dimensional Structure Comparison

1. The PDB (Protein Data Bank) codes for TLR1, TLR2, and TLR10 TIR crystal structures are 1fyv, 1fyw, and 2j67, respectively (*see* **Note 2**).

2.4. Determination of the Structural Validity of a Modelled TIR Domain

1. PDB file for model to be validated.

2.5. Predicting Post-translational Modifications

1. Human TRIF-related adaptor molecule, TRAM, also known as TICAM-2, amino acid sequence (accession number NP_067681).

Table 1
Key to formatted JOY alignments

Structural features	Labelling	Residue format
Alpha helix	Red	x
Beta strand	Blue	x
3_{10} Helix	Maroon	x
Solvent accessible	Lower case	x
Solvent inaccessible	Upper case	X
Hydrogen bond to main-chain amide	Bold	**x**
Hydrogen bond to main-chain carbonyl	Underline	x̲
Disulphide bond	Cedilla	ç
Positive phi torsion angle	Italic	*x*

X: any amino acid; ç: a half-cystine residue.

3. Methods

3.1. Identifying TLR Orthologs

Structurally and functionally important regions of homologous proteins often have high levels of amino acid conservation. Alignment and comparison of the amino acid sequence of homologous proteins from different species (i.e. protein orthologs) can be extremely helpful experimentally through the identification of key functional residues and protein domain boundaries. Here we describe how to identify and align orthologs of TLR4.

3.1.1. Performing a BLAST Search

1. BLAST identifies regions of local similarity between the query and database sequences.
2. Paste the human TLR4 amino acid sequence into the query window of the NCBI-BLAST2 – Protein Database page (http://www.ebi.ac.uk/blastall/index.html).
3. Check that the program selected is blastp and the database is protein and UniProtKB/Swiss-Prot. Run BLAST (*see* **Note 3**).
4. A table of results will be generated showing information about homologous sequences such as protein description and source, length, identity, score, and *E*-value (*see* **Note 4**). From these results it is possible to select TLR4 orthologs identified and download the sequences in a FASTA format (*see* **Note 5**).

3.1.2. Performing a ClustalW Multiple Sequence Alignment

1. Copy the FASTA-formatted orthologs downloaded from the BLAST search (*see* **Subheading 3.1.1**) into the input query field on the EMBL-EBI ClustalW server web page (http://www.ebi.ac.uk/clustalw/index.html).
2. The default parameters can normally be retained. Run ClustalW (*see* **Note 3**).
3. A series of alignment and similarity results will be generated. These include pairwise scores for each sequence aligned, phylogram and cladogram trees, and a multiple sequence alignment (**Fig. 1**).
4. The multiple sequence alignment is especially useful for identifying regions of high and/or low conservation, domain boundaries, and potential substitutions for mutagenic studies.

```
                            10        20        30        40        50
                    ....|....|....|....|....|....|....|....|....|....|
Felis_catus         YDAFVIYSSQDEDWVRNELVKNLEEGVPPFQLCLHYRDFIPGVAIAANII
Equus_caballus      YDAFVIYSSQDEDWVRNELVKNLEEGVPPFQLCLHYRDFIPGVAIAANII
Bos_taurus          YDAFVIYSSQDEDWVRNELVKNLEEGVPPFQLCLHYRDFIPGVAIAANII
Homo_Sapiens        YDAFVIYSSQDEDWVRNELVKNLEEGVPPFQLCLHYRDFIPGVAIAANII
Sus_scrofa          YDAFVIYSSQDEDWVRNELVKNLEEGVPPFHLCLHYRDFIPGVAIAANII
Mus_musculus        YDAFVIYSSQNEDWVRNELVKNLEEGVPRFHLCLHYRDFIPGVAIAANII
Clustal Consensus   **********:***************** *:*******************
                            60        70        80        90       100
                    ....|....|....|....|....|....|....|....|....|....|
Felis_catus         QEGFHKSRKVIVVVSQHFIQSRWCIFEYGIAQTWQFLSSRAGIIFIVLQK
Equus_caballus      QEGFHKSRKVIVVVSQHFIQSRWCIFEYEIAQTWQFLSSRAGIIFIVLHK
Bos_taurus          QEGFHKSRKVIVVVSQHFIQSRWCIFEYEIAQTWQFLSSRAGIIFIVLQK
Homo_Sapiens        HEGFHKSRKVIVVVSQHFIQSRWCIFEYEIAQTWQFLSSRAGIIFIVLQK
Sus_scrofa          QEGFHKSRKVIVVVSQHFIQSRWCIFEYEIAQTWQFLRSHAGIIFIVLQK
Mus_musculus        QEGFHKSRKVIVVVSRHFIQSRWCIFEYEIAQTWQFLSSRSGIIFIVLEK
Clustal Consensus   :**************:************ ******** *::*******.*
                           110       120       130       140       150
                    ....|....|....|....|....|....|....|....|....|....|
Felis_catus         LEKSLLRQQVELYRLLNRNTYLEWEDSVLGRHIFWRRLRKALLDGKPRCP
Equus_caballus      LEKSLLRQQVELYRLLNRNTYLEWEDSVLGRHIFWRRLRKALLDGKPWSP
Bos_taurus          LEKSLLRQQVELYRLLSRNTYLEWEDSVLGRHVFWRRLRKALLAGKPQSP
Homo_Sapiens        VEKTLLRQQVELYRLLSRNTYLEWEDSVLGRHIFWRRLRKALLDGKSWNP
Sus_scrofa          LEKSLLRQQVELYRLLSRNTYLEWEDSVLGRHIFWRRLKKALLDGKPWSP
Mus_musculus        VEKSLLRQQVELYRLLSRNTYLEWEDNPLGRHIFWRRLKNALLDGKASNP
Clustal Consensus   :**:************.*********. ****:*****::*** **.  *
                           160
                    ....|....|....|....
Felis_catus         EGMADAEGS----------
Equus_caballus      AGTADAAESRQHDAETST-
Bos_taurus          EGTADAETNPQ-EATTST-
Homo_Sapiens        EGTVGTGCNWQEATSI---
Sus_scrofa          EGTEDSESNQHDTTAFT--
Mus_musculus        EQTAEEE---QETATWT--
Clustal Consensus
```

Fig. 1. Example ClustalW multiple sequence alignment. The Toll-interleukin-1 receptor (TIR) signalling domains of six of the TLR4 orthologs (host species as labelled in figure panel) identified by a BLAST (Basic Local Alignment Search Tool) search (*see* **Subheading 3.1.1**) were submitted for ClustalW multiple sequence alignment (*see* **Subheading 3.1.2**). The Clustal consensus sequence identifies fully conserved residues (*), strongly similar substitutions (:), weakly similar substitutions (.), and a lack of consensus ( ). The consensus sequence highlights the high degree of conservation in the TLR4 TIR, in contrast the very C-terminus of the protein shows significant variation.

3.2. Finding a Homologue of Known Three-Dimensional Structure

Sequence and structural information can be simultaneously used to improve the homology recognition power and the accuracy of sequence alignments (*see* **Note 6**). Identifying structural homo-logy between a protein sequence of unknown three-dimensional structure and one with known structure provides useful information for understanding protein function. It also provides another layer of information and reflects the high evolutionary pressure for structurally and functionally important residues in a given protein family. In other words, such alignments help identify divergently evolved (homologous) proteins with structural and functional relationships. Furthermore, it allows prediction of the three-dimensional structure through comparative modelling, a technique which is beyond the scope of this chapter (*see* **Note 7**). Here we demonstrate how to use the program FUGUE to identify structural homologues for the TLR4 TIR domain. Unlike the TIR domains of TLR1, 2 and 10, the structure of the TLR4 TIR domain has not yet been solved experimentally.

Annotation of protein sequence alignments with three-dimensional structural features is a useful tool for identifying key structural and functional residues. This can be achieved with a program such as JOY, which provides a modified version of the single-letter amino acid code in order to convey structural information (*see* **Table 1**).

3.2.1. Performing a Sequence-Structure Homology Search with FUGUE

1. Open the FUGUE web page (http://tardis.nibio.go.jp/fugue/prfsearch.html) and enter your e-mail address and the amino acid sequence of the human TLR4 TIR domain (residues 673–839).
2. Keep the default parameters and click on search. The output is sent via e-mail and results can be accessed at.
3. The FUGUE result for human TLR4 TIR domain reveals that the HOMSTRAD (HOMologous STRucture Alignment Database) *(10)* profile hs1fyxa (*see* **Notes 8** and **9**) has the highest *Z*-score. With over 99% confidence, the suggested homology is certain (**Fig. 2A**).
4. The HOMSTRAD family called TIR (*see* **Note 9**), which was built on the crystal structures of human TLR1 and TLR2 TIR domains, is the second best hit. The low *Z*-score of other hits makes them less reliable.
5. Focus only on the two alignments with the highest *Z*-scores by clicking on "alignment" in the results.

3.2.2. Analysing a Sequence-Structure Alignment with JOY

1. The alignments mentioned in **Subheading 3.2.1.5** (**Fig. 2B**) are represented using the JOY annotation described in **Table 1**. In addition to providing a secondary structure prediction for the query sequence, they can also be used to highlight

A

```
##################################################################################
#  Size of fold library: 8465                                                    #
#  Probe sequence ID    : TLR4-TIR                                               #
#  Probe sequence len   : 166                                                    #
#  Probe divergence     : 0.584                                                  #
#  Recommended cutoff   : ZSCORE >=    6.0  (CERTAIN     99% confidence)         #
#  Other cutoff         : ZSCORE >=    4.0  (LIKELY      95% confidence)         #
#  Other cutoff         : ZSCORE >=    3.5  (MARGINAL    90% confidence)         #
#  Other cutoff         : ZSCORE >=    2.0  (GUESS       50% confidence)         #
#  Other cutoff         : ZSCORE <     2.0  (UNCERTAIN)                          #
#                                                                                #
#  PLEN    : Profile length                                                      #
#  RAWS    : Raw alignment score                                                 #
#  RVN     : (Raw score)-(Raw score for NULL model)                              #
#  ZSCORE  : Z-score normalized by sequence divergence                           #
#  PVZ     : P-value based on Z-score jumbling             (Currently Disabled)  #
#  ZORI    : Original Z-score (before normalization)                             #
#  EVP     : E-value based on profile calibration          (Currently Disabled)  #
#  EVF     : E-value based on library search               (Currently Disabled)  #
#  AL      : Alignment algorithm used for Zscore/Alignment calculation           #
#            0 -- Global, 2 -- GloLocSeq (No sequence termini gap penalty)       #
#            3 -- GloLocPrf (No profile termini gap penalty)                     #
##################################################################################

##################################################################################
#--------------------------------------------------------------------------------#
#  Profile              PLEN  RAWS RVN  ZSCORE   PVZ      ZORI    EVP     EVF   AL#
#                                       ******                                   #
#--------------------------------------------------------------------------------#
 hs1fyxa                 145   239 388   27.28 1.0E+03   29.64 1.0E+03 1.0E+03 00
 TIR                     161   232 408   26.80 1.0E+03   29.16 1.0E+03 1.0E+03 00
 hsd1yvei2               225  -213  64    3.63 1.0E+03    5.50 1.0E+03 1.0E+03 00
 hs1lis                  131   -86  54    3.22 1.0E+03    5.23 1.0E+03 1.0E+03 00
 hs2ov8a                 241  -216  65    3.01 1.0E+03    5.14 1.0E+03 1.0E+03 00
 Egg_lysin               131   -92  44    2.82 1.0E+03    4.88 1.0E+03 1.0E+03 00
 hs1gaka                 137   -95  51    2.58 1.0E+03    4.75 1.0E+03 1.0E+03 00
 hs2hiya                 177  -154  49    2.23 1.0E+03    4.68 1.0E+03 1.0E+03 00
 hs2hiya                 177  -154  49    2.23 1.0E+03    4.68 1.0E+03 1.0E+03 00
 LysR_substrate          238  -268  55    2.13 1.0E+03    4.66 1.0E+03 1.0E+03 00
##################################################################################
```

B

```
                                10        20        30        40        50
hs1fyxa        ( 636 )    srniydAFVSYSerdaywVenlMvqeLenfnppFk-LLHkrdfihgkwii
QUERY TLR4                ----YDAFVIYSSQDEDWVRNELVKNLEEGVPPFQLCLHYRDFIPGVAIA
                              bbbb  333aaaaa aaaaaa             aaaa      aa

                                60        70        80        90       100
hs1fyxa        ( 687 )    dnii-dsiekShktVfVLSenFVkseWk-yeldfshfrlfdenndaailI
QUERY TLR4                ANIIHEGFHKSRKVIVVVSQHFIQSRWCIFEYEIAQTWQFLSSRAGIIFI
                          aaaa aaaaa  bbbbbb aaaaaaa                333    bbb

                               110       120       130       140       150
hs1fyxa        ( 736 )    llepIekkaIpqrfkLrkiMntktylewpmdeaqregFWvnLraaIks
QUERY TLR4                VLQKVEKTLLRQQVELYRLLSRNTYLEWEDSVLGRHIFWRRLRKALLDGK
                          b             aaaaaaa    bb     333aaaaaaaaaaaaa

                               160       170
hs1fyxa
QUERY TLR4                SWNPEGTVGTGCNWQEATSI
```

```
                                10        20        30        40        50
1fywa          ( 636 )          srn-iydAFVSYSerdaywVen-lvqeLenf-nppfkLLHkrd
1fyva          ( 625 )    nipleelqrnlqfhAFISySghdsfwVkneLLpnLeke---ggICLhern
QUERY TLR4                ------------YDAFVIYSSQDEDWVRNELVKNLEEGVPPFQLCLHYRD
                                      bbbb     aaaaa

                                60        70        80        90       100
1fywa          ( 679 )    fipgkwiidn-iidsiekShktVfVLSenFvkseW-kyeldfshfrlfde
1fyva          ( 673 )    fvpgksiven-IitCIekSYKSIFVLSpnFVqseWchYelyFahhnlFhe
QUERY TLR4                FIPGVAIAANIIHEGFHKSRKVIVVVSQHFIQSRWCIFEYEIAQTWQFLS
                                aaaa aaaaaaa  bbbbbb aaaaaa

                               110       120       130       140       150
1fywa          ( 728 )    nndaailIllepIekkaIpqrf-kLrkintktylewp-deaqregFWvnL
1fyva          ( 722 )    gsnsLiLIlLepIpqysIpssyhkLksLarrtylewpkekskrglFwanL
QUERY TLR4                SRAGIIFIVLQKVEKTLLRQQVELYRLLSRNTYLEWEDSVLGRHIFWRRL
                               bbbb      333                  bb     333  aaaaaa

                               160       170
1fywa          ( 779 )    raaIks
1fyva          ( 773 )    raaIniklteqak
QUERY TLR4                RKALLDGKSWNPEGTVGTGCNWQEATSI
                          aaaa
```

Fig. 2. Example analysis of a Toll-like receptor 4 (TLR4)-Toll-interleukin-1 receptor (TIR) domain homology search using FUGUE. (**A**) Extract of the output for FUGUE sent via e-mail and (**B**) the sequence-structure alignments of the two top hits based on the JOY annotation.

differences and/or problem areas within the sequence-structure alignments. These could be, for example, insertions or deletions in regions of helical structure; proline residues in regions of predicted helix; the presence, or substitution, of charged residues (e.g. lysine, arginine) for hydrophobic (e.g. phenylalanine, leucine, isoleucine, tyrosine) ones, and vice versa.

2. Analysis of the structural alignments reveals that the core of the TIR domain is well conserved between TLR1 (1fyv), TLR2 (1fyw), and TLR4 (Query). There are however apparent differences. For instance, an extra histidine residue at position 724 in human TLR4 interrupts an alpha helix and is likely to cause some structural distortion. In addition, compared to the structural templates, there are extra residues at the C-terminus of the TLR4 sequence. These are not part of the TIR domain but constitute a tail of unpredictable structure.

3.3. Three-Dimensional Structure Comparison

It can be very helpful to evaluate the degree of three-dimensional structural similarity between either two or more experimentally determined or computer-modelled structures. This can help provide an estimation of structural similarity and/or model/structure reliability. The following method uses the Secondary Structure Matching (SSM) program, available at http://www.ebi.ac.uk/msd-srv/ssm/, to determine the similarity between experimentally determined TLR TIR domains.

3.3.1. Pairwise Structural Comparison

1. Choose the pairwise 3D alignment submission option and select "PDB entry" for both the query and target sequences.
2. Insert the PDB codes for TLR1 (1fyv) and TLR2 (1fyw) TIR domains in the query and target fields, respectively.
3. Retain the default parameters and submit query (*see* **Note 10**).
4. An output table detailing the 3D structural similarity will be generated. In general the higher the number of aligned residues and the lower the rmsd (root mean square deviation of Cα atoms) the greater the degree of structural similarity (*see* **Note 11**). The values for the TLR1 and TLR2 structural comparison suggest a high degree of structural similarity.

3.3.2. Multiple Structural Comparison

1. Choose the multiple 3D alignment submission option and select "PDB entry" as the source.
2. Input the TLR1 TIR PDB code (1fyv) and press the "Actualize" button, followed by the "new entry" button. Repeat for TLR2 (1fyw).
3. Input the TLR10 (2j67) TIR structure files and press "Find Chains"; delete B, Y, and Z from the text box, then press "Actualize". This removes unnecessary information as the TLR10 structure was a dimer. Submit query.

4. The results page will contain information relating to the similarities of the 3D superposition of the structure. This will include rmsd and *Q*-scores, alignment of secondary structure elements, and a structural alignment of input files. The aligned files can be viewed individually, as a superposition, or downloaded.

3.4. Determination of the Structural Validity of a Modelled TIR Domain

There are many computer packages that will produce structural models with little more user input than an amino acid sequence. However, the models produced may contain regions of either poor, or disallowed, stereochemistry. It is always advisable to validate the geometry of any models generated. Two good ways to do this use the programs Verify3D and RAMPAGE. Verify3D assesses the sequence position and structural environment of the model and compares them to databases of known high quality structures. RAMPAGE provides a Ramachandran plot analysis to assess the stereochemical environment of the backbone torsion angles in the modelled structure.

1. Upload, and submit for analysis, the co-ordinate .PDB file of the modelled structure to the servers for Verify3D (http://nihserver.mbi.ucla.edu/Verify_3D/) and RAMPAGE (http://mordred.bioc.cam.ac.uk/~rapper/rampage.php).
2. Sample results for a TLR4 TIR homodimer model are shown in **Fig. 3**.

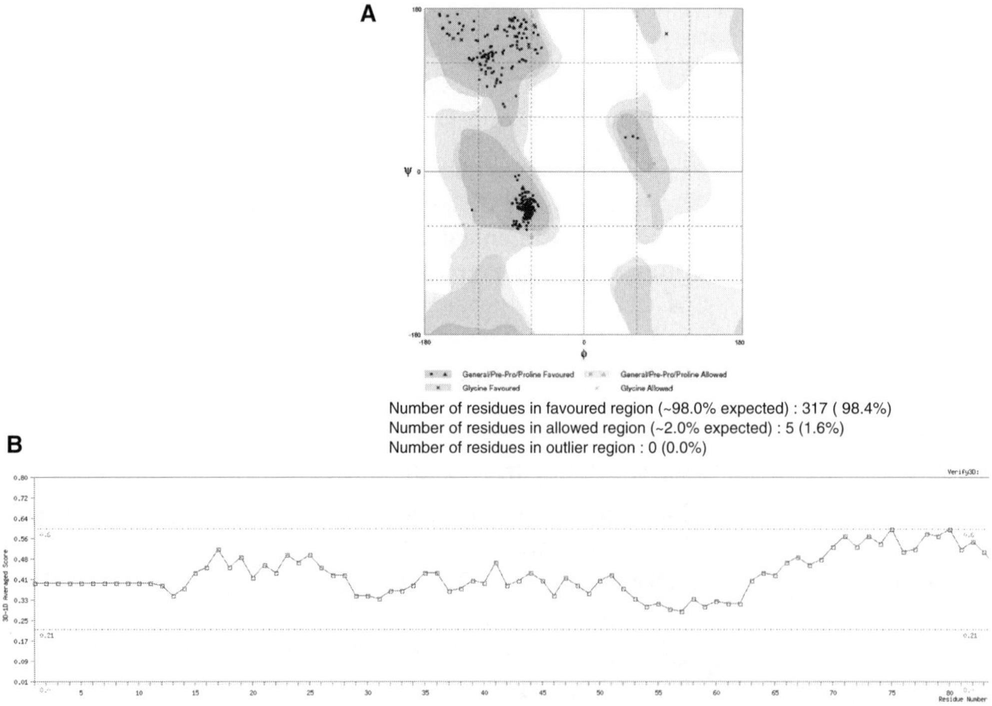

Fig. 3. Example analysis of a modelled Toll-like receptor 4 (TLR4) homodimer using (**A**) RAMPAGE and (**B**) Verify3D.

3. Verify3D scores each residue on a scale of –1 to +1 and a score of >0.2 suggests that the residue is in a structurally favourable environment. Regions with scores below this suggest that those parts of the model must be viewed as less reliable. The region of output shown in **Fig. 3** indicates that the TLR4 model submitted has all Verify3D scores over 0.2 and therefore possesses high quality stereochemistry, with the individual residues being found in structurally favoured environments.
4. RAMPAGE produces a clear graphical output of the Ramachandran plot that identifies the proportion of residues in favoured, allowed, and disallowed regions. This provides a clear indication of the stereochemical quality of the model. For the TLR4 model submitted (**Fig. 3**), over 98% of the residues have torsion angles in the favoured regions, less than 2% in allowed regions, and there are no outliers. This helps confirm the high quality stereochemistry of the model.

3.5. Predicting Post-translational Modifications

Assessing the presence of post-translational modifications in the Toll receptor pathway proteins is critical for understanding the biology of Toll signalling. Many tools exist for this purpose (*see* **Note 12**) and here we use one to identify a protein myristoylation site on the TIR containing adaptor protein TRAM. Myristoylation anchors the adaptor protein to the plasma membrane, where it fulfils its biological role in transferring the signal of activated Toll receptors. The linkage occurs on a consensus sequence consisting of Gly-X-X-X-Ser/Thr-Lys/Arg, where X stands for any amino acid. The 14 carbon fatty acid, myristic acid, is covalently attached by amide linkage to the N-terminal glycine of a protein by an N-terminal myristoyltransferase.

1. Copy the FASTA-formatted TRAM protein sequence into the query field on the NMT server web page (http://mendel.imp.ac.at/myristate/SUPLpredictor.htm).
2. Keep the default parameter of "Eukaryota" as it fits the taxonomy of the sequence.
3. Run the prediction.
4. A reliable myristoylation site is predicted at residue G2 within the sequence GIGKSKINSCPLSLSWG, with an overall score of 0.85 and a probability of false positive prediction of 1.98×10^{-3}.
5. A logical progression would be to confirm the presence and biological relevance of this modification. Indeed site-directed mutagenesis of the predicted myristylation residue (Gly2Ala) and confocal microscopy experiments have determined that wild-type TRAM is myristylated and localises to the plasma membrane. In contrast, a G2A mutant TRAM has a cytoplasmic distribution and is unresponsive to lipopolysaccharide stimulation *(11)*.

4. Notes

1. Multiple TLR4 ortholog sequences can be obtained from a BLAST search (*see* **Subheading 3.1.1**).
2. The full PDB files can be downloaded from the Protein Data Bank (http://www.rcsb.org/pdb/home/home.do).
3. The default search parameters should be fine for these applications. If the user wants further information regarding parameter attributes and variation, it is recommended that they read the related program documentation available through the EMBL-EBI web site (http://www.ebi.ac.uk). The UniProtKB/Swiss-Prot database is the smaller portion of the UniProt database and contains fully annotated sequence information. Using this stops multiple redundant hits being identified. If it was unknown whether orthologs existed then use of the UniProtKB/TrEMBL or UniProt Clusters databases would be more appropriate.
4. The score takes into account the number of gaps and substitutions in the alignment. The greater this number, the better the quality of the alignment. The *E*-value is a measure of the likelihood of the alignment occurring by chance. The smaller this number, the less likely the alignment is a result of chance.
5. The first line of a FASTA-formatted protein sequence starts with a > followed by descriptive text about the sequence. The second, and subsequent, lines contain the protein sequence in single-letter code with no spaces or numbering.
6. A good overview of structural homology modelling can be found in ref. *12*.
7. To find out about the homology modelling approach, go to the Swiss-model (www.expasy.ch/swissmod/SWISS-MODEL.html) and the Modeller (http://www.salilab.org/modeller) web pages.
8. HOMSTRAD is a curated database of structure-based alignments for homologous protein families. Its web site can be found at http://www-cryst.bioc.cam.ac.uk/~homstrad/ or at http://tardis.nibio.go.jp/homstrad/.
9. FUGUE results are given as a list of potentially matching HOMSTRAD profiles. The code hs1fyxa corresponds to the crystal structure of the TLR2 mutant P681H. The 1fyxa relates to the PDB identifier (1fyx; chain A) in the HOMSTRAD "hs" database. The code TIR refers to the HOMSTRAD family containing the TLR1 and TLR2 crystal structures (PDB 1fyv and 1fyw). Clicking of the listed HOMSTRAD profile in the FUGUE results will open the HOMSTRAD entry and show details of its composition.

10. If using a different query sequence and the whole PDB archive as the target then it may be necessary to lower the percentage similarity cut-off for the lowest acceptable target match in order to get any positive hits.
11. Full details of the interpretation of results and scores can be found at http://www.ebi.ac.uk/msd-srv/ssm/ssmresults.html. The higher the *Z* and *Q* score the better.
12. A list of programs for the prediction of post-translation modifications can be found on the Expasy tools web site at http://www.expasy.ch/tools/.

References

1. Akira, S. & Takeda, K. (2004). Toll-like receptor signalling. *Nat Rev Immunol* 4, 499–511.
2. Gay, N. J. & Gangloff, M. (2007). Structure and function of toll receptors and their ligands. *Annu Rev Biochem* 76, 141–65.
3. Gay, N. J., Gangloff, M. & Weber, A. N. (2006). Toll-like receptors as molecular switches. *Nat Rev Immunol* 6, 693–8.
4. Xu, Y., Tao, X., Shen, B., Horng, T., Medzhitov, R., Manley, J. L. & Tong, L. (2000). Structural basis for signal transduction by the Toll/interleukin-1 receptor domains. *Nature* 408, 111–5.
5. Altschul, S. F., Gish, W., Miller, W., Myers, E. W. & Lipman, D. J. (1990). Basic local alignment search tool. *J Mol Biol* 215, 403–10.
6. Altschul, S. F., Madden, T. L., Schaffer, A. A., Zhang, J., Zhang, Z., Miller, W. & Lipman, D. J. (1997). Gapped BLAST and PSI-BLAST: a new generation of protein database search programs. *Nucleic Acids Res* 25, 3389–402.
7. Chenna, R., Sugawara, H., Koike, T., Lopez, R., Gibson, T. J., Higgins, D. G. & Thompson, J. D. (2003). Multiple sequence alignment with the Clustal series of programs. *Nucleic Acids Res* 31, 3497–500.
8. Shi, J., Blundell, T. L. & Mizuguchi, K. (2001). FUGUE: sequence-structure homology recognition using environment-specific substitution tables and structure-dependent gap penalties. *J Mol Biol* 310, 243–57.
9. Mizuguchi, K., Deane, C. M., Blundell, T. L., Johnson, M. S. & Overington, J. P. (1998). JOY: protein sequence-structure representation and analysis. *Bioinformatics* 14, 617–23.
10. Mizuguchi, K., Deane, C. M., Blundell, T. L. & Overington, J. P. (1998). HOMSTRAD: a database of protein structure alignments for homologous families. *Protein Sci* 7, 2469–71.
11. Rowe, D. C., McGettrick, A. F., Latz, E., Monks, B. G., Gay, N. J., Yamamoto, M., Akira, S., O'Neill, L. A., Fitzgerald, K. A. & Golenbock, D. T. (2006). The myristoylation of TRIF-related adaptor molecule is essential for Toll-like receptor 4 signal transduction. *Proc Natl Acad Sci U S A* 103, 6299–304.
12. Nunez Miguel, R., Shi, J. & Mizuguchi, K. (2001). Protein Fold Recognition and Comparative Modeling Using HOMSTRAD, JOY, and FUGUE. In *Protein Structure Prediction: Bioinformatic Approach* (Tsigelny, I. F., ed.). International University Line, La Jolla, CA.

Chapter 6

Expression, Purification, and Crystallization of Toll/Interleukin-1 Receptor (TIR) Domains

Xiao Tao and Liang Tong

Summary

Toll-like receptors (TLRs) and interleukin-1 receptor (IL-1R) play crucial roles in host innate immune response against microbial infections. These receptors share a conserved cytoplasmic domain, the Toll/interleukin-1 receptor (TIR) domain, which is required for signaling through these receptors. Structural information on the TIR domains will be essential for understanding the molecular basis for signal transduction by these receptors.

Key words: Innate immunity, Adaptive immunity, Crystallography, Protein structure and function, Signal transduction, Protein complexes.

1. Introduction

Toll-like receptors (TLRs) have crucial roles in innate immunity *(1–6)*. They recognize various pathogen-associated molecules and initiate host defense responses. Ten TLRs have been identified in humans, and their ligands include cell-wall components from Gram-positive and Gram-negative bacteria (LPS and others), bacterial flagellin, viral dsRNA, unmethylated CpG DNA, and others.

TLRs contain leucine-rich repeat (LRR) and cysteine-rich domains in their extracellular domains. The intracellular region of the TLRs contains a conserved domain of about 150 amino acid residues, which also shares sequence homology with the intracellular region of the interleukin-1 receptor (IL-1R) superfamily. Therefore, this domain is known as the TIR (Toll/interleukin-1

Claire E. McCoy and Luke A.J. O'Neill (eds.), *Methods in Molecular Biology, Toll-Like Receptors, vol. 517*
DOI: 10.1007/978-1-59745-541-1_6

receptor) domain. Upon receptor activation, the TIR domain of the receptor recruits downstream adaptor signaling molecules such as MyD88, Mal, Trif, and others *(7)*. These adapters also contain a TIR domain, suggesting that signal transduction by TLRs and IL-1Rs may be mediated by a homotypic receptor–adaptor TIR domain complex. Mutations in the TIR domain can abrogate this signaling process and cause serious diseases in humans *(8)*. Understanding the molecular basis of TLR and IL-1R signaling will require structural information on the TIR domains *(9–11)*.

This chapter describes protocols about how to express TIR domains of human TLR1 and TLR2 in bacteria, and how to purify and crystallize these TIR domains for structural studies. The protocols described here should be applicable to other TIR domains. In fact, a large number of TIR domains have been purified this way and the TIR domain from IL-1RAPL has also been crystallized *(11)*.

2. Materials

2.1. Protein Expression of the TIR Domains of Human TLR1 and TLR2

1. Kanamycin should be made in a stock solution of 35 mg/ml in water, and filtered through a syringe filter for sterilization.
2. IPTG (0.4 M stock solution) is prepared in water, and filtered to sterilize. It can then be divided into smaller aliquots (1 ml each for example) and stored at –20°C.

2.2. Protein Purification of the TIR Domains of Human TLR1

1. PMSF (100 mM stock solution) is prepared in isopropanol, and should be added to the buffers just prior to use. PMSF is toxic and so should be handled with care.
2. β-Mercaptoethanol (βME) and DTT should be added to the buffers just prior to use, and unused portions of these buffers should be discarded (or supplemented with fresh compounds) in a few days. The oxidized compounds will increase the baseline in the A_{280} readings.
3. Imidazole (0.5 M stock solution) is prepared in water, but make sure to adjust the pH to about 7.
4. Nickel–agarose resins (Qiagen and others) are stored in ethanol. It is good to exchange them into the lysis buffer before adding the bacterial lysate.
5. Lysis buffer: 20 mM Mops (pH 7.0), 300 mM NaCl, 5 mM imidazole (pH 7.0), 5% (v/v) glycerol, 10 mM βME, and 0.3% (v/v) Triton X-100.

6. Nickel column wash buffer: 20 mM Mops (pH 7.0), 300 mM NaCl, 20 mM imidazole (pH 7.0), 5% (v/v) glycerol, and 10 mM βME.
7. Nickel column elution buffer: 20 mM Mops (pH 7.0), 200 mM NaCl, 150 mM imidazole (pH 7.0), 5% (v/v) glycerol, and 10 mM βME.
8. Gel filtration running buffer: 20 mM Mops (pH 7.0), 200 mM NaCl, and 5 mM DTT.
9. A 5 ml cation exchange column can be packed using SP Fast Flow resin (GE Healthcare) following the manufacturer's instructions.
10. The Sephacryl S-300 gel filtration column is from GE Healthcare.

2.3. Protein Purification of the TIR Domains of Human TLR2

1. Lysis buffer: 20 mM Mops (pH 7.0), 200 mM NaCl, 5% (v/v) glycerol, 2 mM DTT, and 0.3% (v/v) Triton X-100.
2. Gel filtration running buffer: 20 mM Mops (pH 7.0), 150 mM NaCl, and 3 mM DTT.

3. Methods

The TIR domains of human TLR1 (residues 625–786, about 20 residues from the transmembrane region) and human TLR2 (residues 626–784, about 16 residues from the transmembrane region) were subcloned into the pET26b vector (Novagen) using the NdeI and XhoI restriction sites, following standard protocols. The resulting protein of TLR1 contains a C-terminal His6 tag (with the sequence LEHHHHHH) while that of TLR2 does not have any tag.

3.1. Protein Expression of the TIR Domains of Human TLR1 and TLR2

1. Transform the desired expression plasmid into competent cells of expression host (BL21 (DE3)) and plate onto an LB-agar plate with 35 μg/ml kanamycin.
2. Incubate the plate at 37°C overnight.
3. Pick a single colony from the plate and inoculate a 50 ml LB culture with 35 μg/ml kanamycin, and shake at 37°C overnight.
4. Dilute the 50 ml overnight culture 100-fold into prewarmed LB media with 35 μg/ml kanamycin and let the cells grow at 37°C for 3–4 h or until the OD_{600} reaches around 0.6.
5. Induce the culture with 0.4 mM IPTG and continue shaking at 20°C overnight (about 16 h).

6. Harvest the bacterial cells by centrifuging at 5,000 × *g* for 15 min.
7. Flash-freeze the cell pellet in liquid nitrogen and store at –80°C for later use, or start the protein purification right away.

3.2. Protein Purification of the TIR Domain of Human TLR1

1. Resuspend the cell pellet in lysis buffer. Use about 30 ml lysis buffer per 700 ml bacterial culture.
2. Add PMSF to a final concentration of 0.5 mM.
3. Lyse the cells by sonicating with 30 bursts on ice, with 50% interval. Let the mixture sit on ice for 30 s to cool. Repeat three more times.
4. Centrifuge at 15,000 × *g* for 30 min at 4°C.
5. In the meantime, prepare the Ni–agarose resin by washing them twice with ddH_2O. Then pre-equilibrate the resin with lysis buffer (*see* **Note 1**).
6. Add the supernatant from **step 4** into the pre-equilibrated Ni–agarose resin.
7. Gently mix the resin and the supernatant at 4°C for about 1 h for protein binding.
8. Load the resin onto a column under gravity flow.
9. Wash the resin with five bed volumes of lysis buffer, followed by ten bed volumes of wash buffer.
10. Elute the bound protein with five bed volumes of elution buffer. The protein should be more than 80% pure already after this step.
11. Dilute the eluate threefold with 20 mM Mops (pH 7.0) to reduce the salt concentration.
12. Load the diluted sample onto a cation exchange column (SP-FF, GE Healthcare) pre-equilibrated with 20 mM Mops (pH 7.0) and 50 mM NaCl (*see* **Note 2**).
13. Wash the column until the UV reading reaches baseline.
14. Elute the protein with a 50–500 mM linear NaCl gradient in ten column volumes.
15. Collect the peak fractions from the column, and concentrate to about 2 ml.
16. Load the concentrated protein onto a gel filtration column (Sephacryl S-300, GE Healthcare) pre-equilibrated with at least one column volume (~120 ml) of running buffer (*see* **Note 3**).
17. Collect the peak fractions from the gel filtration column (keep those fractions with UV reading more than half the maximum UV reading) and concentrate to 30 mg/ml. Exchange the protein buffer to one without salt during the

concentration. The protein concentration was determined by the Bradford (Biorad) method using bovine serum albumin (BSA) as the standard.

18. Divide the sample into small aliquots (20–50 μl each), flash-freeze in liquid nitrogen, and store at –80°C.
19. Analyze the progress of purification by SDS-PAGE, with Coomassie staining. The final protein sample should be more than 95% pure.

3.3. Protein Purification of the TIR Domain of Human TLR2

1. Resuspend the cell pellet in lysis buffer. Use about 30 ml lysis buffer per 700 ml bacterial culture.
2. Add PMSF to a final concentration of 0.5 mM.
3. Lyse the cells by sonicating with 30 bursts on ice, with 50% interval. Let the mixture sit on ice for 30 s to cool. Repeat three more times.
4. Centrifuge at 15,000 × *g* for 30 min at 4°C.
5. Mix the supernatant with 1% (w/v) streptomycin sulfate (Sigma) and incubate at 4°C for half an hour to precipitate DNA.
6. Centrifuge again at 15,000 × *g* for 30 min at 4°C
7. Load the supernatant onto a cation exchange column (SP FF, GE Healthcare) pre-equilibrated with 20 mM Mops (pH 7.0) and 100 mM NaCl (*see***Note 4**).
8. Wash the column extensively with buffer containing 20 mM Mops (pH 7.0) and 250 mM NaCl.
9. Elute the protein with a 250–700 mM NaCl gradient in ten column volumes.
10. Collect the peak fractions from the column, and concentrate to about 2 ml.
11. Load the concentrated protein onto a Sephacryl S-300 gel filtration column (GE Healthcare) pre-equilibrated with at least one column volume (~120 ml) of the running buffer.
12. Collect the peak fractions from the gel filtration column (keep those fractions with UV reading more than half the maximum UV reading) and concentrate to 30 mg/ml.
13. Divide the sample into small aliquots (20–50 μl each), flash-freeze in liquid nitrogen, and store at –80°C.
14. Analyze the progress of purification by SDS-PAGE, with Coomassie staining. The final protein sample should be more than 95% pure.

3.4. Crystallization of the TIR Domains of Human TLR1 and TLR2

Crystals of the TIR domain of human TLR1 were obtained at 21°C, using the hanging-drop vapor diffusion method (*see* **Note 5**). The reservoir solution contained 100 mM Tris (pH 8.0),

Fig. 1. A crystal of the Toll/interleukin-1 receptor (TIR) domain of human Toll-like receptor (TLR).

1.2 M NaH_2PO_4/K_2HPO_4, 5 mM DTT, and 20% (v/v) glycerol. Crystals in the shape of hexagonal bipyramids (Fig. 1) generally appeared in 2 weeks and took another week to grow to full size ($0.3 \times 0.3 \times 0.3\ mm^3$). For cryo-protection, the crystals were transferred to an artificial mother liquid containing 100 mM Tris (pH 8.0), 1.7 M ammonium sulfate (or phosphate), 25% (v/v) glycerol, 20 mM $MgCl_2$, and 10 mM DTT, and flash-frozen in liquid nitrogen (or propane) for data collection at 100 K.

Crystals of the TIR domain of TLR2 were obtained at 4°C, using the hanging-drop vapor diffusion method. The reservoir solution contained 100 mM cacodylate (pH 6.8), 10% (w/v) PEG8000, 20% (v/v) DMSO, 200 mM $MgCl_2$, and 5 mM DTT. The crystals in the same morphology as those of TIR domain of human TLR1 generally appeared overnight and grew to full size ($0.3 \times 0.3 \times 0.3\ mm^3$) in 2–3 days.

4. Notes

1. Generally the amount of nickel–agarose used for purification is just sufficient for binding all the His-tagged proteins. This should prevent the nonspecific binding of contaminant proteins to the resin. For TLR1 and TLR2 TIR domains, generally about 1 ml slurry of the resin is used for each 700 ml of bacterial culture. The nickel–agarose is supplied as a 50% slurry in ethanol.

2. The TIR domain of TLR1 has a high pI, and therefore can bind a cation exchange column at pI 7. This cation exchange step allows further purification of the protein.
3. The gel filtration column both allows further purification of the protein and gives an indication of the solution behavior (mono-disperse or aggregated) of the protein. The TIR domains of TLR1 and TLR2 consistently migrated as monomers on the gel filtration column, suggesting that TIR domains alone have weak affinity for self-association.
4. The TIR domain of TLR2 has a rather high pI value of 8.5. It binds tightly to the cation exchange column even at pH 7, and can be eluted only at higher salt (about 300 mM NaCl). Therefore, the TIR domain is mostly pure after the single cation exchange purification step.
5. In a hanging-drop vapor diffusion crystallization setup, the precipitant solution is placed in a well (typically on a 24-well plate). The protein solution is mixed (generally 1 µl + 1 µl) with the precipitant solution on a cover slip. The cover slip is then inverted and placed over the well, forming an airtight enclosure (grease is applied to the top edge of the well). Over time, equilibration between the drop and reservoir leads to changes in the drop and crystallization of the protein.

Acknowledgments

We thank Yingwu Xu and Javed Khan for their contributions to the TIR domain project. This research was supported in part by a grant from the National Institutes of Health (AI49475 to LT).

References

1. Dunne, A. & O'Neill, L. A. J. (2003) The interleukin-1 recepor/Toll-like receptor superfamily: signal transduction during inflammation and host defense. *Sci. STKE* 2003, re3.
2. Underhill, D. M. & Ozinsky, A. (2002) Toll-like receptors: key mediators of microbe detection. *Curr. Opin. Immunol.* 14, 103–110.
3. Tong, L. (2005) Structural basis of bacterial pathogenesis. (Waksman, G., Caparon, M. G., & Hultgren, S., eds.), ASM Press, Washington, DC, pp. 241–263
4. Kawai, T. & Akira, S. (2007) TLR signaling. *Semin. Immunol.* 19, 24–32.
5. Hoebe, K., Jiang, Z., Georgel, P., Tabeta, K., Janssen, E., Du, X., & Beutler, B. (2006) TLR signaling pathways: opportunities for activation and blockade in pursuit of therapy. *Curr. Pharm. Des.* 12, 4123–4134.
6. Pasare, C. & Medzhitov, R. (2005) Toll-like receptors: linking innate and adaptive immunity. *Adv. Exp. Med. Biol.* 560, 11–18.
7. O'Neill, L. A. J. & Bowie, A. G. (2007) The family of five: TIR-domain-containing adaptors in Toll-like receptor signaling. *Nat. Rev. Immunol.* 7, 353–364.
8. Poltorak, A., He, X., Smirnova, I., Liu, M.-Y., Huffel, C. V., Du, X., Birdwell, D., Alejos, E., Silva, M., Galanos, C., Freudenberg, M., Ricciardi-Castagnoli, P., Layton, B., & Beutler, B. (1998) Defective LPS signaling in C3H/HeJ and C57BL/10ScCr mice: mutations in Tlr4 gene. *Science* 282, 2085–2088.

9. Xu, Y., Tao, X., Shen, B., Horng, T., Medzhitov, R., Manley, J. L., & Tong, L. (2000) Structural basis for signal transduction by the Toll/interleukin-1 receptor domains. *Nature* 408, 111–115.
10. Tao, X., Xu, Y., Zheng, Y., Beg, A., & Tong, L. (2002) An extensively-associated dimer in the structure of the C713S mutant of the TIR domain of human TLR2. *Biochem. Biophys. Res. Commun.* 299, 216–221.
11. Khan, J. A., Brint, E. K., O'Neill, L. A. J., & Tong, L. (2004) Crystal structure of the Toll/interleukin-1 receptor (TIR) domain of IL-1RAPL. *J. Biol. Chem.* 279, 31664–31670.

Part II

Methods to Analyze Signal Transduction Downstream of Toll-Like Receptor Stimulation

Chapter 7

Proteomic Analysis of Protein Complexes in Toll-Like Receptor Biology

Kiva Brennan and Caroline A. Jefferies

Summary

Purification of protein complexes and identification of the constituent components therein have been made relatively simple by the recent advances in proteomics. Uniting good biochemical and protein chemistry techniques with protein identification by mass spectrometry (MS) has resulted in advances in this field that are unprecedented. Our knowledge of Toll-like receptor (TLR) biology has been considerably advanced through the use of such techniques, with key intermediates such as TRAF3, TANK, RIP1 all being identified using proteomic strategies. Applying these techniques to key questions in TLR biology will undoubtedly serve to further advance the field.

Key words: TLR, immunoprecipitation, Pull-down, Trypsin digestion, Mass spectrometry.

1. Introduction

Proteomics is an extremely powerful and ever-growing technology. It is the use of protein chemistry and mass spectrometry (MS) in tandem. The success of proteomics depends almost entirely on good experimental design. This chapter is specifically addressing the use of proteomic techniques to isolate and identify protein complexes. There are many different experimental approaches that can be employed to achieve this with a variety of complexity (reviewed in **refs.** *1–4*). Tandem affinity purification is a highly sophisticated approach and requires epitope-tagging of the protein of interest, stably expressing it in a suitable cell-line and two separate purification techniques *(5)*. It has been

Claire E. McCoy and Luke A.J. O'Neill (eds.), *Methods in Molecular Biology, Toll-Like Receptors, vol. 517*
DOI: 10.1007/978-1-59745-541-1_7

used with great success by a number of groups, and variations on this approach have yielded interesting and novel information regarding Toll-like receptor signalling such as the identification of TRAF3 *(6, 7)*.

However, if such a technique is unavailable, several more simple alternatives can be employed. One of the simpler approaches is to use biotinylated recombinant peptides against the protein of interest, or His-tagged full-length proteins and use them to fish out any interacting proteins from whole cell lysates. Purification of complexes by immunoprecipitation is also a useful method and again has yielded important information regarding TLR signalling intermediates and their post-translational modification *(8–10)*. Once the complexes have been isolated, the next step is to resolve the proteins in the complex by one-dimensional gel electrophoresis, excise the bands, digest the proteins (detailed in **ref.** *11)*, and analyse by mass spectrometry (MS). At this stage, access to a mass spectrometry facility that can separate out the peptides by liquid chromatography (LC) prior to tandem mass spectrometry (MS/MS) analysis is essential. There are many considerations when isolating protein complexes: the nature of the target protein – how abundant it is, how stable it is, and the reagents available.

Here we give an example of how to perform a pull-down experiment where the target protein is recombinantly produced as a His-tagged protein in *Escherichia coli* and purified. Once it has been coupled to nickel (Ni^{2+}) agarose beads, the protein-conjugated beads are incubated with cell lysates and the abilities of proteins to associate with the protein of interest, pre- and post-stimulation with a TLR ligand, are analysed. Once the protein complexes are isolated, the proteins are eluted off the beads and separated out by one-dimensional SDS-PAGE. The SDS-PAGE gel is stained and potential interacting proteins are excised and analysed by LC-MS/MS. This type of analysis is extremely powerful and can yield enormous amounts of information about the protein of interest such as binding partners, substrates, regulators, and even post-translational modifications. This protocol can be modified to look at a target protein (tagged or untagged) over-expressed in a cell-line using an antibody specific to the over-expressed protein (or its tag), or looking at endogenous protein complexes if the protein is in high abundance and the antibodies against it are both specific and of high affinity. In all cases, the main concerns or points to consider are how to reduce the degree of non-specific binding of proteins to the antibodies, proteins, peptides, or beads that are being used. This can be achieved by increasing the stringency of lysis or wash buffers by varying salt and detergent concentration. Performing a number of small-scale trial experiments to address these points is essential.

2. Materials

2.1. Cell Culture and Lysis

1. RPMI (Biosera, East Sussex, UK) supplemented with 10% fetal bovine serum (FBS), 100 units/ml penicillin/100 µg/ml streptomycin.
2. Lipopolysaccharide (LPS) dissolved in sterile, endotoxin-free H_2O at a concentration of 1 mg/ml, stored at 4°C. LPS must be sonicated for 5 min in a sonicating water bath just prior to use and is used at a concentration of 1 µg/ml (concentration needs to be optimised for each cell-line).
3. Ice-cold phosphate-buffered saline (PBS).
4. Cell lysis buffer: 0.1% NP40, 0.05% sodium deoxycholate, 150 mM NaCl, 100 mM Tris-HCl (pH 7.4), 50 mM imidazole, protease inhibitors (*see* **Note 1**) (added just prior to use). For immunoprecipitations and peptide pull-downs, there are some specific differences in lysis buffers (*see* **Note 2**).
5. A 21 gauge needle and 1–5-ml syringe.

2.2. Pull-Down

1. Ni^{2+} agarose beads (Qiagen, Valencia, CA), washed in PBS three times and resuspended in PBS as a 50% v/v slurry. For immunoprecipitation and peptide pull-downs, different types of beads are used (*see* **Note 3**).
2. Cell lysis buffer (five times the volume needed above).
3. A 1X SDS sample buffer: 50 mM Tris-HCl (pH 6.8), 100 mM dithiothreitol, 2% (w/v) SDS, 0.01% bromophenol blue, 10% glycerol (store sample buffer at room temperature and add DTT fresh from a 1 M stock before use or add DTT, aliquot and store at –20°C).
4. A 5X SDS sample buffer: 250 mM Tris-HCl (pH 6.8), 500 mM dithiothreitol, 10% (w/v) SDS, 0.05% bromophenol blue, and 50% glycerol (add DTT as above).

2.3. SDS-Polyacrylamide Gel Electrophoresis (SDS-PAGE)

1. Running buffer: 1.5 M Tris-HCl, pH 8.8, store at room temperature (can make larger volumes and store at 4°C).
2. Stacking buffer: 1 M Tris-HCl, pH 6.8, store at room temperature (can make larger volumes and store at 4°C).
3. Ten per cent SDS solution, store at room temperature.
4. Ten per cent (w/v) solution of ammonium persulphate (APS) in water (stable at room temperature for up to a month … more stable at 4°C).
5. Thirty per cent acrylamide:bisacrylamide solution 37.5:1 ratio (neurotoxin when unpolymerised, handle with care).
6. *N*,*N*,*N*′,*N*′-Tetramethylethylenediamine (TEMED) (bad odour, best used in fume hood).

7. Water-saturated butanol: Shake equal volumes of water and butanol in a glass bottle and allow to separate into upper (butanol) and lower (water) phases. Store at room temperature.
8. Running buffer (10X): 250 mM Tris, 1.92 M glycine, 1% (w/v) SDS. Store at room temperature.
9. Pre-stained molecular weight markers (New England Biolabs UK Ltd., Hertfordshire, UK).

2.4. Gel Staining and Trypsin Digest

1. Colloidal Coomassie gel staining kit (Invitrogen, Carlsbad, CA) (*see* **Note 4**).
2. Methanol.
3. Millipore water.
4. Fifty per cent acetonitrile (ACN).
5. One hundred per cent acetonitrile.
6. One hundred millimolar NH_4HCO_3 made up fresh on day of digest (0.79 g/10 ml water).
7. Trypsin V1115A (Promega Ltd., Southampton, UK).
8. Fifty per cent acetonitrile with 1% trifluoroacetic acid (TFA).
9. One per cent formic acid (FA).
10. One per cent trifluoroacetic acid.

3. Methods

3.1 Preparation of Samples

1. THP1 cells (human monocytic leukaemia cell-line) are set up the night before at a density of 5×10^5 /ml to ensure that they are in log phase of growth on the day of the experiment. Setting them up at a lower density means that the volume of cells to spin down on the day of the experiment is unmanageable and also increasing the cell density cuts down on the amount of RPMI used and hence the cost of the experiment.
2. To minimise keratin contamination, ensure powder-free gloves are worn at all times, changed regularly, and that long hair is tied back and up. It is recommended that clean lab coats be worn, wearing of wool on the day be avoided, and all samples used be filtered. Taking steps to minimise keratin contamination is essential. As keratin is so abundant that contamination can mask any "hits" that may occur, as the keratin peptides are predominant in what the mass spectrometer sees.

3. The recombinant protein is pre-coupled to the Ni^{2+} agarose beads by incubating 5 μg of recombinant protein with 50 μl of Ni^{2+} agarose beads in PBS, rotating overnight at 4°C. Differences between this protocol and that for immunoprecipitation or peptide pull-down are highlighted below (*see* **Note 5**).
4. On the day of the experiment, cells are spun down and resuspended at a density of 5×10^7 cells/ml (5×10^8 cells in a 50 ml falcon tube) in fresh medium. This is perfect for short stimulations (anything up to 60 min) but for longer stimulations, the cell density is going to affect the performance of the cells and lower cell densities should be used (normally around 1×10^6 cells/ml for overnight stimulations).
5. LPS is sonicated for 5 min prior to use and used to stimulate the cells for the required time at a concentration of 1 μg/ml. Following stimulation cells are spun down at 800 × *g* at 4°C and washed with an equal volume of ice-cold PBS and the cells centrifuged again. The resulting cell pellet is resuspended in 5 ml (1 ml/1×10^8 cells) of ice-cold PBS and divided between five 1.5 ml Eppendorf tubes. The cells are centrifuged again and PBS removed.
6. The cells are lysed by the addition of 1 ml of ice-cold lysis buffer with freshly added inhibitors. Cells are lysed at 4°C with rotation for 30 min. Each millilitre of lysate is then subjected to ten strokes with a 21 gauge needle to shear DNA and ensure adequate lysis of cells. The lysates are cleared by centrifugation in a minifuge at the top speed for 10 min. The resulting lysate should be clear and easily removed from the pellet to a fresh Eppendorf tube and kept on ice.

3.2. Protein-Complex Isolation

1. Eighty microlitres of the lysates should be taken for later analysis and 20 μl 5X SB added to each.
2. The lysates can be pre-cleared at this stage by incubating with Ni^{2+} agarose beads on their own for 30 min. This removes any proteins that will bind non-specifically to the beads by virtue of their charge (*see* **Note 6**).
3. The cleared lysates are then incubated with the pre-coupled recombinant protein for 1 h at 4°C.
4. Following the 1 h incubation, the protein complexes are washed three times with 1 ml cell lysis buffer (with freshly added inhibitors). Each wash involves addition of lysis buffer, centrifugation in a minifuge for 1 min at 800 × *g*, and removal of most of supernatant (the low speed centrifugation prevents crushing of the beads; it is recommended that beads are allowed to settle for 1 min after centrifugation and prior to removal of the wash buffer).

5. After the last wash, all supernatant is removed (*see* **Note 7**) and beads are resuspended in 50 μl 1X SB
6. Before loading onto a one-dimensional SDS-PAGE gel, samples are boiled for 5 min and centrifuged in a minifuge at top speed for 10 min.

3.3. Sds-page

1. SDS-PAGE is carried out using the Laemmli method *(12)* and these instructions assume the use of the Atto Electrophoresis Dual Mini Slab system. Gel plates must be washed using detergent and rinsed after use. Before use (particularly for gels for subsequent proteomic analysis), gel plates should be dipped in nitric acid (*see* **Note 8**). After nitric acid application, gel plates should be rinsed with distilled water and 70% ethanol and allowed to air-dry.
2. Gel plates are assembled according to the manufacturer's instructions with the plastic gaskets lying ridge up on the spacer plate and the backing plate placed directly on top. The clips supplied are used to hold the plates and gasket in position.
3. Different percentage gels are made up according to **Table 1** but for an overall view of lysate proteins, a 10% gel is a good starting point as it shows proteins with molecular weights between 20 and 200 (*see* **Note 9**).
4. Pour the resolving gel solution between the plates, leaving enough space at the top for the stacking gel, the length of the teeth of the comb plus 1 cm. Overlay the resolving gel with ~200 μl water-saturated butanol and allow to set for 20 min (*see* **Note 10**).
5. When the gel has set, wash off the water-saturated butanol and rinse the top of the gel with distilled water.
6. Prepare the stacking gel solution, according to **Table 2**, and pour on top of the resolving gel. Very quickly, place the

Table 1
Resolving gel (volumes given to make two gels)

	6%	8%	10%	12%	15%
Water (ml)	7.9	6.9	5.9	4.9	3.4
1.5 M Tris-HCl (pH 8.8) (ml)	3.8	3.8	3.8	3.8	3.8
Acrylamide:bisacrylamide (ml)	3	4	5	6	7.5
10% SDS (ml)	0.15	0.15	0.15	0.15	0.15
10% Ammonium persulphate (APS) (ml)	0.15	0.15	0.15	0.15	0.15
TEMED (μl)	6	6	6	6	6

Table 2
Stacking gel (volumes given to make two gels)

H_2O	4.1 ml
1 M Tris-HCl (pH 6.8)	0.75 ml
Acrylamide:bisacrylamide	1 ml
10% SDS	60 μl
10% APS	60 μl
TEMED	60 μl

comb, with its teeth downward, into the stacking gel solution between the gel plates.

7. Once the stacking gel has set carefully, remove the comb and rinse out the wells of the gel with distilled water. Remove the clips and gasket and rinse around the gel with distilled water.
8. Pour the running buffer into the electrophoresis rig to a depth of ~3 cm and place the gel into the gel rig (*see* **Note 11**) with the open face of the wells facing toward the middle of the rig. Gels are secured in place using the clamps provided.
9. The central reservoir created between two gels is then filled with running buffer and the samples applied to the gel using gel loading tips (*see* **Note 12**).

3.4. Staining and De-Staining of Gel

1. Shake gel in staining solution (**Table 3**) for a minimum of 3 h and a maximum of 12 h (*see* **Note 13**).
2. Decant staining solution and replace with a minimum of 200 ml of deionised water per gel. Shake gel in water for at least 7 h. Gel will have a clear background after 7 h in water (*see* **Note 14**).
3. For long-term storage (over 3 days), keep the gel in a 20% ammonium sulphate solution at 4°C.
4. Scanning of the gel is best achieved by placing the destained gel between the leaves of a plastic A4 pocket that has been wiped clean with 50% methanol and 1% TFA solution. A flat bed scanner is ideal for this, or a gel doc system. It is a good idea at this stage to mark out the bands that are going to be excised on the plastic pocket, label them, and take a scan of that also so that there is a permanent record of the bands and the labelling of the gel (*see* **Note 15**).

Table 3
Staining solution (volumes given to be added in this order to stain two gels)

Solution	
Deionised water	110 ml
Methanol	40 ml
Stainer B (shake well before addition)	10 ml
Stainer A	40 ml

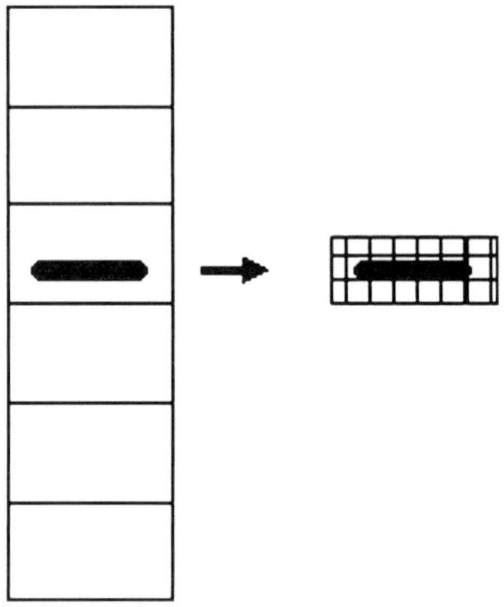

Fig. 1. Illustration of how to excise a protein of interest from the gel.

3.5. Digestion

1. After staining, identify bands of interest. Using a fresh blade, cut out the band and as much as possible avoid taking gel around the protein band.
2. Slice the gel piece lengthwise and across so that slices are approximately 1 × 3 mm pieces (*see* **Fig. 1**) – gives the trypsin a bigger surface area to get into the gel.
3. Use the scalpel to place into clean Eppendorf tubes – if unsure how clean Eppendorf tubes are rinse with 50% acetonitrile (ACN) H_2O with 1% TFA in it – this will get rid of any impurities including plastic polymers that can contaminate samples (*see* **Note 16**).
4. Samples are ready to be washed, using 500 μl per gel piece for 30 min per wash. Wash in water first. If gel slices do not

destain after the first water wash (rarely do) alternate the water and 50% ACN washes – the acetonitrile dehydrates slightly, the water rehydrates, and in the process the Coomassie comes out. It is important to completely destain the bands as the stain interferes with efficient digestion. Once destained perform two final 15 min washes with 50% ACN.

5. Once the final ACN wash has been performed, add 200 µl of 100 mM NH_4HCO_3 for 5 min – the purpose of this wash is to adjust the pH so that the trypsin is optimally active.
6. This following step is perhaps the most important: remove NH_4HCO_3 wash, and add 100% ACN (250 µl) – this dehydrates the gel pieces. They should shrink and turn completely white within 5 min. If this does not occur, remove the ACN and add fresh 100% for a further 10 min. Remove the ACN and speedyvac to dry – pierce the lids of the tubes (make sure the sample numbers are written on sides of tubes as well as on lids). Complete dehydration of the gel pieces is critical – as soon as the trypsin is added to the gel pieces, the solution is absorbed by the gel and trypsin taken into the gel to start the digest.
7. Make up the trypsin – 20 µl of the resuspension buffer to the 20 µg in the tube. Then dilute 1:100 in 25 mM NH_4HCO_3 to give sufficient trypsin solution for the number of experimental points. Add 1 mM $CaCl_2$ to this if possible as it aids with the digestion – some water purifying systems remove the calcium salts which are important for optimal trypsin activity. Add 30 µl of the trypsin solution per sample to cover the gel pieces – gel pieces will swell and take up the trypsin. If not completely rehydrated using this volume (can check after 15–20 min), add the same volume of 25 mM NH_4HCO_3 to the tube. In the majority of cases, this is not necessary. The gel pieces are then left overnight to digest at 37°C.

3.6. Extraction of Peptides

1. Add 200 µl 50% ACN, 1% TFA to the tubes, and agitate for 30–60 min. Remove supernatant and keep (start speedyvacing this extraction – no need to close lids at this stage as all digested and no risk of contamination).
2. Re-extract and then combine supernatants and speedyvac to dryness – store peptides at –80°C until ready to send to be analysed by mass spectrometry. The peptides are analysed using LC-MS/MS (liquid chromatography-tandem mass spectrometry) where the peptides are separated out on a C18 column and as they are eluted, they are analysed directly by the mass spectrometer.

4. Notes

1.

Stocks	Functions
Aprotinin	Serine protease inhibitor, inhibits chymotrypsin, plasmin, kallikrein, trypsin *(13)*
100 mM PMSF	Inhibits serine proteases and acetylcholinesterase
	Inhibits trypsin, chymotrypsin, thrombin, papain
	Prepared in anhydrous isopropanol, it is insoluble in water
200 mM Na_3VO_4	Inhibits ATPases and phosphate-transferring enzymes, alkaline phosphatase and tyrosine phosphatase
	pH adjusted to 10. To ensure presence of monomers solution is boiled until translucent, allowed to cool and pH adjusted to 10. Repeat process until colour stays clear and no orange/yellow colour is noticed
1 M KF/NaF	Serine and threonine phosphatase inhibitor *(14)*
5 mg/ml Pepstatin A	Inhibits acid proteases pepsin, rennin, cathepsin D, chymosin
5 mg/ml leupeptin	Inhibits serine and cysteine proteases

2. As a general rule of thumb, the less stringent a lysis buffer used the better. Protein–protein interactions are much more likely to survive in low salt (100–150 mM NaCl) conditions and where a non-ionic detergent (0.05–1% NP40 or Triton X-100) is used.

   CHAPS is a zwitterionic detergent, which at very low concentration, i.e. 0.05%, allows for solubilisation of membrane proteins without disruption of protein–protein interaction. Examples of some of the buffers commonly used are found at the end of this note. It is important to remember that lysis buffers need to be optimised to fit the needs of each particular experiment *(15–17)*.

   Imidazole is an essential component of buffers for His-tagged recombinant protein pull-downs as it limits non-specific interactions. Nickel beads will bind any protein that is negatively charged but with less strength than they will bind to histidine. In order to deplete non-specific interaction, imidazole is used to dislodge negatively charged molecules. The concentration of imidazole is important and is used to remove His-tagged proteins from Ni^{2+} columns.

Immunoprecipitation lysis buffer: 50 mM HEPES, 150 mM NaCl, 2 mM EDTA, 1% NP40/ IGEPAL, 10% glycerol (*18*).

His-tagged peptide pull-down lysis buffer: 25 mM Tris-HCl, pH 8.0, 140 mM NaCl, 4 mM EDTA, 0.1% NP40, 50–150 mM imidazole.

Biotinylated peptide pull-down lysis buffer: 50 mM HEPES, 150 mM NaCl, 1.5 mM $MgCl_2$, 1 mM EDTA, 10 mM $Na_4P_2O_7$, 10% glycerol, 1% TritonX-100 *(19)*.

3. Immunoprecipitation: Protein G Sepharose (Amersham International plc, Buckinghamshire, UK) washed in PBS three times and resuspended in PBS as a 50% v/v slurry. Peptide pull-down: Streptavidin agarose resin (Pierce, IL 61105, USA) washed in PBS three times and resuspended in PBS as a 50% v/v slurry.
4. The Colloidal Blue Staining Kit from Invitrogen provides nanogram-level detection of proteins and water clear backgrounds without destaining. Using this stain you can detect <10 ng of BSA on a 4–20% 1.0 mm Tris-Glycine gel in an hour. The Colloidal Blue Staining Kit is based on the colloidal properties of Coomassie® Blue dyes created in aqueous or methanolic solutions containing inorganic acids and high salt concentrations (*20*). The free dye in solution is greatly reduced due to the hydrophobic effect, resulting in low background staining and high affinity binding of the dye to the proteins fixed in the gel.
5. In immunoprecipitation, 2–4 μg of antibody is pre-coupled overnight at 4°C to 50 μl Protein A/G beads which have been washed with cold immunoprecipitation lysis buffer *(21)*. The choice of beads depends on the type of immunoglobulin. In most cases, it is suggested to use Protein G as it covers a broader range of immunoglobulins; however Protein A, in some cases, has a higher affinity for a particular immunoglobulin. It is important to use a control antibody on one set of lysates. The control antibody for a mouse monoclonal should be a monoclonal of the same subclass and for rabbit or goat serum should be pre-immune serum from the same animal. For a purified rabbit or goat polyclonal, the control antibody should be another purified rabbit or goat polyclonal.
6. Pre-clearing using Ni^{2+} agarose beads should be preceded by trial experiments where the non-specific interaction of proteins in your lysate is determined initially. Including low concentrations of imidazole at this step significantly reduces non-specific interactions and can be assessed by simply pre-clearing with the Ni^{2+} agarose beads with a range of con-

centrations of imidazole in the lysis buffer and running the beads on a Coomassie gel to assess non-specific binding. Once this has been determined, it is important to run one of the pre-cleared samples as a control on the gel along with the pull-down samples to determine specificity of binding. Pre-clearing for immunoprecipitations will be done using Protein A/G beads for 30 min. This removes any proteins that will bind non-specifically to the beads by virtue of their affinity for Protein A/G.

7. Take a gel loading tip and crimp the end between a pair of thick tweezers. Remove most of supernatant, leaving ~100 μl to be removed using crimped gel loading tip. This will stop the beads from being sucked up into the tip and prevent loss of protein/immunocomplexes.
8. Nitric acid can be used to thoroughly clean plates before use. Nitric acid achieves this through leeching of ions from the surface of the plates. Plates should be immersed in a 30% HNO_3 solution for approximately 30 min. This eliminates contaminants including any residual acrylamide, detergents and dust.
9. It is essential to ensure gel solutions are filtered to remove keratin because most keratin is introduced at gel running and staining stages.
10. The total volume of gel solution prepared will not be used; keep the end of the solution to estimate when the gel sets.
11. To avoid air bubbles, place the gel into the running buffer at a 45° angle and lower gently into place allowing bubbles to be displaced. Bubbles can affect the smooth migration of proteins though the gel.
12. One of the common problems with this type of experiment is contamination with nuclear proteins, i.e. pull-down of a cytosolic protein results in mostly splicing factors, RNA helicases, etc. This mostly results from incomplete shearing of the DNA from the whole cell lysates. Either add DNase to the lysate or sonicate for longer. If there is a problem in pelleting the cell lysate at the clearance stage and the lysate is a bit stringy then it is clear there is a lot of DNA still present. It is not worth proceeding with the experiment until this problem has been resolved.
13. It is important here to use designated staining boxes that have not been used in Western blotting experiments. A common contaminant of proteomic experiments is BSA or alpha-casein – both introduced by using staining boxes that had previously been used to block membranes using either BSA or Marvel. If in doubt, rinse the gel box in concentrated HCl for 15 min, followed by three washes in dH_2O to remove any contaminants.

It is essential during staining and destaining steps that the box remains covered and any handling of the gel takes place with gloves on.

14. Staining intensity does not vary significantly if gels are left in stain for a minimum of 3 h and a maximum of 12 h. Gels can be left in water for up to 3 days without significant change in band intensity and background clarity.
15. Keratin contamination is commonly introduced at this stage of the experiment. It often results from the eager experimentalist peering into the gel box to monitor development of the bands. Keep the gel box covered at all times – keratin is in the air and will settle on a gel exposed to the air remarkably quickly.
16. When pipetting the TFA or formic acid take a glass beaker (smallest available and clean), pour a small amount of the acid into it. Take the pipette that is to be used with tip, pipette up and down (leaches out any plastics – and it is from TFA and FA that most of these make their way into the samples). Now using a fresh clean beaker, pour a small amount of the acid into it again and pipette straight into the solution that it is to be added to.

References

1. Purcell, A. W. & Gorman, J. J. (2004). Immunoproteomics: mass spectrometry-based methods to study the targets of the immune response. *Mol Cell Proteomics* 3, 193–208.
2. Aebersold, R. & Mann, M. (2003). Mass spectrometry-based proteomics. *Nature* 422, 198–207.
3. Liu, H., Lin, D. & Yates, J. R., 3rd (2002). Multidimensional separations for protein/peptide analysis in the post-genomic era. *Biotechniques* 32, 898, 900, 902 passim.
4. Zhou, M. & Veenstra, T. D. (2007). Proteomic analysis of protein complexes. *Proteomics* 7, 2688–97.
5. Puig, O., Caspary, F., Rigaut, G., Rutz, B., Bouveret, E., Bragado-Nilsson, E., Wilm, M. & Seraphin, B. (2001). The tandem affinity purification (TAP) method: a general procedure of protein complex purification. *Methods* 24, 218–29.
6. Hacker, H., Redecke, V., Blagoev, B., Kratchmarova, I., Hsu, L. C., Wang, G. G., Kamps, M. P., Raz, E., Wagner, H., Hacker, G., Mann, M. & Karin, M. (2006). Specificity in Toll-like receptor signalling through distinct effector functions of TRAF3 and TRAF6. *Nature* 439, 204–7.
7. Wang, T., Gu, S., Ronni, T., Du, Y. C. & Chen, X. (2005). In vivo dual-tagging proteomic approach in studying signaling pathways in immune response. *J Proteome Res* 4, 941–9.
8. Aki, D., Mashima, R., Saeki, K., Minoda, Y., Yamauchi, M. & Yoshimura, A. (2005). Modulation of TLR signalling by the C-terminal Src kinase (Csk) in macrophages. *Genes Cells* 10, 357–68.
9. Kong, H. J., Anderson, D. E., Lee, C. H., Jang, M. K., Tamura, T., Tailor, P., Cho, H. K., Cheong, J., Xiong, H., Morse, H. C., 3rd & Ozato, K. (2007). Cutting edge: autoantigen Ro52 is an interferon inducible E3 ligase that ubiquitinates IRF-8 and enhances cytokine expression in macrophages. *J Immunol* 179, 26–30.
10. Ivison, S. M., Khan, M. A., Graham, N. R., Bernales, C. Q., Kaleem, A., Tirling, C. O., Cherkasov, A. & Steiner, T. S. (2007). A phosphorylation site in the Toll-like receptor 5 TIR domain is required for inflammatory signalling in response to flagellin. *Biochem Biophys Res Commun* 352, 936–41.
11. Cramer, R., Saxton, M. & Barnouin, K. (2004). Sample preparation of gel electrophoretically separated protein binding partners for analysis by mass spectrometry. *Methods Mol Biol* 261, 499–510.

12. Laemmli, U. K. (1970). Cleavage of structural proteins during the assembly of the head of bacteriophage T4. *Nature* 227, 680–5.
13. Sherman, M. P. & Kassell, B. (1968). The tyrosine residues of the basic trypsin inhibitor of bovine pancreas. Spectrophotometric titration and iodination. *Biochemistry* 7, 3634–41.
14. Jaumot, M. & Hancock, J. F. (2001). Protein phosphatases 1 and 2A promote Raf-1 activation by regulating 14-3-3 interactions. *Oncogene* 20, 3949–58.
15. Winters, M., Dabir, B., Yu, M. & Kohn, E. C. (2007). Constitution and quantity of lysis buffer alters outcome of reverse phase protein microarrays. *Proteomics* 7, 4066–68.
16. Bowie, A., Kiss-Toth, E., Symons, J. A., Smith, G. L., Dower, S. K. & O'Neill, L. A. J. (2000). A46R and A52R from vaccinia virus are antagonists of host IL-1 and toll-like receptor signaling. *Proc Natl Acad Sci U S A* 97, 10162–7.
17. Brint, E. K., Xu, D., Liu, H., Dunne, A., McKenzie, A. N. J., O'Neill, L. A. J. & Liew, F. Y. (2004). ST2 is an inhibitor of interleukin 1 receptor and Toll-like receptor 4 signaling and maintains endotoxin tolerance. *Nat Immunol* 5, 373.
18. Jefferies, C. A., Doyle, S., Brunner, C., Dunne, A., Brint, E., Wietek, C., Walch, E., Wirth, T. & O'Neill, L. A. J. (2003). Bruton's Tyrosine Kinase is a Toll/interleukin-1 receptor domain-binding protein that participates in nuclear factor {kappa} B activation by Toll-like receptor 4. *J. Biol. Chem.* 278, 26258–64.
19. Rogers, N. C., Slack, E. C., Edwards, A. D., Nolte, M. A., Schulz, O., Schweighoffer, E., Williams, D. L., Gordon, S., Tybulewicz, V. L., Brown, G. D. & Reis e Sousa, C. (2005). Syk-dependent cytokine induction by Dectin-1 reveals a novel pattern recognition pathway for C-type lectins. *Immunity* 22, 507.
20. Neuhoff, V., Arold, N., Taube, D. R., Ehrhardt, W. (1988). Improved staining of proteins in polyacrylamide gels including isoelectric focusing gels with clear background at nanogram sensitivity using Coomassie Brilliant Blue G-250 and R-250. *Electrophoresis* 9, 255–62.
21. Doyle, S. L., Jefferies, C. A., Feighery, C. & O'Neill, L. A. J. (2007). Signaling by Toll-like receptors -8 and -9 requires Bruton's tyrosine kinase. *J Biol Chem* 282, 36953–60.

Chapter 8

2D-DIGE: Comparative Proteomics of Cellular Signalling Pathways

Nadia Ben Larbi and Caroline Jefferies

Summary

Two-dimensional (2-D) gel electrophoresis concerted with protein identification by mass spectrometry (MS) is an extremely powerful method for comparative expression profiling of complex protein samples such as cell lysates. The highly resolutive 2-D electrophoresis allows the separation of heterogeneous protein samples on the basis of isoelectric point (p*I*), molecular mass (*Mr*), solubility, and relative abundance (*(1)* J Biol Chem 250: 4007–4021, 1975; *(2)* Electrophoresis 14: 1067–1073, 1993). Consequently, it provides a comprehensive view of a proteome state (*(3)* Electrophoresis 21: 1037–1053, 2000), where variations in protein expression levels, isoforms, or post-translational modifications (e.g. phosphorylation) can be highlighted and investigated (*(4)* Electrophoresis 21: 2196–2208, 2000). Furthermore, this allows the identification of biological markers that characterize a specific physiological or pathological background of a cell or a tissue (*(5)* Proteomics 1: 397–408, 2001; *(6)* J Bacteriol 179: 7595–7599, 1997). In this way one can compare the effects of a stimulus or drug on cells or tissue, or more importantly, analyse the effects of disease on the expression level of proteins. Relatively recently, conventional 2-D gel electrophoresis has been combined with protein labelling strategies using up to three different fluorescent dyes to allow comparative analysis of different protein samples within a single 2-D gel platform. In this technique, termed differential in-gel electrophoresis (DIGE), samples are labelled separately then combined and run on the same 2D gel minimizing experimental variation and greatly facilitating spot matching. When three CyDyes (Cy2, Cy3, and Cy5) have been used, three images of the gel are captured then superposed to localize the differentially regulated spots on the 2-D gel using image analysis software. This is an extremely powerful tool in comparative proteomics as these dyes provide a linear response to protein concentration up to five orders of magnitude and great sensitivity with detection down to 125 pg of a single protein, which is less than needed for MS identification. In this chapter, we describe the basic methods for protein labelling, optimization of the isoelectrofocusing parameters for the first dimension (where proteins are separated according to their isoelectric point (p*I*)), sodium-dodecyl sulphate-polyacrylamide gel electrophoresis (SDS-PAGE) separation for the second dimension (based on molecular weight (MW)), and different post-staining protocols of the 2-D gel and protein preparation for mass spectrometry identification.

Key words: Proteomic difference analysis, Two-dimensional (2-D) gel electrophoresis, Sample preparation, Fluorescent labelling, Internal standard, Protein detection, Protein identification.

Claire E. McCoy and Luke A.J. O'Neill (eds.), *Methods in Molecular Biology, Toll-Like Receptors, vol. 517*
DOI: 10.1007/978-1-59745-541-1_8

1. Introduction

The genomic era has heralded exciting and challenging new techniques (such as transcriptomics) and new information that has greatly facilitated our understanding of cellular complexity. However, this genomic study is limited as the fundamental knowledge of protein expression, stability, and post-translational modifications in response to biological or physical signals or a physio-pathological state is not available. Moreover, the mRNA level in a cell or tissue does not necessarily reflect the level of protein expression. A protein cannot be produced without an mRNA, but a protein can exist even if its coding mRNA is not present and vice versa; the mRNA can be massively transcribed with no protein being translated and some of mRNAs are not translated at all.

Often, proteins undergo numerous co- and post-translational modifications such as proteolysis, phosphorylation, glycosylation, acetylation, isoprenylation, etc., to reach their functional active form. Interestingly, these modifications represent an essential source of information that cannot be deduced from the sequence of the corresponding gene. Consequently, it is now clearly established that only the direct study of proteins or proteomics can provide a comprehensive and quantitative description of the protein expression and changes under biological perturbations like disease or drug treatment.

In this way, proteomics is a dynamic description of gene regulation with two wide application areas: (i) cartography of the protein content in different subcellular organelles, an inventory of the proteins in such compartments, information on protein–protein interactions *(7)*, and intracellular localization (i.e. addition or elimination of a chemical group on a polypeptidic chain can, e.g., represent a localization signal in a cellular compartment); (ii) knowledge of the levels of expression of proteins and fallouts of disturbing events.

Despite of attractive alternatives (e.g. multidimensional protein identification, stable isotope labelling, protein arrays) *(8–18)* two-dimensional (2-D) electrophoresis remains a routinely employed technique for the differential protein expression profiling and quantitative proteomics. It relies on the powerful combination of 2-D electrophoresis with mass spectrometry (MS) techniques. Two-dimensional electrophoresis can jointly separate more than 5,000 proteins (~2,000 habitually) and can detect less than 1 ng of protein per spot in conformity with the gel size and pH gradient used. Also, one of its best forces is its competency to easily locate post-translationally modified proteins (PTM), as they often appear as distinct rows of spots in the horizontal and/or vertical axis of the 2-D gel due to differences in molecular weight and charge as a result of the addition of the modification.

Up to several hundred PTMs including phosphorylation, ubiquitination, glycosylation, acetylation, lipidation, sulphation, or limited proteolysis have been lately reported and reviewed *(19–23)*. Protein phosphorylation is a fundamental PTM, vital for the balance of numerous regulatory signalling pathways, enzyme activities, and protein degradation, whilst glycosylation is implicated in biochemical modifications, developmental evolution and pathogenesis, e.g. tumorigenesis.

1.1. Introduction to Two-Dimensional Differential In-Gel Electrophoresis (2-D DIGE)

In conventional 2-D technique, protein samples are separated on single gels, stained, and analysed by image comparison using image analysis software. Unfortunately, as different gels images are hardly perfectly superimposable, the comparative 2-D analysis is often extremely time consuming. This limitation is due to the high degree of gel-to-gel variation in spot patterns particularly between samples. As a result, it is difficult to distinguish any true biological variation from an experimental variation, notably when subtle protein modifications are investigated. To circumvent variations, several replicate gels should be run for each sample to create electronic "average" gels, which can then be matched. This tedious procedure has been dramatically simplified as a result of developments by Unlu et al. *(24)*, from which the two-dimensional differential in-gel electrophoresis (2-D DIGE) concept has emerged. The main difference between conventional 2-D electrophoresis and 2-D DIGE technique is that the latter will enable up to three different protein samples to be run on a single 2-D gel, where the differently fluorescently labelled protein samples are mixed before the first dimension run. The multiplexing procedure of the 2-D DIGE methodology allows the integration of the same internal standard on every 2-D gel. This internal standard is a mix of all the samples within the experiment, and therefore includes each protein from each sample. It is useful for matching the protein patterns over gels thereby considerably minimizing the problem of inter-gel variation, which is common in standard "one sample per gel" 2-D electrophoresis experiments. Accurate quantitation of differences between samples, with an associated statistical significance, is then possible and the real biological change is easily distinguished from system variation and inherent biological variation. Therefore, 2-D DIGE technique offers appropriate standardized quantitation in comparative proteomics.

The CyDye DIGE Fluor minimal dyes (Cy™2, Cy™3, and Cy™5) are the three fluorescent cyanine minimal dyes used to label proteins in the 2-D DIGE technique. They are spectrally resolvable (differing in their excitation and emission wavelengths) and matched for mass and charge. As a result, the same protein labelled with any of the CyDye DIGE Fluor minimal dyes will migrate to the same location on the 2-D gel. This multiplexing methodology minimizes intra-gel variation. These dyes furnish great sensitivity with detection down to 125 pg of a single protein and a linear response to protein concentration up to five orders of magnitude. In contrast, silver stain detects 1–60 ng of protein with a range of less than two orders of magnitude. The CyDye DIGE Fluor minimal dyes labelled samples are captured separately but because they have been resolved on the same gel, the images can be superimposed and straightly compared without warping, using the DeCyder Extended Data Analysis (EDA) software that has been designed for the Ettan DIGE system from GE Healthcare (**Fig. 1**). This software contains algorithms that carry out co-detection of differently labelled samples on the same gel. This process reduces the number of gels that must be run as gels can be standardized using one sample as an internal control.

The EDA software also achieves automated detection, background subtraction, quantitation, normalization, internal standardization, and inter-gel matching. The advantages are low user operation, high output, and low experimental variation by offering innovative statistical analysis in a simple-to-use version, uncovering patterns in expression data, and relationship using multivariate analysis and sophisticated clustering combinations. Note that the Progenesis software can also be used for 2-D DIGE analysis.

1.2. The Chemistry of Labelling Proteins with CyDye DIGE Fluor Minimal Dyes

CyDye DIGE Fluor minimal dyes are conceived to form a covalent bond between their NHS ester reactive group and the epsilon amino group of lysine in proteins via an amide connection

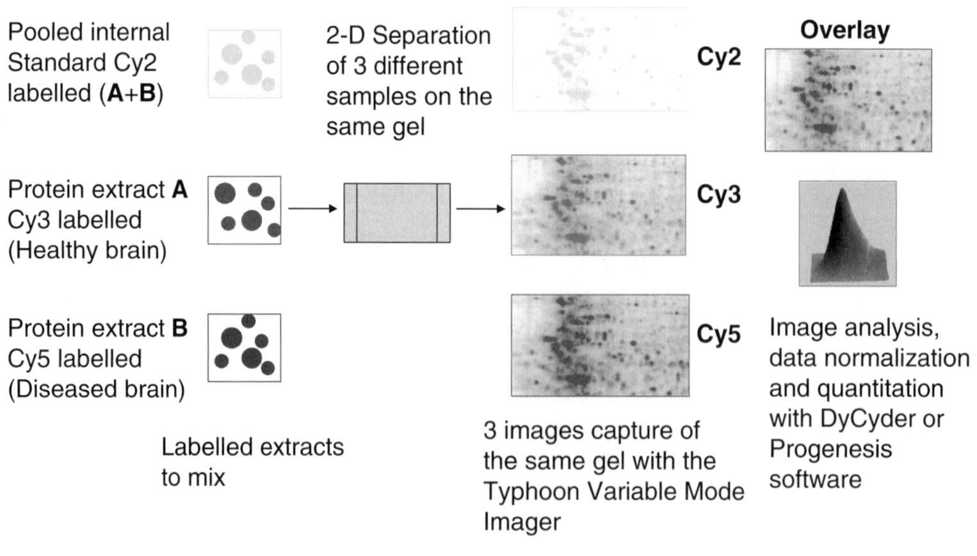

Fig. 1. Minimal Two-dimensional differential in-gel electrophoresis (2-D DIGE) concept: three CyDye DIGE Fluor minimal dyes are separated on a single gel (adapted from GE Healthcare).

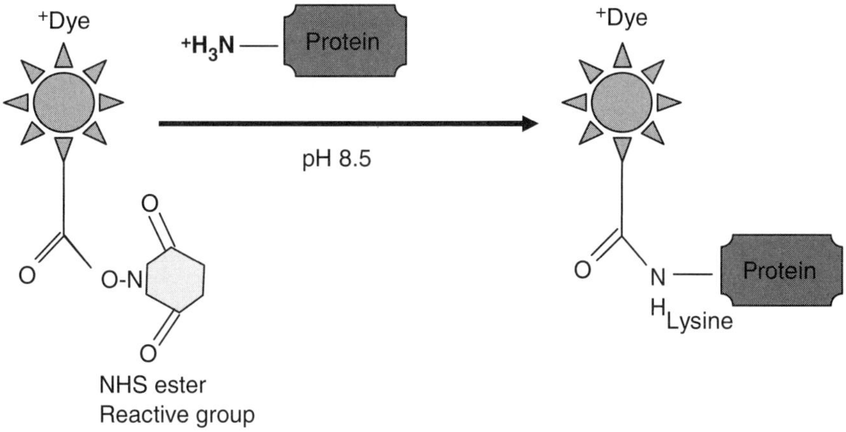

Fig. 2. Diagram representing the minimal labelling reaction. The CyDye differential in-gel electrophoresis (DIGE) Fluor minimal dye contains an NHS ester active group covalently binding the lysine residue of a protein through an amide linkage (adapted from GE Healthcare).

(**Fig.2**). The amount of dye supplied to the protein is limiting within the labelling reaction. As a consequence, the dyes label only 1–2% of lysine residues. Thereby, each labelled protein carries only one dye label and appears as a single protein spot on the 2D-DIGE gel. The lysine amino acid in proteins presents an intrinsic +1 charge at neutral or acidic pH. When coupled to the lysine, the CyDye DIGE Fluor minimal dyes substitute the protein lysine's +1 charge with their own, so the p*I* of the protein is not significantly altered. When conjugated to a protein, each CyDye will add just about 500 Da to the mass of the protein. This mass shift is not visible on a 2-D gel.

1.3. Gel Analysis and Spot Picking

As only a small proportion of the total protein in a sample is labelled, unlabelled species for each protein are also obtained. This can shift slightly the labelled species caused by the addition of the CyDye and is more pronounced for lower molecular weight proteins. Therefore, the centre of the spot may not correspond to the zone of the highest protein content. For this reason, it is more useful to post-stain the 2-D gel in order to localize the exact position of the total protein. The standard post-stain used for this step is SYPRO™ Ruby (a toxic dye that requires proper handling and disposal measures) or Deep Purple from GE Healthcare (non-toxic, therefore easier to handle). Other post-staining protocols, such as Coomassie™ and silver staining, can be employed, but EDA software is set up for a fluorescent signal output, so these stains necessitate more complex analyses. An analytical gel can be post-stained and utilized directly for spot picking. However, the use of a separate preparative gel loaded with 500 μg of sample (pH 3–10 NL 24 cm Immobiline DryStrip) maximizes the amount of protein needed for MS identification. The SYPRO Ruby and CyDye DIGE images are superimposed thus allowing the spots for picking to be located and linking the selected spots from the post-stained gel image with the analytical data. The investigator should optimize the loading for each strip lengths and pH ranges.

In this chapter, we describe basic protocols for 2-D DIGE, hoping to give the reader an overview on how ideal is this technique. However, as the sample preparation is a crucial step for the following separations, it needs to be tested and optimized for each biological sample used *(25–27)*.

2. Materials

2.1. Lysate Preparation for 2-D DIGE

1. Wash buffer: 0.35 M sucrose (isotonic) (*see* **Note 1**).
2. Lysis buffer: 30 mM Tris-HCl, 7 M urea, 2 M thiourea, 4% CHAPS (USB Corporation), adjust the pH to 8.5 and store at –20°C up to 3 months (*see* **Note 2**).
3. Ion exchange resin: MTO-Amberlite MB-1 (Supelco).
4. Whatman filter system, cellulose nitrate membrane filters 0.45 μm.
5. pH indicator paper strips pH 8.0–9.7 (Whatman International Ltd.).

2.2. Labelling of Samples for 2D-DIGE

1. Ten millimolar of L-Lysine monohydro-chloride min 98% TLC (MW 182.6, Sigma). Store at –20°C. Stable for 6 months.
2. Ultrapure anhydrous dimethylformamide (DMF) 99.8% (DMF, Sigma).
3. Five or ten or twenty-five nanomoles of CyDye DIGE Fluor Cy2 minimal dye, 5 or 10 or 25 nmol CyDye DIGE Fluor Cy3 minimal dye and 5 or 10 or 25 nmol CyDye DIGE Fluor Cy5 minimal dye (GE Healthcare). Store at –20°C.

2.3. Preparation of the Labelled Protein Samples for the First Dimension

1. A 2X sample buffer: 7 M urea (USB), 2 M thiourea (USB), 2% Pharmalyte broad range pH 3–10 (GE Healthcare), 2% (20 mg/mL or 130 mM) DTT (USB), 4% CHAPS Ultrapure (USB) (*see* **Note 3**).
2. A 2X rehydration buffer: 7 M urea, 2 M thiourea, 1% pharmalyte broad range pH 3–10, 0.2% (2 mg/mL or 13 mM) DTT, 4% CHAPS and few granules of bromophenol blue (*see* **Note 3**).
3. Immobiline DryStrip pH 3–10, 24 cm and Immobiline DryStrip reswelling tray (GE Healthcare).

4. PlusOne™ DryStrip Cover Fluid (GE Healthcare).
5. The Ettan IPGphor 3 or the Multiphor II Isoelectric Focusing (IEF) System (GE Healthcare).
6. Paper wicks.

2.4. Second Dimension: SDS-PAGE

1. Water-saturated butanol: 50 mL butan-2-ol and 50 mL or more distilled water until two layers become visible. Shake to mix and once completely separated, use the top layer to overlay gels. Store at room temperature. Stable for 6 months.
2. Ettan DALT*six* or *twelve* Large Vertical System, low-fluorescence plates with integral spacers for Ettan DALT, Ettan DALT Cassette Rack, and Ettan DALT Gel Caster. The BIO-RAD PROTEAN® Plus Dodeca Cell system can also be used as a separation unit.
3. Two-dimensional DIGE SDS-PAGE (sodium-dodecyl sulphate-polyacrylamide gel electrophoresis) 12.5% gel composition (900 mL for 12 gels): 281.25 mL acrylamide/bis 40% (w/v), 225 mL Tris-HCL (1.5 M, pH 8.8), 9 mL SDS (10%), 9 mL ammonium persulphate (APS, 10%), 1.24 mL *N,N,N,N′*-tetramethyl-ethylenediamine (TEMED) make up to 900 mL with distilled water. Prior to addition of APS and TEMED, filter complete solution through a 0.2 μm filter into a clean bottle. Allow solution to warm to room temperature before adding APS and TEMED and before pouring the gel.
4. SDS electrophoresis running buffer: 25 mM Tris, 192 mM glycine, and 0.1% SDS. Make up to 23 L with distilled water. Store at room temperature. Stable for 3 months.
5. One per cent of agarose overlay solution: 100 mL electrophoresis running buffer, 1 g low melting point agarose prep and few grains bromophenol blue. Mix components in a 250-mL conical flask and heat on a low setting in the microwave for 1 min. Ensure all the agarose has melted. Allow the solution to cool slightly before use. Store at room temperature. Do not keep for more than 1 month.
6. SDS equilibration buffer stock solution: 100 mM Tris-HCl, pH 8.0, 6 M urea, 30% glycerol and 2% SDS. Make up to 200 mL with deionized water. This stock solution can be stored at room temperature. Stable for 6 months.
7. Equilibration solution 1: 0.5% DTT in SDS equilibration buffer stock solution. Should be used immediately. Do not store.
8. Equilibration solution 2: 4.5% iodoacetamide in SDS equilibration buffer stock solution. Should be used immediately. Do not store.

2.5. Cleaning and Bind-Silane Treating Glass Plates

1. A 0.5 M NaOH and 10% acetic acid.
2. Bind-silane working solution: 8 mL ethanol, 200 μL glacial acetic acid, 10 μL bind-silane, and 1.8 mL double-distilled H_2O.
3. Reference markers used for spot picking from the preparative gel (GE Healthcare).

2.6. Staining of Gels

2.6.1. Silver Staining

1. Fixation solution F1: 450 mL deionized H_2O, 500 mL absolute ethanol (50%), and 50 mL acetic acid (5%).
2. Fixation solution F2: 500 mL deionized H_2O and 500 mL absolute ethanol (50%).
3. Sodium thiosulphate (ST) solution: 2 mL sodium thiosulphate 10% (0.02% final) and 11 mL deionized H_2O.
4. Silver staining solution Ag: 2 mL NO_3Ag 50% (0.1% final) and 11 mL deionized H_2O.
5. Developing solution cNa: 40 g sodium carbonate (2% final), 2.16 mL formaldehyde 37% (0.04% final), and 2 L deionized H_2O (2 L = 2 baths per gel).
6. Stop solution Ac 5%: 50 mL acetic acid (5%) and 950 mL deionized H_2O.
7. Storing solution Ac 1%: 10 mL acetic acid (1%) and 990 mL deionized H_2O.

2.6.2. Coomassie Blue G250 Staining

1. Fixation buffer: 500 mL absolute ethanol (50%), 30 mL *O*-phosphoric acid (instead of H_3PO_4) (3%) and complete to 1 L with deionized H_2O.
2. Staining buffer: 170 g ammonium sulphate (1.3 M), 350 mL deionized H_2O (heat to dissolve), 340 mL methanol (34%), 20 mL *O*-phosphoric acid (instead of H_3PO_4) (2%), and complete to 1 L with deionized H_2O.

2.6.3. SYPRO Ruby Staining

1. Propylene, polycarbonate, or polyvinyl chloride tray.
2. Fixation solution: 30% methanol, 7.5% acetic acid.
3. SYPRO Ruby stain (Invitrogen – Molecular Probes).

2.7. Spot Picking and Preparation for MS

The following buffers should be prepared freshly as they are highly volatile.

1. Acetonitrile 100%.
2. Digestion buffer: 100 mM NH_4HCO_3, 5 mM $CaCl_2$ (stock $CaCl_2$ 100X).
3. Formic acid (FAc) 5%.
4. Trypsin (Gold Promega).
5. ZipTip C18® (Millipore™)

3. Methods

3.1. Experimental Design

Before commencing practical work, the investigator should carefully think about how to design the 2D-DIGE experiments. In fact, it is very important to distinguish the 2-D analysis experiences variation from the biologically induced differential regulation (generated by a pathological state/drug treatment/metabolism cycle, etc.). For example, differences between first dimension strips or second dimension gels, gel distortions, user-to-user variation, and user-specific editing and interpretation represent a source of system variation that can be effectively minimized in the 2D-DIGE gels by including an internal standard within each gel and using the EDA software. This allows automated detection, background subtraction, accurate quantitation, normalization, internal standardization, and increased confidence in matching between gels. Thereby, reliable data and limited subjective editing are offered by the EDA flexibility of statistical analysis.

Furthermore, differences between animals, plants, or cultures must be carefully taken into account as they can be a source of inherent biological differences. To circumvent this type of variation, we strongly recommend the use of biological replicates (multiple cultures, tissues, etc.) in every set of experiments. By the use of the statistical facilities, the differential regulations are accurately separated from the inherent biological variation. Swapping of samples can also be used to confirm the efficiency of the experimental practice.

3.2. Preparation of Cell Lysate Samples for 2D-DIGE Labelling

1. Pellet the cells at +4°C. Discard the media and wash the cells at +4°C with 1 mL of wash buffer at 12,000 × *g* for 5 min. Repeat these washes 4–6 times.
2. Remove the entire cell wash buffer completely.
3. Resuspend the washed cell pellet in 1 mL of lysis buffer (or dilute the pellet 1 in 10) and leave on ice for 10 min.
4. Leave the cells on ice and sonicate intermittently with low-intensity 30 s pulses until the lysate turned clear.
5. Spin the cells at 4°C for 10 min at 12,000 × *g*.
6. Make sure that the pH of the lysate is still at pH 8.0–9.0 using a pH indicator strip. If the pH of the cell lysate is below pH 8.0, then you will need to adjust the lysates pH before labelling (*see* **Note 4**).
7. Determine the protein concentration with a protein assay, compatible with detergents and thiourea (i.e. Bradford).
8. The cell lysate can now be stored in aliquots, at –80°C.

3.3. Resuspension of the CyDye DIGE Fluor Minimal Dyes in DMF for Protein Labelling

As these are standard protocols, I have chosen the GE Healthcare protocols as a model and modified them as appropriate. The following sections are taken from GE Healthcare manual, which are extremely comprehensive and easy to follow. Rather than direct the reader to the manual, I have added them to this chapter for easy reference and have included my own modifications and notes where appropriate.

The resuspension and storage of CyDye DIGE Fluor minimal dyes is critical to the success of sample labelling. If the dimethylformamide (DMF) is of a low quality or the CyDye DIGE Fluor minimal dyes are improperly stored, protein labelling will not be efficient. If the protein labelling is not optimal, problems can occur later in the experiment during gel scanning and image analysis.

CyDye DIGE Fluor minimal dye is supplied as a powder and is resuspended in DMF at the concentration of 1 nmol/µL. After reconstitution in DMF, the dye will give a deep colour: Cy2-yellow, Cy3-red, and Cy5-blue.

1. Allow the CyDye DIGE Fluor minimal dye to defrost for 5 min at room temperature without opening the tube to avoid exposure of the dye to condensation, which can cause hydrolysis.
2. Take a small volume of DMF from the original container and put it into a microfuge tube (*see* **Note 5**).
3. Deduct from this tube the specified volume of DMF (see manufacturer instructions supplied with the CyDye DIGE Fluor minimal dye) and add it to each new vial of dye. For example, add 25 µL DMF to 25 nmol of dye.
4. Vortex vigorously for 30 s to dissolve the dye (*see* **Note 6**).
5. Spin the microfuge tube for 30 s at 12,000 × *g*. The dye stock solution (1 nmol/µL) will need to be further diluted to the working dye solution prior to use in the labelling reaction (*see* **Note 7**).

3.4. Preparation of a Working Dye Solution to Label a Protein Lysate

It is recommended to use 400 pmol of dye to label 50 µg of protein. This dye:protein ratio must be maintained for all samples in the same experiment when more than 50 µg of protein are used for labelling. Different dye:protein ratios can be employed but must be optimized for the sample by checking the labelling efficiency on a one-dimensional (1-D) gel.

1. Spin down the dye stock solution in a microcentrifuge.
2. To make 400 pmol/µL of working dye solution in 5 µL, take a fresh microfuge tube and add 3 µL of DMF.
3. To the DMF, add 2 µL of reconstituted dye stock solution. Pipette up and down several times to remove all dye from the tip into the working dye solution (*see* **Note 8**).

3.5. Labelling of the Protein Sample

The dye labelling reaction is simple and should take about 45 min to perform. The ratio of dye to protein should be maintained at 400 pmol dye: 50 μg protein.

1. Add 50 μg of protein sample to a microfuge tube. Mass labelling reactions can be performed by scaling up as needed.
2. Add 1 μL of working dye solution to the microfuge tube containing the protein sample (i.e. 50 μg of protein is labelled with 400 pmol of dye).
3. Vortex substantially the mix of dye with protein sample.
4. Spin briefly in a microcentrifuge to collect the solution at the bottom of the tube. Leave on ice for 30 min in the dark.
5. Add 1 μL of 10 mM lysine to stop the reaction. Mix and spin briefly and leave on ice for 10 min in the dark. Labelling is now finished.

   The labelled samples can be analysed immediately or stored for at least 3 months at –70°C in the dark.
6. It is recommended to test whether labelling of the proteins is successful before the samples are processed through the 2-D electrophoresis. To do so, a small sample of the freshly labelled lysate is run on a 1-D SDS-PAGE gel along with a control lysate known to label successfully. The gel is then scanned at the appropriate wavelength. The total fluorescent signal of each labelled sample is then compared. Protein lysates that contain previously untested chemical compounds should also be assessed (**Fig.3a**).

3.6. Preparation of the Labelled Protein Samples for the First Dimension

1. After the protein samples have been CyDye labelled, add an equal volume of 2X sample buffer to each sample and leave on ice for at least 10 min.
2. Combine the two or three differentially labelled samples into a single microfuge tube and mix. One of these samples should be the pooled internal standard.

   The samples are now ready for the isoelectric focusing step. Once 2X sample buffer has been added, the sample should be run immediately on an Immobiline DryStrip.

3.7. Rehydration of the Immobiline DryStrip in the Presence of Protein Sample

The differentially CyDye DIGE Fluor minimal dyes labelled protein samples are mixed together to be focused on the same single Immobiline DryStrip. This makes sure that the protein samples, labelled with different dyes, undergo exactly the same electrophoretic running conditions. Immobiline DryStrips must be rehydrated prior to the first separation with the rehydration buffer. They can be rehydrated without samples, followed by sample application by cup-loading (for sample under 100 μL) or with labelled protein already dissolved in rehydration buffer (sample in-gel rehydration, more appropriate for dilute protein samples and preparative gels >350 μg). The volume of rehydration buffer

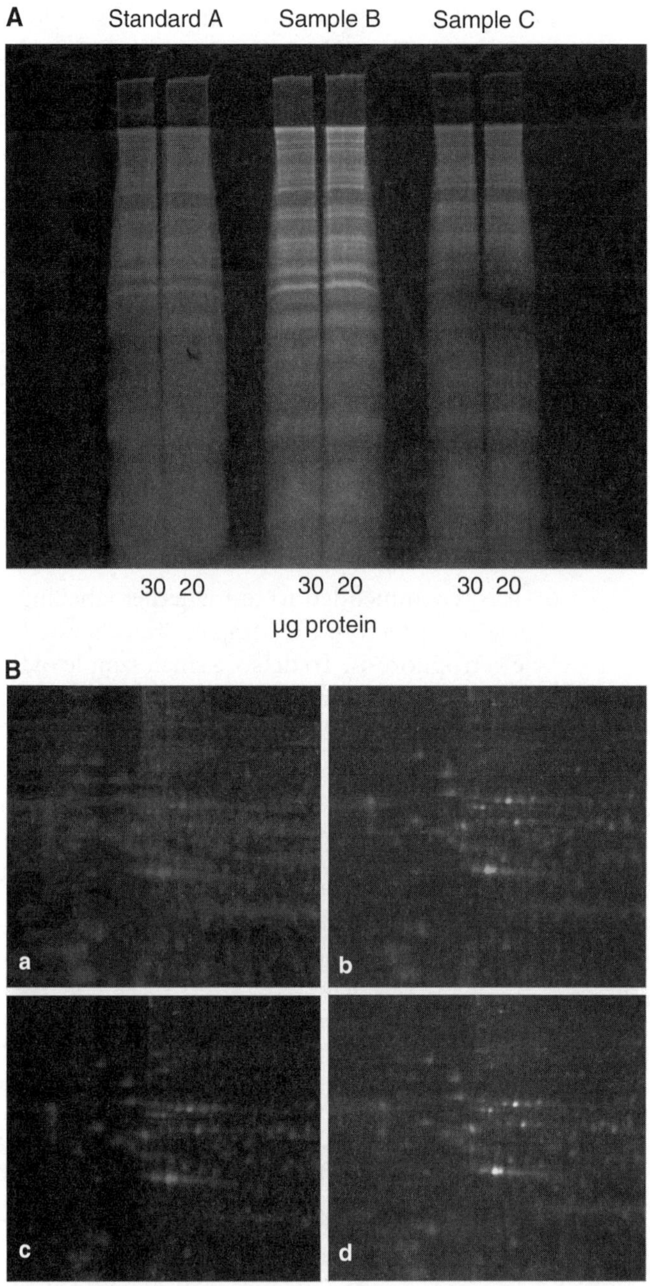

Fig. 3. (**A**) Testing the efficiency labelling for new protein lysates. Freshly labelled lysates from control and LPS-treated cells were run on a one-dimensional sodium-dodecyl sulphate-polyacrylamide gel electrophoresis (1-D SDS-PAGE) gel. Sample A, Cy2-labelled internal standard (Pool of control and LPS sample); Sample B, Cy3-labelled LPS-treated sample; and Sample C; Cy5-labelled control sample. (**B**) Two-dimensional differential in-gel electrophoresis (2-D DIGE) gel images. Cell lysates samples were separately pre-labelled with Cy2 or Cy3 or Cy5, and mixed prior to two-dimensional polyacrylamide gel electrophoresis (2-D PAGE) on a 3–10 NL 24 cm strip. The gel was immediately scanned with the Typhoon Variable Mode Imager using Cy2 or Cy3 or Cy5 excitation wavelengths. The images are then merged and differences between them can be determined by the EDA or Progenesis software. (a) Internal standard (pool of control and LPS-treated cells), (b) control sample, (c) LPS-treated sample, and (d) is the merge of the three captured images (A-B-C) (*See Color Plates*).

Table 1
Rehydration volumes required per Immobiline DryStrips

Immobiline DryStrip length (cm)	Total volume strip (μL)
7	125
11	200
13	250
18	350
24	450

must not go beyond the designated volumes for the Immobiline DryStrip size as shown in **Table 1**:

1. Make up the total volume of labelled protein to the required volume for each Immobiline DryStrip using the rehydration buffer.
2. Apply the labelled protein solution slowly to the centre of the slot in the Immobiline DryStrip reswelling tray. Remove any large bubbles by bursting them using a thin needle (*see* **Note 9**).
3. Remove the protective plastic cover from the Immobiline DryStrip.
4. Place the Immobiline DryStrip with the gel side down and the acidic end of the strip against the end of the slot closest to the spirit level. Put down the Immobiline DryStrip onto the solution. Softly lift and lower the strip along the surface of the solution to cover the Immobiline DryStrip. Avoid trapping bubbles under the Immobiline DryStrip.
5. Overspread each Immobiline DryStrip with 2 mL of PlusOne™ DryStrip Cover Fluid to prevent evaporation and urea crystallization.
6. Put the lid onto the Immobiline DryStrip reswelling tray and allow the Immobiline DryStrips to rehydrate at room temperature overnight. A minimum of 10 h is required for rehydration (*see* **Note 10**).

After the Immobiline DryStrip has been rehydrated for at least 10 h in the reswelling tray, it can be transferred to the ceramic Ettan IPGphor 3 strip holder. Paper electrode wicks should be used with the electrodes and changed many times during the focusing to minimize protein streaking caused by salts effects.

7. Make sure that the Ettan IPGphor 3 is level.
8. Place the electrode paper wicks (5 mm × 15 mm pieces) on a clean dry surface and soak with distilled water. Remove excess

water using filter paper so the wicks are damp and not wet as excess of water may cause streaking.

9. Fill the slot of the ceramic Ettan IPGphor 3 strip holder with PlusOne DryStrip Cover Fluid.
10. Remove the Immobiline DryStrip from its slot by sliding a pair of forceps along the sloped end of the slot and into the depression under the Immobiline DryStrip. Hold the end of the strip with the forceps and lift the strip out of the tray.
11. Position the Immobiline DryStrip gel side up with the basic end superposed on the negative end of the ceramic IPGphor 3 strip holder.
12. Put a humidified electrode paper wick onto the acidic and basic extremities of the gel.
13. Clip down the electrodes firmly onto the electrode pads. Check if the contact between the Immobiline DryStrip and the metal on the outside of the strip holder is good.
14. Lay out at least 4 mL of PlusOne DryStrip Cover Fluid to the IPGphor 3 strip holder, which should completely cover the Immobiline DryStrip and the paper wicks.
15. Close the light excluding lid onto the strip holder. The strip is now ready to focus on the Ettan IPGphor 3 IEF unit.

3.8. Isoelectric Focusing Parameters

Instructions are also available in the Ettan IPGphor 3 Isoelectric Focusing System User Manual for programming the instrument. The investigator should optimize focusing parameters for different pH gradients of Immobiline DryStrips and different protein loadings. However, there are some universal rules that can be taken into account:

1. Use higher total power, measured in voltage hours (Vh), when more protein is loaded onto an Immobiline DryStrip to allow complete focus of the protein sample.
2. Focusing a same amount of protein loaded on Immobiline DryStrips of pH 3–10 range will require fewer Vh than focusing on a narrow range Immobiline DryStrip such as a pH 5.5–6.5.
3. Long Immobiline DryStrips (e.g. pH 3–10, 24 cm) will require more Vh to entirely focus a protein sample than the 18, 13, or 7 cm shorter strips.
4. The following IEF program has been successfully used in our hands for a NL 3–10 and a 4–7, 24 cm DryStrips (*see* **Note 11**):
   - Ten to twelve hours rehydration at 20° C
   - IEF at 20° C and 50 μA maximum per strip (75,000 Vh).
   - S1 Step 3,500 V

- S2 Gradient 8,000 V, 10 min
- S3 Step 8,000 V, 1 h
- S4 100 V, 72 h

3.9. Casting Homogenous Gels in DALTsix Gel Caster

It is recommended to prepare the gels at least 1 day in advance to ensure reproducible results and homogenous polymerization to occur. They can be prepared during the period when the strip is focusing. The following casting instructions assume the use of the Ettan DALT gel system (*see* **Note 12**).

3.9.1. Preparing Glass Cassettes

1. Ensure that all low-fluorescent glass cassettes are clean and dry. If not, the glass plates should be cleaned with deionized water and ethanol using lint-free tissue, such as KleanTech paper. It is important that all glass cassettes are dust-free.
2. Plates should be taped along the sides with an electrical insulating tape (*see* **Note 13**).

3.9.2. Assembling the Caster

The caster should be placed on a level bench or on a levelling table so that the gel tops are level.

1. Remove the faceplate of the DALT*six* caster.
2. Place a separator sheet against the back wall and then alternate gel cassettes with separator sheets. The short glass plate should be next to the pouring channel. Finish with a separator sheet and fill any remaining space with blank cassettes and thicker spacer sheets.
3. Brush a light coating of GelSeal steadily onto the gasket in the faceplate.
4. Insert the faceplate onto the caster with the V-shaped bottom slots resting on their respective screws. Screw these into the holes at the bottom of the faceplate and tighten all screws.
5. Lock the faceplate with the six clamps provided.

3.9.3. Casting the Gels

1. Place the DALT*six* caster in a tray to prevent the acrylamide solution to spread on the bench in case of leak. Prepare the monomer solution in a beaker, adding the APS and TEMED last.
2. Pour the acrylamide solution regularly into the channel at the back of the caster until the level of solution is about 0.5 cm below the level of the shorter glass plate.
3. Apply approximately 1.5 mL of water-saturated butanol onto the top of each gel. This can be washed off with deionized water after 2–3 h.
4. Ideally gels should be left to polymerize overnight. Cover the top of gels with deionized water and a plastic film. Store at +4°C.

3.10. 2-D DIGE Second Dimension: SDS-PAGE

After the Immobiline DryStrips have been focused and equilibrated, the strips can be run immediately on the 2-D gel Ettan DALT*six* or *twelve* electrophoresis systems.

3.10.1 Equilibration of Focused Immobiline DryStrips

Before equilibrating the Immobiline DryStrips, it is strongly recommended to check the quality of the gels polymerization to make sure that they can be used, as once equilibrated the strips cannot be re-frozen and must be run immediately on SDS gels. If the gels are not good to use, freeze your non-equilibrated strips at –80°C and cast new gels (*see* **Note 14**).

1. Prepare SDS equilibration solutions 1 and 2. Reserve 5 mL per strip for each equilibration solution.
2. Pull out the Immobiline DryStrip from the IPGphor 3 strip holder or the Multiphor II using forceps. If the Immobiline DryStrips have been focused and frozen, leave the strips to defrost entirely beforehand.
3. Place the Immobiline DryStrips in individual equilibration tubes or in individual 10 mL disposable pipettes in which the upper extremity has been removed and the lower tight extremity has been stoppered with parafilm.
4. Add 5 mL of the DTT-containing equilibration solution 1 to each tube and stopper the upper extremities with parafilm.
5. Incubate the strips for 10–15 min with gentle agitation. Longer equilibration times may cause proteins diffusion out of the strip during this step.
6. During equilibration, prepare the gel cassettes for loading by rinsing the top of the gels with deionized water, then with SDS electrophoresis running buffer, and pouring off all buffer from the top of the gel. Before loading the Immobiline DryStrips, ensure that the gel surface and plates are dry.
7. Drain away the equilibration solution 1 and add 5 mL of equilibration solution 2. Incubate the strips for 10–15 min with gentle agitation. Pour off the solution and drain thoroughly.
8. During the equilibration step, prepare the agarose overlay solution.

3.10.2. Applying the Focused Immobiline DryStrips to SDS Gels

1. Place the gels in the Ettan DALT cassette rack.
2. Discard the equilibration solution 2 and succinctly rinse the Immobiline DryStrips by submerging them in the 10 mL pipettes with SDS electrophoresis running buffer, which contains 0.1% SDS.
3. Use forceps to carefully slide the Immobiline DryStrip in-between the glass plates across the top of the gel with the plastic side of the strip facing the back. Push against the plastic backing of the Immobiline DryStrip (not the gel itself) using a thin small

spatula until the IPG strip comes into contact with the gel and just lies on its surface. Avoid piercing the second dimension gel with the strip. The IPG strip will slide more easily if the glass plates are slightly wet with 1X SDS electrophoresis buffer. However, make sure that the excess of buffer is drained off before sealing with agarose otherwise the agarose will not set properly. Do not trap air bubbles between strip and the gel. The acidic or positive end of the Immobiline DryStrip is positioned to the left when the shorter plate is facing you. The gel face of the strip should not touch the opposite glass.

4. Pipette the molten (cooled) agarose on top of the IPG strip. Once the agarose solution has completely set, the gel should be run in a separation unit as soon as possible.

3.10.3. Running Gels into the Ettan DALT Electrophoresis Buffer Tank

1. Check the levelling of the separation unit, then fill the lower buffer tank with 1X SDS electrophoresis running buffer. Switch on the control unit, turn on the pump, and set the temperature to 15°C. Make sure that there is enough water in this unit.
2. Once the running buffer has reached the temperature of interest, insert the loaded gel cassettes with the Immobiline DryStrips into the gaskets. Place blank cassette inserts into any free slots. Gel cassettes and blank cassettes should be lubricated with 1X running buffer which makes them slide much more easily into the unit. The buffer level should come up slightly below the level of the buffer seal gaskets, once the 12 slots are occupied.
3. Fill the top of the buffer tank with SDS running buffer to the fill line.
4. Place the lid firmly onto the buffer tank, program the run parameters into the control unit, and start electrophoresis (*see* **Note 15**).

3.10.4. Running Parameters

Gels are run at 0.2 W per gel during the first hour and then at 1–2 W per gel overnight. This can be increased the following morning to 15 W per gel. The run ends when the dye front (which moves in a horizontal line) reaches the end of the gel. The following examples give an idea on how long a gel run can take depending on the power used:

1. For 16 h (overnight run) duration, 15°C, 2 W per gel
2. For 8 h duration, 15°C, 4 W per gel
3. For 4 h duration, 15°C, 8 W per gel (*see* **Note 16**)

3.11. Gels Scanning

1. After the gels have run, they can now be scanned immediately using the Typhoon Variable Mode Imager. Keep the gels between the glass plates but remove the yellow electrical tape along the sides (*see* **Note 17**).

2. Scan the gels as soon as possible to avoid proteins spot diffusion. If gels capture cannot be performed immediately store in SDS electrophoresis running buffer (kept moist), at 4°C and in the dark. However, the gels should be allowed to reach room temperature before scanning. Gels scanned more than a day after running might show significant protein diffusion.
3. Do not fix the gels before they are scanned as this may affect EDA software quantitation of CyDye DIGE Fluor minimal dye labelled proteins (**Fig. 3b**).

3.12. Preparing Gels for the Spot Picking

Once the gel has been analysed by EDA software and the differentially regulated spots have been localized on it, a preparative gel loaded with 500 μg of sample is prepared to identify these differentially regulated proteins using MS. These protein spots are picked from the preparative gel using the Ettan Spot Picker. The Ettan Spot Picker interacts with the EDA software to specifically pick spots from gels as large as the Ettan DALT gels. Spot picking of the correct protein spots detected in EDA software is made possible by attaching two reference markers under the gel. The gel also has to be bound to the glass plate to make sure that the gel does not deform during the staining, imaging, and picking operations.

3.12.1. Gel Preparation

The gels have to be immobilized on a backing to undergo spot picking. It is recommended that the glass plate is treated with a bind-silane solution to lay down/block the gel onto the glass plate (*see* **Note 18**).

Cleaning and Bind-Silane Treating Glass Plates

The following protocol describes the cleaning technique of glass plates on which gels have been bound by a PlusOne bind-silane treatment.

1. Submerge the glass plate in a 0.5 M sodium hydroxide solution for several hours until the gel comes unstuck.
2. Rinse with deionized water.
3. Soak the glass plate for 15 min in 10% acetic acid.
4. Rinse abundantly with deionized water.
5. Dry the plate using a lint-free tissue drenched with ethanol or leave to air-dry in a dust-free environment. Store in a dust-free environment if not to be used immediately.
6. Prepare the bind-silane working solution.
7. Place 2–4 mL (depending on plate size) of the bind-silane solution over the surface of the plate and wipe it over with a lint-free tissue until it is dry. Cover the plate with a lint-free tissue to avoid dust deposit and leave on the bench for 1.5 h (minimum 1 h) for excess bind-silane to evaporate (*see* **Note 19**).

Positioning the Reference Markers

The reference markers must be attached to the treated glass plate before gel pouring. It is important that the markers are properly placed on the treated surface of the bind-silanized plate. Make sure that the markers are not bound where they can interfere with the pattern of protein spots in the gel:

1. Measure the length of the edge.
2. Position the marker at around halfway along the treated plate edge near the spacer, but it should not touch the spacer. Ensure that the markers are tightly stuck to the plate by pushing down with a lint-free tissue or powder-free glove. The markers should be in positions similar to those shown below (**Fig.**4).
3. Repeat **steps 1** and **2** for the other treated backing plate.
4. When finished, pour gels as described above..

3.12.2. Loading Immobiline DryStrips onto a Preparative Picking Gel

The positioning of the Immobiline DryStrip and the reference markers are delicate as it dictates matching of the preparative gel to the analytical gels in EDA software, and ensures that the correct spots are picked from the gel when the picklist is exported to the Ettan Spot Picker:

1. Position the preparative gel adhered to the back of the glass plate (with suitable reference markers), so the front of the reference markers is facing the investigator.
2. Load the Immobiline DryStrip after it has been equilibrated, making sure that the acidic end of the Immobiline DryStrip is on the left-hand side of the gel as shown in the **Fig.**5 and run the gel.

3.12.3. Gel Post-staining for Spot Picking

The gel must be post-stained, e.g. with SYPRO Ruby to visualize spots of interest. This guarantees that the majority of the unlabelled protein is picked for MS identification. Other post-staining methods, such as Coomassie and silver staining, can be used, but these stains will require more complex analysis. The migration differences, between the unlabelled and labelled proteins due to the addition of a single CyDye DIGE Fluor minimal dye molecule to the labelled protein, are more pronounced for

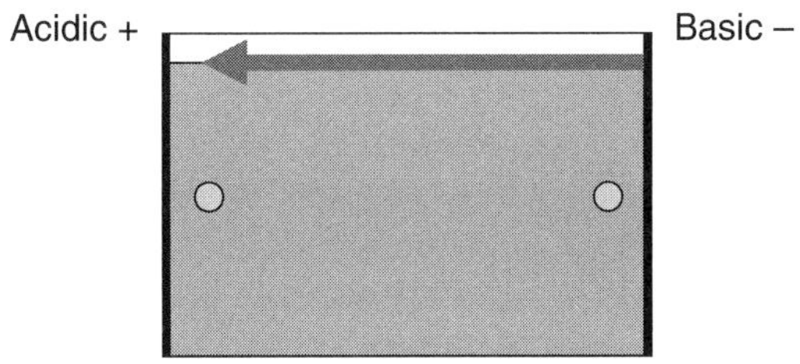

Fig. 4. Model showing the optimum position of reference markers on the gel backing (adapted from GE Healthcare).

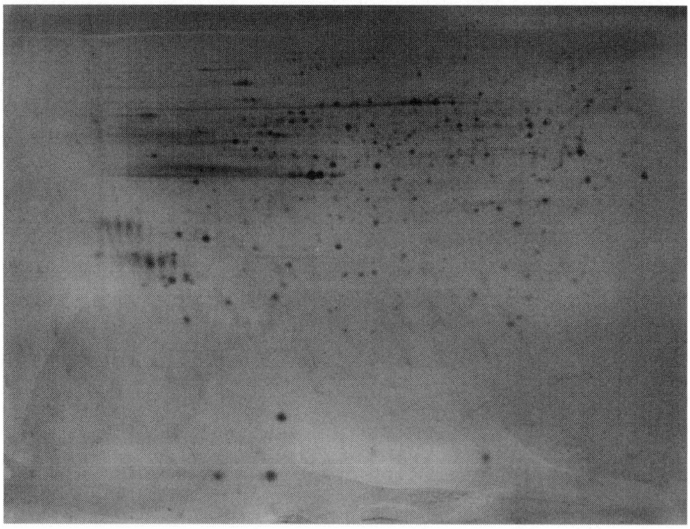

Fig. 5. Silver-stained two-dimensional (2-D) gel. The cell lysate sample was separated on a pH 4–7, 24 cm strip and was post-stained using the PlusOne Silver Kit (GE Healthcare) for focusing optimization purposes.

lower molecular weight proteins. SYPRO Ruby staining allows visualization of the majority of unlabelled protein which needs to be picked for MS identification.

Gel Fixing and SYPRO Ruby Staining

1. Fix the proteins on the gel to prevent the spots diffusing and washing away across the post-staining process using 30% methanol/7.5% acetic acid. Use 1 L of this solution to ensure that the gel is completely covered for a minimum of 2 h (*see* **Note 20**).
2. Once fixed, place the gel directly into a light excluding tray and submerge with the 600 mL of the SYPRO Ruby stain.
3. Incubate the gel for 5 h or overnight with gentle shaking and keep away from the light.
4. Discard the SYPRO Ruby solution and wash in deionized water for 2 h with gentle shaking. Four changes of deionized water should be used during this step.
5. Reassemble the gel for scanning.

Scanning of Fluorescent Post-stained Gels

The steps described below are specifically indicated when a fluorescent post-electrophoresis stain has been used. Work as quickly as you can as some fluorescent post-stains have poor photostability.

1. Once the staining process is finished, clean and rinse with double-distilled water, then dry the back of the glass plate to which the gel is bound. Take care to not damage the gel.
2. Position the gel (glass side down) onto a clean surface with the wells to one side.
3. Damp the farthest edge of the gel with distilled water.
4. Take a clean piece of low-fluorescence glass and place one edge along the farthest edge of the gel.
5. Lower gradually the new glass plate on the gel without forming bubbles. To keep clear of bubbles, ensure there is plenty of water on the gel.
6. Once the new plate is flat on the gel, pick the gel up and let any excess water drain away. Dry off the outside of the plates. The gel is ready to scan.
7. Place the gel onto the Gel Alignment Guide and place into a Typhoon Variable Mode Imager.
8. Scan the gel choosing the appropriate filter set and exposure times and ensure that both reference markers are visible in the gel image. If not, re-scan the gel until they appear as circles. Set the image resolution for the analytical and preparative gels at the same level and at least at 100 μm.
9. When scanning is finished, remove the top plate and store the gel into the fix solution.

Fixing and Gel Silver Staining *(See **Note 21**)*

Use a very clean airtight container (be careful with protein contaminations) to fix and colour the gels, handle always with powder-free gloves. Identify each container for each gel:

1. Fix the gel for 20 min with the F1 solution.
2. Wash for 10 min with the F2 solution.
3. Wash for 10 min with deionized H_2O.
4. Etching for 1 min with the ST solution.
5. Proceed by two washes for 10 s with deionized H_2O.
6. Stain for 20 min with the Ag solution at 4°C.
7. Proceed by two washes for 1 min with deionized H_2O. Develop quickly (turns yellow) with the cNa solution, discard this first bath, and repeat again with a second bath.
8. Stop the development with the Ac solution 5% for 30 min.
9. Store the gel at +4°C in the Ac solution 1%.
10. Scan the gel (**Fig.5**). Clean immediately all the instruments and containers with water and soap, then rinse with demineralized water, deionized water, and leave them to air-dry on absorbent paper before to rank.

Fixing and Coomassie Blue G250 Staining

1. Fix the gel for 2 × 1 h with the fixation buffer under agitation.
2. Wash for 3 × 30 min with water.
3. Incubate 1 h with 500 mL of staining buffer with agitation.
4. Add 333 mg of Coomassie G250 dissolved in methanol.
5. Leave it to stain for 4 days at room temperature with agitation.
6. Wash 3 × 20 min at least with big volumes of deionized water.

3.13. Spot Picking and Preparation for MS

Work on clean surfaces with sterile or clean material (deionized water and ethanol).

Never vortex to avoid protein adsorption on the eppendorf tube walls. As the acetonitrile is highly volatile, it is recommended to aliquot it into small glass containers protected from light.

3.13.1. Washes

1. Pick up the spot of interest (+ blank) with a cut tip depending on the size of the spot.
2. Split the gel spot into small pieces and place them in an Eppendorf.
3. Wash 2 × 10 min with deionized water and discard the supernatant each time.
4. Dehydrate for 10 min with acetonitrile, eliminate supernatant.
5. Rehydrate for 10 min with NH_4HCO_3.
6. Add identical volume of acetonitrile for 10 min, discard supernatant.
7. Speed Vac for 3–5 min.

3.13.2. Digestion (+4°C)

1. Prepare freshly the digestion buffer + Trypsin: 400 µL of digestion buffer and 6 µL of Trypsin (gold 1 µg/µL) aliquoted and stored at −80°C.
2. Rehydrate the gel pieces with digestion buffer + Trypsin for 45 min on ice (150 ng per spot).
3. Add a sufficient volume of digestion buffer to completely cover the pieces of the gel.
4. Incubate over the night at 25°C.

3.13.3. Extraction

1. Recover the digestion supernatant.
2. Add acetonitrile for 10 min and recover the supernatant into the same tube.
3. Add NH_4HCO_3 for 10 min, add the same amount of acetonitrile for 10 min, and recover the supernatant always into the same tube.

4. Repeat again the previous step.
5. Add FAc for 10 min, add the same volume of acetonitrile for 10 min supernatant.
6. Repeat again the previous step.
7. Dry the peptides recovered through speed vac (1.5 h–2 h).
8. Samples can be desalted using ZipTip C18® (Millipore™) according to the manufacturers instructions.

4. Notes

1. The cell wash buffer should not lyse the cells, but it should dilute and discard any growth media or compounds that might affect the labelling reaction. Also, it should be free of any primary amines such as ampholytes as they are able to compete with the proteins for CyDye DIGE Fluor minimal dyes. A poor output of dye labelled proteins might affect the data after scanning and spot detection.
2. Make sure that the pH remains between pH 8.0 and 9.0 by incorporating a buffer such as Tris, HEPES, or bicarbonate in the protein solution at a concentration of approximately 30 mM. Higher buffer concentrations may interfere with isoelectric focusing. If a suitable buffer is not included, the pH of the solution may fall below pH 8.0 resulting in little or no protein labelling. As the lysis buffer is needed to work at 4°C, the pH should be determined when the solution is cold. The protein solution should be free of any added primary amines before labelling as these will compete with the proteins for dyes.
3. Both of the 2X sample buffer and rehydratation buffer used have the same concentration of urea, thiourea, and CHAPS. It is worthwhile to deionize the urea with an ion-exchange resin before adding the other compounds, because in aqueous solution urea is present in equilibrium with ammonium cyanate which can react with protein amino groups and bring in charge artefacts, resulting in additional spots on the IEF gel. The DTT and pharmalytes are added to both stock buffers and are aliquoted (0.5 and 1 mL) for storage at −20°C (stable for 6 months). Pharmalytes are added to improve protein solubility but also as a cyanate scavenger. Once the solutions have been defrosted, they are unstable and must be used the same day, discarding any unused material.
4. Do not proceed with labelling if the pH of the protein sample is outside the scale of pH 8.0–9.0,. If the lysate pH is too

low, e.g. pH is at 7.5, proceed as following: prepare an identical lysis buffer, without cells, at pH 9.5. Then add increasing volumes of the new lysis buffer to the protein sample. This will increase the pH of the protein sample as more lysis buffer is added. Stop when the pH of the protein sample is at pH 8.5. The pH can also be adjusted by the careful addition of a sodium hydroxide solution (50 mM).

5. A new bottle of DMF should be opened every 3 months or for every new CyDye powders.

   The DMF must be high quality anhydrous (specification: ≤0.005% H_2O, ≥9.8% pure) and should not be contaminated with water. DMF will start to degrade as soon as it is opened generating amine compounds, which will react with the CyDye DIGE Fluor minimal dyes and decrease the concentration of dye available for protein labelling.

6. Make sure that the dye solution gives an intense colour. The dye powder may spread around the inside surface of the tube including the lid. If the colour is not deep, pipette the solution around the tube and lid to ensure complete resuspension of dye. Vortex and spin down.

7. After use, store the CyDye DIGE Fluor minimal dyes in a light excluding container, and bring back to the –20°C freezer as quickly as possible. Once resuspended, the dye stock solution is stable for 2 months or until the expiry date on the container.

8. The working dye solution is only stable for 2 weeks at –20°C.

9. The optimal application point is dictated by the characteristics and nature of the sample. Finest results are obtained when the samples are delivered at the pH extremes (near the anode or the cathode). When the proteins of interest have acidic p*I*s or when SDS has been included in sample preparation, it is preferable to deliver the sample near the cathode. Anodic sample application is necessary with pH 6–11 gradients and preferred when pH 3–10 gradients are used as it proved to be superior to cathodic application in most cases. When working with basic pH gradients such as IPGs 6–10, 6–12, or 9–12, anodic delivery is mandatory for all types of samples. Empirical choice of the optimal application point is best.

10. Rehydrate overnight at around 20°C. Higher temperatures (>+37°C) risk to promote protein carbamylation, whereas lower temperatures (<+10°C) result in urea crystallization on the IPG gel. Improved entry of higher (M_r > 100 kDa) into the IPG gel matrix is helped by active rehydration, by applying low voltages (30–50 V) during reswelling.

11. Settings are commonly limited to 50 μA per IPG strip and 150 V to bypass Joule heating, because the conductivity is at the start elevated due to salts. As the run progresses, the salt ions move to the electrodes, decreasing the conductivity and allowing high voltages to be applied. Samples containing high salt concentrations can be desalted directly in the IPG gel by limiting the voltage to 50–100 V over the first 4–5 h with many changes of the electrode paper wicks. Restricted voltage to 100 V overnight can also be employed for large sample volumes (micropreparative runs and/or narrow IPGs) before continuing IEF at higher voltages (>3,500 V). Final settings (up to 8,000 V) are especially helpful for zoom-in gels and alkaline pH gradients. Specific protocols describing optimum focusing parameters for several wide and narrow pH range IPGs have been published by Görg et al. *(3, 26)* and are also available at http://www.wzw.tum.de/proteomik.
12. Low-fluorescence glass plates must be used for gel electrophoresis within Ettan DIGE system as they provide the lowest background pixel values of scanned images. Scratched low-fluorescence glass plates should not be used as the scratches will appear on the image. Make sure that the casting system is clean, dry, and free of any polymerized acrylamide. For best results, filter the acrylamide solution before adding APS and TEMED as any dust in the gel will fluoresce during scanning and will interfere with the quantitative data from EDA software.
13. This is especially useful when the BIO-RAD® separation unit is used. Indeed, as the gel plates are run on their side, the electrophoresis proceeds from right to left instead of top to bottom. When the gasket in the BIO-RAD® tank becomes less tight with time, a current leakage across the unhinged side of the glass plate might happen. This results in a trailing dye front where there is leaking rather than a straight blue line
14. If the Immobiline DryStrip is not run immediately on the second dimension gel, after it has been focused, it can be stored for up to 3 months at −80°C in a sealed container or plastic bag. The container has to be rigid because a frozen Immobiline DryStrip is quite brittle and can easily be spoiled. Do not equilibrate Immobiline DryStrips before storage; this must be carried out immediately prior to the second dimension separation.
15. The BIO-RAD® PROTEAN® Plus Dodeca Cell system is a very interesting separation unit to use as it contains only one large chamber so that all leak problems are completely discarded. It includes electrophoresis buffer tank with built-in ceramic cooling core, lid, buffer recirculation pump with

tubing, and two gel releasers. The investigator interested in this system can follow these steps:

(1) Prepare sufficient 1X SDS running buffer. Depending on the number of gels run, this tank can require up to 23 L. Switch the Multi Temp unit on and set to 15°C. Ensure that there is enough water in this unit.

(2) Check that the separation unit is level and that the outlet/waste tap at the rear of the unit is closed.

(3) Fill halfway through the tank with buffer to avoid overfilling the tank once all plates and blanks are in place.

(4) Lubricate the gels and blank cassettes with 1X buffer before inserting into gaskets (rubber slots into which the glass cassette is placed).

(5) For correct orientation the IPG strip should be to the right side of the gaskets and the hinge side should be at the bottom.

(6) The running buffer (1X) should come up just to the start of the glass spacer, so that all of the gel is submerged with buffer – but don't overfill the tank.

(7) Place the lid firmly onto the unit.

(8) Turn on the pump on and set it to the top setting (this mixes the buffer in the separation tank) and start the run. The run ends when the blue dye (which moves in a vertical line from right to left) reaches the end of the gel.

16. These run-times are recommended for 12 gels. They can become shorter if the number of gels per run decreases as this allows increased watts per gel (up to a maximum of 10 W per gel), which reduces run-times.

17. To scan a gel with fluorescently labelled proteins (either pre-labelled or post-stained), avoid using commonly available plastics as gel backing. Plastic materials will fluoresce intensely at the wavelengths used for scanning.

18. Ettan Spot Picker can perform spot picking from 1 to 1.5 mm thick, 8% to 18% polyacrylamide gels.

19. Enough time should be left to the bind-silane to dry efficiently prior to assemble the glass plates for casting. Otherwise, the solution will evaporate off the treated plate and coat the facing glass surface. As a result, the gel will stick to both plates and will stay attached to bind-silane treated glass during electrophoresis, staining procedures, scanning, and storage.

20. A stronger fix (higher concentration of methanol and/or acetic acid) can make the gel peel away from the backing or

crack. When using a weaker fix, the gels must be incubated for a longer time to become fully fixed.

21. When new samples are tested for the optimal focusing conditions, the silver staining is recommended to visualize the quality of the separation pattern on the 2D gel. This allows the investigator to save the CyDyes (as they are quite expensive) for the decisive experiments.

Acknowledgements

We would like to thank the European Commission support for funding this work through the Marie Curie Actions Human Resources and Mobility Activity program.

References

1. O'Farrell, P.H. (1975) High resolution two-dimensional electrophoresis of proteins. *J Biol Chem* 250, 4007–4021.
2. Burstin, J., Zivy, M., de Vienne, D. & Damerval, C. (1993) Analysis of scaling methods to minimize experimental variations in two-dimensional electrophoresis quantitative data: application to the comparison of maize inbred lines. *Electrophoresis* 14, 1067–1073.
3. Görg, A., Obermaier, C., Boguth, G., Harder, A., Scheibe, B., Wildgruber, R., & Weiss, W. (2000) The current state of two-dimensional electrophoresis with immobilized pH gradients. *Electrophoresis* 21, 1037–1053.
4. Ducret, A., Desponts, C., Desmarais, S., Gresser, M.J. & Ramachandran, C. (2000) A general method for the rapid characterization of tyrosine-phosphorylated proteins by mini two-dimensional gel electrophoresis. *Electrophoresis* 21, 2196–2208.
5. Friso, G., Kaiser, L., Raud, J. & Wikstrom, L. (2001) Differential protein expression in rat trigeminal ganglia during inflammation. *Proteomics* 1, 397–408.
6. Lambert, L.A., Abshire, K., Blankenhorn, D. & Slonczewski, J.L. (1997) Proteins induced in *Escherichia coli* by benzoic acid. *J Bacteriol* 179, 7595–7599.
7. Blackstock, W.P. & Weir, M.P. (1999) Proteomics: quantitative and physical mapping of cellular proteins. *Trends Biotechnol* 17, 121–127.
8. Aebersold, R. (2003) Constellations in a cellular universe. *Nature* 422, 115–116.
9. Aebersold, R. (2003) Quantitative proteome analysis: methods and applications. *J Infect Dis* 187 Suppl 2, S315–320.
10. Ducret, A., Van Oostveen, I., Eng, J.K., Yates, J.R., 3rd & Aebersold, R. (1998) High throughput protein characterization by automated reverse-phase chromatography/electrospray tandem mass spectrometry. *Protein Sci* 7, 706–719.
11. Figeys, D., Gygi, S.P., McKinnon, G. & Aebersold, R. (1998) An integrated microfluidics-tandem mass spectrometry system for automated protein analysis. *Anal Chem* 70, 3728–3734.
12. Gygi, S.P., Rist, B., Gerber, S.A., Turecek, F., Gelb, M.H., & Aebersold, R. (1999) Quantitative analysis of complex protein mixtures using isotope-coded affinity tags. *Nat Biotechnol* 17, 994–999.
13. Haynes, P.A. & Yates, J.R., 3rd (2000) Proteome profiling-pitfalls and progress. *Yeast* 17, 81–87.
14. Link, A.J. Internal standards for 2-D. (1999) *Methods Mol Biol* 112, 281–284.
15. Link, A.J. Autoradiography of 2-D gels. (1999) *Methods Mol Biol* 112, 285–290.
16. Link, A.J. & Bizios, N. (1999) Measuring the radioactivity of 2-D protein extracts. *Methods Mol Biol* 112, 105–107.
17. Link, A.J., Eng, J., Schieltz, D.M., Carmack, E., Mize, G.J., Morris, D.R., Garvik, B.M., & Yates, J.R. 3rd. (1999) Direct analysis of protein complexes using mass spectrometry. *Nat Biotechnol* 17, 676–682.

18. Washburn, M.P., Wolters, D. & Yates, J.R., 3rd (2001) Large-scale analysis of the yeast proteome by multidimensional protein identification technology. *Nat Biotechnol* 19, 242–247.
19. Kalume, D.E., Molina, H. & Pandey, A. (2003) Tackling the phosphoproteome: tools and strategies. *Curr Opin Chem Biol* 7, 64–69.
20. Mann, M. & Jensen, O.N. (2003) Proteomic analysis of post-translational modifications. *Nat Biotechnol* 21, 255–261.
21. Packer, N.H., Ball, M.S. & Devine, P.L. (1999) Glycoprotein detection of 2-D separated proteins. *Methods Mol Biol* 112, 341–352.
22. Peters, E.C., Brock, A. & Ficarro, S.B. (2004) Exploring the phosphoproteome with mass spectrometry. *Mini Rev Med Chem* 4, 313–324.
23. Yan, J.X., Packer, N.H., Gooley, A.A. & Williams, K.L. (1998) Protein phosphorylation: technologies for the identification of phosphoamino acids. *J Chromatogr A* 808, 23–41.
24. Unlu, M., Morgan, M.E. & Minden, J.S. (1997) Difference gel electrophoresis: a single gel method for detecting changes in protein extracts. *Electrophoresis* 18, 2071–2077.
25. Chevallet, M., Santoni, V., Poinas, A., Rouquié, D., Fuchs, A., Kieffer, S., Rossignol, M., Lunardi, J., Garin, J., & Rabilloud, T. (1998) New zwitterionic detergents improve the analysis of membrane proteins by two-dimensional electrophoresis. *Electrophoresis* 19, 1901–1909.
26. Gorg, A., Drews, O. & Weiss, W. (2004) *Purifying Proteins for Proteomics*, Simpson, R. J. (Ed.), Cold Spring Harbor Laboratory Press, New York, USA.
27. Rabilloud, T., Adessi, C., Giraudel, A. & Lunardi, J. (1997) Improvement of the solubilization of proteins in two-dimensional electrophoresis with immobilized pH gradients. *Electrophoresis* 18, 307–316.

Chapter 9

MAPPIT (Mammalian Protein–Protein Interaction Trap) Analysis of Early Steps in Toll-Like Receptor Signalling

Peter Ulrichts, Irma Lemmens, Delphine Lavens, Rudi Beyaert, and Jan Tavernier

Summary

The mammalian protein–protein interaction trap (MAPPIT) is a two-hybrid technique founded on type I cytokine signal transduction. Thereby, bait and prey proteins are linked to signalling deficient cytokine receptor chimeras. Interaction of bait and prey and ligand stimulation restores functional JAK (Janus kinase)–STAT (signal transducers and activators of transcription) signalling, which ultimately leads to the transcription of a reporter or marker gene under the control of the STAT3-responsive rPAP1 promoter. In the subsequent protocol, we describe the use of MAPPIT to study early events in Toll-like receptor (TLR) signalling. We here demonstrate a "signalling interaction cascade" from TLR4 to IRAK-1.

Key words: MAPPIT, Two-hybrid, JAK-STAT, Cytokine signal transduction, TLR.

1. Introduction

Monitoring interaction partners of a given protein is often a major step in revealing its biological role. Therefore, a wide spectrum of biochemical and genetic methods for studying protein–protein interactions has been developed. For a recent general review, we refer to Lievens et al. *(1)*. The mammalian protein–protein interaction trap (MAPPIT) *(2)* was used in this protocol to map Toll-like receptor (TLR) signalling from TLR4 to IRAK-1. MAPPIT is a two-hybrid assay based on type I cytokine signal transduction. Ligand-induced activation and reorganisation of type I cytokine receptors lead to cross-phosphorylation and activation of receptor-associated cytosolic Janus kinases (JAKs). Subsequently, these

Claire E. McCoy and Luke A.J. O'Neill (eds.), *Methods in Molecular Biology, Toll-Like Receptors, vol. 517*
DOI: 10.1007/978-1-59745-541-1_9

activated JAKs phosphorylate conserved tyrosine motifs in the cytoplasmic tail of the receptor, which become docking sites for signalling molecules like signal transducers and activators of transcriptions (STATs). Upon recruitment, STATs are phosphorylated by the JAKs and those activated STATs translocate as dimers to the nucleus, where they induce specific gene transcription (**Fig. 1A**). In MAPPIT, a C-terminal fusion of a given "bait" protein with a leptin receptor (LR) that is deficient in STAT3 recruitment is made. The "prey" protein on the other hand is linked to a series of six functional STAT3 recruitment sites of the gp130 chain. Association of bait and prey and ligand stimulation leads to phosphorylation of the prey chimeras resulting in STAT3 recruitment, activation, and translocation to the nucleus leading to induction of a STAT3-responsive luciferase reporter (rPAPI-Luci) (**Fig. 1B**).

The mammalian cell context of the assay is a major asset compared to the classical yeast-two-hybrid method. This near-optimal physiological environment encompasses the need for posttranscriptional modification, often crucial in mammalian

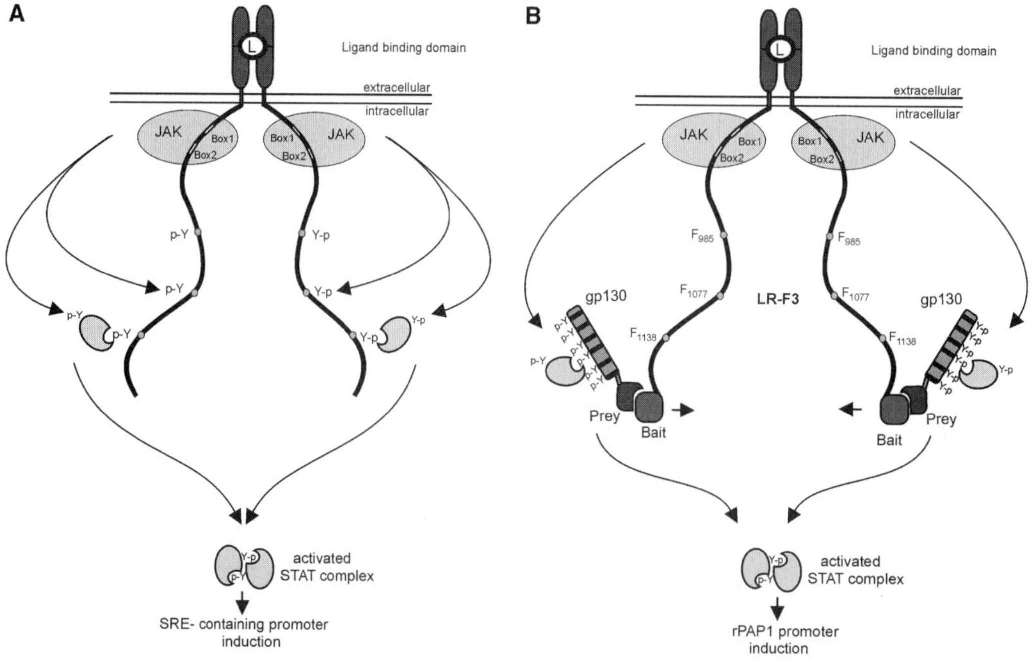

Fig. 1. (**A**) Schematic overview of the JAK (Janus kinase)–STAT (signal transducers and activators of transcription) signal transduction pathway (see main text). L, ligand; SRE, STAT-responsive element. (**B**) Mammalian protein–protein interaction trap (MAPPIT) principle. A given "bait" is C-terminally fused with a leptin receptor (LR) that is deficient in STAT3 recruitment. The extracellular domain of either the erythropoietin (EpoR) or of the leptin receptor (LR) can be used. The "prey" protein is linked to a series of six functional STAT3 recruitment sites of the gp130 chain. Association of bait and prey and ligand stimulation leads to STAT3 activation and induction of a STAT3-responsive luciferase reporter (rPAPI-luci).

protein–protein interactions. For example, JAK2 target sites in the bait protein were found to be phosphorylated upon receptor activation, further underscoring the relevance of using MAPPIT to study signalling pathways. Serine phosphorylation-dependent interactions could also readily be detected using heteromeric MAPPIT *(3)*, whereby kinase and substrate were fused to one of both heteromeric receptor chains.

Intrinsic to the MAPPIT setup, interactor and effector zones are physically separated, respectively in the cytosol and the nucleus since signal readout is mediated by endogenous STAT molecules. In that way, direct interference of bait and prey chimeras with transcription of the reporter gene is avoided. Moreover, signals are cytokine-stimulation dependent, allowing us to discriminate for ligand-independent false positives.

The potency of the analytical use of the MAPPIT technology is reflected by a growing number of studies on different type I cytokine receptors. For instance, aspects of erythropoietin *(4)*, leptin *(5, 6)*, and growth hormone *(7)* receptor signalling have been extensively studied. Next to these type I cytokine receptors, this method can also be used to study some aspects of Toll-like receptor signalling *(8)*, as described in this protocol.

Based on this MAPPIT technology, a reverse two-hybrid system has been developed (reverse MAPPIT) *(9)*, which allows relatively easy discovery and analysis of disruptor molecules. Using reverse MAPPIT in a TLR context could lead to the development of therapeutics alleviating the consequences of uncontrolled TLR stimulation like excessive inflammation, autoimmunity, and septic shock.

2. Materials

2.1. MAPPIT Vectors

1. pXP2d2-rPAP1-luci (pXP2d2 is a gift from Dr. S. Nordeen, Colorado Health Sciences Center, Department of Pathology, Denver, CO, 80262, USA, steve.nordeen@uchsc.edu).
2. pCEL4f, pCLL2f (derived from pcDNA5/FRT (Flp recombination target) vector; Flp-In System, Invitrogen), and pMG2 vectors (derived from pMet7 *[2]*).

2.2. Analytical Application of MAPPIT

2.2.1. Seeding of Cells

1. Hek293T cell line (available from ATCC).
2. Growth medium: Dulbecco's modified Eagle's medium (DMEM) supplemented with 10% fetal calf serum.
3. Gentamycin and penicillin/streptomycin solution (Invitrogen).

2.2.2. Transient $Ca_3(PO4)_2$ Transfection of Hek293T Cells

1. A 2.5 M $CaCl_2$ solution. Prepare in dH_2O. Filter sterilise by passage through a 0.45 mM nitrocellulose membrane and store at −20°C.
2. A 2X HEPES-buffered saline (HeBS) solution: 280 mM NaCl, 1.5 mM Na_2HPO_4, 50 mM HEPES. Adjust pH to 7.05 with NaOH. Filter sterilise by passage through a 0.45 mM nitrocellulose membrane and store at −20°C.

2.2.3. Cell Transfer and Stimulation

1. Ca^{2+}- and Mg^{2+}-free phosphate-buffered saline (PBS) (Invitrogen) and cell dissociation agent (Invitrogen).
2. Mouse leptin (R&D systems). Dilute to 20 µg/ml in growth medium. Store aliquots at −20°C.

2.2.4. Luciferase Reporter Gene Assays

1. Cell culture lysis reagent: 25 mM Tris-phosphate (pH 7, 8), 2 mM DTT, 2 mM 2,2 diaminocyclohexane-*N,N,N′,N′*-tetra-acetate (DCTA), 10% glycerol, 1% Triton X-100.
2. Luciferase substrate: 20 mM Tricine, 1.07 mM $(MgCO_3)_4Mg(OH)_2.5H_2O$, 2.67 mM $MgSO_4$, 3333 mM DTT, 0.1 mM EDTA, 270 µM CoA, 530 µM ATP, 470 µM D-luciferin.
3. Luminescence counter (e.g. Topcount, Packard).

2.3. Assaying Prey Expression

1. Modified RIPA lysis buffer: 200 mM NaCl, 50 mM Tris-HCl pH 8, 0.05% SDS, 2 mM EDTA, 1% NP40, 0.5% desoxycholate, 1 mM NaVanadate, 1 mM NaF, 20 mM b-glycerophosphate. Prepare 50 ml in dH_2O. Add one tablet of Complete® proteinase inhibitor cocktail just prior to use. Store at 4°C.
2. Separating buffer: 1.5 M Tris-HCl, pH 8.8, 0.1% SDS.
3. Stacking buffer: 0.5 M Tris-HCl, pH 6.8, 0.1% SDS.
4. Thirty per cent acrylamide/bis solution (29:1) (Bio-Rad) (caution: this agent is a neurotoxin when unpolymerised).
5. *N,N,N,N′*-Tetramethyl-ethylenediamine (TEMED, Bio-Rad).
6. Ammonium persulphate (10% solution).
7. Prestained molecular weight marker (e.g. All Blue Standards, Biorad).
8. Running buffer (5X): 125 mM Tris, 960 mM glycine, 0.5% (w/v) SDS.
9. Loading buffer (2X): 5% beta-mercaptoethanol, 2% SDS, 8% glycerol, 62 mM Tris-HCl pH 6.8, 1% bromine-phenolblue.
10. Nitrocellulose membrane (e.g. Hybond-C, Amersham Biosciences) and 3MM Whatman paper.
11. Transfer buffer: 25 mM Tris base, 0.2 M glycine, 20% methanol.

12. A 10X Tris-buffered saline (TBS) solution: 0.2 M Tris-HCl, 1.37 M NaCl. Adjust pH to 7.6 with HCl.
13. Wash buffer (1X TBS-T): 1X TBS, 0.1% Tween-20.
14. Blocking buffer (LI-COR® Biosciences).
15. Antibody dilution buffer: Blocking buffer (LI-COR® Biosciences), 0.1% Tween-20.
16. Primary antibody: Anti-Flag M2 (Sigma).
17. Secondary antibody: Goat anti-mouse IRDye® 680 or 800CW (LI-COR® Biosciences).
18. Odyssey® infrared imaging system (LI-COR® Biosciences).

3. Methods

The standard MAPPIT assay makes use of the human embryonic 293T (HEK293T) cells. Other cell lines can also be used. Examples include the haematopoietic TF1 *(10)* and the neuronal N38 cell lines *(11)*. Briefly, the cells are transfected with the bait plasmid, the prey plasmid, and a STAT-dependent reporter gene (pXP2d2-rPAP1-luci). The stimulation of the reporter gene is used as a measure of the interaction between the bait and prey proteins.

To study TLR signalling using MAPPIT, we cloned the intracellular part of some TLRs as a bait and TLR adaptors and downstream signalling molecules both as bait and prey. As proof-of-principle, we studied the signalling cascade starting from TLR4. Tested interactions are depicted in **Fig. 2** (*see* **Note 1**).

We first examined the adaptor recruitment by TLR4, using the intracellular part of TLR4 as bait (TLR4ic) and different adaptors as prey (**Fig. 3A**). Clear interaction could be detected between TLR4ic and Tram or Mal. However, co-transfection of the TLR4ic bait and the MyD88-prey did not lead to any luciferase induction, although the role of MyD88 in TLR4 signalling is well documented *(12)*. We therefore examined if this interaction could be indirect. When co-transfecting a Mal expression vector together with TLR4ic-bait and MyD88-prey, a clear MAPPIT signal was detected, indicating that Mal bridges MyD88 to TLR4. These data are consistent with the phenotype of Mal-deficient mice, which is analogous to MyD88-deficient mice in terms of TLR2 and TLR4 signalling *(13)*, and with a recent report, describing delivery of MyD88 to TLR4 as the primary function of Mal *(14)*. These data prove that indirect interactions can be detected using MAPPIT assay.

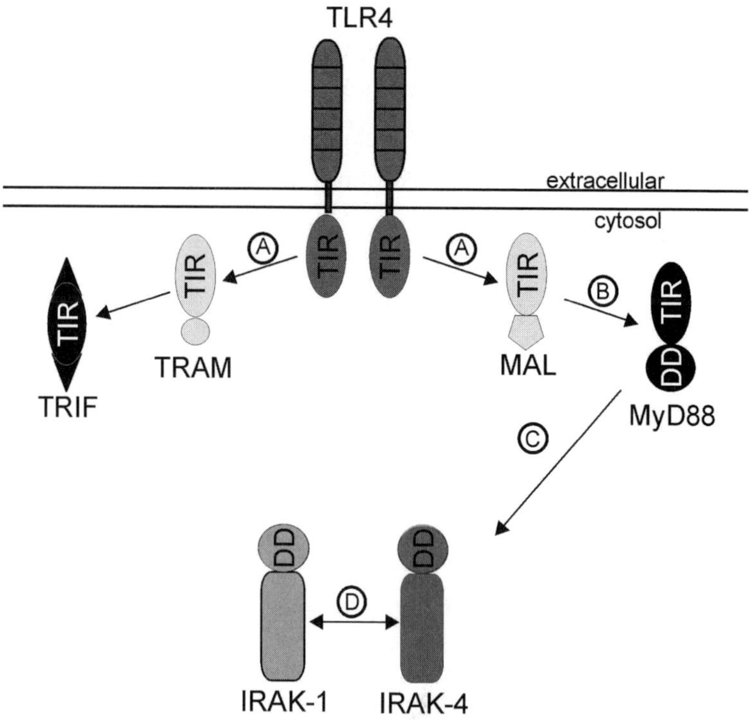

Fig. 2. Initial steps in TLR4 signalling. All tested mammalian protein–protein interaction trap (MAPPIT) interactions, as shown in **Fig. 3**, are annotated with an encircled capital.

As expected, heterodimerisation of Mal and MyD88 was readily detected in a reciprocal way, using these adaptors either as bait or prey (**Fig. 3B**). Further downstream, the interaction of MyD88 and IRAK-4 was also observed (**Fig. 3C**) and association of Mal with IRAK-4 was strongly increased upon MyD88 co-expression, in analogy with the indirect TLR4-MyD88 binding (**Fig. 3C**, lower panel). Finally, dimerisation of IRAK-1 and IRAK-4 could be demonstrated (**Fig. 3D**). Taken together, these data illustrate that MAPPIT can be used to walk down the TLR signalling pathway from the receptor to the downstream IRAK kinases. MAPPIT thus provides a unique tool to study these molecular interactions in great detail in intact mammalian cells.

The design of prey and bait vectors and a typical protocol for an analytical MAPPIT assay are described below.

3.1. MAPPIT Vectors

The plasmid vectors pMG2, pCEL4f, and pCLL2f were developed for analytical MAPPIT applications. The structure of these vectors and of the protein chimeras they encode is shown in **Fig. 4**. Subcloning is performed using standard methods.

The protein used as a prey is cloned C-terminal to the gp130 fragment. The pMG2 plasmid is derived from the mammalian

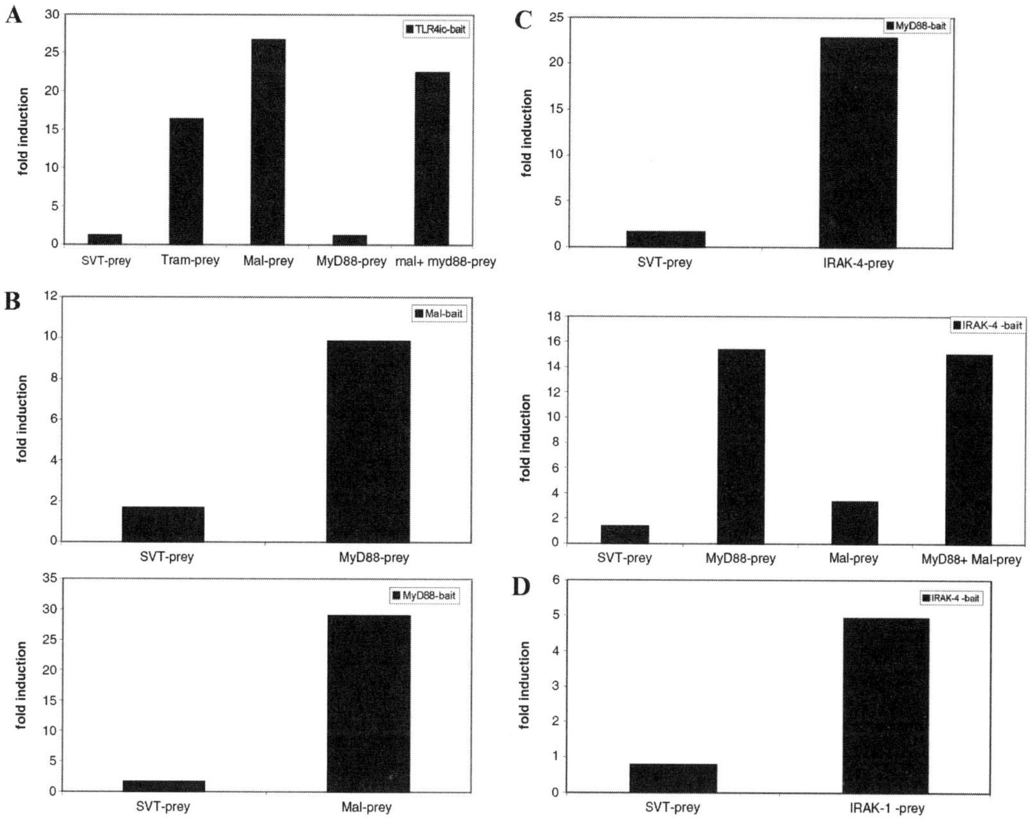

Fig. 3. (**A**) Adaptor recruitment by TLR4. Hek293T cells were transiently co-transfected with the mammalian protein–protein interaction trap (MAPPIT) bait plasmid pCLL-TLR4ic, various TLR-adaptor prey constructs (or a SVT-prey as negative control (*see* **Note 6**)), and the STAT3-responsive rPAPI-luci reporter. The effect of Mal expression on MyD88 association was assayed by co-transfection of a Mal expression vector (pcDNA5-Mal). Twenty-four hours after transfection the transfected cells were stimulated with leptin (100 ng/ml) for another 24 h or were left untreated (NS). Luciferase measurements were performed in triplicate. Data are expressed as mean fold induction (ratio stimulated/NS). (**B**) Heterodimerisation of Mal and MyD88. Hek293T cells were transiently co-transfected with the MAPPIT bait plasmid pCLL-Mal and the prey vector pMG2-MyD88 (*upper panel*) or the reciprocal situation (*lower panel*) together with the rPAPI-luci reporter. Experimental setup was as in (**A**). (**C**) MyD88 and IRAK-4 association. Interaction of MyD88 and IRAK-4 was assayed using the MyD88-bait plasmid pCLL-MyD88 and the pMG2-IRAK4 prey vector (*upper panel*) or with the reciprocal setup (*lower panel*). Indirect association of Mal and IRAK-4 was demonstrated by co-transfecting a MyD88 expression vector (pMet7-MyD88) (*lower panel*). Experimental setup was as in (**A**). (**D**) Dimerisation of IRAK-1 and IRAK-4. Using IRAK-4 as a bait (pCLL-IRAK-4) and IRAK-1 as a prey (pMG2-IRAK-1), heterodimerisation of both kinases was monitored. Experimental setup was as in (**A**).

expression vector pMET7, which contains the strong SRα promoter. This vector encodes the FLAG-tagged gp130 fragment (aa 760–918), of which aa 905–918 were duplicated, with a glycine–glycine–serine (GGS) amino acid linker region preceding the stuffer.

To exchange the stuffer with the prey-encoding sequence, the following restriction sites can be used: EcoRI in combination with NotI, XhoI, or XbaI (**Fig. 4A**).

The pCEL4f and pCLL2f vectors express the receptor-bait chimera. Both vectors originate from the pcDNA5-FRT plasmid (Flp-In system, Invitrogen). Bait expression is controlled by the human cytomegalovirus (CMV) promoter. The extracellular portion of either the human erythropoietin receptor (EpoR, pCel4f) or the murine leptin receptor (LR, pCLL2f) is used (*see* **Note 2**). Transmembranary and intracellularly, both bait-chimeras consist of the LR in which conserved tyrosine residues have been mutated to phenylalanine (LR-F3). Bait proteins can be cloned after the C-terminal GGS hinge sequence using SacI or BamHI and NotI restriction sites (**Fig. 4B**). These vectors also contain an Flp recombination target (FRT) site followed by a hygromycin resistance cassette to permit Flp recombinase-mediated integration in suited cell types (Invitrogen).

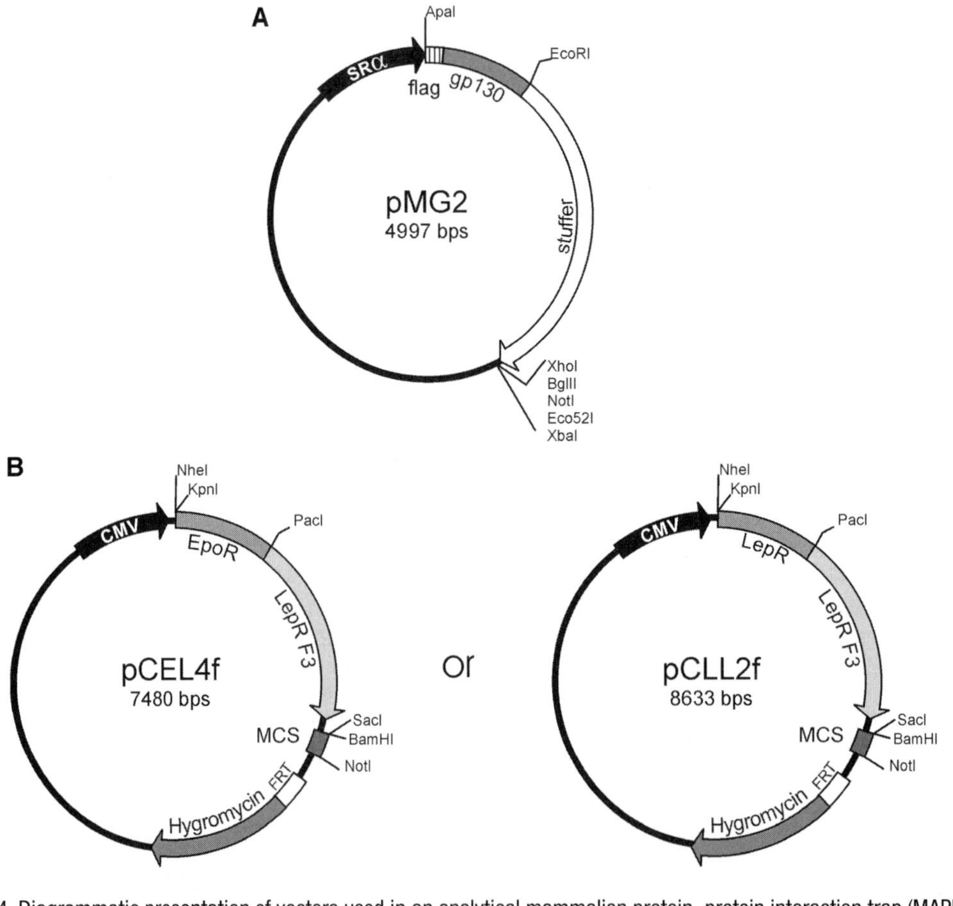

Fig. 4. Diagrammatic presentation of vectors used in an analytical mammalian protein–protein interaction trap (MAPPIT) approach. (**A**) The pMG2 vector is used to express the prey. An N-terminal flag-tag was added to check for prey expression, which is controlled by the strong SRα promoter. The pMG2 vector contains aa 760–918 of the gp130 chain, of which aa 905–918 were duplicated. (**B**) The pCEL4f and pCLL2f vectors encode the chimeric bait receptor under control of the cytomegalovirus (CMV) promoter. pCEL4f and pCLL2f contain the extracellular domains of the erythropoietin receptor (EpoR) or LepR, respectively. Both vectors contain an Flp recombination target (FRT) sequence, which can be used for recombination-assisted integration into the genome of suited cell lines, e.g. T-Rex cell lines (Invitrogen).

3.2. Analytical Application of MAPPIT

3.2.1. Seeding of Cells

1. For each tested condition, plate 4×10^5 subconfluent HEK293T cells in one 10 cm^2 well (6-well plate). Sufficient cells should be plated to allow the experiments to be performed in triplicate and to monitor expression of the bait and prey proteins.
2. Grow overnight in a humidified atmosphere at 37°C and 8–10% CO_2.

3.2.2. Transient $Ca_3(PO4)_2$ Transfection of Hek293T Cells

1. For each well, make a DNA/$CaCl_2$ mixture containing 0.5 µg of pCLL2f bait construct, 0.5 µg of pMG2 prey construct, 200 ng pXP2d2-rPAP1-luci, and 15 µl 2.5 M $CaCl_2$ in a total volume of 150 µl dH_2O in a 1.5 ml tube (*see* **Note 3**). Gently drop 125 µl of the DNA mixture into 125 µl 2X HeBS solution in a 1.5 ml tube, while vortexing.
2. After vortexing the mixture for five additional seconds, leave precipitation mixture for 10–20 min at room temperature (*see* **Note 4**).
3. Add the precipitation mixture to one of the wells of a 6-well plate containing the HEK293T cells.
4. Leave precipitate on cells overnight in the incubator.
5. Remove medium and wash cells once with 1 ml PBS (*see* **Note 5**).
6. Add 2 ml fresh growth medium.
7. Incubate the cells for 8 h in a humidified atmosphere at 37°C and 8–10% CO_2.

3.2.3. Cell Transfer and Stimulation

1. Remove the growth medium and wash once with 1 ml PBS.
2. Add 200 µl cell dissociation agent to each well and incubate for 5 min.
3. Gently tap plate to detach all cells.
4. Add 2 ml growth medium and triturate using a 1 ml pipette to break cell clusters.
5. Plate 50 µl of cell suspension in a black-well plate. We recommend three wells for every stimulation condition. The remainder of the cells are left in the 6-well and can be used for Western blot analysis to verify prey expression (*see* **Subheading 3.3**).
6. Add 50 µl growth medium or 50 µl growth medium containing leptin to each well of the black-well plate. Final leptin concentration should be 100 ng/ml.
7. Grow both the black-well plate and the 6-well plate with the remainder of the cells overnight in a humidified atmosphere at 37°C and 8–10% CO_2.

3.2.4. Luciferase Reporter Gene Assays

1. Remove growth medium from the black 96-well plates.
2. Add 50 μl of cell culture lysis reagent to each well and incubate for 15 min. At this time point, plates can be stored at –20°C.
3. Add 35 μl of luciferase substrate to each well and instantly measure the plate in a luminescence counter.

3.3. Assaying Prey Expression

For a correct analysis of the data generated with the MAPPIT technique, the expression of the prey chimera should be assayed. Using standard SDS-PAGE and Western blotting techniques, prey expression can be checked using the N-terminal flag-tag present on every prey.

1. Discard growth medium of the remainder of the transfected cells in the 6-well plate (*see* **Subheading 3.2.3**) and wash once with 1 ml PBS.
2. Add 150 μl modified RIPA to each well and incubate on ice for 5 min.
3. Transfer lysates into a 1.5 ml tube and centrifuge at full speed to pellet the nuclei.
4. Use 15 μl of the supernatant and add 15 μl 2X loading buffer.
5. Boil samples for 5 min.
6. After cooling down to room temperature, samples are ready for SDS-PAGE.
7. Load samples on a 1.5 mm thick, 10% polyacrylamide gel. The gel can be run at 80 V through the stacking gel and at 120 V through the separating gel until the dye fronts run-off the gel.
8. Next, the separated samples are transferred to a nitrocellulose membrane electrophoretically. Transfer can be obtained at 120 V for 1.5 h or overnight at 30 V.
9. After transfer, the membrane is incubated in 10 ml blocking buffer for 1 h at 4°C on a rocking platform.
10. The blocking buffer is discarded and a 1:4,000 dilution of the anti-Flag antibody (Sigma) is administered. Incubate for 1 h at 4°C on a rocking platform.
11. Discard the primary antibody and wash the membrane five times for 5 min each with 10 ml TBS-T.
12. Incubate the membrane for another hour with a 1:5,000 dilution of the secondary antibody.
13. Discard the secondary antibody and wash the membrane five times for 5 min each with 10 ml TBS-T.

14. Once the final wash is removed from the blot, it can be stored at –20°C or scanned directly for prey expression with the Odyssey® infrared imaging system (LI-COR® Biosciences).

4. Notes

1. No MAPPIT signals were detected using TRIF as bait or as prey. Expression of the TRIF-prey was limited to a perinuclear compartment. Since MAPPIT measures interactions in the sub-membranary space, this finding provides a likely explanation for the lack of TRIF-dependent signals in MAPPIT experiments. Of note, mutagenesis studies in conjunction with the MAPPIT readout and immunofluorescence studies can be used to identify the localisation signals involved in targeting TRIF to this cellular sub-compartment.
2. Generally, a stronger signal is obtained when using the extracellular part of the LR, which can be explained by the fact that after binding of its ligand the LR undergoes higher order clustering, while the EpoR forms homodimers. Throughout this protocol, only the pCLL-2f vector is used.
3. Different ratios of bait and prey plasmids may yield better results in some settings.
4. Do not let the mixture incubate much longer since this will reduce the transfection efficiency. The precipitate can be checked microscopically; the particles should look like small speckles (almost invisible at a 100× magnification) to obtain optimal transfection efficiencies.
5. Do not leave the PBS too long on the cells since this can cause detachment.
6. The pMG2-SVT vector is routinely used as a negative control. This vector encodes for a SV40 large T prey, which lacks its nuclear localisation signal. Bait expression can be monitored by using a prey capable of interacting with the LR-F3 portion of the bait, e.g. the SH2β-prey (data not shown).

Acknowledgements

This work was supported by grants from the Flanders Institute of Science and Technology (GBOU 010090 grant, and to P.U.), from Ghent University (GOA 01G00606), and from the VIB/J&J-COSAT fund.

References

1. Lievens, S., Lemmens, I., Montoye, T., Eyckerman, S., and Tavernier, J. (2006) Two-hybrid and its recent adaptations, *Drug Discov. Today: Technol.* 3, 317–324.
2. Eyckerman, S., Verhee, A., der Heyden, J. V., Lemmens, I., Ostade, X. V., Vandekerckhove, J., and Tavernier, J. (2001) Design and application of a cytokine-receptor-based interaction trap, *Nat. Cell Biol.* 3, 1114–1119.
3. Lemmens, I., Eyckerman, S., Zabeau, L., Catteeuw, D., Vertenten, E., Verschueren, K., Huylebroeck, D., Vandekerckhove, J., and Tavernier, J. (2003) Heteromeric MAPPIT: a novel strategy to study modification-dependent protein–protein interactions in mammalian cells, *Nucl. Acids Res.* 31, e75.
4. Montoye, T., Lemmens, I., Catteeuw, D., Eyckerman, S., and Tavernier, J. (2005) A systematic scan of interactions with tyrosine motifs in the erythropoietin receptor using a mammalian 2-hybrid approach, *Blood* 105, 4264–4271.
5. Lavens, D., Montoye, T., Piessevaux, J., Zabeau, L., Vandekerckhove, J., Gevaert, K., Becker, W., Eyckerman, S., and Tavernier, J. (2006) A complex interaction pattern of CIS and SOCS2 with the leptin receptor, *J. Cell Sci.* 119, 2214–2224.
6. Montoye, T., Piessevaux, J., Lavens, D., Wauman, J., Catteeuw, D., Vandekerckhove, J., Lemmens, I., and Tavernier, J. (2006) Analysis of leptin signalling in hematopoietic cells using an adapted MAPPIT strategy, *FEBS Lett.* 580, 3301-3307.
7. Uyttendaele, I., Lemmens, I., Verhee, A., De Smet, A. S., Vandekerckhove, J., Lavens, D., Peelman, F., and Tavernier, J. (2007) MAPPIT analysis of STAT5, CIS and SOCS2 interactions with the growth hormone receptor, *Mol. Endocrinol.* 21, 2821–2831.
8. Ulrichts, P., Peelman, F., Beyaert, R., and Tavernier, J. (2007) MAPPIT analysis of TLR adaptor complexes, *FEBS Lett.* 581, 629–636.
9. Eyckerman, S., Lemmens, I., Catteeuw, D., Verhee, A., Vandekerckhove, J., Lievens, S., and Tavernier, J. (2005) Reverse MAPPIT: screening for protein–protein interaction modifiers in mammalian cells, *Nat. Methods* 2, 427–433.
10. Montoye, T., Piessevaux, J., Lavens, D., Wauman, J., Catteeuw, D., Vandekerckhove, J., Lemmens, I., and Tavernier, J. (2006) Analysis of leptin signalling in hematopoietic cells using an adapted MAPPIT strategy, *FEBS Lett.* 580, 3301–3307.
11. Wauman, J., De Smet, A.S., Catteeuw, D., Belsham, D., and Tavernier J. (2008) Insulin receptor substrate 4 couples the leptin receptor to multiple signaling pathways, *Mol. Endocrinol.* 22, 965–977.
12. Kawai, T., Adachi, O., Ogawa, T., Takeda, K., and Akira, S. (1999) Unresponsiveness of MyD88-deficient mice to endotoxin, *Immunity* 11, 115–122.
13. Yamamoto, M., Sato, S., Hemmi, H., Sanjo, H., Uematsu, S., Kaisho, T., Hoshino, K., Takeuchi, O., Kobayashi, M., Fujita, T., Takeda, K., and Akira, S. (2002) Essential role for TIRAP in activation of the signalling cascade shared by TLR2 and TLR4, *Nature* 420, 324–329.
14. Kagan, J. C. and Medzhitov, R. (2006) Phosphoinositide-mediated adaptor recruitment controls Toll-like receptor signaling, *Cell* 125, 943–955.

Chapter 10

Analysis of the Functional Role of Toll-Like Receptor-4 Tyrosine Phosphorylation

Andrei E. Medvedev and Wenji Piao

Summary

Toll-like receptors (TLRs) are principal innate immune sensors critically involved in the recognition of evolutionary conserved microbial and viral structures called "pathogen-associated molecular patterns" (PAMPs). Although recognition patterns of many TLRs have been characterized, molecular mechanisms that initiate TLR signaling are poorly understood. Since posttranslational modifications of many receptor systems are important in initiating signaling, we studied whether tyrosine phosphorylation of TLR4, the principal sensor of Gram-negative bacterial lipopolysaccharide (LPS) plays a role in TLR4 signal-transducing functions. We found that LPS induced TLR4 tyrosine phosphorylation and mutations of tyrosine residues in the Toll-IL-1R signaling domain markedly suppressed TLR4-mediated activation of JNK and p38 MAP kinases and transcription factors NF-κB, RANTES, and IFN-β. This chapter summarizes a combination of methodological approaches that can be used to demonstrate an indispensable role of TLR4 tyrosine phosphorylation in receptor signaling, including transient transfections, site-directed mutagenesis, immunoprecipitation and immunoblot analyses, and analyses of transcription factor activation in reporter assays.

Key words: Toll-like receptors, Posttranslational modifications, Signal transduction, Cell activation, Transcription factors.

1. Introduction

Toll-like receptors (TLRs) are evolutionary conserved innate immune sensors expressed on macrophages, monocytes, dendritic cells, epithelial, and endothelial cells *(1)*. TLRs sense invariant structures of bacteria and viruses (pathogen-associated molecular patterns, PAMPs), activating innate immunity and initiating adaptive immune responses *(2)*. The seminal discovery of Drosophila

Claire E. McCoy and Luke A.J. O'Neill (eds.), *Methods in Molecular Biology, Toll-Like Receptors, vol. 517*
DOI: 10.1007/978-1-59745-541-1_10

Toll as the receptor critically involved in antifungal immune defense in adult flies *(3)* gave rise to searches of mammalian innate signaling sensors of bacteria and viruses. Very shortly, three important breakthroughs were reported: (1) human Toll homolog was cloned and constitutively active human CD4-TLR4 was found to signal cytokine production and up-regulation of co-stimulatory molecules *(4)*; (2) positional cloning identified a P→H substitution in *Tlr4* as the mutation responsible for lipopolysaccharide (LPS) refractoriness of the mouse C3H/HeJ strain *(5)*; and (3) TLR4 was confirmed as the principal signaling receptor of Gram-negative bacterial LPS by the demonstration of profound LPS refractoriness of TLR4$^{-/-}$ mice *(6)*. Subsequently, other mammalian TLRs involved in the recognition of a variety of bacterial and viral PAMPs were cloned *(7)*, defining the TLR-IL-1R family. Eleven mammalian TLRs identified to date share a similar structural organization, expressing leucine-rich ectodomains that sense PAMPs, transmembrane domains, and cytoplasmic tails with conserved TIR domains mediating signal transduction *(1, 2, 8)*. TLR2 senses lipoproteins, lipopeptides, and lipoteichoic acids expressed by Gram-positive bacteria, mycoplasma, and mycobacteria *(9, 10)*. TLR5 responds to bacterial flagellins *(11)*, TLR9 recognizes unmethylated CpG motifs present in bacterial, fungal, and viral DNA *(12–15)*, and TLR3 and TLR7/8 are sensors of viral single-stranded and double-stranded RNA, respectively *(16–18)*. Mouse TLR11 was shown to sense uropathogenic *Escherichia coli* and profilin-like molecules from *Toxoplasma gondi (19, 20)*, while the ligands for TLR10 are still unknown. TLR recognition of PAMPs leads to hetero-dimerization (TLR2-TLR1 or TLR2-TLR6) or homo-dimerization (TLR4) (rev. in **ref.** 8). This is thought to initiate conformational changes within the TIR domains, creating docking platforms for downstream adapters and kinases and eliciting their recruitment to the receptor complex. As a result, downstream MAP kinases and transcription factors are activated and mediate expression of inflammatory cytokines and chemokines, up-regulation in the expression of adhesion, MHC, co-stimulatory molecules and antimicrobial effectors, and maturation of dendritic cells *(8, 21)*.

Despite advances in the characterization of TLR ligand specificity and downstream signaling, molecular events that signify TLR activation necessary for its signal-transducing functions are poorly understood. Posttranslational modifications, including tyrosine phosphorylation, have been reported to play a significant role in initiating signal transduction by many receptor systems, including B-cell R, FcR, TLR2, TLR3, and TLR5 *(22–27)* and regulating signalosome complex formations amongst adapters (e.g., MyD88) and kinases (e.g., IL-1R-associated kinase (IRAK)-1) *(28)*. We found that TLR4 undergoes tyrosine phosphorylation upon activation with LPS that could be detected by immunoblot analyses of immunoprecipitated TLR4 proteins with commercially available antibodies detecting phosphotyrosine residues *(29)*. Alternatively, TLR4 tyrosine phosphorylation was also examined by immunoprecipitation of total phosphotyrosine proteins followed by the detection of TLR4 proteins with antibodies reacting to native endogenous or epitope-tagged TLR4 variants. Functional significance of TLR4 tyrosine phosphorylation was further shown by demonstrating signaling deficiencies of TLR4 variants that express mutations of tyrosine residues within their TIR domains following transient transfection of wild-type and tyrosine-deficient TLR4 species and analyses of MAP kinase phosphorylation and transcription factor activation in reporter assays (**ref. 29** and this chapter).

2. Materials

2.1. Cell Culture

1. Cell lines: the human embryonic fibroblast (HEK) 293T cell line was obtained from ATCC (Manassas, VA). HEK293 cells stably transfected with native TLR4 and Flag-tagged MD-2 were generated as described *(30)* and kindly provided by Dr. Douglas T. Golenbock (University of Massachusetts School of Medicine, Worcester, MA).
2. Cell culture medium: HEK293T cells were cultured in Dulbecco's Modified Eagle's Medium (DMEM) (Invitrogen, Carlsbad, CA) supplemented with 10% fetal bovine serum (FBS, HyClone, Ogden, UT), 1 mM L-glutamine, 100 U/ml penicillin, and 100 μg/ml streptomycin (Invitrogen) (complete (c) DMEM). HEK293/TLR4/MD-2 cells were cultured in cDMEM containing 1 mg/ml G418 (Invitrogen).
3. Solution of trypsin (0.25%) and ethylenediamine tetraacetic acid (EDTA) (1 mM) (Invitrogen).
4. T225 and T75 cm^2 Corning tissue culture (TC) flasks (Sigma-Aldrich, St. Louis, MO).

2.2. Transformation of Bacterial Cells and Plasmid Generation

1. Luria Bertani (LB) broth (in solution) and LB agar (Invitrogen).
2. Ampicillin (Sigma-Aldrich) is dissolved in distilled water as 1 mg/ml stock solution, aliquoted, and stored at −20°C.
3. LB agar-ampicillin dishes: 10 g Select peptone 140, 5 g Select yeast extract, 5 g sodium chloride, and 12 g Select agar per 1,000 ml of water solution. Autoclave, cool down to ~45–50°C, add ampicillin to a final concentration of 50 μg/ml, and quickly cover plates with ~20 ml of LB/ampicillin solution. After ~10 min,

agar is solidified, and plates are stored for up to 1 month at 4°C.

4. Bacteriological petri dishes (BD Biosciences, Bedford, MA) and 15, 50, and 250-ml high-speed centrifuge tubes (Nalgene, Rochester, NY).
5. DH5-α cells and EndoFree Plasmid Maxi Kits (Qiagen, Valencia, CA).
6. Restriction enzymes and 100x bovine serum albumin (BSA) (10 mg/ml) (New England BioLabs, Ipswich, MA).
7. Ultrapure agarose, 10x TBE buffer, ethidium bromide (10 mg/ml aqueous solution) (Invitrogen).
8. 5x nucleic acid sample loading buffer and 1 kb DNA ladder (Bio-Rad).

2.3. Transient Transfection

1. Highly purified LPS, free of TLR2 contaminants, and the synthetic TLR2 agonist, Pam3Cys (*S*-[2,3-bis(palmitoyloxy)-(2-*RS*)-propyl]-*N*-palmitoyl-(*R*)-Cys-(*S*)-Ser-Lys4-OH trihydrochloride) (Invivogen, San Diego, CA).
2. Superfect transfection reagent and plasmid Maxiprep kits (Qiagen).
3. Plasmids and expression vectors: pCDNA3-YFP-TLR4, pCDNA3-huCD14, and pEFBOS-His/HA-huMD-2 were kindly provided by Dr. Douglas T. Golenbock. Expression vectors pFlag-CMV-1 encoding WT CD4-TLR4 *(31)* were from Dr. Stephen T. Smale (Howard Hughes Medical Institute, UCLA, Los Angeles, CA). P714H, Y674A, and Y680A mutations were introduced into the TIR domain of CD4-TLR4 or YFP-TLR4 by site-directed mutagenesis, using Quick-Change Site-Directed Mutagenesis kit (Stratagene, La Jolla, CA) as described below.
4. Corning TC dishes (150 and 100 mm) (Sigma-Aldrich) and cell scrapers (18 cm) (BD Biosciences).

2.4. Site-Directed Mutagenesis and Sequencing

1. QuikChange® II Site-Directed Mutagenesis Kit containing Turbo-Pfu high-fidelity DNA polymerase, *Dpn* I endonuclease, dNTP mix, and XL1-blue supercompetent cells (Stratagen).
2. QIAprep® Spin Miniprep kit and EndoFree Plasmid Maxi Kit (Qiagen).
3. SmartSpec™Plus spectrophotometer (Bio-Rad).
4. BigDye Terminator Cycle Sequencing Ready Reaction V3.1 Kit and Edge Biosystems spin columns (Applied Biosystems, Foster City, CA).

2.5. Immunoprecipitation and Immunoblotting

1. The following antibodies were used: polyclonal antibody (Ab) to human TLR4 (H80), IκB-α, tubulin, β-actin and IRAK-1 (Santa Cruz Biotechnology, CA), anti-phospho (p)-p38 and anti-p-JNK Abs (Promega), anti-Flag monoclonal Ab (M2) and M2-horseradish peroxidase (HRP) conjugate (Sigma-Aldrich), and anti-p-tyrosine Ab PY20 (BD Biosciences). Anti-TLR4 antiserum (AS) was kindly provided by Dr. Ruslan Medzhitov (Yale University School of Medicine, New Haven, CT).
2. Lysis buffer: 20 mM HEPES (pH 7.4), 0.5% Triton X-100, 150 mM NaCl, 20 mM β-glycerophosphate, 50 mM NaF, 1 mM dithiothreitol, 5 mM *p*-nitrophenyl phosphate, 1 mM sodium orthovanadate, 2 mM EDTA, 1 mM phenylmethylsulfonyl fluoride, and Complete protease inhibitor cocktail (Roche Applied Science, Indianapolis, IN).
3. Four to twenty percent gradient polyacrylamide gels (Invitrogen).
4. A 10x running buffer: 250 mM Tris, pH 8.3, 1920 mM glycine, 1% SDS.
5. A 10x Tris-glycine buffer: 250 mM Tris, pH 8.3, 1920 mM glycine.
6. Laemmli sample buffer: 62.5 mM Tris-HCl, pH 6.8, 2% SDS, 25% glycerol, 0.01% bromophenol blue (Bio-Rad, Hercules, CA).
7. Transfer buffer is prepared by addition of methanol to make up 1X Tris-glycine buffer containing 20% methanol.
8. A 10X Tris-buffered saline 10x Tween 20, and nonfat dry milk (Bio-Rad).
9. Bovine serum albumin (fraction V) (Sigma).
10. Stripping buffer: 100 mM 2-mercaptoethanol, 2% SDS, 62.5 mM Tris-HCl, pH 6.7.
11. Prestained SDS-PAGE Standards, low range (Bio-Rad).
12. Benchmark prestained protein ladder (Invitrogen).
13. ECL and ECL + detection reagents (Amersham Biosciences, Piscataway, NJ).
14. Bio-Rad *Dc* Protein Assay Kit and BSA (10 mg/ml) (Bio-Rad).
15. Kodak Biomax films (Sigma).
16. Protein G-agarose beads (Roche Applied Science).
17. Immobilon-P membranes (Millipore, Billerica, MA).
18. Model 680 Microplate Reader (Bio-Rad).

2.6. Reporter Assays

1. pELAM-luciferase was obtained from Dr. Douglas T. Golenbock (University of Massachusetts School of Medicine, Worcester, MA), pGL3-RANTES-luciferase reporter plasmid was kindly provided by Dr. John Hiscott (McGill University, Montreal, Canada), and pCMV1-β-galactosidase (Clontech, Mountain View, CA).
2. Superfect transfection reagent (Qiagen), passive lysis buffer, and Luciferase Assay System (Promega, Madison, WI), and Galacto-Light™ Chemiluminescent Reporter Gene Assay System (Applied Biosystems).
3. Plastic 12-well plates (BD Biosciences).
4. A Berthold LB 9507 luminometer and luminometer tubes (Berthold Technologies, Bad Wildbad, Germany).

3. Methods

To delineate a functional role for TLR4 tyrosine phosphorylation in signal transduction, two methodological approaches can be used. First, it is important to show ligand (LPS)-inducible phosphorylation of the receptor (*see* **Subheading 3.1**). To this end, TLR4-negative HEK293T cells are transfected with expression vectors encoding untagged, full-length human TLR4, along with overexpression of MD-2 and CD14 to impart LPS sensitivity (*see* **Subheading 3.1.1**). Cells are treated with medium or stimulated with LPS over a time course (*see* **Subheading 3.1.2**), transfected TLR4 is immunoprecipitated and subjected to immunoblot analyses with antibodies recognizing phosphotyrosine residues (PY20) (*see* **Subheadings 3.1.3–3.1.5**). As a modification of this approach, total phospho-tyrosine proteins can be first immunoprecipitated, followed by detection of TLR4 with an anti-TLR4 Ab. Second, mutagenesis of several tyrosine residues within the TIR domain of TLR4 is carried out to study their effects on TLR4 signal-transducing functions (*see* **Subheading 3.2**). For these studies, two TLR4 expression vectors could be employed: pCMV1-CD4-TLR4 encoding constitutively active TLR4 because of the replacement of the ectodomain with the CD4 dimerization tag; and pCDNA3-YFP-TLR4 encoding full-length YFP-TLR4 that is not constitutively active since YFP is located at the C-terminus of TLR4. Site-directed mutagenesis of TLR4-encoding expression plasmids is used to introduce mutations into tyrosine residues within the TIR domain of TLR4 (*see* **Subheadings 3.2.1–3.2.3**). HEK293T cells are transfected with the expression plasmids encoding WT or mutant TLR4 variants (*see* **Subheading 3.2.4**), and the effect of tyrosine deficiencies on signaling cascades driven by overexpressed, constitutively active CD4-TLR4, or activated as a result of LPS engagement of full-length YFP-TLR4 (replicating the behavior of native TLR4) is studied. TLR4-mediated signaling is examined by assessing IRAK-1 and p38 phosphorylation, IκB-α degradation by Western blot analyses (*see* **Subheading 3.2.5**), and NF-κB, RANTES, and IFN-β activation in reporter assays (*see***Subheading 3.2.6**). The description of the methods given below follows these two approaches.

3.1. Studies of LPS-Inducible Tyrosine Phosphorylation of Full-Length YFP-TLR4

3.1.1. Cell Culture and LPS Stimulation

1. HEK293T are maintained in cDMEM, and subcultured twice a week (upon reaching confluence) by aspiration of spent culture medium, addition of trypsin-EDTA (3 ml/75 cm^2-tissue culture flask), and incubation for 2–5 min at 37°C. Flasks are slapped, cell detachment is ensured by microscopic examination, cDMEM is added to neutralize trypsin (7 ml per flask), and cells are transferred to fresh T75 cm^2 flasks containing 10 ml of fresh cDMEM to make 1:20; 1:10, and 1:5 splitting ratios (0.5, 1, and 2 ml of cell suspensions per flask, respectively).
2. The remaining cells are plated onto 150 mm tissue culture dishes (1×10^7 cells per dish), and incubated for 24 h before transfection.
3. Mix the following reagents (all reagent amounts are given per dish):
   - Serum-free (SF) DMEM to make up a total volume of 600 μl, including the amount of added plasmids and Superfect transfection reagent (see below)
   - Ten micrograms of pCDNA3 (control), or 10 μg pCDNA3-TLR4, or 10 μg pCMV-Flag-TLR2 (*see* **Note 1**)
   - Eight micrograms of pCDNA3-CD14
   - Two micrograms of pEFBos-HA-MD2
   - Ninety microliters of Superfect
4. Vortex and incubate the tube at room temperature for 30 min.
5. Add serum-containing cDMEM to the mixture to a total volume of 15 ml and mix five times by pipetting up and down. Aspirate spent culture medium from cell monolayers, add transfection mixture to cells, and incubate for 3 h at 37°C in a CO_2-incubator (5% CO_2).
6. At the end of incubation, aspirate the transfection mixture, add 20 ml of fresh cDMEM per dish and recover cells for 20 h at 37°C in a CO_2-incubator (5% CO_2).
7. Prepare LPS and Pam3Cys solutions in cDMEM at 1.1 μg/ml concentration and warm the solutions to 37°C. Prepare large plastic trays with ice, and ensure to have ice-cold PBS and a lysis buffer ready to use.

8. Stimulate cells with 100 ng/ml LPS (TLR4-expressing cells) and 1 μg/ml Pam3Cys (TLR2-expressing cells) for 0, 1, 3, 5, 15, and 30 min by adding 2 ml of stimuli stock (1.1 μg/ml) to 20 ml of medium volume in a 150 mm TC dish.

3.1.2. Preparation of Whole Cell Lysates

1. Immediately after stimulation, place cells on ice, aspirate medium, and wash cell monolayers with ice-cold PBS for three times. After addition of 10 ml of ice-cold PBS per dish, cells are scrapped off into 50-ml tubes, centrifuged for 10 min at 400 × *g*, 4°C and lysed in a lysis buffer (0.5–1 ml per tube).
2. Cells are pipetted up and down in a lysis buffer (at least five times), using a 1-ml pipetman, transferred into 1.5-ml Eppendorf tubes, and rotated for 30 min at 4°C.
3. Centrifuge lysates for 20 min at ≥15,000 × *g*, 4°C to remove insoluble debris and carefully transfer supernatants into fresh Eppendorf tubes without disturbing the pellet.
4. The protein concentrations of cell extracts are measured by Bio-Rad *Dc* Protein Assay Kit as recommended by the manufacturer, and samples for immunoblotting are prepared by resuspending protein at a 1 μg/μl final concentration in Laemmli sample buffer containing β-ME, using safe-lock Eppendorf tubes. Samples are boiled for 5 min and could be frozen at –20°C until use. Alternatively, samples can be loaded on a gel and subjected to immunoblotting immediately.

3.1.3. Protein Measurement

1. The protein concentration of the whole cell lysate is measured by Bio-Rad *Dc* Protein Assay Kit (Bio-Rad).2. Protein standards are prepared by diluting BSA stock solution with PBS to obtain 1,000, 800, 600, 400, 200, and 100 μg/ml concentrations. 3. The lysates are diluted with lysis buffer at a 1:5 ratio, and BSA standards, lysis buffer (a negative control), or the diluted lysates are added to triplicate wells of a 96-well plate (5 μl per well). 4. Prepare reagent A by adding 20 μl of reagent A to each 980 μl of reagent B, and add 25 μl of reagent A to each well. 5. Thereafter, reagent B is added (200 μl per well) using a multichannel pipette, the content of the wells is gently mixed with a multichannel pipette avoiding bubble formation, and optical density (OD) is measured on a Microplate Reader at the wavelength of 595 nm.

3.1.4. Immunoprecipitation

1. Prepare protein samples, resuspending at least 1 mg of total protein in a lysis buffer (a total volume is 1 ml).
2. Samples are then precleared with protein G-agarose beads by addition of 10 μl beads per 1,000 μl lysate, rotated for 4 h at 4°C, and beads are removed by centrifugation at 12,000 × *g* for 1 min.

3. The precleared lysate is incubated overnight with 1 μg/ml of the respective Ab in a lysis buffer to immunoprecipitate either TLR4 (H80), Flag-TLR2 (α-Flag), or phosphotyrosine proteins. Protein G-agarose beads (45 μl per sample) are added and incubation continued for additional 4 h, beads are then extensively washed five times with ice-cold lysis buffer to remove nonbound proteins (*see* **Note 2**).
4. After the final wash, the settled protein G-agarose beads (25 μl pellet volume) are resuspended in 25 μl of 2x Laemmli sample buffer containing β-ME, and boiled for 10 min.

3.1.5. Immunoblotting

1. Prepare running buffer and fill the XCell Sure Lock unit. Remove the gel cassette from the pouch and rinse with distilled water. Peel off the tape covering the slot on the back of the gel cassette and pull the comb out of the cassette. Rinse the wells with 1x running buffer and fill the sample wells with running buffer.
2. Insert the XCell SureLock™ Assembly in its unlocked position into the center of the cell base. Place one cassette on each side of both buffer cores, and lock the XCell SureLock™ Assembly by moving the tension lever to the locked position, squeezing the gels together, and creating leak-free seals. Fill the created upper chamber with 1x running buffer.
3. Protein samples (20 μl of whole cell lysate sample or immunoprecipitates) and protein standards are loaded in each pocket of the 4–20% gel. Place the lid on the assembled XCell Sure Lock ™ Midi-Cell. The lid firmly seats if the (–) and (+) electrodes are properly aligned. With the power off, connect the electrode cords to power supply. Turn on the power and perform electrophoresis for ~3 h at 120 mV constant voltage (the dye front should reach the slot mark of the gel).
4. Remove the lid, unlock the XCell SureLock™ Assembly by moving the tension lever to the unlocked position (indicated on the XCell SureLock™ Assembly), and remove gel cassettes. Insert the gel knife's beveled edge into the narrow gap between the two plates of the cassette; push up and down gently on the knife's handle to separate the plates. Open the cassette, remove, and discard the plate without the gel, and carefully transfer the gel to a tray with a transfer buffer.
5. Cut the Immobilon-P membrane and the filter paper to the dimensions of the gel. Soak the membrane, filter paper, and fiber pads in transfer buffer (at least 15 min).
6. Prepare the gel sandwich by placing the cassette, with the dark side down, in a tray with transfer buffer. Place one prewetted fiber pad on the black side of the cassette and roll out any bubbles with a plastic cut-to-size tube (cutting a 5-ml plastic pipette is a good example). Place a sheet of filter paper on

the fiber pad, roll out bubbles, and put the equilibrated gel on the filter paper. Place the prewetted membrane on the gel and gently roll out air bubbles, and place a piece of prewetted filter paper on the membrane. Add the last fiber pad and roll out air bubbles again.

7. Close the cassette and lock the white latch and repeat for another cassette. Place the cassettes in the tank (dark sides of the cassettes facing the dark side of the tank) sliding them through the slots and place the assembled unit into the TransBlot Electrophoretic Transfer Cell (Bio-Rad). Place a plastic ice pack and a stirrer into the cell, completely fill the tank with transfer buffer, insert the lid, and attach electrodes to a power supply.
8. Place the TransBlot unit into an ice box (ensure that ice covers the unit) and place the entire assembly (the icebox with the TransBLot unit) onto a stirring plate. Carry out the transfer for 2–3 h at 100 mV.
9. Once the transfer is complete, remove the cassette out of the tank and carefully disassemble, with the top sponge and blotting paper removed. Remove the membranes that should clearly show the prestained molecular weight markers and incubate in 50-ml blocking buffer for 1 h at room temperature on a rocking platform.
10. The blocking buffer is discarded and the membrane washed twice in TBS-T washing buffer prior to addition of a 1:1,000 dilution of the α-TLR4 H80Ab, α-Flag Ab, α-β-actin Ab in TBS-T containing 5% nonfat milk, or α-phosphotyrosine PY20 Ab in TBST/4% BSA for 18–24 h at 4°C on a rocking platform.
11. The primary antibody is removed and the membrane washed five times for 10 min each with 100-ml TBS-T.
12. The secondary antibody is freshly prepared for each experiment as 1:10,000-fold dilution in the respective blocking buffers (see above) and added to the membrane for 60 min at room temperature on a rocking platform.
13. The secondary antibody is discarded and the membrane washed five times for 10 min each with TBS-T.
14. During washes, the ECL + reagent is warmed separately to room temperature and reagents A and B are mixed at a 1:40 ratio (2-ml reagent A and 50-μl reagent B), the membranes are held vertically with a forceps to remove TBS-T, placed in plastic pouches (VWR) and the reagent mix is added to the membranes, and the pouches are rotated for 1 min on a rocking platform.
15. The membranes are then placed in an x-ray film cassette and all remaining exposure steps are performed in a darkroom under safe light conditions. Films are placed in the cassette for suitable exposure times, typically 1, 10, and 30 s, and 1,

5, 30, 60, and 240 min. An example of the results produced is shown in **Fig. 1**.

3.2. The Effect of Mutagenesis of Tyrosine Residues Within the TIR Domain of TLR4 on TLR4 Signal-Transducing Functions

3.2.1. Site-Directed Mutagenesis

QuikChange® II Site-Directed Mutagenesis Kit (Stratagene) is used to make the P714H substitution in the TIR domains of constitutively active human CD4-TLR4 by PCR, using a pair of mutagenic oligonucleotide primers containing the sequence resulting in the desired mutation. These primers are incorporated and extended by *PfuTurbo* DNA polymerase, followed by treatment with *Dpn* I endonuclease, which cleaves methylated and hemimethylated parental plasmid DNA obtained from in vivo grown Dam⁺*E. coli* strains but does not cleave mutant vectors

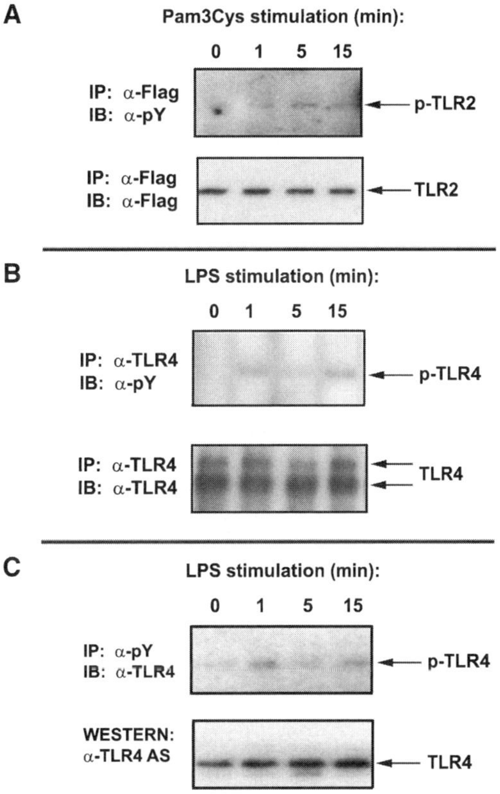

Fig. 1. Toll-like receptor-2 (TLR2) and TLR4 agonists induce tyrosine phosphorylation of respective TLRs. HEK293T cells were transiently transfected with pCDNA3-huCD14, pEF-BOS-HA-huMD-2, along with either (**A**) pCMV-1-FLAG-huTLR2 or (**B**) pCDNA3-TLR4. Cells were recovered for 24 h, followed by stimulation with 1 μg/ml Pam3Cys (**A**) or (**B** and **C**) 100 ng/ml LPS. Flag-TLR2 and TLR4 were immunoprecipitated from cell extracts with α-FLAG and α-TLR4 H80 antibody (Ab), respectively, and subjected to immunoblot analysis for total TLR2 (α-FLAG Ab) or TLR4 (H80 Ab) expression, as well as TLR tyrosine phosphorylation (α-phosphotyrosine Ab PY20). In (**C**), total phosphotyrosine proteins were immunoprecipitated from cell extracts with PY 20 Ab, followed by immunoblotting with α-TLR4 Ab (H80). The bottom blot shows TLR4 total expression analyzed by immunoblotting with α-TLR4 antiserum (AS). The results demonstrate that TLR2 and TLR4 undergo tyrosine phosphorylation upon cell stimulation with TLR2 (pam3Cys) and TLR4 (LPS) agonists.

produced by in vitro PCR. The mutant containing synthesized DNA with staggered nicks is transformed into XL1-blue supercompetent cells. The following procedures modified from the manufacturer's manual are recommended.

1. Design mutagenic oligonucleotide primers for the desired P714H, Y674A, and Y680A mutations. The mutations should be in the middle of the primer, flanked with ~15–20 bases of unmodified nucleotide sequence on both sides; the primers should have minimum 40% GC content and should terminate in more than one G/C bases. The melting temperature (T_m) of the primers should be ≥78°C. For calculating T_m, use the following formula $T_m = 81.5 + 0.41$ (% GC) – 675/N – % mismatch (N is the primer length in bases; values for % GC and % mismatch are whole numbers). Using these rules, the following primers could be used: to generate P714H mutation in TLR4 protein: 5′-CTACAGAGACTTTATTCACGGTGTGGCCATTGCTGC-3′, forward; 5′-GCAGCAATGGCCACACCGTGAATAAAGTCTCTGTAG-3′, reverse; to generate Y674A mutation: 5′-GGTAGAGGTGAAAACATCGCTGATGCCTTTGTTATC-3′, forward; 5′-GATAACAAAGGCATCAGCGATGTTTTCACCTCTACC-3′, reverse; to generate Y680A mutation: 5′-GATGCCTTTGTTATCGCCTCAAGCCAGGATGAGG-3′, forward; 5′-CCTCATCCTGGCTTGAGGCGATAACAAAGGCATC-3′, reverse.
2. Mix the following reagents in a PCR tube: 5 μl of 10× reaction buffer; 5–20 ng of pCMV1-CD4-TLR4 or pCDNA3-YFP-TLR4; 125 ng of each of oligonucleotide primers; 1 μl of dNTP mix; H_2O (PCR quality) to a final volume of 50 μl, then add 1 μl of *PfuTurbo* DNA polymerase (2.5 U/μl). Put the tube in a PCR thermocycler, denature at 95°C, 45 s, and then the reaction is cycled for 16 cycles using denature at 95°C for 30 s; annealing at 55°C for 1 min; and extension at 68°C for 25 min; with the final extension at 68°C for 60 min.
3. Add 1 μl of the *Dpn* I restriction enzyme (10 U/μl) directly to each amplification reactions, thoroughly mix by gently pipetting, and incubate the mixture at 37°C for 1 h. The digested reaction product could be either used for immediate transformation or stored at –20°C until use.
4. On ice, slowly thaw 45 μl XL1-blue supercompetent cells, which should be originally thawed and aliquoted in Falcon® 2059 polypropylene tubes. Transfer 2 μl of the *Dpn* I-treated PCR product to the cells, gently mix by swirling, and incubate on ice for 30 min.
5. Heat-pulse the tube at 42°C for 30–45 s, transfer on ice for 2 min, followed by addition of 0.5-ml preheated (42°C) SOC

medium and incubated at 37°C for 1 h with horizontal shaking at 200 × *g*.

6. Spread cells on two LB-ampicillin agar plates (0.25 ml per plate), and incubate at 37°C for overnight.
7. Pick up the colonies (at least 10) into Falcon sterile plastic tubes containing 2-ml LB-ampicillin broth, and incubate at 37°C for >16 h, followed by isolation of plasmid DNA with Miniprep kits (Qiagen) according to the manufacturer's instruction.

3.2.2. Restriction Analysis of Plasmids

Plasmid DNA isolated from different colonies is screened for the presence of the TLR4 insert with the correct size by restriction analyses of pCMV1-TLR4s with Cla I and Kpn I, and pCDNA3-YFP-TLR4s with BamH I and Xho I. In our hands, this step is necessary, as despite the antibiotic selection, we found that ~2–5% colonies did not contain plasmid DNA or contained TLR4 insert with altered sizes, most likely, due to errors attributable to PCR.

1. Into PCR tubes, mix 2 μg plasmid DNA with the corresponding restriction enzymes (Cla I and Kpn I for pCMV1-CD4-TLRs, and BamH I and Xho I for pCDNA3-YFP-TLR4s,1 unit per microliter each) in a final volume of 20 μl of the appropriate 1× NEBuffer containing 100 μg/ml BSA (NEBuffer 4 for pCMV1-CD4-TLR4 and BamH I Buffer for pCDNA3-YFP-TLR4s).
2. Reactions are incubated for 2 h at 37°C.
3. During incubation, prepare 1% agarose gel by weighing 0.5 g agarose, placing it into in 50 ml 1× TBE buffer containing 1 μl ethidium bromide, microwaving, and casting into a plastic tray. A comb for loading samples is inserted, and following incubation for ~30 min, the comb is removed and the tray is placed into a gel electrophoresis apparatus.
4. Following restriction, add 5 μl 5× nucleic acid sample loading buffer to each PCR tube with 20 μl samples, load the entire mixture into the wells, load 1 kb ladder as a standard, and run a gel for 2 h at 100 V.
5. Place a gel onto a transilluminator and photograph under UV-light.

3.2.3. DNA Sequencing

The presence of the mutations in plasmid minipreps containing the TLR4 inserts with correct sizes is confirmed by DNA sequencing using the BigDye Terminator Cycle Sequencing Ready Reaction V3.1 Kit.

1. The following reagents are mixed in a PCR tube: 250 ng plasmid (pCMV1-CD4-TLR4 wild type and P714H, Y674A, or Y680A mutants; and pCDNA3-YFP-TLR4 WT, P714H, and Y674A or Y680A mutants); 2 μl 5× BigDye sequenc-

ing buffer; 2 μl 2.5× Ready Reaction premix; 1 μl sequencing primer (5 pmol/μl); add nuclease-free water up to 10 μl. The following primers are used to verify the presence of the mutations: 5′-GTGCTGAGTTTGAATATCACCTG-3′ (for pCMV1-CD4-TLR4s), 5′-TAATACGACTCACTATAGGG-3′; 5′-CTCACAC CAGAGTTGCTTTC-3′, and 5′-TAGAAGGCACAGTCGAGG-3′ (for pCDNA3-YFP-TLR4s).

2. The PCR tubes are placed in a thermocycler with top-heating and cycling performed using the following parameters: denaturation for 5 min at 96°C, followed by 25 cycles each consisting of 96°C for 1 min; 50°C for 15 s; 60°C for 4 min; and one final extension at 60°C for 10 min.
3. Remove unincorporated dyes by gel filtration on Edge Biosystems spin columns: spin down columns at 850 × *g* for 3 min, add 10 μl of nuclease-free water to the sequence reaction, and load 20 μl of the sequencing reaction to the central column bed without touching the gel surface. Spin down the columns at 850 × *g* for 3 min, collecting the elution fluid into Eppendorf tubes.
4. Analyze the cleaned sequencing product on the ABI PRISM® 3100 Genetic Analyzer, select proper DNA analyzing software (CHROMAS, Sequencher, etc.), and perform BLAST analysis of the sequence against the NCBI nucleotide data base.
5. Expression vectors encoding wild-type or mutant TLR4 variants are then transformed into DH5-α bacterial cells, grown, and plasmid DNAs are isolated with Maxiprep kits (Qiagen) as specified by the manufacturer. After isolation, determine the DNA concentration by UV spectrophotometry at 260 nm and by quantitative agarose gel electrophoresis, using low-mass DNA ladder.

3.2.4. Transient Transfection

1. Cell culture and seeding of HEK293T cells into 150-mm TC dishes are exactly as described in **Subheading 3.1.1**. Two separate experiments are recommended to carry out for two different sets of expression vectors (pCMV-CD4-TLR4s and pCDNA3-YFP-TLR4s).
2. In the first experiment, CD4-TLR4 variants are expressed that elicit constitutive signaling due to replacement of the ectodomain with the dimerization CD4 tag; therefore, there is no need for LPS stimulation and co-expression of CD14 and MD-2 (to enable LPS sensitivity to TLR4). Use 50-ml TC tubes to mix the following reagents in SF DMEM to make up a total volume of 600 μl, including the amount of added plasmids and Superfect transfection reagent: 20 μg pCDNA3 (control, first tube), 20 μg pCMV1-CD4-TLR4

WT (second tube), 20 μg pCMV1-CD4-TLR4 P714H (third tube); 20 μg pCMV1-CD4-TLR4 Y674A (fourth tube), and 20 μg pCMV1-CD4-TLR4 Y680A (fifth tube). To each tube, add 90 μl Superfect per tube, vortex, and incubate for 20–30 min at room temperature.

3. Add serum-containing cDMEM to the mixture to a total volume of 15 ml and mix five times by pipetting up and down. Aspirate spent culture medium from cell monolayers, add transfection mixture to cells, and incubate for 3 h at 37°C in a CO_2-incubator (5% CO_2).
4. At the end of incubation, aspirate the transfection mixture, add 20 ml of fresh cDMEM per dish and recover cells for 20 h at 37°C in a CO_2-incubator (5% CO_2). Thereafter, cells are placed on ice, washed in ice-cold PBS, collected by centrifugation, and cell lysates prepared as described in the **Subheading 3.1.2**.
5. In the second experiment, full-length YFP-TLR4 variants are overexpressed, which do not mediate constitutive cell activation because the YFP tag is located at the C-terminus of TLR4. The purpose of this experiment is to study the effect of mutations on LPS-inducible cell activation, therefore, co-expression of CD14 and MD-2 co-receptors are required to enable LPS sensitivity to TLR4. In separate 50-ml tubes, the following reagents are mixed in SF DMEM to make up a total volume of 600 μl:
   - Ten micrograms of pCDNA3 or pCDNA3-YFP-TLR4 WT, P714H, Y674A, or Y680A
   - Eight micrograms of pCDNA3-CD14
   - Two micrograms of pEFBOS-HA-MD-2
   - Ninety microliters of Superfect
6. The remaining steps of the transient transfection procedure are exactly as described in **Subheading 3.1.1**.
7. After recovery, cells are stimulated with LPS (100 ng/ml final concentration), TC dishes are placed on ice, cells are collected, and cell lysates prepared as described in **Subheading 3.1.2**.
8. Samples for immunoblotting are prepared as described in **Subheading 3.1.2**.

3.2.5. Immunoprecipitation and Immunoblotting

1. Cell lysates (at least 1 mg total protein) are precleared with protein G-agarose as described in **Subheading 3.1.4**, and incubated with α-CD4 Ab (experiment 1) or with α-GFP Ab (experiment 2) to immunoprecipitate CD4-TLRs or YFP-TLR4s (both antibodies are used at 1 μg per sample) for 18 h at 4°C on a rotation platform. Analyses of protein

levels of transfected TLR4s are critical to ensure comparable expression of wild-type and mutant TLR4 variants, as differences in receptor expression can cause differences in receptor-initiated signaling cascades. Although it is possible to directly perform immunoblot analyses of TLR4 total protein expression in cell lysates, in our hands the immunoprecipitation step allowed for increased sensitivity and significantly improved the detection of epitope-tagged, transfected TLR4 proteins.

2. Thereafter, 45 μl prewashed protein G-agarose beads (50% slurry) are added, incubation continued for additional 4°C on a rotator, and beads bound to TLR4 immune complexes are extensively washed at least five times with ice-cold lysis buffer. After the last wash (*see* **Note 2**), 25 μl 2 × Laemmli sample buffer is added to the beads, and samples are boiled for 10 min.
3. Sample loading, immunoblotting, and electrotransfer are performed as described in **Subheading 3.1.5**.
4. After the electrotransfer, membranes are placed in 50-ml TBS-T/5% nonfat milk, and incubated for 1 h at room temperature on a rotating platform.
5. The blocking buffer is discarded and the membrane washed twice in TBS-T prior to addition of the following antibodies: α-Flag-HRP Ab (experiment 1, *see* **Note 3**) for the detection of CD4-TLR4s (1:400 dilution), α-GFP Ab for the detection of YFP-TLR4s (experiment 2, 1:2,000 dilution), α-IRAK-1 Ab (1:1,000 dilution), all these dilutions are made in TBS-T/5% nonfat milk, and α-β-actin Ab (used at a 1:1,000 dilution in TBST/4% normal rabbit serum, *see* **Note 4**).
6. The membranes are incubated with the respective antibodies for 18–24 h at 4°C on a rocking platform and washed; secondary antibodies are added and the rest of the procedure is performed as described in **Subheading 3.1.4**. **Figure 2** depicts data from a representative experiment showing the effect of P714H, Y674A, and Y680A mutations on constitutive (experiment 1, **Fig. 2A**) or LPS-inducible (experiment 2, **Fig. 2B**) TLR4-mediated activation of JNK, p38, and NF-κB (*see* **Note 5**).

3.2.6. Reporter Assays

1. HEK293T cells are seeded into 12-well plates (1 × 10^5 cells per well) in cDMEM and grown for 24 h before transfection. Two separate experiments are performed to examine the effect of the P714H (signaling-deficient TLR4 used as a control to achieve maximal inhibition), Y674A, and Y680A mutations on activation of transcription factors NF-κB, RANTES, and IFN-β mediated by constitutively

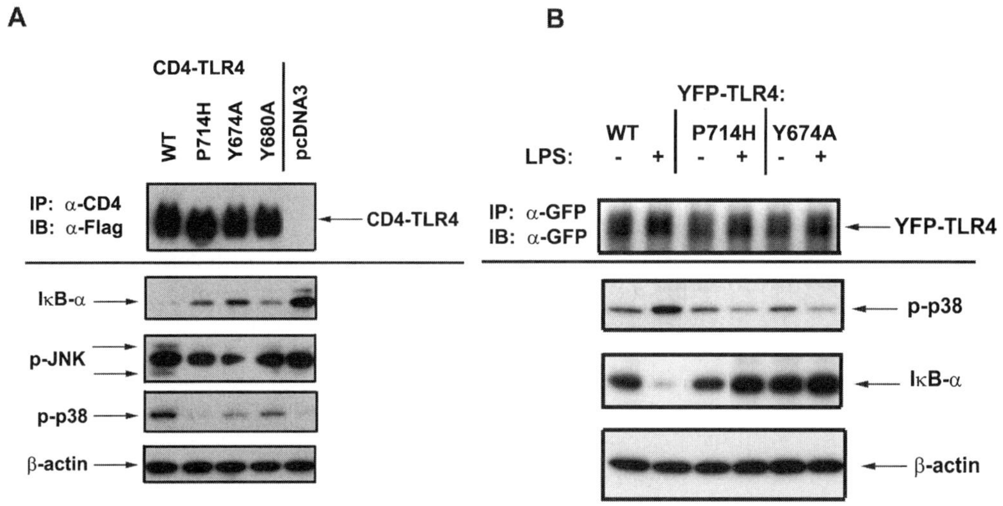

Fig. 2. Tyrosine–alanine substitutions in the TIR domain impair Toll-like receptor-4 (TLR4)-mediated IL-1R-associated kinase (IRAK)-1, JNK and p38 phosphorylation and IκB-α degradation. (**A**) WT, P714H, Y674A, or Y680A CD4-TLR4 or (**B**) YFP-TLR4 species were overexpressed in HEK293T cells. In control cultures, pcDNA3 transfection was carried out. In (**B**), CD14 and MD-2 were co-expressed to enable LPS sensitivity upon TLR4. CD4-TLR4s were immunoprecipitated with α-CD4 antibody (Ab) and subjected to immunoblotting with α-FLAG Ab (*see* **Note 6**) (**A**); YFP-TLR4s were immunoprecipitated with α-GFP Ab and analyzed by immunoblotting with α-GFP Ab (total TLR4 expression). (**A** and **B**) IκB-α degradation and phosphorylation of JNK and p38 MAP kinases were assessed by immunoblot analyses of whole cell lysates with antibodies against IκB-α or phosphorylated species of JNK and p38. The medium lane in the JNK immunoblot represents a nonspecific band. β-Actin immunoblot shows equal protein loading. The data demonstrate that Y674A and Y680A mutations, similar to P714H mutation, significantly inhibit TLR4 signal-transducing capacities.

active CD4-TLR4s or LPS-induced transcription factor induction elicited by YFP-TLR4s.

2. In experiment 1, the following reagents are mixed in 15-ml plastic tubes in a total volume of 75-μl SF DMEM to carry out transient transfections to measure the effect of tyrosine deficiencies on CD4-TLR4 constitutive activation of NF-κB (pELAM-luciferase), RANTES (pGL3-RANTES-luciferase), and IFN-β (p125-luciferase) reporters (all amounts of reagents are given per well of 12-well plates):
   - One hundred nanograms of pCMV-β-galactosidase reporter
   - Six hundred nanograms of pCDNA3 (control) or pCMV1-CD4-TLR4 WT, or P714H, Y674A, or Y680A mutants (each expression vector is added to separate tubes)
   - Three hundred nanograms of pELAM-luciferase
   - Five hundred nanograms of pCDNA3 to adjust the total amount of plasmid DNA to 1.5 μg
   - A 7.5-μl Superfect

The same experimental setup is used when pGL3-RANTES-luciferase or p125-luciferase is utilized instead of pELAM-luciferase reporter.

3. In experiment 2, the following reagents are mixed in 15-ml plastic tubes in a total volume of 75-μl SF DMEM to carry out transient transfections to measure the effect of tyrosine deficiencies on LPS-inducible, YFP-TLR4-mediated activation of NF-κB (pELAM-luciferase), RANTES (pGL3-RANTES-luciferase), and IFN-β (p125-luciferase) reporters (all amounts of reagents are given per well of 12-well plates):
   - One hundred nanogram pCMV-β-galactosidase reporter
   - Four hundred nanogram pCDNA3 (control transfection) or pCDNA3-YFP-TLR4 WT, P714H, Y674A, or Y680A variants (each TLR4 variant is placed to a separate tube)
   - Two hundred nanogram pCDNA3-CD14
   - One hundred nanogram pEFBOS-HA-MD2
   - Three hundred nanogram pELAM-luciferase (NF-κB) reporter
   - Four hundred nanogram pCDNA3 to adjust the total amount of plasmid DNA to 1.5 μg
   - A 7.5-μl Superfect

   The same experimental setup is used when pGL3-RANTES-luciferase or p125-luciferase is utilized instead of pELAM-luciferase reporter.
4. The transfection mixture of plasmid DNA, Superfect, and serum-free (SF) DMEM is vortexed for 10–15 s, incubated at room temperature for 30 min, followed by addition of cDMEM (up to a total volume of 750 μl per well) and thorough mixing.
5. Following aspiration of spent medium, transfection mixture is added into each well of 12-well plates according to the experiment schedule (750 μl total volume per well), and transfection is carried out for 3 h at 37°C in a CO_2-incubator (5% CO_2).
6. After transfections, cells are washed twice, cDMEM (0.5 ml per well) is added to each well, and cells are recovered for 20 h at 37°C in a CO_2-incubator (5% CO_2).
7. In experiment 1, cells are washed with PBS and processed to prepare cell lysates as indicated in **step 8**. In experiment 2, medium is replaced in wells with fresh cDMEM (0.5 ml per well), and cells are treated with medium (addition of 0.5 ml cDMEM per well) or 100 ng/ml LPS (addition of 0.5 ml 200 ng/ml LPS per well) for 6 h at 37°C in a CO_2-incubator (5% CO_2).

8. In all experiments, cells are washed twice with PBS, and 1x passive lysis buffer (5x lysis buffer is diluted with distilled water 1:4) are added to wells (0.25 ml per well). At this point, it is possible to stop experiments by freezing plates at –20°C until further processing.
9. Cells are scrapped off wells with a 1-ml pipette tip (separate tips for different treatments), pipetted up and down, and transferred into 0.5-ml Eppendorf tubes. Following centrifugation (10,000 × *g*, 30 s, room temperature), supernatants are transferred into a fresh set of 0.5-ml Eppendorf tubes,

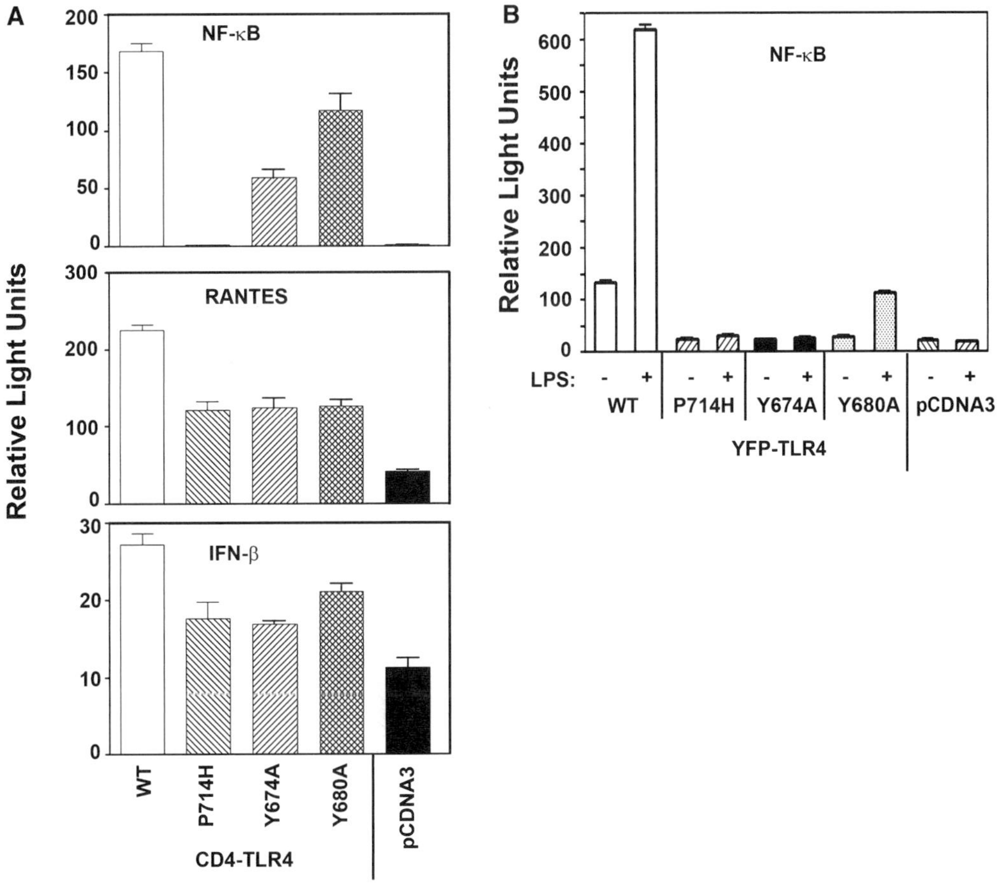

Fig. 3. Tyrosine–alanine substitutions within the TIR domain suppress Toll-like receptor-4 (TLR4)-mediated activation of transcription factors NF-κB, RANTES, and IFN-β. HEK293T cells were transiently transfected with expression vectors encoding either (**A**) CD4-TLR4 WT, CD4-TLR4 P714H, Y674A and Y680A mutants or (**B**) pCDNA3-YFP-TLR4 species encoding P714H, Y674A, and Y680A mutations. As a control, pCDNA3 was transfected. (**A**) NF-κB (pELAM-luciferase), RANTES (pGL3-RANTES-luciferase), and IFN-β (p125-luciferase) reporters were co-expressed along with pCMV–galactosidase reporter, cells were recovered for 24 h, and luciferase versus β-galactosidase activities were measured and normalized to β-galactosidase activities (**A**, NF-κB and **B**). Alternatively, luciferase activity was measured and expressed per protein amount (**A**, RANTES and IFN-β). These results demonstrate that Y674A and Y680A mutations markedly impair the capacity of both CD4-TLR4 (constitutively active) and YFP-TLR4 (LPS-inducible activation) to mediate activation of NF-κB, RANTES, and IFN-β reporters. Their inhibitory effects are comparable to that of signaling-inactive P714H TLR4 mutants used as controls.

and either processed for measuring protein, luciferase, and β-galactosidase activities as described in **step 10** or frozen down at –80°C to stop experiments.

10. Luciferase and β-galactosidase activities are measured using Luciferase Assay and Galacto-Light™ Systems, respectively. To measure luciferase activities, the Luciferase Reaction Buffer is prepared first by resuspending the Luciferase Assay Substrate with supplied Assay Buffer, and the solution is used to prime an injector of a Berthold tube luminometer. Twenty microliters of cell lysates from **step 9** are added to luminometer tubes, tubes are placed into the luminometer holder, and firefly luciferase luminescence is read for 5–10 s per tube following injection of 100 μl Luciferase Reaction Buffer.
11. For measuring β-galactosidase activity with Galacto-Light™ system, Reaction Buffer is first prepared by diluting Galacton substrate with Reaction Buffer Diluent at a ratio 1:100. Two hundred microliters of Reaction Buffer are added to luminometer tubes containing 1–2 μl cell extracts from **step 9**; the mixtures are vortexed and incubated in the dark for 60 min. Following incubation, the tubes are placed into the luminometer (primed with light emission accelerator), injected with 300 μl of light emission accelerator, and luminescence is read for 2–5 s.Protein levels of the cell extract are also controlled by measuring protein concentrations by Bio-Rad *Dc* Protein Assay Kit as described in **Subheading 3.1.3**.
12. Data are normalized by dividing firefly luciferase activity with that of β-galactosidase or by protein amounts (e.g., luciferase activity per 100 μg protein, *see* **Note 6**). **Figure 3** depicts the results of a representative experiment 1 (**Fig. 3A**) and a representative experiment 2 (**Fig. 3B**) in HEK293T cells.

4. Notes

1. As a control for ligand-inducible TLR tyrosine phosphorylation, it is recommended to perform transient transfection of pCMV1-Flag-TLR2 in order to see TLR2 tyrosine phosphorylation following cell stimulation with a TLR2 ligand, Pam3Cys. The same setup as shown for TLR4 is used, with the exception of replacing pCDNA3-TLR4 with pCMV1-Flag-TLR2 in the transfection mixture.
2. When aspirating, it is critical not to aspirate the beads to avoid differences in loading. After the last wash, leave ~25–35 μl

above the bead pellet and use manual removal of this volume of the lysis buffer with a pipette tip, using a Pipettman-200 (Gilson, Middleton, WI, USA).

3. pCMV-CD4-TLR4 plasmids encode two tags: Flag and CD4; therefore, it is recommended to immunoprecipitate with Ab against CD4 and use α-Flag Ab for immunoblot detection of CD4-Flag-TLR4 proteins. In our hands, α-Flag-HRP conjugate gives fewer nonspecific bands and lower background compared to the use of α-Flag Ab followed by addition of secondary α-mouse IgG-HRP.
4. TBST/4% normal rabbit serum is used as a blocking buffer because other variants of blocking buffers (e.g., TBS-T/4% BSA or TBS-T/5% milk) in our hands significantly increased the background noise wherever secondary anti-goat-HRP antibody was used.
5. TLR4-mediated activation of JNK and p38 can be judged by their phosphorylation that is a necessary step in the induction of kinase activity of MAP kinases. Activation of NF-κB classic pathway is reflected by degradation of an inhibitory protein, IκB-α, in the cytoplasm that results in the release of NF-κB dimers, unmasking their nuclear localization sequence, and their translocation to the nucleus.
6. In our hands, we observed that β-galactosidase activities vary in different treatment regiments. For measuring NF-κB reporter activation, this effect does not cause significant influence due to high inducibility of the NF-κB reporter (8–18-fold). However, in our hands, RANTES and IFN-β reporters were induced by LPS to a lower extent (2–4-fold); therefore, variations in β-galactosidase activities could cause difficulties in determining transcription factor activation. For this reason, we recommend to normalize luciferase activities to protein, e.g., luciferase light units per 100 ng protein.

References

1. Pasare, C., and Medzhitov, R. (2005) Toll-like receptors: linking innate and adaptive immunity. *Adv. Exp. Med. Biol.* 560, 11–18.
2. Kaisho, T., and Akira, S. (2006) Toll-like receptor function and signaling. *J. Allergy Clin. Immunol.* 117, 979–987.
3. Lemaitre, B., Nicolas, E., Michaut, L., Reichhart, J. M., and Hoffmann, J. A. (1996) The dorsoventral regulatory gene cassette spätzle/Toll/cactus controls the potent antifungal response in Drosophila adults. *Cell* 86, 973–983.
4. Medzhitov, R., Preston-Hurlburt, P., and Janeway, C. A. Jr. (1997) A human homologue of the Drosophila Toll protein signals activation of adaptive immunity. *Nature* 388, 394–397.
5. Poltorak, A., He, X., Smirnova, I., Liu, M. Y., Van Huffel, C., Du, X., Birdwell, D., Alejos, E., Silva, M., Galanos, C., Freudenberg, M., Ricciardi-Castagnoli, P., Layton, B., and Beutler, B. (1998) Defective LPS signaling in C3H/HeJ and C57BL/10ScCr mice: mutations in Tlr4 gene. *Science* 282, 2085–2088.
6. Hoshino, K., Takeuchi, O., Kawai, T., Sanjo, H., Ogawa, T., Takeda, Y., Takeda, K., and Akira, S. (1999) Cutting edge: Toll-like receptor 4 (TLR4)-deficient mice are

hyporesponsive to lipopolysaccharide: evidence for TLR4 as the Lps gene product. *J. Immunol.* 162, 3749–3752.

7. Rock, F. L., Hardiman, G., Timans, J. C., Kastelein, R. A., and Bazan, J. F. (1998) A family of human receptors structurally related to Drosophila Toll. *Proc. Natl. Acad. Sci. U. S. A.* 95, 588–593.
8. Doyle, S. L., and O'Neill, L. A. (2006) Toll-like receptors: from the discovery of NF-κB to new insights into transcriptional regulations in innate immunity. *Biochem. Pharmacol.* 72, 1102–1113.
9. Lien, E., Sellati, T. J., Yoshimura, A., Flo, T. H., Rawadi, G., Finberg, R. W., Carroll, J. D., Espevik, T., Ingalls, R. R., Radolf, J. D., and Golenbock, D. T. (1999) Toll-like receptor 2 functions as a pattern recognition receptor for diverse bacterial products. *J. Biol. Chem.* 274, 33419–33425.
10. Means, T. K., Lien, E., Yoshimura, A., Wang, S., Golenbock, D. T., and Fenton, M. J. (1999) The CD14 ligands lipoarabinomannan and lipopolysaccharide differ in their requirement for Toll-like receptors. *J. Immunol.* 163, 6748–6755.
11. Hayashi, F., Smith, K. D., Ozinsky, A., Hawn, T. R., Yi, E. C., Goodlett, D. R., Eng, J. K., Akira, S., Underhill, D. M., and Aderem, A. (2001) The innate immune response to bacterial flagellin is mediated by Toll-like receptor 5. *Nature.* 410, 1099–10103.
12. Hemmi, H., Takeuchi, O., Kawai, T., Kaisho, T., Sato, S., Sanjo, H., Matsumoto, M., Hoshino, K., Wagner, H., Takeda, K., and Akira, S. (2000) A Toll-like receptor recognizes bacterial DNA. *Nature* 408, 740–745.
13. Hochrein, H., Schlatter, B., O'Keeffe, M., Wagner, C., Schmitz, F., Schiemann, M., Bauer, S., Suter, M., and Wagner, H. (2004) Herpes simplex virus type-1 induces IFN-alpha production via Toll-like receptor 9-dependent and -independent pathways. *Proc. Natl. Acad. Sci. U. S. A.* 101, 11416–11421.
14. Abe, T., Hemmi, H., Miyamoto, H., Moriishi, K., Tamura, S., Takaku, H., Akira, S., and Matsuura, Y. (2005) Involvement of the Toll-like receptor 9 signaling pathway in the induction of innate immunity by baculovirus. *J. Virol.* 79, 2847–2858.
15. Bellocchio, S., Montagnoli, C., Bozza, S., Gaziano, R., Rossi, G., Mambula, S. S., Vecchi, A., Mantovani, A., Levits, S. M., and Romani, L. (2004) The contribution of the Toll-like/IL-1 receptor superfamily to innate and adaptive immunity to fungal pathogens in vivo. *J. Immunol.* 172, 3059–3069.
16. Alexopoulou, L., Holt, A. C., Medzhitov, R., and Flavell, R. A. (2001) Recognition of double-stranded RNA and activation of NF-kappaB by Toll-like receptor 3. *Nature* 413, 732–738
17. Heil, F., Hemmi, H., Hochrein, H., Ampenberger, F., Kirschning, C., Akira, S., Lipford, G., Wagner, H., and Bauer, S. (2004) Species-specific recognition of single-stranded RNA via toll-like receptor 7 and 8. *Science* 303, 1526–1529.
18. Lund, J. M., Alexopoulou, L., Sato, A., Karow, M., Adams, N. C., Gale, N. W., Iwasaki, A., and Flavell, R. A. (2004) Recognition of single-stranded RNA viruses by Toll-like receptor 7. *Proc. Natl. Acad. Sci. U. S. A.* 101, 5598–5603.
19. Zhang, D., Zhang, G., Hayden, M. S., Greenblatt, M. B., Bussey, C., Flavell, R. A., and Ghosh, S. (2004) A toll-like receptor that prevents infection by uropathogenic bacteria. *Science* 303, 1522–1526.
20. Yarovinsky, F., Zhang, D., Andersen, J. F., Bannenberg, G. L., Serhan, C. N., Hayden, M. S., Hieny, S., Sutterwala, F. S., Flavell, R. A., Ghosh, S., and Sher, A. (2005) TLR11 activation of dendritic cells by a protozoan profilin-like protein. *Science* 308, 1626–1629.
21. Gay, N. J., and Gangloff, M. (2007) Structure and function of toll receptors and their ligands. *Annu. Rev. Biochem.* 76, 141–165.
22. Hsueh, R. C., and Scheuermann, R. H. (2000). Tyrosine kinase activation in the decision' between growth, differentiation, and death responses initiated from the B cell antigen receptor. *Adv. Immunol.* 75, 283–316.
23. Mustelin, T., and Tasken, K. (2003) Positive and negative regulation of T-cell activation through kinases and phosphatases. *Biochem. J.* 371, 15–27.
24. Park, J. G., Murray, R. K., Chien, P., Darby, C., and Schreiber, A. D. (1993) Conserved cytoplasmic tyrosine residues of the subunit are required for a phagocytic signal mediated by Fc gamma RIIIA. *J. Clin. Invest.* 92, 2073–2079.
25. Arbibe, L., Mira, J. P., Teusch, N., Kline, L., Guha, M., Mackman, N., Godowski, P. J., Ulevitch, R. J., and Knaus, U. G. (2000) Toll-like receptor 2-mediated NF-κB activation requires a Rac1-dependent pathway. *Nat. Immunol.* 1, 533–540.
26. Sarkar, S. N., Peters, K. L., Elco, C. P., Sakamoto, S., Pal, S., and Sen, G. C. (2004) Novel roles of TLR3 tyrosine phosphorylation and PI3 kinase in double-stranded RNA signaling. *Nat. Struct. Mol. Biol.* 11, 1060–1067.

27. Ivison, S. M., Khan, M. A., Graham, N. R., Bernales, C. Q., Kaleem, A., Tirling, C. O., Cherkasov, A., and Steiner, T. S. (2007) A phosphorylation site in the Toll-like receptor 5 TIR domain is required for inflammatory signalling in response to flagellin. *Biochem. Biophys. Res. Commun.* 352, 936–941.
28. Wesche, H., Henzel, W. J., Shillinglaw, W., Li, S., and Cao, Z. (1997) MyD88: an adapter that recruits IRAK to the IL-1 receptor complex. *Immunit.* 7, 837–847.
29. Medvedev, A. E., Piao, W., Shoenfelt, J., Rhee, S. H., Chen, H., Basu, S., Wahl, L. M., Fenton, M. J., and Vogel, S. N. (2007) Role of TLR4 tyrosine phosphorylation in signal transduction and endotoxin tolerance. *J. Biol. Chem.* 282, 16042–53.
30. Visintin, A., Latz, E., Monks, B. G., Espevik, T., and Golenbock, D. T. (2003) Lysines 128 and 132 enable lipopolysaccharide binding to MD-2, leading to Toll-like receptor-4 aggregation and signal transduction. *J. Biol. Chem.* 278, 48313–48320.
31. Ronni, T., Agarwal, V., Haykinson, M., Haberland, M. E., Cheng, G., and Smale, S. T. (2003) Common interaction surfaces of the toll-like receptor 4 cytoplasmic domain stimulate multiple nuclear targets. *Mol. Cell. Biol.* 23, 2543–55.

Chapter 11

Analysis of Ubiquitin Degradation and Phosphorylation of Proteins

Pearl Gray

Summary

Both ubiquitination and phosphorylation are crucial mediators involved in controlling the functions of numerous proteins belonging to the Toll-like receptor (TLR) signaling pathways. Altering the aforementioned post-translational events can be detrimental to the host survival. Therefore, the importance of these modifications cannot be overestimated. This chapter describes techniques used to examine if a protein is ubiquitinated and/or phosphorylated. In addition, a method is provided to identify the modified amino acids. We have previously shown using these techniques that the protein MyD88 adapter-like (Mal) is phosphorylated and ubiquitinated following activation of the TLR2 and TLR4 signaling pathways. Both post-translational modifications are essential for the activation and degradation of Mal, and thus are crucial steps, in regulating these TLR signaling cascades and consequently the innate immune response.

Key words: Ubiquitination, Degradation, Phosphorylation, TLR.

1. Introduction

Toll-like receptors (TLRs) are vital instigators in eliciting an effective and precise immune response against invading pathogens. From the initiation of the TLR signaling pathways to the expression of pro-inflammatory genes, a profuse number of proteins are targeted for phosphorylation and/or ubiquitination. Both processes have evolved to tightly control TLR signal transduction and thereby promote host survival. In particular, these post-translational modifications are essential for activating, down-regulating, and terminating TLR signaling.

Claire E. McCoy and Luke A.J. O'Neill (eds.), *Methods in Molecular Biology, Toll-Like Receptors, vol. 517*
DOI: 10.1007/978-1-59745-541-1_11

Several members of the TLR family have been shown to be modified by phosphorylation and ubiquitination. In particular, it has been reported that TLR2 *(1)*, TLR3 *(2, 3)*, TLR4 *(4, 5)*, TLR5 *(6, 7)*, and TLR9 *(8)* are phosphorylated. Furthermore, with the exception of the adapter protein sterile alpha and HEAT/Armadillo motif (SARM), all the TLR signaling adapters, namely myeloid differentiation factor 88 (MyD88) *(9)*, MyD88 adapter-like (Mal) *(10, 11)*, Toll/IL-1 receptor domain-containing adapter-inducing interferon-β (TRIF) *(12)*, and TRIF-related adapter molecule (TRAM) *(13)*, have been shown to undergo phosphorylation. Moreover, it has emerged that TLR2 *(14)*, TLR4 *(15)*, TLR9 *(14)*, and the signaling adapters Mal *(16)* and MyD88 *(17)* are ubiquitinated and subsequently degraded.

Apart from the aforementioned TLRs and their adapter molecules, numerous downstream proteins, partaking in the TLR signaling pathways, are modified by phosphorylation and/or ubiquitination. However, further work is required to identify all the proteins involved in the TLR signaling cascade, which are altered by these post-translational events and the exact site(s) of modification that are vital in regulating these proteins and the subsequent host immune response to pathogenic microorganisms.

This chapter describes techniques that have been employed to determine if a particular protein is phosphorylated and/or ubiquitinated. The first part focuses on the methods used to investigate if a protein is ubiquitinated. Primarily, the protein of interest is immunoprecipitated and any attached ubiquitin molecules are immunodetected. However, certain challenges may arise during this technique. In particular, the immunoprecipitating antibody which has most likely been raised against the entire unmodified protein of interest, or a portion of it, may not be able to efficiently bind to the modified form of that protein. In that case, or if the amounts of the immunoprecipitated endogenous protein are not sufficient to detect ubiquitination, it is recommended that studies are conducted by overexpressing tagged forms of the protein of interest. Given that the structure of the tag should not be altered during the post-translational modification of the protein to which it is attached, commercially available antibodies raised to specific epitopes of this tag should be more efficient at immunoprecipitating the modified target protein. Furthermore, as ubiquitinated proteins are prone to rapid deubiquitination, it is imperative that certain procedures are conducted while performing these experiments. Principally, all steps must be carried out at 4°C and inhibitors of deubiquitinating enzymes must be included during the lysis of the cells.

Further classification of the mode of degradation of the target protein can be determined via the use of specific inhibitors that block either the lysosomal or the proteasomal degradation pathways.

Finally, having determined that the protein is ubiquitinated, the specific lysine residue(s) which are modified can be ascertained by mutational analysis. Mutant protein(s) are generated and their ability to be ubiquitinated, compared to the wild-type target protein, is assessed. The mutants that cannot be modified may subsequently be used to investigate the role of this post-translational event in the TLR signaling pathways. In recent years, mass spectrometry has also been used to identify the site(s) of ubiquitination; however, this particular technique is beyond the scope of this chapter.

The next part of the chapter describes protocols that can be used to investigate if a protein of interest is phosphorylated. Firstly, in order to examine if the protein is a phospho-acceptor, two-dimensional sodium dodecyl sulfate polyacrylamide gel electrophoresis (SDS-PAGE), followed by immunoblotting, is conducted *(18–20)*. The negative charge of the phosphate(s) attached to a protein will alter its p*I*, causing an electrophoretic shift thereby indicating that a phosphorylation event may have occurred. A post-translational modification may even cause a shift in the apparent molecular weight of a protein on one-dimensional SDS-PAGE. To investigate if the observed altered mobility of the target protein is the result of phosphorylation, the migration pattern of the protein, with and without phosphatase treatment, is examined. Such treatment will result in the disappearance of the phosphoforms of a protein that exhibit altered mobility, establishing that a phosphorylation event had occurred. It is nonetheless important to note that phosphorylation does not always induce a mobility shift in the molecular weight of a protein. In this case, dephosphorylation assays will be ineffective. However, phosphatase treatment can be conducted on proteins analyzed by two-dimensional SDS-PAGE, if a p*I* shift is observed.

Lastly, having determined that the protein under investigation is phosphorylated, the next step is to identify the phospho-accepting residues. This can be achieved by mutational analysis of threonine, serine and tyrosine residues (as is the case with determining the sites of ubiquitination, phosphorylated residues can be detected by mass spectrometry – a technique which is not discussed in this chapter). Mutant protein(s) are generated and it is determined, by overexpression studies or by in vitro kinase assays, if these proteins are phosphorylated to the same level as the wild-type target protein. Having determined the site(s) of phosphorylation, mutant proteins that do not contain one or more of the phospho-accepting serine(s), threonine(s), or tyrosine(s) can be used to evaluate the function of the specific phosphorylation of these residue(s) in TLR signal transduction.

2. Materials

2.1. Immunoprecipitation of Endogenous or Overexpressed Ubiquitinated Protein

1. Plastic cell scrapers (Sarstedt).
2. Phosphate-buffered saline (PBS), pH 7.4, 1x: 154 mM NaCl, 1.9 mM NaH_2PO_4, 8.1 mM Na_2HPO_4. A 10x stock solution can be prepared.
3. Suggested lysis buffer: 50 mM Tris-HCl, pH 7.5, 150 mM NaCl, 1 mM EDTA, 1% (v/v) Triton X-100. Store at 4°C. Immediately before use, the following inhibitors should be added: 2 mM Na_3VO_4, 10 mM NaF, 10 mM iodoacetamide (Sigma), 10 mM β-glycerol phosphate, aprotinin 2 μg/ml, leupeptin 5–10 μg/ml, Phenylmethanesulfonyl fluoride 1 mM.
4. Protein A/G Sepharose beads (Amersham).
5. Immunoprecipitating antibody.
6. Control for immunoprecipitating antibody.
7. Anti-ubiquitin antibody (P4D1) (Santa Cruz).
8. Horse radish peroxidase (HRP)-conjugated secondary antibody (Promega, Santa Cruz).
9. Reducing SDS-PAGE sample buffer, 5x: 125 mM Tris-HCl, pH 6.8, 2% (w/v) SDS, 10% (v/v) glycerol, 0.2% (w/v) bromophenol blue. Store at room temperature. Immediately prior to use, dithiothreitol (DTT) is added to a final concentration of 50mM.
10. Polyvinylidene diflouride (PVDF) or nitrocellulose membranes (Amersham, Invitrogen).
11. Enhanced chemiluminescent reagent (ECL) (Amersham).

2.2. Protein Concentration Measurement – Bradford Method

Bradford reagent: 0.01% (w/v) Coomassie Brilliant Blue G-250, 4.7% (v/v) ethanol, 8.5% (v/v) orthophosphoric acid. Care must be taken as orthophosphoric acid is a corrosive reagent. Protect solution from light, filter through Whatman no. 1 filter paper, and store at 4°C.

2.3. Stripping the Membrane

Stripping buffer: 0.2 M NaOH.

2.4. Using Inhibitors to Determine if the Protein of Interest Is Degraded by Either the Proteasome or the Lysosome

1. Proteasomal inhibitors (Calbiochem, Boston Biochem, Sigma):

   MG-132 and lactacystin: soluble in Dimethyl sulfoxide, once reconstituted store at – 20°C for up to 1 month.

   Proteasome inhibitor I: soluble in DMSO, once reconstituted store at – 20°C for up to 3 months.
2. Lysosomal inhibitors (Calbiochem, Boston Biochem, Sigma):

Chloroquine: soluble in H_2O, store at room temperature.

NH_4Cl: soluble in H_2O, once reconstituted store at 4°C for up to 3 months.

Epoxomicin: protect from light; soluble in DMSO, once reconstituted store at – 20°C for up to 3 months.

3. Antibody that recognizes the protein of interest.

2.5 Two-Dimensional Electrophoretic Analysis of Phosphoproteins (All Chemicals Should Be of Electrophoresis Grade)

1. Immobilized pH gradient (IPG) strips of the appropriate length and pH gradient range (Amersham Biosciences, Biorad). Store at – 20°C.
2. IPG rehydration tray (Amersham Biosciences).
3. Mineral oil (Sigma).
4. Pre-stained molecular weight markers. Store at – 20°C (NEB).
5. Electrode wicks (Biorad, Amersham).
6. Sample solubilization solution: 8 M urea, 50 mM DTT, 4% (w/v) CHAPS, 0.2% (v/v) carrier ampholytes (corresponding to the required pH gradient range), 0.002% (w/v) bromphenol blue. Store at – 20°C.
7. SDS-PAGE equilibration buffer I (with DTT): 6 M urea, 0.375 M Tris-HCl, pH 8.8, 2% (w/v) SDS, 20% (v/v) glycerol, 2% (w/v) DTT.
8. SDS-PAGE equilibration buffer II (with iodoacetamide): 6 M urea, 0.375 M Tris-HCl, pH 8.8, 2% (w/v) SDS, 20% (v/v) glycerol, 2.5% (w/v) iodoacetamide.
9. Reagents for resolving gel: *see* Chapter 7, **Subheading 3.3**.
10. Water-saturated isobutanol.
11. Sealing agarose solution: 25 mM Tris base, 192 mM glycine, 0.1% (w/v) SDS, 0.5% (w/v) low melting point agarose, 0.002% (w/v) bromophenol blue. Store at room temperature.
12. SDS-PAGE running buffer: 25 mM Tris base, 192 mM glycine, 0.1% (w/v) SDS. A 10x stock solution can be prepared. Store at room temperature.

2.6. In Vitro Dephosphorylation of Phosphoproteins with Calf Intestinal Alkaline Phosphatase (CIP)

1. CIP (NEB).
2. Phosphatase buffer: 50 mM Tris-HCl, pH 7.5, 1 mM $MgCl_2$.

2.7. Site-Directed Mutagenesis

1. Molecular biology grade water (DNase, RNase, and nucleic acid free).
2. Two complementary oligonucleotide primers with the appropriate nucleotide base(s) altered to create the required amino acid in the protein of interest.

3. *Pfu*Turbo® DNA polymerase (Stratagene).
4. 10x cloned *Pfu* DNA polymerase reaction buffer (Stratagene).
5. dNTP mix-12.5 mM each of dATP, dTTP, dGTP, and dCTP in molecular biology grade water (Invitrogen).
6. Dpn I restriction enzyme (NEB).
7. Competent bacterial cells (Stratagene, Invitrogen).
8. Luria Broth (LB) (Sigma). The solution is prepared and stored according to manufacturer's instructions.
9. LB agar (Sigma) plates with the appropriate antibiotic: the LB agar solution is prepared according to manufacturer's instructions and autoclaved at 121°C for 15 min. When it cools down to 60°C, the appropriate antibiotic is added, and the solution is poured into 10 cm petri dishes. Store plates at 4°C for the length of time which is appropriate for the antibiotic used.

3. Methods

3.1. Immunoprecipitation of Endogenously Ubiquitinated Proteins

3.1.1. Pre-coupling the Immunoprecipitating Antibody to Protein A or G Sepharose Beads

1. For each sample to be immunoprecipitated, 40 µl of a 50% (v/v) suspension of either protein A or G Sepharose beads (*see* **Table 1** to determine the appropriate type for the antibody being used) is transferred to a microcentrifuge tube (*see* **Note 1**). An extra sample that will serve as an experimental control must be included.
2. The beads are washed once with ice-cold PBS and centrifuged at 0.5 × *g* for 1 min. The supernatant is subsequently aspirated and the beads are resuspended in 750 µl of ice-cold PBS per sample (*see* **Note 2**).
3. The desired volume of immunoprecipitating antibody is added to each sample. If using a purified monoclonal or polyclonal antibody, a starting concentration of 1–5 µg per sample is recommended. For polyclonal serum, 0.5–5 µl of antibody may be used. As a control, a separate sample with an equivalent amount of the appropriate species and isotype-matched antibody or pre-immune serum is required.
4. Samples are rotated for 1–2 h at 4°C (during this incubation step, one may proceed to **Subheading 3.1.2** for preparation of the cell lysates).
5. Once the incubation period is completed, the samples are centrifuged at 0.5 × *g* for 1 min at 4°C. The beads are then washed twice with 1 ml of lysis buffer per sample and the

Table 1
Binding affinity of either protein A or protein G to mammalian immunoglobulins. W, M, and S indicate weak, medium, and strong binding, respectively

Species	Subclass	Protein A binding	Protein G binding
Mouse	Total IgG	S	S
	IgG1	W	S
	IgG2a	S	S
	IgG2b	S	S
	IgG3	S	S
Rabbit	Total IgG	S	M
Rat	Total IgG	W	M

supernatant is aspirated. Each sample of beads is resuspended to give a 50% (v/v) suspension with lysis buffer (*see* **Note 3**).

3.1.2. Preparation of Cell Lysates

1. The amount of cells required must be empirically determined for each protein under study. As a starting point, 10^7 cells per immunoprecipitation is recommended. In order to increase the amount of certain target proteins, cells may be pre-treated with either a proteasomal or lysosomal inhibitor (*see* **Subheading 3.3**, **step 2**). If required, cells are stimulated with specific TLR ligands for various lengths of time. As a control, an unstimulated sample is included to confirm that ubiquitination of the protein of interest is stimulus-dependent.
2. (a) If adherent cells are used, the media is aspirated and the cells are rinsed once with ice-cold PBS, taking care not to dislodge the cells. The PBS is then removed and the plates are left briefly tilted on ice so that the remaining liquid can be aspirated. Proceed to **step 3**.

   (b) If cells that grow in suspension are used, the samples are centrifuged at 125 × *g* for 3 min at 4°C. The media is then aspirated, without disturbing the cell pellet. The cells are subsequently rinsed with ice-cold PBS. The samples are centrifuged at 125 × *g* for 3 min at 4°C and the remaining liquid is removed by aspiration.
3. (a) For adherent cells, 1 ml of lysis buffer/10^7 cells, containing freshly added iodoacetamide, is dispensed onto each plate and the cells are dislodged with a cell scraper (*see* **Note 4**). Following the addition of the lysis buffer, the cell lysate is transferred to pre-chilled microcentrifuge tubes. Proceed to **step 4**.

(b) For suspension cells, the pelleted cells are resuspended in 1 ml of lysis buffer/10^7 cells, containing freshly added iodoacetamide (*see* **Note 4**).

4. The samples are rotated at 4°C using an end-over-end rotator, or left on ice for 10–30 min.
5. Subsequently, the samples are centrifuged at maximum speed in a microcentrifuge for 10–20 min at 4°C and the supernatant is transferred to a new pre-chilled microcentrifuge tube.

3.1.3. Cell Lysate Pre-clearance (Optional)

1. Protein A or G Sepharose beads are pre-coupled to the control antibody as described in **Subheading 3.1.1**. Forty microliters of the 50% (v/v) suspension of antibody-coated protein A or G beads are transferred to each cell lysate.
2. The samples are rotated at 4°C using an end-over-end rotator for 30 min.
3. Once this incubation is complete, the samples are centrifuged at 0.5 × *g* for 2 min at 4°C.
4. The cell lysates are then transferred to new pre-chilled microcentrifuge tubes.

3.1.4. Protein Concentration Measurement – Bradford Method (21)

1. Using bovine serum albumin (BSA), a set of standard concentrations ranging from 0 to 20 μg/ml of lysis buffer are prepared. Each sample is prepared in triplicate.
2. The cell lysates to be measured are diluted 1:20 with distilled water. As a blank, lysis buffer alone is diluted in the same way. Twenty microliters of each sample to be measured is then transferred, in triplicate, to individual wells in the microtiter plate.
3. Two hundred microliters of the Bradford reagent, equilibrated to room temperature, are dispensed into each well. The plate is tapped gently on the side to mix the samples adequately.
4. The reaction is allowed to develop for 5 min at room temperature.
5. If air bubbles are present in the wells, a syringe needle may be used to burst them (*see* **Note 5**).
6. The absorbance of the standards and the samples is subsequently measured at 595 nm.
7. Once the absorbance readings are obtained, a standard curve is prepared by plotting the absorbance readings of the BSA standards on the *y*-axis and the known BSA concentration on the *x*-axis. The absorbance values obtained for the samples can then be used to determine the protein concentration of each cell lysate. The values obtained for the blank and the dilution factor that was used must also be taken into account.

In addition, in order to correctly determine the protein concentrations, the absorbance values of the samples must be in the linear range of the standard curve.

3.1.5. Immunoprecipitation of the Protein of Interest and Detection of Ubiquitin

1. Having determined the protein concentrations, equal amounts of protein from each cell lysate are transferred to the microcentrifuge tubes, which contain 40 μl of the previously prepared protein A or G Sepharose beads (*see* **Subheading 3.1.1**), pre-coated with either the immunoprecipitating antibody or the control.
2. Samples are rotated from 1 h to overnight at 4°C. The duration of this incubation step is dependent on the concentrations of the protein of interest and of the antibody, as well as the affinity of the antibody for the protein.
3. Samples are then centrifuged at 0.5 × *g* for 1 min. The supernatant is subsequently aspirated and the beads are washed three times with 800 to 1000 μl of lysis buffer.
4. After the final wash, it is important to ensure that the supernatant is completely aspirated without any loss of beads.
5. Thirty microliters of 5x reducing SDS-PAGE sample buffer are added to each sample, and mixed by gently tapping the side of the tube (*see* **Note 6**).
6. The samples are heated at 85–100°C for 3–5 min and subsequently centrifuged briefly in a microcentrifuge. The supernatants can be analyzed directly by SDS-PAGE or stored at – 20°C for future use.
7. Following SDS-PAGE, the samples are transferred electrophoretically to a PVDF or nitrocellulose membrane (*see* **Chapters** 7, **10**, and **13**).
8. To determine if the protein of interest is ubiquitinated, immunodetection is carried out with an anti-ubiquitin antibody (*see* **Notes** 7 and **8**, and see chapters 7,10 and 13 for immunoblotting technique).
9. Once ubiquitination of the protein has been detected to a satisfactory level, the membrane is stripped of the antibodies (*see* **Subheading 3.1.7**) so that it can be reprobed with an antibody that recognizes the protein of interest.

3.1.6. Stripping the Membrane

1. Following detection of the ubiquitinated protein by ECL, the membrane is washed with distilled water for 5 min at room temperature with gentle rocking.
2. The membrane is then covered with a sufficient amount of stripping buffer and incubated for 5 min at room temperature with rocking (*see* **Note 9**).

3. Once the stripping incubation is complete, the membrane is washed once with distilled water for 5 min.
4. (Optional) To ensure that the membrane has been stripped completely of the previous antibodies, ECL detection is repeated.
5. The membrane is then blocked again by immersing it in blocking reagent for 1 h at room temperature with rocking and is reprobed with the antibody that recognizes the protein of interest.

3.2. Detection of Ubiquitin on Over-expressed Tagged Proteins (See Note *10)*

1. The cell line of choice is transiently co-transfected (*see* Chapter 10, **Subheading 3.1.1**) with a plasmid that encodes for a tagged form of the protein of interest together with a plasmid that encodes for a tagged form of ubiquitin *(22)* (*see* **Note 11**). As controls, the expression plasmids that encode for the protein of interest or ubiquitin should be transiently transfected separately.
2. In order to detect the interaction of the protein of interest with ubiquitin, it may be necessary to inhibit certain protein degradation pathways. Therefore, 18 h after the cells have been transiently transfected, they can be pre-treated with an appropiate inhibitor for a pre-determined time (*see* **Subheading 3.3, step 2**). If required, the cells may be stimulated with TLR agonists, for various lengths of time. An unstimulated control must be included.
3. Immunoprecipitation is performed as described in **Subheading 3.1**, using the antibody that recognizes the tag of the protein of interest.
4. Samples are then analyzed by SDS-PAGE and immunoblotting is carried out with an antibody that recognizes the tagged ubiquitin. The membrane is subsequently stripped as described in **Subheading 3.1.7** and reprobed with the antibody that recognizes the tagged protein of interest. See **Fig. 1** as an example of the results that can be obtained.

3.3. The Use of Inhibitors to Determine if the Protein of Interest Is Degraded by the Proteasome or the Lysosome

1. The required amount of cells must be empirically determined for each protein of study. If the endogenous protein of interest is expressed at detectable levels, proceed to **step 2**. However, if the protein is not easily detected, the cells can be transiently transfected with a plasmid that expresses the target protein (*see* Chapter 10, **Subheading 3.1.1**). Eighteen hours post-transfection, proceed to **step 2**.
2. Separate samples of cells are treated with the following inhibitors: 5–50 μM MG-132; 1–20 μM lactacystin; 3 μM proteasome inhibitor I; 10–100 μM chloroquine; 10–50 mM NH_4Cl; 10–50 μM epoxomicin. The concentration of each

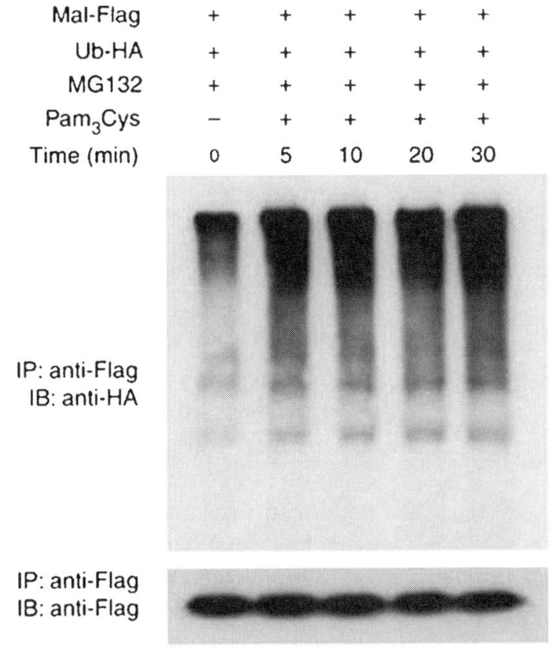

Fig. 1. Analysis of ubiquitination of exogenously expressed MyD88 adapter-like (Mal). Cells were made responsive to Pam$_3$Cys by co-transfecting HEK293 cells with 100 ng of TLR2 in conjugation with HA-ubiquitin and Flag-Mal, incubated for 24 h and pretreated for 4 h with 20 μM of MG-132 before stimulation with 100 ng/ml of Pam$_3$Cys. IP: immunoprecipitation; IB: immunoblot (obtained from Mansell et al., 2006 [16]).

inhibitor and the length of treatment time must be determined, as these factors are dependent on the half-life of the protein of interest. Many of the inhibitors mentioned above are reconstituted in DMSO (*see* **Subheading 2.4**); therefore, as a control the cells must also be treated with DMSO alone.

3. If required, the cells are then stimulated with TLR agonist(s).
4. (a) If using adherent cells, the media is aspirated and the cells are washed once with ice-cold PBS. The PBS is then removed; the plates are left tilted briefly on ice and any remaining liquid is aspirated. Proceed to **step 5**.

   (b) If cells grown in suspension are used, the samples are centrifuged at 125 × *g* for 3 min at 4°C. The media is aspirated and the cells are washed once with ice-cold PBS. The samples are centrifuged as before and the remaining liquid is aspirated without disturbing the cell pellet.
5. The cells are subsequently lysed by using a suitable volume of lysis buffer (e.g., 1 ml of lysis buffer/10^7 cells). For adherent

cells, a cell scraper is used to dislodge the cells from the plate; the cell lysates are then transferred to pre-chilled microcentrifuge tubes and left on ice, or rotated, for 10–30 min at 4°C.

6. Once the cells are lysed, the samples are centrifuged at maximum speed for 15 min at 4°C in a microcentrifuge. The supernatants are then transferred to new pre-chilled microcentrifuge tubes.
7. The protein concentration of each sample is determined using the Bradford method (*see* **Subheading 3.1.4**).
8. If necessary, the protein of interest can be immunoprecipitated (*see* **Subheading 3.1**). Otherwise the cell lysate may be analyzed directly by SDS-PAGE followed by immunoblotting (*see* **Notes 12** and **13**). Samples can be stored at – 80°C for future analysis.

3.4. Two-Dimensional Electrophoretic Analysis of Phosphoproteins (See Note *14)*

3.4.1. Sample Preparation

1. The predicted isoelectric point of the protein under investigation is determined by analyzing its amino acid sequence using appropriate computer software (e.g., http://www.expasy.ch/tools/pi_tool.html) (*see* **Note 15**).
2. If the endogenous protein of interest is easily detected, proceed to **step 3**. Otherwise a plasmid encoding for the target protein is transiently transfected into a suitable cell line (*see* Chapter 10, **Subheading 3.1.1**). As a control, a sample that is mock transfected must be included.
3. If required, the cells are treated with specific TLR ligands, in order to induce phosphorylation of the protein of study.
4. Cell extracts are prepared and the target protein is immunoprecipitated as described in **Subheading 3.1** up to **Subheading 3.1.5**, **step 4** (the use of ubiquitin isopeptidase inhibitors may not be required) (*see* **Note 16**). Immunocomplexes are washed once more with ice-cold PBS and the supernatant is aspirated.
5. Immunocomplexes are resuspended in 400 μl of sample solubilization solution. Samples are left at room temperature for 20 min (*see* **Note 17**).

3.4.2. Rehydration of IPG Strips

1. A rehydration tray, of an appropriate size for the length of the IPG strip used, is thoroughly washed. A suitable implement (e.g., toothbrush) should be used to ensure sufficient cleaning. The tray is rinsed with high purity water and air-dried. It is essential that the rehydration tray is adequately cleaned before its use, in order to prevent contamination of the samples with proteins from earlier experiments.

2. A suitable volume of the sample (according to the instructions provided by the manufacturers of the IPG strips) is transferred in a line along the length of the individual strip holder in the rehydration tray.
3. Before rehydration, the plastic cover from the IPG strip is carefully removed with a pair of forceps. While still holding the strip with the forceps, the opposite end is gently placed, gel side down, onto the sample in the strip holder. The strip holder is then examined to ensure that no air bubbles are trapped beneath the gel strip and that even coverage of the sample has been accomplished. If air bubbles do exist, the IPG strip is gently raised with a forceps and then lowered again onto the sample. In this way, the air bubbles are removed and the strip is evenly rehydrated.
4. (Optional) The IPG strip is allowed to absorb the sample for 1 h.
5. Mineral oil is carefully added, drop wise, directly onto the plastic backing of the IPG strip. The complete strip is covered so that evaporation of the sample is prevented. The rehydration tray is covered with its supplied lid and is positioned on a level surface.
6. The IPG strips are allowed to rehydrate overnight or for a minimum of 10 h.

3.4.3. First Dimension: Isoelectric Focusing (IEF)

1. The strip is removed from the rehydration tray with forceps, held vertically, and rinsed with high purity water. It is then gently placed on its side on a piece of filter paper in order to drain. Care needs to be taken to ensure that the gel is not detached from its plastic backing.
2. The IPG strip is then transferred to an electrophoresis unit, as recommended by the manufacturers. Taking into account the length of the IPG strip, program parameters are set and the electrophoresis is started as indicated in the instruction manual of the apparatus.
3. Following IEF, the IPG strip may be stored at – 80°C for future analysis or directly equilibrated for second-dimension electrophoresis.

3.4.4. Equilibration of IPG Strips

Before equilibrating the focused IPG strips, the resolving gel is prepared for SDS-PAGE analysis. The resolving gel (the percentage of which is dependent on the molecular weight of the protein of interest) is prepared, as in the case of one-dimensional SDS-PAGE (*see* **Chapter 7, Subheading 3.3**). It is then poured to within 5 mm of the top of the short plate and overlaid with water-saturated isobutyl alcohol. A stacking gel is not required.

1. If the focused IPG strips were stored at – 80°C, they are allowed to thaw at room temperature for no longer than 20 min, as proteins may begin to diffuse (*see* **Note 18**).
2. Each focused IPG strip is transferred, gel side up, to a clean strip holder on the rehydration tray, which contains a sufficient amount of SDS-PAGE equilibration buffer I to cover the entire IPG strip. The strip is then incubated for 10 min at room temperature with gentle rocking and the buffer is subsequently removed.
3. An adequate amount of equilibration buffer II is then added to cover the IPG strip, which is subsequently left for 10 min at room temperature with gentle rocking.

3.4.5. Second-Dimension SDS-PAGE

1. The sealing agarose solution is melted and allowed to cool down to approximately 60°C (or until the container of agarose can be comfortably held).
2. The water-saturated isobutyl alcohol is decanted from the resolving gel and the top of the gel is then rinsed with high purity water to remove any remaining alcohol.
3. A suitable volume of molecular weight protein solution (see supplier's recommendations) is dispensed onto an IEF sample application piece. With the use of a forceps, the application piece is placed on top of the gel in one corner.
4. Approximately 1 ml of sealing agarose solution is applied to the top of the gel. The following two steps must be performed in quick succession before the agarose begins to solidify.
5. To lubricate the IPG strip, so that it will easily slide between the gel plates, it is briefly dipped into SDS-PAGE buffer.
6. The IPG strip is subsequently placed, using a forceps, onto the top of the resolving gel. To ensure that the IPG strip has made direct contact with the gel and that no air bubbles exist between them, the top edge of the plastic backing is gently pushed with a thin piece of plastic (e.g., gel-loading micropipette tips) until the whole length of the strip touches the resolving gel. The sealing agarose solution will rise to the top, thereby covering the gel strip and holding it in place during electrophoresis.
7. Once the agarose has solidified, electrophoresis is carried out in the same way as one-dimensional SDS-PAGE, followed by immunoblotting. See **Fig. 2** as an example of the results that can be obtained.

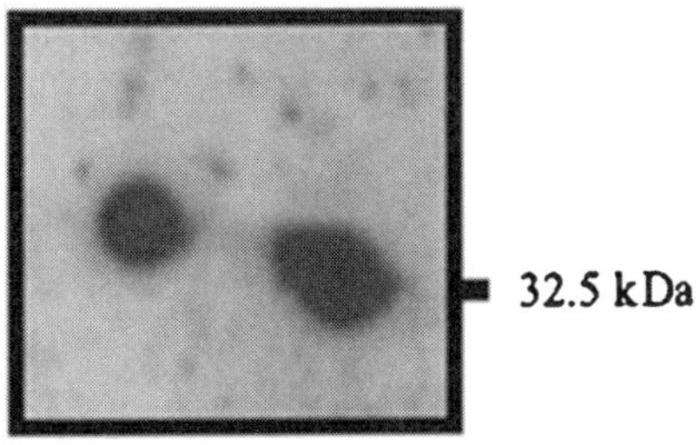

Fig. 2. Analysis of the phosphorylation status of MyD88 adapter-like (Mal) using two-dimensional electrophoresis. HEK293 cells were transiently transfected with a plasmid encoding for HA-Mal. Cell lysates were prepared, and Mal was immunoprecipitated with an anti-HA antibody. The sample was analyzed by two-dimensional sodium dodecyl sulfate polyacrylamide gel electrophoresis (SDS-PAGE), followed by immunoblotting with an anti-HA antibody. The sample was electrofocused at a pH gradient of 4 (*left*) to 7 (*right*) (obtained from Gray et al. 2006 [*10*]).

3.5. In Vitro Dephosphorylation of Phosphoproteins with Calf Intestinal Alkaline Phosphatase (CIP) (See Note *19)*

1. In order to detect the protein of interest, the amount of cells required must be empirically determined. If the endogenous protein is expressed at detectable levels, proceed to **step 2**. If the protein is not easily detected, the cells may be transiently transfected with a plasmid that encodes for that protein (*see* **Chapter 10**, **Subheading 3.1.1**) and 18 h post-transfection, **step 2** is carried out.
2. If required, the cells are stimulated with specific TLR ligands that may induce a shift in the target protein's predicted molecular mass. Unstimulated cells must be used as a control.
3. The protein of interest is isolated in the absence of phosphatase inhibitors (*see* **Note 20**):
   (a) For immunoprecipitations, the candidate phosphoprotein is isolated according to **Subheading 3.1** up to **Subheading 3.1.5**, **step 4** (the use of ubiquitin isopeptidase inhibitors may not be required). After the immunocomplexes are washed with cell lysis buffer, in the absence of phosphatase inhibitors, they are washed twice with 400 μl of phosphatase digestion buffer. The buffer is aspirated without disturbing the bead pellet.
   (b) If using cell lysates directly, *see* **Subheading 3.1.2** for the preparation of the samples. It is important that the cells are lysed in the absence of phosphatase inhibitors (the use of ubiquitin isopeptidase inhibitors may not be required). The protein concentration of the samples is then determined as described in **Subheading 3.1.4**.

4. For immunocomplexes, 30 μl of phosphatase buffer is added to each sample. If using cell lysates, 50–200 μg of total protein is added to 100 μl of phosphatase buffer.
5. As a control, dephosphorylation is inhibited in one sample by the addition of the following phosphatase inhibitors at the given final concentrations: 10 mM NaF, 1.5 mM Na_2MoO_4, 1 mM β-glycerolphosphate, 0.4 mM Na_3VO_4, and 0.1 μg/ml okadaic acid. As an additional control, a sample that contains no CIP can be included.
6. The samples are incubated at 37°C for 15 min.
7. Ten to one hundred units of CIP are added to each sample (*see* **Note 21**).
8. Reactions are incubated at 37°C. It is recommended that a time course (15 min–3 h) is employed to determine the time required to dephosphorylate the protein of interest.
9. The reactions are terminated by the addition of the appropriate volume of SDS-PAGE sample buffer and heating at 85–100°C for 3–5 min. The samples are either directly resolved by SDS-PAGE and immunoblotting is performed or stored at – 20°C for future analysis (*see* **Note 22**). **Figure 3** offers an example of the results that can be obtained.

3.6. Site-Directed Mutagenesis

3.6.1. Primer Design

1. Optimally the length of the mutagenic primers should be between 25 and 45 nucleotides.
2. Both primers should be designed to anneal to the same sequence on complementary strands of the plasmid. In addition, the chosen altered nucleotide base(s) should be located in the middle region of the sequences.

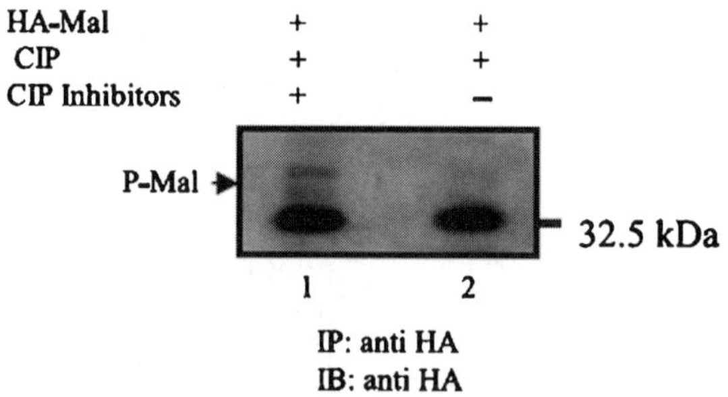

Fig. 3. Phosphatase treatment results in disappearance of the slower migrating forms of MyD88 adapter-like (Mal). HEK293 cells were transiently transfected with a plasmid encoding HA-Mal. Cell lysates were prepared, and Mal was immunoprecipitated with an anti-HA antibody. Immunoprecipitates were incubated at 37°C for 3 h with 100 U of calf intestinal alkaline phosphatase (CIP) in the presence (*lane 1*) or absence (*lane 2*) of calf intestinal alkaline phosphatase inhibitors. Samples were then analyzed by sodium dodecyl sulfate polyacrylamide gel electrophoresis (SDS-PAGE) and immunoblotted with an anti-HA antibody (obtained from Gray et al. 2006 *[10]*).

3. The melting temperature should be approximately 78°C. This value can be estimated for each primer using the following equation:

   81.5 + 0.41(% GC) −675/N −% mismatch

   where N is the number of bases in the primer.

4. If possible, a GC content of approximately 40% is desired. It is also preferred to have the primer ending in either one or more G/C bases.

5. Once the primer is successfully designed, it is synthesized by the company of choice.

3.6.2. Pcr

1. The following reagents are added to a pre-chilled thin-walled PCR tube:

   Five microliters of 10x reaction buffer

   X microliters (5–50 ng) of dsDNA template

   X microliters (125 ng) of mutagenic oligonucleotide primer 1

   X microliters (125 ng) of mutagenic oligonucleotide primer 2

   One microliter of dNTP mix (250 μM final concentration)

   molecular biology grade water is then added to a final volume of 50 μl.

2. Lastly, 1 μl of *Pfu*Turbo DNA polymerase (2.5 U/μl) is added into the reaction tube (*see* **Note 23**). The end of each reaction tube is tapped gently to mix the contents and the tubes are briefly centrifuged.

3. If the thermocycler does not have a heated lid, each reaction is overlaid with approximately 30–50 μl of PCR grade mineral oil.

4. The following cycling parameters are recommended: the initial denaturation step is 95°C for 60 s. If the mutation of interest requires one base change, 12 cycles for the reaction are recommended. However, if two or three base changes are introduced, 16 cycles can be initially used. The second denaturation step is 95°C for 30 s. The annealing step is 55°C for 60 s and the extension step is 68°C for 60 s/kb of plasmid length.

5. Once the above settings are programmed into the thermocycler, the PCR is commenced. On completion, the samples can be digested with Dpn I or stored at – 20°C.

3.6.3. Dpn I digestion

1. If samples were frozen, they must be thawed on ice, otherwise on completion of the PCR reaction, the samples are cooled to approximately 37°C. Ten units of Dpn I are

added per sample. If mineral oil was applied, ensure that the enzyme is added below it and directly into the sample. The reaction is mixed by pipetting the solution up and down and the samples are centrifuged briefly.

2. The samples are incubated for 1 h at 37°C. This procedure will digest the parental-methylated plasmid DNA, so that only the mutated non-methylated DNA remains.
3. The samples can be stored at – 20°C or alternatively transformations can be performed directly.

3.6.4. Transformation

1. Competent bacterial cells are thawed on ice. Fifty microliters of these cells are dispensed into each 14 ml pre-chilled polypropylene tube.
2. Four microliters of the Dpn I-digested PCR are added to the competent cells. The tube is then gently tapped to ensure adequate mixing.
3. The samples are incubated on ice for 30–60 min.
4. The bacteria are heat shocked at 42°C; the duration time for this step is dependent on the strain of the bacterial cells used (manufacturer's recommendations must be followed).
5. The samples are subsequently incubated on ice for 2 min.
6. Nine hundred microliters of LB are added to each tube. Alternatively, other growth media may be recommended by the manufacturer to increase the transformation efficiency.
7. The samples are incubated at 37°C with shaking at 250 rpm for 1 h.
8. Once this incubation period is complete, the samples are centrifuged at 1,000 × *g* for 10 min. Eight hundred microliters of supernatant are removed from each tube and the pellet is resuspended in the remaining growth medium.
9. It is then transferred to LB agar plates, which contain the appropriate antibiotic for selection of the transformants. A plastic spreader is used to evenly distribute the samples.
10. Plates are inverted and incubated at 37°C for up to 16 h or until bacterial colonies are clearly visible.
11. Plasmid DNA is then prepared from a single bacterial colony and sequenced to confirm that the desired mutation is present.

4. Notes

1. If multiple samples are being immunoprecipitated, a master mix may be prepared.

2. If a master mix is used, the beads are resuspended in an adequate volume of PBS, in an appropriate vessel, to permit sufficient mixing.
3. If a master mix is used, 40 μl of the beads are aliquoted into pre-chilled microcentrifuge tubes.
4. Ubiquitin isopeptidase inhibitors are added during cell lysis to inhibit deubiquitination of the protein of interest. One millimolar *N*-ethylmaleimide is also routinely used alone, or in combination with iodoacetamide.
5. If air bubbles are not removed, the reading will be inaccurate.
6. It is recommended that pipette tips are not used, as beads tend to stick to the inside of the tip thereby resulting in unequal sample loading.
7. When immunoblotting, the membrane should be blocked in BSA, as ubiquitinated proteins can be present in milk.
8. If the protein of interest is ubiquitinated, a shift in its gel electrophoretic mobility should be observed. For monoubiquitinated proteins, the altered gel mobility may be represented by an 8 kDa molecular weight change. A smear of higher molecular weight species should be detected following immunodetection of polyubiquitinated proteins.
9. Longer incubation times may be required (up to 20 min) depending on the affinity of the antibody for the protein of interest. Stripping buffers are also commercially available.
10. If ubiquitination of the protein of interest is not detected at it's endogenous levels, it is recommended that overexpression studies are conducted, as the immunoprecipitating antibody for the target protein may not recognize its ubiquitinated form.
11. The protein of interest and ubiquitin should contain different tags.
12. In most cases, polyubiquitination targets proteins for degradation by the 26S proteasome and monoubiquitination facilitates the internalization and degradation of membrane proteins by the endosome–lysosome pathway.
13. An increase in the expression level of the target protein following pre-treatment with MG-132, lactacystin, and proteasome inhibitor I indicates that it is degraded through the proteasome degradation pathway. If the protein of interest is degraded in the lysosome, chloroquine, NH_4Cl, and epoxomicin will inhibit protein degradation.
14. Not all proteins are amenable to two-dimensional analysis. In particular, hydrophobic proteins or those of high molecular weight are generally not easily detected.

15. Due to the addition of phosphates, the isoelectric point of a phosphorylated protein is more acidic than the unphosphorylated form of the same protein.
16. Additional samples may be prepared up to this point and analyzed by one-dimensional SDS-PAGE using 30 μl of reducing SDS-PAGE sample buffer per sample. Subsequently, immunoblotting with an anti-phosphotyrosine antibody (4G10, Upstate) may be conducted to determine if the protein of interest is tyrosine phosphorylated. Antibodies detecting phosphoserine and phosphothreonine residues are also commercially available, but many are not as highly specific and are unable to always provide conclusive results.
17. Samples containing urea should not be heated because urea can be degraded at temperatures above 30°C, forming isocyanate. This compound may carbamylate the protein of interest, and thus change its isoelectric point.
18. Once frozen IPG strips have reverted to their transparent state and no white insoluble urea is visible, they can be equilibrated.
19. It is important to note that not all phosphorylated proteins display changes in their electrophoretic mobility when analyzed by SDS-PAGE. Therefore, if a change in the mobility of a protein is not detected, this does not necessarily mean that it is not a phosphoprotein.
20. Cell lysis must be conducted in the absence of phosphatase inhibitors, as even trace amounts are liable to inhibit protein dephosphorylation by CIP.
21. This protocol describes the use of CIP as the general phosphatase employed. However, numerous phosphatases specific for either tyrosine, or serine or threonine dephosphorylation are commercially available which can also be used. In that way, the amino acid that the phosphate group is attached to, can be determined.
22. Dephosphorylation of the target protein will eliminate the differently migrating phosphoform(s) thereby confirming that the protein contained phosphates.
23. It is important that the *Pfu* DNA polymerase is added last, as this enzyme contains 3' to 5' exonuclease activity that may result in primer degradation, which may cause non-specific amplification and reduce the yield of the desired product.

Acknowledgment

The author would like to sincerely thank Dr. Constantinos Brikos for his advice and help in the writing of this chapter.

References

1. Arbibe, L., Mira, J. P., Teusch, N., Kline, L., Guha, M., Mackman, N., Godowski, P. J., Ulevitch, R. J. & Knaus, U. G. (2000). Toll-like receptor 2-mediated NF-kappa B activation requires a Rac1-dependent pathway. *Nat Immunol* 1, 533–40.
2. Sarkar, S. N., Elco, C. P., Peters, K. L., Chattopadhyay, S. & Sen, G. C. (2007). Two tyrosine residues of Toll-like receptor 3 trigger different steps of NF-kappa B activation. *J Biol Chem* 282, 3423–7.
3. Sarkar, S. N., Peters, K. L., Elco, C. P., Sakamoto, S., Pal, S. & Sen, G. C. (2004). Novel roles of TLR3 tyrosine phosphorylation and PI3 kinase in double-stranded RNA signaling. *Nat Struct Mol Biol* 11, 1060–7.
4. Chen, L. Y., Zuraw, B. L., Zhao, M., Liu, F. T., Huang, S. & Pan, Z. K. (2003). Involvement of protein tyrosine kinase in Toll-like receptor 4-mediated NF-kappa B activation in human peripheral blood monocytes. *Am J Physiol Lung Cell Mol Physiol* 284, L607–13.
5. Medvedev, A. E., Piao, W., Shoenfelt, J., Rhee, S. H., Chen, H., Basu, S., Wahl, L. M., Fenton, M. J. & Vogel, S. N. (2007). Role of TLR4 tyrosine phosphorylation in signal transduction and endotoxin tolerance. *J Biol Chem* 282, 16042–53.
6. Ivison, S. M., Khan, M. A., Graham, N. R., Bernales, C. Q., Kaleem, A., Tirling, C. O., Cherkasov, A. & Steiner, T. S. (2007). A phosphorylation site in the Toll-like receptor 5 TIR domain is required for inflammatory signalling in response to flagellin. *Biochem Biophys Res Commun* 352, 936–41.
7. Yu, Y., Nagai, S., Wu, H., Neish, A. S., Koyasu, S. & Gewirtz, A. T. (2006). TLR5-mediated phosphoinositide 3-kinase activation negatively regulates flagellin-induced proinflammatory gene expression. *J Immunol* 176, 6194–201.
8. Sanjuan, M. A., Rao, N., Lai, K. T., Gu, Y., Sun, S., Fuchs, A., Fung-Leung, W. P., Colonna, M. & Karlsson, L. (2006). CpG-induced tyrosine phosphorylation occurs via a TLR9-independent mechanism and is required for cytokine secretion. *J Cell Biol* 172, 1057–68.
9. Ojaniemi, M., Glumoff, V., Harju, K., Liljeroos, M., Vuori, K. & Hallman, M. (2003). Phosphatidylinositol 3-kinase is involved in Toll-like receptor 4-mediated cytokine expression in mouse macrophages. *Eur J Immunol* 33, 597–605.
10. Gray, P., Dunne, A., Brikos, C., Jefferies, C. A., Doyle, S. L. & O'Neill, L. A. (2006). MyD88 adapter-like (Mal) is phosphorylated by Bruton's tyrosine kinase during TLR2 and TLR4 signal transduction. *J Biol Chem* 281, 10489–95.
11. Kubo-Murai, M., Hazeki, K., Sukenobu, N., Yoshikawa, K., Nigorikawa, K., Inoue, K., Yamamoto, T., Matsumoto, M., Seya, T., Inoue, N. & Hazeki, O. (2007). Protein kinase Cdelta binds TIRAP/Mal to participate in TLR signaling. *Mol Immunol* 44, 2257–64.
12. Sato, S., Sugiyama, M., Yamamoto, M., Watanabe, Y., Kawai, T., Takeda, K. & Akira, S. (2003). Toll/IL-1 receptor domain-containing adaptor inducing IFN-beta (TRIF) associates with TNF receptor-associated factor 6 and TANK-binding kinase 1, and activates two distinct transcription factors, NF-kappa B and IFN-regulatory factor-3, in the Toll-like receptor signaling. *J Immunol* 171, 4304–10.
13. McGettrick, A. F., Brint, E. K., Palsson-McDermott, E. M., Rowe, D. C., Golenbock, D. T., Gay, N. J., Fitzgerald, K. A. & O'Neill, L. A. (2006). Trif-related adapter molecule is phosphorylated by PKC{epsilon} during Toll-like receptor 4 signaling. *Proc Natl Acad Sci U S A* 103, 9196–201.
14. Chuang, T. H. & Ulevitch, R. J. (2004). Triad3A, an E3 ubiquitin-protein ligase regulating Toll-like receptors. *Nat Immunol* 5, 495–502.
15. Husebye, H., Halaas, O., Stenmark, H., Tunheim, G., Sandanger, O., Bogen, B., Brech, A., Latz, E. & Espevik, T. (2006). Endocytic pathways regulate Toll-like receptor 4 signaling and link innate and adaptive immunity. *Embo J* 25, 683–92.
16. Mansell, A., Smith, R., Doyle, S. L., Gray, P., Fenner, J. E., Crack, P. J., Nicholson, S. E., Hilton, D. J., O'Neill, L. A. & Hertzog, P. J. (2006). Suppressor of cytokine signaling 1 negatively regulates Toll-like receptor signaling by mediating Mal degradation. *Nat Immunol* 7, 148–55.
17. Naiki, Y., Michelsen, K. S., Zhang, W., Chen, S., Doherty, T. M. & Arditi, M. (2005). Transforming growth factor-beta differentially inhibits MyD88-dependent, but not TRAM- and TRIF-dependent, lipopolysaccharide-induced TLR4 signaling. *J Biol Chem* 280, 5491–5.
18. O'Farrell, P. H. (1975). High resolution two-dimensional electrophoresis of proteins. *J Biol Chem* 250, 4007–21.
19. Gorg, A., Postel, W., Gunther, S., Weser, J., Strahler, J. R., Hanash, S. M., Somerlot, L. & Kuick, R. (1988). Approach to stationary

two-dimensional pattern: influence of focusing time and immobiline/carrier ampholytes concentrations. *Electrophoresis* 9, 37–46.

20. Westermeier, R., Postel, W., Weser, J. & Gorg, A. (1983). High-resolution two-dimensional electrophoresis with isoelectric focusing in immobilized pH gradients. *J Biochem Biophys Methods* 8, 321–30.
21. Bradford, M. M. (1976). A rapid and sensitive method for the quantitation of microgram quantities of protein utilizing the principle of protein-dye binding. *Anal Biochem* 72, 248–54.
22. Ellison, M. J. & Hochstrasser, M. (1991). Epitope-tagged ubiquitin. *A new probe for analyzing ubiquitin function. J Biol Chem* 266, 21150–7.

Chapter 12

The Generation of Highly Purified Primary Human Neutrophils and Assessment of Apoptosis in Response to Toll-Like Receptor Ligands

Lisa C. Parker, Lynne R. Prince, David J. Buttle, and Ian Sabroe

Summary

Neutrophils are crucial components of our defence against microbial assault. They are short-lived cells, with regulation of their lifespan being a primary mechanism involved in the regulation of their function. Delay of apoptosis facilitates their clearance of pathogens, whilst appropriate induction of cell death facilitates wound healing. A variety of methods are available to study neutrophil function: purification of human neutrophils and analysis of their lifespan are described here.

Key words: Granulocytes, Phoshotidylserine, Annexin V, To-Pro-3, Caspase, Flow cytometry.

1. Introduction

Apoptosis, or programmed cell death, of the neutrophil is an event that is accepted as essential for the resolution of inflammation *(1)*. In the course of acute inflammation, neutrophil apoptosis must first be delayed until its essential functions, such as pathogen ingestion and clearance, are completed. Thus, a central component of the neutrophils' response to pathogens is the prolongation of their lifespan. Following this, the cells promptly undergo apoptosis to abrogate inflammation and limit tissue damage. A range of factors, including pathogen-derived factors

Claire E. McCoy and Luke A.J. O'Neill (eds.), *Methods in Molecular Biology, Toll-Like Receptors, vol. 517*
DOI: 10.1007/978-1-59745-541-1_12

that are recognised by Toll-like receptors (TLRs), exquisitely regulate the balance between neutrophil survival and death by apoptosis *(2–4)*. It is well documented that human neutrophils express TLRs on their cell surface *(5–8)*, and that these receptors can regulate neutrophil survival. For example, we and others have shown that spontaneous apoptosis can be delayed by activation of TLR4 *(9,10)*, TLR2 *(11)*, and TLR9 *(12)*. TLR4-mediated neutrophil survival is also further enhanced by the presence of monocytes, through the release of monocyte-derived neutrophil survival factors *(9)*.

In vitro studies of neutrophil function typically involve the isolation of human neutrophils from peripheral blood followed by ex vivo study in culture. Here we describe three techniques suitable for the measurement of apoptosis in neutrophils. The first uses light microscopy to look at the characteristic morphological changes observed in apoptosis, namely the formation of apoptotic bodies that comprise condensed nuclear material within an intact cell membrane *(13)*. The second utilises flow cytometry to detect externalisation of phosphatidylserine (PS), an event indicative of a loss of membrane symmetry and thus apoptosis *(14)*. The third involves the fluorometric analysis of effector caspase activity, measuring the ability of caspase 3, an apoptotic effector caspase involved in cleavage of DNA, chromatin, and cytoskeletal proteins *(15)*.

2. Materials

2.1. Isolation and Purification of Neutrophils from Peripheral Blood

1. Ensure all reagents are kept sterile and handled in a sterile hood.
2. Dextran T500 (Sigma, Poole, UK) is used at 6% dissolved in 0.9% saline and pre-warmed to 37°C prior to use. Dextran takes a few minutes to dissolve into warm saline (with vigorous mixing); working aliquots of 50–100 ml may be stored at 4°C and pre-warmed to room temperature prior to use.
3. Percoll (GE Healthcare, Chalfont St. Giles, UK) is stored at −20°C in 45 ml stock aliquots. Working aliquots are diluted by addition of 5 ml 0.9% saline (generating a 90% Percoll solution), stored at 4°C, and pre-warmed to room temperature prior to use.
4. Hanks-balanced salts solution (HBSS) with Ca^{2+} and Mg^{2+}, or without Ca^{2+} and Mg^{2+} (both from Invitrogen, Paisley, UK), are stored at 4°C and pre-warmed to 37°C prior to use. A 10X concentrated stock solution can be diluted with sterile water to a 1X working solution on the day, or made up in advance and stored for short periods at 4°C.

5. Equipment for the magnetic separation (where performed): 0.3 in. negative selection column, 3-way tap, blunt-ended needle, and side syringe are from StemCell Technologies (Vancouver, Canada). Magnetic columns from other manufacturers (e.g. Miltenyi Biotech Ltd., Surrey, UK) are also suitable.
6. Cell purity can be improved by depletion of contaminating cells. Our lab utilises a custom made granulocyte antibody cocktail (comprising of antibodies to CD36, CD2, CD3, CD19, CD56, and glycophorin A) and magnetic colloid from StemCell Technologies (Vancouver, Canada). These are stored at −20°C in aliquots of 100 and 60 μl, respectively.
7. Column buffer: HBSS without Ca^{2+} and Mg^{2+} + 2% FCS.

2.2. Neutrophil Culture and TLR Stimulation

1. RPMI 1640 media containing 2 mM L-glutamine (Invitrogen, Paisley, UK) supplemented with 10% low endotoxin (<0.5 EU/ml) fetal calf serum (FCS) (BioWhitaker, Verviers, Belgium), 100 U/ml penicillin, and 100 μg/ml streptomycin (Invitrogen, Paisley, UK).
2. Ninety-six well polyvinyl chloride Flexiwell plates (Falcon, Becton Dickinson, UK).
3. Examples of TLR agonists include purified lipopolysaccharide (LPS) from *Escherichia coli* serotype R515 (Alexis Corporation, Nottingham, UK), which is stored at 4°C in 10 μl aliquots, and the synthetic bacterial-like lipopeptide Pam_3Cys-$SerLys_4$ (Pam_3CSK_4) (EMC Microcollections, Tübingen, Germany), which is stored at −20°C in 10 μl aliquots.

2.3. Assessment of Apoptosis

2.3.1. Light Microscopy

1. Menzel-Gläser microscope slides (76 × 26 mm^2 ground edge 90°) and cover glasses (22 × 22 mm^2) are from Thermo Fisher Scientific (Loughborough, UK).
2. The Reastain Quick-Diff kit containing Reastain Quick-Diff RED (0.12% Eosin Y, <0.1% sodium azide) and Reastain Quick-Diff BLUE (0.09% Azur II, 5% glycerol, <0.1% sodium azide) are from Reagena Ltd (Finland).

2.3.2. PS Detection

1. A 10X Annexin binding buffer (BD Biosciences, Oxford, UK): 0.1 M HEPES, 1.4 M NaCl, 25 mM $CaCl_2$ is stored at 4°C. Prior to use, dilute to 1X in distilled water at room temperature. Annexin V-PE (BD Biosciences, Oxford, UK) is stored at 4°C, whilst To-Pro-3 iodide (Invitrogen, Paisley, UK) is stored protected from light at −20°C in 10 μl aliquots.
2. Ethylenediamine tetraacetic acid (EDTA: Invitrogen, Paisley, UK) is used at 5 mM diluted in sterile phosphate-buffered saline (PBS) (Invitrogen, Paisley, UK) and stored at room temperature.

2.3.3. Caspase-3 Activity

1. Caspase lysis buffer: 100 mM HEPES, pH 7.5, 10% (w/v) sucrose, 0.1% CHAPS, 5 mM DTT (added fresh each time), and is filtered before use. Caspase lysis buffer is stored at 4°C for up to a week.
2. The fluorogenic caspase substrate set II is from Merck Chemicals Ltd (Beeston, UK), stored at –20°C and protected from light and moisture. The anti-Fas (Clone CH11) monoclonal antibody is from Upstate (Hampshire, UK) and is aliquoted and stored at –20°C. Z-Val-DL-Asp-fluoromethylketone (Z-VD-FMK) is from Bachem (St Helens, UK) and is aliquoted and stored at –20°C.

3. Methods

3.1. Isolation and Purification of Neutrophils from Peripheral Blood

Perform all procedures aseptically in a sterile environment. All procedures involving human subjects require appropriate ethics and informed consent.

3.1.1. Isolation

1. Draw blood through a needle of adequate bore (19–21 gauge) to avoid activation of the cells upon venesection (*see* **Note 1**) and add anticoagulate using 3.8% tri-sodium citrate (4.4 ml 3.8% tri-sodium citrate + 35.6 ml blood).
2. Immediately spin the blood in 50 ml polypropylene tubes at 270 × *g* with low brake for 20 min at room temperature to separate all the blood cells from the plasma and platelets. Without disturbing the interface, carefully remove the plasma upper phase to a fresh tube and spin this upper phase at 1,200 × *g* to give platelet-poor plasma (PPP). Retain this PPP in a fresh tube. To the cell-rich lower phase of the first centrifuge step, add 6 ml of 6% Dextran (*see* **Note 2**) in saline and top up to 50 ml with 0.9% saline. Gently invert the closed tubes to mix the contents, remove air bubbles from around the lid, replace with a fresh lid that is not contaminated with blood, and leave for 25–30 min to allow the erythrocytes to sediment.
3. When a clear interface has formed, aspirate the upper white cell-rich phase, transfer to a new 50 ml tube, and spin at 200–300 × *g* for 6 min at room temperature to pellet the white blood cells. During this time, prepare the plasma-Percoll (*see* **Note 3**) gradient by carefully overlaying an upper phase (42% Percoll/58% PPP) onto a bottom phase (51% Percoll/ 49% PPP), taking care to avoid mixing the two phases. Gradients should be made in 15 ml polypropylene centrifuge tubes. Add 3 ml of the denser Percoll solution and then overlay, slowly using a plastic disposable Pasteur pipette, the lighter layer. One gradient tube per 40 ml of starting volumes of blood is required.

4. Carefully resuspend the white cell pellet (obtained in **step 3**) in approximately 1.5 ml PPP and gently layer over the upper phase of the Percoll gradient. Spin at 225 × *g* for 11 min at room temperature with low brake.
5. The gradient yields three populations of cells: a top band of mononuclear cells (monocytes, lymphocytes, and basophils), a middle band of granulocytes (comprising of neutrophils and eosinophils), and a red cell pellet. Individually remove the top two layers with separate Pasteur pipettes, taking care to avoid contamination between the populations. Excessive red blood cell contamination of the desired granulocyte layer can be removed using an additional hypotonic shock lysis if required (*see* **Note 4**).
6. Resuspend the neutrophil cell layer (and/or mononuclear cell layer, *see* **Note 5**) in 30 ml (the volume can be adjusted depending on the size of the cell pellet) HBSS without Ca^{2+} and Mg^{2+} (containing 25% PPP). Remove 10 µl and use to count cell numbers. Spin the cell suspension at 300–400 × *g* for 6 min at room temperature.
7. Take cells requiring further purification (*see* **Note 6**) at this stage (*see* **Subheading 3.1.2**), whilst those not requiring purification can be washed a second time with HBSS without Ca^{2+} and Mg^{2+}, followed by a third and final wash in HBSS containing Ca^{2+} and Mg^{2+}, before culture (*see* **Subheading 3.2**). Repeated washing can result in cell losses; thus a further cell count prior to using the cells experimentally may be required.

3.1.2. Purification

Neutrophils can be further purified by negative magnetic selection. These instructions assume the use of a 0.3 in. StemSep® column for negative magnetic selection with gravity feed. Columns from Miltenyi Biotech Ltd (Surrey, UK) can be used according to the manufacturer's instructions.

1. Resuspend the granulocytes obtained in **step 7** (*see* **Subheading** 3.1.1) in column buffer (HBSS without Ca^{2+} and Mg^{2+} + 2% FCS) at 8 × 10^7/ml and incubate with 100 µl/ml custom granulocyte antibody cocktail (comprising of antibodies to CD36, CD2, CD3, CD19, CD56, and glycophorin A) for 15 min at room temperature with periodic swirling.
2. Incubate the cells with magnetic colloid (60 µl/ml) for a further 15 min at room temperature with periodic swirling. During this incubation period, prepare the column (*see* **Fig. 1**):
   - Take the column, a 3-way tap, and a 23 gauge blunt-ended needle out of their packaging; attach the 3-way tap to the column and the needle (keeping the cover on) to the end of the tap vertically below the column.

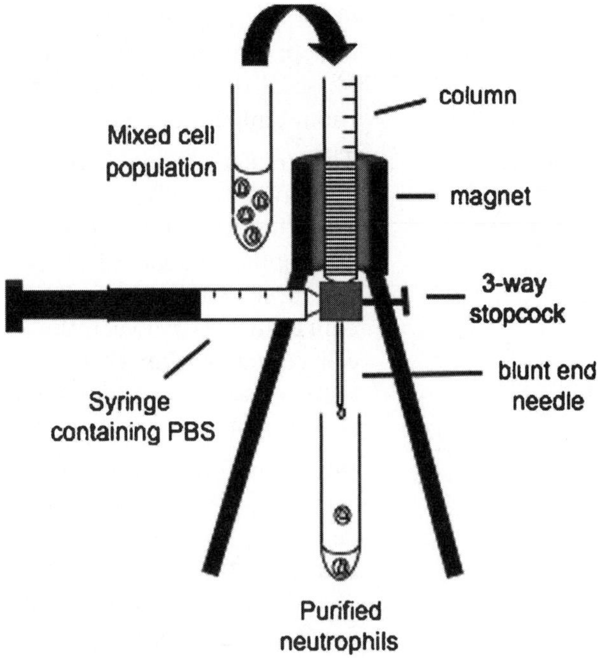

Fig. 1. Purification of neutrophils. Unwanted cells are immunomagnetically labelled and bound to a magnetic column. The purified neutrophils are collected in the column flow-through; they have not had antibody bound to their surface and are therefore suitable for further studies.

- Place the column into the magnet and remove the gauze from the top of the column.
- Take a sterile syringe and fill it with PBS before attaching it to the side connection of the 3-way tap, taking care not to introduce air bubbles. Set the tap to allow flow from the side connection and deliver PBS up the column to just above the stainless steel matrix (if air bubbles enter the mesh matrix sharply tapping the column should remove them).
- Wash the column by adding column buffer to the top of the column, place a container below the blunt end needle to collect waste, and then remove the cover from the needle. Allow flow through the column by turning the 3-way tap to the down position and continue to add column buffer until three column volumes worth (8 ml) has passed through. Take care not to allow the column to run dry at any point.
- Stop the flow of column buffer when the fluid level is just above the mesh matrix by turning the tap to the closed position. Remove the waste pot and replace with a sterile collection tube. The column is now prepared for separation of the cell population.

3. Load the cell suspension onto the column; turn the tap to allow the sample to run into the mesh matrix. Wash the cells through the column with three volumes (8 ml) of column buffer and allow to drip into the sterile collection tube. Again take care not to allow the column to run dry at any point.
4. Spin the collected cells at 300–400 × *g* for 6 min at room temperature. Resuspend the cell pellet in HBSS with Ca^{2+} and Mg^{2+}, and remove a 10 μl aliquot to count cell numbers before culture (*see* **Subheading 3.2**).

3.2. Neutrophil Culture and TLR Stimulation

1. Resuspend the neutrophils (normal or highly purified) at 5 × 10^6/ml in media (RPMI 1640 + 1% pen/strep and 10% low endotoxin FCS) and culture in 96 well Flexiwell plates (which minimise adhesion) at a final density of 2.5 × 10^6/ml. Keep the final well volume (including test substance and vehicle) at a constant volume (usually 100 μl) and culture the wells in duplicate. Incubation conditions are 37°C, 5% CO_2 in a water-jacketed tissue culture incubator. Neutrophils are terminally differentiated cells which should be immediately used for experiments; it is important to note that they rapidly undergo apoptosis once placed in culture, with typically >80% showing apoptotic characteristics at 24 h.
2. Purified LPS, a TLR4 ligand, is vortexed for 10 min at room temperature before dilution in RPMI media before addition to culture wells as required. Pam_3CSK_4, a TLR2/1 ligand, is heated to 42°C for 5 min and briefly vortexed before dilution in RPMI media before addition to culture wells as required.

3.3. Assessment of Neutrophil Apoptosis

3.3.1. Light Microscopy

During apoptosis, neutrophils exhibit characteristic morphological changes *(13)*, which can be clearly visualised by light microscopy. Cytocentrifuge slides (cytospins) are a way of preserving the image of a cell to assess these morphological changes. This method has been shown to closely correlate with other measurements of neutrophil apoptosis, including annexin V binding *(16)* and shedding of CD16 *(17)*:

1. Prepare the cytospins by spinning 100 μl cell suspension (at 2.5 × 10^6/ml) in assembled cytocentrifuge chambers for 3 min at 10 × *g* (*see* **Note 7**) in a Shandon Cytospin 3 cytocentrifuge (Thermoscientific, UK). This method allows the cells to be deposited on a defined area of a glass slide and allows absorption of any excess fluid into the chamber's filter card.
2. Allow the slides to air-dry, fix by dropping methanol onto the cells, and again air-dry.
3. Stain the cells using the Reastain Quick-Diff kit in staining chambers. Dip the slides into the Reastain Quick-Diff RED solution for 15–30 s to stain the cytoplasm, remove and blot

off excess, place slides into the Reastain Quick-Diff BLUE solution for 30–60 s to stain the nuclei, rinse the slides in distilled water, and air-dry (*see* **Note 8**).

4. Once dry, mount the slides with a coverslip and view by light microscopy under an oil immersion lens (×1,000) with 300 neutrophils scored from each slide of a duplicate pair. It is usually advisable to have the individual who is counting the slides blinded to the experimental conditions, or have cytospins recounted in a blinded fashion, to avoid the possibility of inadvertent introduction of experimental bias.
5. Assess the cell populations based on their cellular morphological characteristics. Cells displaying shrunken cell size and pyknotic nuclei should be recorded as apoptotic, whilst those retaining normal cell size with a multi-lobed nucleus and chromatin bridges still visible should be recorded as non-apoptotic (*see* **Fig. 2**). These data are used to calculate percentage apoptosis.

3.3.2. PS Detection

One of the earliest features of apoptosis is the exposure of the membrane aminophospholipid PS on the outer leaflet of the plasma membrane *(14)*, which occurs as the cells lose their membrane symmetry. PS exposure can be detected using the phycoerythrin-labelled PS probe annexin V-PE *(16)*, which binds to PS with high affinity in a calcium-dependent manner. The vital dye, To-Pro-3 iodide, is used to measure cell viability. Cells that lose their membrane integrity (necrotic or late apoptotic) allow entry of To-Pro-3, which intercalates with DNA and fluoresces in the FL4 channel of a flow cytometer:

1. Transfer 0.5 × 10^6 neutrophils (from duplicate wells) into 1.5 ml Eppendorf tubes, spin at 500 × *g* for 2 min in a refrigerated microcentrifuge, wash once in cold PBS, and resuspend

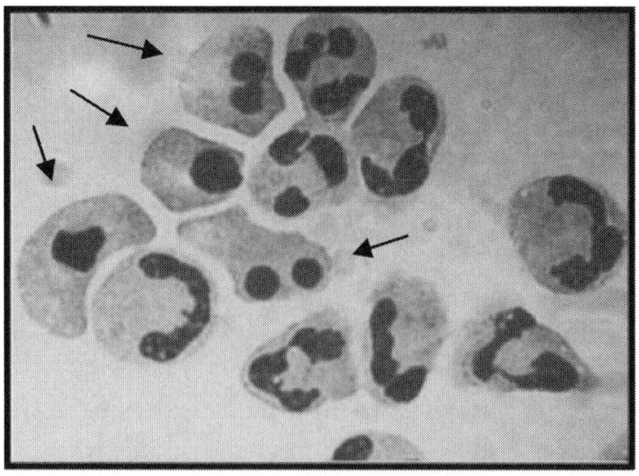

Fig. 2. Apoptotic and non-apoptotic neutrophils. The arrowed cells represent the apoptotic population, and while membrane integrity remains intact, cell shrinkage and chromatin condensation are apparent. Non-apoptotic cells retain segmented nuclear structure.

in 96.5 μl annexin binding buffer. Prepare a 1:100 dilution of To-Pro-3 in annexin binding buffer.

2. Incubate cells with 3.5 μl annexin V-PE for 20 min at room temperature, wrapped in foil. Add a further 197 μl of annexin V binding buffer to the tubes then add 3 μl To-Pro-3 iodide (final dilution 1:10,000). The To-Pro-3 should be added shortly before analysis: if many samples are to be acquired on the flow cytometer, it is best to add this compound a few minutes before each sample is acquired, by taking an aliquot of diluted To-Pro-3 (wrapped in foil) to the cytometer (*see* **Note 9**).
3. The following controls are used to set up quadrants.
   - Unstained untreated cells
   - Untreated cells stained with annexin V-PE alone (no To-Pro-3)
   - Untreated cells stained with To-Pro-3 alone (no annexin-V-PE)

     Other staining controls often include
   - Apoptotic positive control (dual-stained UV-treated cells) – expose cells to UV light for 20 min and then place in incubator for 2–4 h.
   - Annexin V/To-Pro-3 positive control (dual-stained aged cells) – allow untreated cells to age for 24 h and thus enter late apoptosis/necrosis.
   - Annexin V non-specific binding control (dual-stained cells) – cells are resuspended in 5 mM EDTA (instead of annexin binding buffer) prior to staining to prevent the Ca^{2+}-dependent binding of PS to the phospholipid binding protein annexin-V.
4. Analyse by flow cytometry immediately (*see* **Note 9**). We used a FACSCalibur flow cytometer, Becton Dickinson Immunocytometry Systems, San Jose, CA. Annexin V and To-Pro-3 are detected using FL-2 and FL-4 parameters, respectively (**Fig. 3**).

3.3.3. Caspase-3 Activity

Caspases are key players in neutrophil apoptosis *(15, 18)* and caspase activity is often used as a marker for measuring apoptosis in vitro *(19–21)*. Caspase activity is measured using selective small synthetic substrates that release a chromophore or fluorophore upon cleavage. In this way, the amount of substrate cleavage can be measured quantitatively *(22)*.

Caspase 3 is known as an effector caspase and is responsible for the cleavage of structural proteins (e.g. laminins) and cytoskeletal proteins.

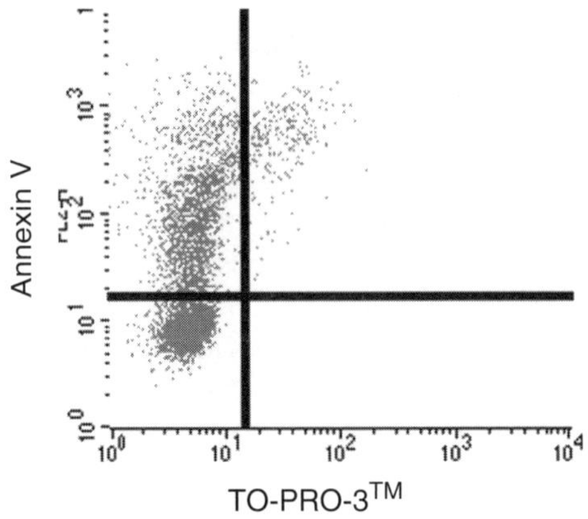

Fig. 3. An example of confirmation of apoptosis by annexin V staining. Neutrophils were treated with a proapoptotic toxin, stained with annexin V and To-Pro-3™ and assayed using a FACSCalibur FACScan flow cytometer. This figure shows an illustrative histogram with normal cells (healthy, *lower left*), annexin V positive cells (apoptotic, *upper left*) and dual annexin V/To-Pro-3 positive cells (late apoptotic/necrotic, *upper right*).

1. Caspase-3 activity is determined by measuring an enzymatically cleaved-labelled substrate such as Ac-DEVD-AMC (AMC = 7-(4-methyl)coumarylamide). For these experiments, a continuous assay is employed, which collects kinetic data over time.
2. Remove neutrophils from culture, wash in PBS, and resuspend in caspase lysis buffer at a concentration of 100 × 10^6/ml. Use the equivalent of five million cells in each run (see below). Neutrophil lysates can be frozen at −80°C until required.
3. Set the fluorimeter (we used a Perkin-Elmer LS-50 B and the FLUSYS software package *[23]*) to 37°C, and use an excitation wavelength between 365 and 380 nm and emission between 430 and 460 nm.
4. Run a known amount of free aminomethyl coumarin as a reference standard to calibrate the fluorimeter. This is typically a 100-fold lower concentration than the caspase substrate concentration so that all measurements are taken under conditions where there is no significant substrate depletion. In this instance, 200 nM aminomethyl coumarin in a final volume of 3 ml lysis buffer was used.
5. Blank with 2,800 μl caspase lysis buffer in a 3 ml cuvette ($t = 0$) and allow to run until the readings are stable. After 5 min ($t = 5$), add 50 μl of 20 μM caspase 3 substrate Ac-DEVD-AMC in DMSO to the cuvette and leave to run until the activity has stabilised (approximately 5 min). Add 50 μl of neutrophil lysate

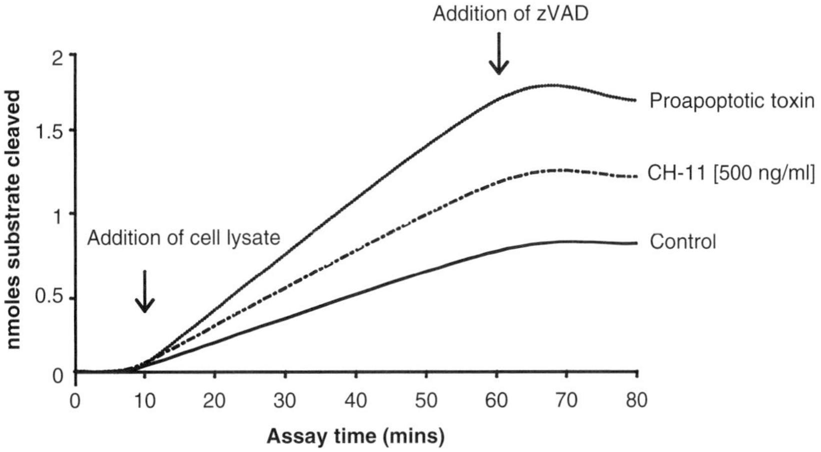

Fig. 4. An example of the increased caspase-3 activity observed after treatment with a proapoptotic toxin. This figure shows an example trace from a fluorometric analysis of effector caspase activity. Neutrophils were treated with either a proapoptotic toxin, anti-Fas antibody, CH-11 (500 ng/ml) or media alone (control) for 2 h. Cell lysates were prepared as described in **Subheadings** 2 and **3**. Ten million cells were used in each assay run. Z-VD-FMK (100 μM) was added after 60 min to confirm the specificity of substrate cleavage.

(equivalent to 5×10^6 neutrophils) and leave the assay to run for the required amount of time (e.g. 1–5 h (*see* **Note 10**).

6. Control samples should include cells treated with media alone, whilst an agonistic antibody of the Fas receptor can be used to induce cell death and caspase activation in samples whose lysates are then used as a positive control. The anti-Fas (clone CH11) mAb (500 ng/ml) is suitable for this. To ensure cleavage of the substrate is caspase-specific, add the cell-permeable pan-caspase inhibitor Z-VD-FMK (100 μM) (individual caspase inhibitors can be used to verify specificity of activity). This should terminate the reaction and prevent any further increase in fluorescence.
7. Caspase activity is proportional to the amount of free aminomethyl coumarin produced (1 U of caspase activity is defined as the amount of caspase required to produce 1 pmol of aminomethyl coumarin from Ac-DEVD-AMC/min at saturating substrate conditions) (**Fig. 4**).

4. Notes

1. To avoid activation of the neutrophils, all pipetting, resuspending, and general handling procedures must be carried out with great care.

2. Dextran consists of multiple glucose molecules joined into chains of differing lengths to form a complex polysaccharide, and functions by binding to erythrocytes and causing them to sediment to the bottom of the falcon tube.
3. Percoll comprises polyvinylpyrrolidone (PVP)-coated colloidal silica particles of 15–30 nm diameter and is a useful reagent for density separation. The gradients utilised here allow the heavier red cells to pass completely through both Percoll densities to settle as a pellet, granulocytes pass through the upper lighter Percoll layer and reside between the two densities, whilst the lighter mononuclear cells remain above the upper Percoll layer.
4. Red cell contamination of the granulocyte layer can be reduced using hypotonic shock lysis. Centrifuge the granulocytes at 300–400 × *g* for 6 min at room temperature, resuspend the cell pellet in 0.2% saline (1 ml) using a Pasteur pipette, then rapidly top up to 25 ml with 0.2% saline and gently mix so that the total exposure time of the cells to the hypotonic environment is about 45 s. At the end of this period, add an equal volume of 1.6% saline and mix (rapidly but gently) to restore isotonicity. Centrifuge the cells at 300–400 × *g* for 6 min at room temperature to obtain a cell pellet and continue from **Subheading 3.1.1**, **step 6**.
5. If monocytes are required they can be obtained from the same method of preparation by following the same protocol as the neutrophil separation procedure. The upper mononuclear cell layer is washed in HBSS (*see* **Subheading 3.1.1**, **steps 6** and 7) and then resuspended in media and cultured as described (*see* **Subheading 3.2**). If further purification by negative magnetic selection is required, an almost identical protocol to the neutrophil purification is carried out (*see* **Subheading 3.1.2**, **steps 1–4**) but with one modification, a different antibody cocktail in **step 1** (comprising of monoclonal antibodies to CD2, CD3, CD19, CD56, CD66b, Glycophorin A).
6. Granulocyte populations are typically comprised of >95% neutrophils, with the contaminating cells being mostly eosinophils and a few mononuclear cells (<2%). Further purification by negative magnetic selection typically produces cells that are >99% granulocytes, where >95% neutrophils and the remainder are eosinophils. We have found that small numbers of contaminating PBMCs (<5%) can greatly enhance the survival of neutrophils in response to TLR agonists, most likely through the release of survival cytokines *(5,9,24)*. Thus, for experiments investigating neutrophil apoptosis, it is recommended that highly purified neutrophils are utilised,

with the option of adding back known numbers of PBMCs if required.

7. If preparing cytospins of fresh cells ($t = 0$), increase the spin speed to $18 \times g$. Cytospins of monocytes should be prepared with 50% FCS to prevent cellular disintegration upon contact with the slides.
8. The staining solutions are kept in separate staining chambers and should be changed periodically (after staining approximately 250 slides).
9. Vital dyes such as To-Pro-3 iodide are cell-impermeant, passing through the plasma membrane of dead cells only; however, if left on the cell for a length of time they can permeate the cell and thus give a false positive result. If access to a flow cytometer cannot be immediate, the annexin V-PE stain can be left for up to 1 h; however, the To-Pro-3 dye should only be added immediately before reading.
10. Due to the prolific protease activity of neutrophils, once lysates are prepared they must be kept at 4°C or lower and freeze thaws avoided. In the event of a low caspase signal, more neutrophil lysate can be added during the data collection period.

References

1. Bianchi, S. M., Dockrell, D. H., Renshaw, S. A., Sabroe, I., and Whyte, M. K. (2006) Granulocyte apoptosis in the pathogenesis and resolution of lung disease. *Clin. Sci. (Lond).* 110, 293–304.
2. Colotta, F., Re, F., Polentarutti, N., Sozzani, S., and Mantovani, A. (1992) Modulation of granulocyte survival and programmed cell death by cytokines and bacterial products. *Blood* 80, 2012–2020.
3. Lee, A., Whyte, M. K., and Haslett, C. (1993) Inhibition of apoptosis and prolongation of neutrophil functional longevity by inflammatory mediators. *J. Leukoc. Biol.* 54, 283–288.
4. Kobayashi, S. D., Braughton, K. R., Whitney, A. R., Voyich, J. M., Schwan, T. G., Musser, J. M., and DeLeo, F. R. (2003) Bacterial pathogens modulate an apoptosis differentiation program in human neutrophils. *Proc. Natl. Acad. Sci. U.S.A.* 100, 10948–10953.
5. Sabroe, I., Jones, E. C., Usher, L. R., Whyte, M. K., and Dower, S. K. (2002) Toll-like receptor (TLR)2 and TLR4 in human peripheral blood granulocytes: a critical role for monocytes in leukocyte lipopolysaccharide responses. *J. Immunol.* 168, 4701–4710.
6. Hayashi, F., Means, T. K., and Luster, A. D. (2003) Toll-like receptors stimulate human neutrophil function. *Blood* 102, 2660–2669.
7. Kurt-Jones, E. A., Mandell, L., Whitney, C., Padgett, A., Gosselin, K., Newburger, P. E., and Finberg, R. W. (2002) Role of toll-like receptor 2 (TLR2) in neutrophil activation: GM-CSF enhances TLR2 expression and TLR2-mediated interleukin 8 responses in neutrophils. *Blood* 100, 1860–1868.
8. Parker, L. C., Whyte, M. K., Dower, S. K., and Sabroe, I. (2005) The expression and roles of Toll-like receptors in the biology of the human neutrophil. *J. Leukoc. Biol.* 77, 886–892.
9. Sabroe, I., Prince, L. R., Jones, E. C., Horsburgh, M. J., Foster, S. J., Vogel, S. N., Dower, S. K., and Whyte, M. K. (2003) Selective roles for Toll-like receptor (TLR)2 and TLR4 in the regulation of neutrophil activation and life span. *J. Immunol.* 170, 5268–5275.
10. Radsak, M. P., Salih, H. R., Rammensee, H. G., and Schild, H. (2004) Triggering receptor expressed on myeloid cells-1 in neutrophil inflammatory responses: differential regulation of activation and survival. *J. Immunol.* 172, 4956–4963.

11. Lotz, S., Aga, E., Wilde, I., van Zandbergen, G., Hartung, T., Solbach, W., and Laskay, T. (2004) Highly purified lipoteichoic acid activates neutrophil granulocytes and delays their spontaneous apoptosis via CD14 and TLR2. *J. Leukoc. Biol.* 75, 467–477.
12. Jozsef, L., Khreiss, T., and Filep, J. G. (2004) CpG motifs in bacterial DNA delay apoptosis of neutrophil granulocytes. *FASEB J.* 18, 1776–1778.
13. Savill, J. S., Wyllie, A. H., Henson, J. E., Walport, M. J., Henson, P. M., and Haslett, C. (1989) Macrophage phagocytosis of aging neutrophils in inflammation. Programmed cell death in the neutrophil leads to its recognition by macrophages. *J. Clin. Invest.* 83, 865–875.
14. Fadok, V. A., Voelker, D. R., Campbell, P. A., Cohen, J. J., Bratton, D. L., and Henson, P. M. (1992) Exposure of phosphatidylserine on the surface of apoptotic lymphocytes triggers specific recognition and removal by macrophages. *J. Immunol.* 148, 2207–2216.
15. Chang, H. Y., and Yang, X. (2000) Proteases for cell suicide: functions and regulation of caspases. *Microbiol. Mol. Biol. Rev.* 64, 821–846.
16. Homburg, C. H., de Haas, M., von dem Borne, A. E., Verhoeven, A. J., Reutelingsperger, C. P., and Roos, D. (1995) Human neutrophils lose their surface Fc gamma RIII and acquire Annexin V binding sites during apoptosis in vitro. *Blood* 85, 532–540.
17. Dransfield, I., Buckle, A. M., Savill, J. S., McDowall, A., Haslett, C., and Hogg, N. (1994) Neutrophil apoptosis is associated with a reduction in CD16 (Fc gamma RIII) expression. *J. Immunol.* 153, 1254–1263.
18. Miller, D. K. (1997) The role of the caspase family of cysteine proteases in apoptosis. *Semin. Immunol.* 9, 35–49.
19. Los, M., Walczak, H., Schulze-Osthoff, K., and Reed, J. C. (2000) Fluorogenic substrates as detectors of caspase activity during natural killer cell-induced apoptosis. *Methods Mol. Biol.* 121, 155–162.
20. Niwa, M., Hara, A., Kanamori, Y., Matsuno, H., Kozawa, O., Yoshimi, N., Mori, H., and Uematsu, T. (1999) Inhibition of tumor necrosis factor-alpha induced neutrophil apoptosis by cyclic AMP: involvement of caspase cascade. *Eur. J. Pharmacol.* 371, 59–67.
21. Wilkie, R. P., Vissers, M. C., Dragunow, M., and Hampton, M. B. (2007) A functional NADPH oxidase prevents caspase involvement in the clearance of phagocytic neutrophils. *Infect. Immun.* 75, 3256–3263.
22. Stennicke, H. R., and Salvesen, G. S. (2000) Caspase assays. *Methods Enzymol.* 322, 91–100.
23. Rawlings, N. D., and Barrett, A. J. (1990) FLUSYS: a software package for the collection and analysis of kinetic and scanning data from Perkin-Elmer fluorimeters. *Comput. Appl. Biosci.* 6, 118–119.
24. Sabroe, I., Prince, L. R., Dower, S. K., Walmsley, S. R., Chilvers, E. R., and Whyte, M. K. (2004) What can we learn from highly purified neutrophils? *Biochem. Soc. Trans.* 32, 468–469.

Chapter 13

Cellular Expression of A20 and ABIN-3 in Response to Toll-Like Receptor-4 Stimulation

Kelly Verhelst, Lynn Verstrepen, Beatrice Coornaert, Isabelle Carpentier, and Rudi Beyaert

Summary

Although Toll-like receptor (TLR)-induced expression of several proinflammatory genes is required to provoke an efficient immune response, excessive or prolonged activation of TLR signaling can contribute to the development of septic shock and several inflammatory diseases. Given this inherent danger of unrestrained TLR signaling to the organism, it is not surprising that many negative feedback mechanisms have evolved to hold TLR signaling in check. In this context, TLR stimulation induces several negative regulators of TLR-induced signaling to nuclear factor (NF)-κB dependent gene expression. Here we describe the use of Western blotting and reverse transcriptase polymerase chain reaction (RT-PCR) to study respectively the cellular protein and mRNA expression levels of the NF-κB inhibitory proteins A20 and ABIN-3 in response to TLR4 stimulation by lipopolysaccharide (LPS).

Key words: A20, ABIN, Lipopolysaccharide, Negative feedback, NF-κB, RT-PCR, Toll-like receptor, Western blotting.

1. Introduction

An effective immune system requires the rapid and appropriate activation of an inflammatory response to terminate the spread of an infection as quickly as possible. However, as the inflammatory response evolved at the cost of self-tissue damage, hyperactivation or failure of resolution can have detrimental effects to the organism. Because of this inherent danger of inflammation, the crucial receptor systems and signaling pathways that shape

Claire E. McCoy and Luke A.J. O'Neill (eds.), *Methods in Molecular Biology, Toll-Like Receptors, vol. 517*
DOI: 10.1007/978-1-59745-541-1_13

an inflammatory response are tightly regulated. In this context, TLR-induced activation of the transcription factor nuclear factor κB (NF-κB) plays a pivotal role in initiating and propagating the inflammatory response *(1)*. Hence, many regulatory proteins control the NF-κB signaling pathways induced by microbial and other TLR ligands. These checkpoints act at different levels in NF-κB signaling and apply various strategies such as interfering with protein–protein interactions, protein phosphorylation, or protein ubiquitination *(2, 3)*. In many cases, the expression of these NF-κB inhibitory proteins is induced by TLR stimulation, which is itself regulated by NF-κB, thus inducing a negative feedback of NF-κB activation. Knowledge on the mRNA and protein expression of these proteins under different conditions is of significant interest for the understanding of TLR signaling and may eventually contribute to the development of new therapeutic approaches for several inflammatory diseases.

In this chapter, we describe two simple methods to measure the TLR4-inducible expression of A20 and the A20-binding protein ABIN-3, which both negatively regulate NF-κB activation in response to a wide range of proinflammatory stimuli, including lipopolysaccharide (LPS), tumor necrosis factor (TNF), and interleukin-1 (IL-1) *(4–7)*. A20 has been proposed to inhibit TLR-induced NF-κB activation by de-ubiquitinating crucial TLR signaling molecules such as RIP1, TRAF6, and IKKγ (also known as NEMO) *(3, 5, 8–9)*. Although the molecular mechanism of action of ABIN-3, as well as of the related A20-binding inhibitors of NF-κB activation ABIN-1 and ABIN-2, is still largely unclear, ABINs have been suggested to mediate at least part of the effects of A20 by functioning as an adaptor between A20 and other proteins *(10)*. However, evidence for an A20-independent mechanism of action of ABINs has also been published *(6, 11–13)*. In contrast to ABIN-1 and ABIN-2, which are constitutively expressed in most cell types, expression of A20 and ABIN-3 relies on specific NF-κB sites in their promoter region and can in many cases only be observed upon treatment of cells with NF-κB activating stimuli such as LPS and TNF (*14, 15*). Inducible expression of A20 and ABIN-3 can be detected at both the mRNA and protein level using various techniques. Here we describe the use of reverse transcriptase polymerase chain reaction (RT-PCR) as a simple and highly sensitive technique for the detection of ABIN-3 mRNA expression in different cell lines. For the detection of TLR4-inducible changes in A20 and ABIN-3 protein expression, we document the use of sodium dodecyl sulfate–polyacrylamide gel electrophoresis (SDS-PAGE) and Western blotting of cell extracts, followed by immunodetection with A20 and ABIN-3-specific antibodies that have recently become available.

2. Materials

2.1. Cell Culture and Lysis

1. Culture medium for HeLa (human cervical carcinoma cell line) and HepG2 (human hepatoma cell line): DMEM (Gibco/BRL, Bethesda, MD) supplemented with 10% fetal bovine serum (Greiner, Belgium), 2 mM L-glutamine, 0.4 mM sodium pyruvate, 0.1 mM nonessential amino acids, 100 IU/mL penicillin, and 0.1 mg/mL streptomycin.
2. Culture medium for THP-1 (human monocytic cell line), Jurkat (human T-cell line), and U937 (human monocyte leukemia cell line): RPMI1640 (Gibco/BRL, Bethesda, MD) supplemented with 10% fetal bovine serum (Greiner, Belgium), 0.4 mM sodium pyruvate, 2 mM L-glutamine, 4 μM β-mercaptoethanol, 100 IU/mL penicillin, and 0.1 mg/mL streptomycin.
3. Culture medium for A549 (human lung epithelial cell line): DMEM supplemented with 10% newborn calf serum (Gibco/BRL, Bethesda, MD), 2 mM L-glutamine, 100 IU/mL penicillin, and 0.1 mg/mL streptomycin.
4. Trypsin/EDTA buffer: 400 mL 0.04% EDTA (1 mM EDTA, 0.1 mM NaCl, 2.5 mM KCl, 7 mM $Na_2HPO_4.12H_2O$, and 2 mM KH_2PO_4) and 100 mL trypsin solution (0.1 mM trypsin, 0.1 M NaCl, 2.5 mM KCl, 66 mM Tris-HCl, 0.7 mM Na_2HPO_4, and 1% phenolred), pH 7.6. Store at 4°C.
5. LPS from *Salmonella abortus* (Alexis, San Diego, CA) is dissolved at 1 mg/mL in ultra pure water and stored in single aliquots at −20°C. The working solutions are prepared by diluting in DMEM or RPMI1640.
6. Recombinant human TNF was expressed in *Escherichia coli* in our laboratory, purified to at least 99% homogeneity and has a specific biological activity of 2.3×10^7 IU/mg purified protein, as determined with the international standard (code 87/650; National Institute for Biological Standards and Control, Potters Bar, UK). Store in single aliquots at −70°C.
7. Phosphate-buffered saline (PBS) (10x): 95.44 g DPBS (Lonza, Verviers, Belgium) in 1 L water. Store at 4°C.
8. Modified laemmli buffer for cell lysis (5xLL): 250 mM Tris-HCl pH 8, 10% sodium dodecyl sulfate (SDS), 50% glycerol, 0.005% bromphenol blue, 25% β-mercaptoethanol. Dilute with water if necessary.
9. E1A buffer for cell lysis: 50 mM HEPES pH 7.6, 250 mM NaCl, 0.5% NP-40 (w/v), 5 mM EDTA, supplemented with phosphatase inhibitors (200 mM sodiumorthovanadate, 20 mM 2-glycerophosphate, and 10 mM sodiumfluoride) and

protease inhibitors (0.3 mM aprotinin, 1 mM leupeptin, and 1 mM Pefabloc).

2.2. SDS–Polyacrylamide Gel Electrophoresis (SDS-PAGE)

1. Separating buffer (4x): 1.5 M Tris-HCl, pH 8.8, 0.4% SDS. Store at 4°C.
2. Stacking buffer (4x): 0.5 M Tris-HCl, pH 6.8, 0.4% SDS. Store at 4°C.
3. Forty percent acrylamide/bis solution (37.5:1, Biosolve, Valkenswaard, The Netherlands) and *N,N,N,N'*-tetramethylethylenediamine (TEMED, Merck, VWR, Belgium). These solutions are stored at 4°C.
4. Ammonium persulfate (APS): prepare 10% solution in water and store at 4°C (*see* **Note 1**).
5. Water-saturated isobutanol (*see* **Note 2**).
6. Running buffer (10x): 250 mM Tris, 192 mM glycine, and 10% SDS. Store at room temperature.
7. Prestained molecular weight markers (Bio-Rad, Hercules, CA).

2.3. Semi-dry Western Blotting/Detection of A20

1. Transfer buffer: 20% methanol, 39 mM glycine, 48 mM Tris-HCl pH 9.2, 0.0375% SDS.
2. Nitrocellulose membrane (0.45 μM) from Schleicher & Schuell, MA and chromatography paper from Whatman, Maidstone, UK.
3. Tris-buffered saline with Tween-20 (TBS-T): prepare 10x TBS stock solution (200 mM Tris-HCl pH 7.6, 1.37 M NaCl). To make 1x TBS-T, dilute 100 mL of 10x TBS with 900 mL water and add 0.1% Tween-20.
4. Blocking buffer: 5% nonfat dry milk in TBS-T (*see* **Note 3**).
5. Primary antibody: mouse antihuman A20 monoclonal antibody clone 59A426 (eBioscience, San Diego, CA).
6. Secondary antibody: antimouse IgG conjugated to horseradish peroxidase (Amersham GE Healthcare, UK).
7. Enhanced chemiluminescent (ECL) reagents from Perkin & Elmer (Waltham, MA) and Hyperfilm (Amersham GE Healthcare, UK).

2.4. Semi-dry Western Blotting/Detection of ABIN-3

1. PBS-T: dilute 100 mL 10x PBS with 900 mL water and add 0.1% Tween-20.
2. Blocking buffer: 5% nonfat dry milk in PBS-T (*see* **Note 3**).
3. Primary antibody: rabbit anti-ABIN-3 polyclonal antibody clone 949 raised against a human ABIN-3-specific peptide (NH2-CDVQHKANGLSSVKKVHP-COOH) coupled to keyhole limpet hemocyanin.

4. Secondary antibody: antirabbit IgG conjugated to horseradish peroxidase (Amersham GE Healthcare, UK).

2.5. Total RNA Isolation

1. TRIzol®Reagent (Invitrogen, Carlsbad, CA).
2. Chloroform (Merck, NJ).
3. Isopropyl alcohol (Fiers, Belgium).
4. Seventy-five percent ethanol (Merck, NJ).
5. RNAse-free water: add 0.01% diethylpyrocarbonate (DEPC; Sigma-Aldrich, Belgium) let stand/stir overnight and autoclave.

2.6. Semi-quantitative RT-PCR for Detection of ABIN-3 mRNA

1. Superscript™ First-Strand Synthesis System for RT-PCR (Invitrogen, Carlsbad, CA).
2. Taq DNA polymerase (Invitrogen, Carlsbad, CA).
3. Sense primer human ABIN-3: 5′-ACTGGACGCCGCG-GAAAGAT-3′.
4. Antisense primer human ABIN-3: 5′-TGGCGGAAGCT-GGTCAAGAG-3′.

3. Methods

3.1. Seeding of Cells

1. For A20 and ABIN-3 protein expression analysis, plate 8×10^5 subconfluent THP-1 cells in 2 mL of the appropriate medium in 6-well plates.
2. For ABIN-3 mRNA analysis, plate 8×10^5 subconfluent THP-1, U937 or Jurkat cells, or 2×10^5 subconfluent HeLa, HepG2, or A549 cells in 2 mL of the appropriate medium in 6-well plates.
3. Grow all cells overnight in a humidified atmosphere at 37°C and 5% CO_2.

3.2. Stimulation of Cells and Preparation of Samples for Expression Analysis of Human A20 and ABIN-3 by Western Blotting

1. Stimulate THP-1 cells with 100 ng/mL LPS or 1,000 IU/mL human TNF for different times.
2. Collect and centrifuge cells at $1{,}000 \times g$ for 5 min, remove the growth medium, and wash cells with 1 mL PBS by centrifugation at $1{,}000 \times g$ for 5 min.
3. Add 100 μL E1A lysis buffer to the cell pellet and incubate for 15 min on ice.
4. Centrifuge the sample for 15 min at $14{,}000 \times g$ to remove insoluble material and collect the supernatant.
5. Determine the protein concentration of each sample and add 25 μL 5xLL to 100 μL lysate. Boil the samples for 10 min at 95°C.

3.3. SDS-PAGE

1. These instructions assume the use of a BioRad mini gel system (type Mini-Protean 3). Use clean glassplates and rinse them with 95% ethanol and let air-dry.
2. Prepare a 1 mm thick, 10% gel by mixing 2.5 mL of 4x separating buffer, with 2.5 mL acrylamide/bis solution, 5 mL water, 62.5 μL APS, and 6.25 μL TEMED. Pour the gel, leaving space for a stacking gel (±1.5 cm). Overlay with water-saturated isobutanol to obtain a smooth surface. The gel should polymerize within 30 min.
3. Remove the isobutanol, rinse the top of the gel twice with water, and dispose all water with Whatman paper.
4. Prepare the stacking gel by mixing 2.5 mL 4x stacking buffer with 0.75 mL acrylamide/bis solution, 3 mL water, 15 μL APS, and 5 μL TEMED. Pour the stacking gel and insert the comb immediately. The stacking gel should polymerize within 30 min.
5. Once the stacking gel has set, remove the comb and fill the wells with running buffer.
6. Place the gel in the tank filled with running buffer and load 25 μg of each cell extract/well for expression of human A20 or ABIN-3. Include one well with prestained molecular weight marker.
7. Complete the assembly of the gel unit and connect to the power supply. The gel can be run at 100–200 V for 1–1.5 h.

3.4. Western Blotting for Human A20

1. The samples that have been separated by SDS-PAGE are transferred to supported nitrocellulose membranes using a semi-dry blotting system (type OWL; VWR, Belgium).
2. Three sheets of Whatman paper of the size of the separating gel are soaked in transfer buffer and laid on the surface of the blotting device.
3. The gel unit is disconnected from the power supply and disassembled. The stacking gel is removed and discarded. Then, the separating gel is laid on top of the Whatman papers on the blotting device. A nitrocellulose sheet of the size of the separating gel is soaked in transfer buffer and is placed on the surface of the separating gel on the blotting device. On top of this device, add another three Whatman papers, soaked in transfer buffer.
4. The assembly of the blotting cassette is completed and transfer can be obtained at 150 mA (dependent on the surface area of the blot; count 1 mA/cm^2) for 1.5 h.
5. After transfer, the membrane is incubated in 5 mL blocking buffer for 1 h at room temperature on a rocking platform.

6. The blocking buffer is discarded and replaced with a 1:1,000 dilution of the anti-A20 antibody in fresh blocking buffer. Incubate 2 h at room temperature on a rocking platform.
7. Discard the primary antibody (*see* **Note 4**) and wash the membrane three times for 10 min each with 10 mL blocking buffer.
8. Incubate the membrane for 1 h with a 1:2,500 dilution of the secondary antibody in blocking buffer at room temperature.
9. Discard the secondary antibody (*see* **Note 4**) and wash the membrane three times for 10 min each with 10 mL blocking buffer. Wash away the residual milk by rinsing the blot a few times with water.
10. Once the final wash is removed from the blot, 1 mL of each portion of the ECL reagents is mixed and then immediately added to the blot, which is then rotated by hand for 1 min to ensure equal coverage.
11. The ECL reagents are removed and the blot is placed between two transparent protector plastic leaves in an x-ray film cassette and exposed to an x-ray film (Hyperfilm; Amersham GE Healthcare, UK) for a suitable exposure time, typically a few minutes. An example of the results obtained in a typical experiment is shown in **Fig.1**.

3.5. Western Blotting for Human ABIN-3

1. Follow **Steps 1–5** as in **Subheading 3.4**.
2. The blocking buffer is discarded and replaced with a 1:2,000 dilution in fresh blocking buffer of the anti-ABIN-3 antibody. Incubate 1 h at room temperature on a rocking platform.
3. Discard the primary antibody (*see* **Note 4**) and wash the membrane three times for 10 min each with 10 mL blocking buffer.

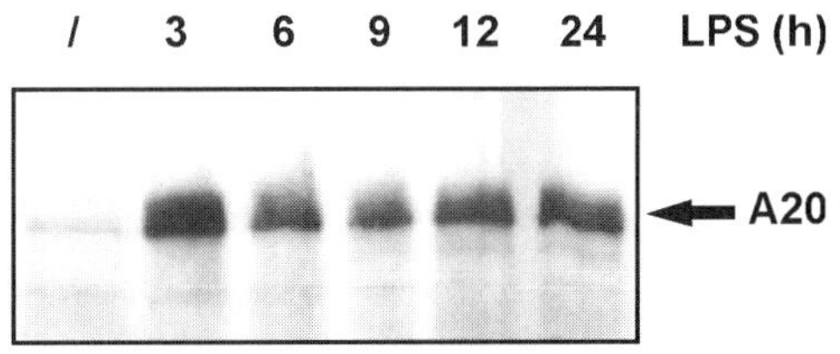

Fig. 1. Immunodetection of lipopolysaccharide (LPS)-induced A20 protein expression in THP-1 cells. Cells were left untreated or stimulated for 3, 6, 9, 12, or 24 h with 100 ng/mL LPS. Cell extracts were prepared and equal amounts of protein were separated by sodium dodecyl sulfate–polyacrylamide gel electrophoresis (SDS-PAGE), followed by Western blotting and immunodetection with anti-A20 antibodies.

4. Incubate the membrane for 1 h with a 1:2,500 dilution of the secondary antibody in blocking buffer at room temperature.
5. Discard the secondary antibody (*see* **Note 4**) and wash the membrane three times for 10 min each with 10 mL blocking buffer.
6. Follow **steps 10** and **11** as in **Subheading 3.4**. An example of the results obtained in a typical experiment is shown in **Fig. 2**.

3.6. Total RNA Isolation

1. Stimulate THP-1, U937, Jurkat, HeLa, HepG2, and A549 cells with 100 ng/mL LPS or 1,000 IU/mL human TNF.
2. Subsequently, total cellular RNA is isolated by the TRIzol®-method (Invitrogen, Carlsbad, CA). Work in a fume hood.
3. Cells are pelleted by centrifugation and lysed in 500 μL of TRIzol®Reagent by repetitive pipetting.
4. Following homogenization, the samples are incubated at room temperature for 5 min (*see* **Note 5**). Add 100 μL chloroform and shake the samples vigorously by hand for 15 s and incubate them at room temperature for 2 min.
5. Centrifuge the samples at 12,000 × *g* for 15 min at 4°C.
6. The aqueous phase is transferred to a fresh tube (*see* **Note 6**). RNA is precipitated by mixing the aqueous phase with 250 μL isopropyl alcohol. Incubate the samples at room temperature for 10 min and then centrifuge at 12,000 × *g* for 10 min at 4°C.
7. Remove the supernatant and wash the RNA pellet once with 500 μL 75% ethanol (*see* **Note 7**).

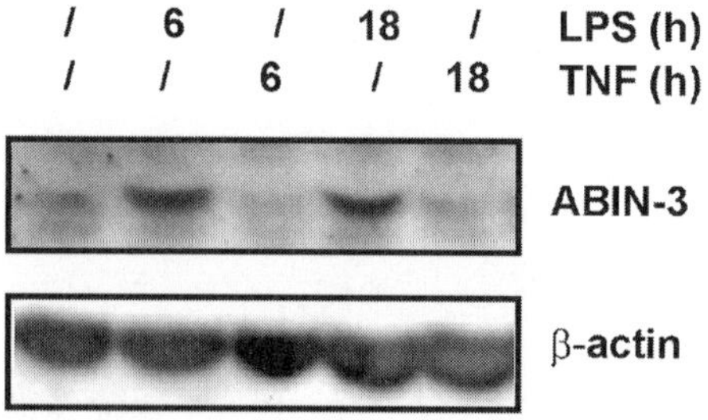

Fig. 2. Immunodetection of lipopolysaccharide (LPS)- and tumor necrosis factor (TNF)-induced ABIN-3 protein expression in THP-1 cells. Cells were left untreated or stimulated for 6 or 18 h with 100 ng/mL LPS or 1,000 IU/mL TNF as indicated. Cell extracts were prepared and equal amounts of protein were separated by sodium dodecyl sulfate–polyacrylamide gel electrophoresis (SDS-PAGE), followed by Western blotting and immunodetection with anti-ABIN-3 antibodies. Immunodetection of the same blot with anti-β-actin was used as a control for equal loading in each lane.

8. Vortex the sample and centrifuge at 7,500 × *g* for 5 min at 4°C.
9. Briefly air-dry the RNA pellet. Dissolve the pellet in 20 μL RNAse-free water and incubate the samples for 10 min at 55°C.

3.7. Semi-quantitative RT-PCR for Detection of ABIN-3 mRNA

1. First-strand cDNA is synthesized by the "Superscript™ First-Strand Synthesis System for RT-PCR" kit (Invitrogen, Carlsbad, CA).
2. Mix 1 μg total RNA, 10 mM dNTP mix, 0.5 μg oligo(dT), and add RNAse-free water to 10 μL. Incubate this mix at 65°C for 5 min, then place it on ice for at least 1 min.
3. Add 2 μL 10x RT buffer, 4 μL 25 mM $MgCl_2$, 2 μL 0.1 M DTT, and 1 μL RNaseOUT™ Recombinant RNAse Inhibitor, mix gently, and recollect by brief centrifugation.
4. Incubate at 42°C for 2 min.
5. Add 1 μL (50 units) of SuperScript™ II RT to each tube, mix, and incubate at 42°C for 50 min.
6. Terminate the reactions at 70°C for 15 min. Place it on ice.
7. Collect the reactions by brief centrifugation. Add 1 μL of RNase H to each tube and incubate for 20 min at 37°C.
8. Add 1 μL of the synthesized cDNA, 5 μL 10x PCR buffer minus Mg, 1.5 μL $MgCl_2$, 1 μL 10 mM dNTP mix, 1 μL 10 μM sense primer, 1 μL 10 μM antisense primer, 0.4 μL Taq DNA polymerase (5 U/μL), and 38.1 μL autoclaved distilled water.
9. Mix gently.
10. Perform PCR: 5 min at 94°C; 1 cycle: 30 s at 94°C, 30 s at 62°C, and 1 min at 72°C; 1 cycle: 30 s at 94°C, 30 s at 59°C,

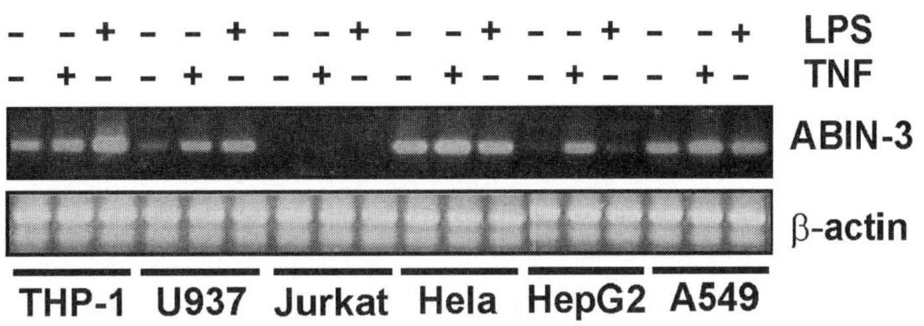

Fig. 3. Reverse transcriptase polymerase chain reaction (RT-PCR) of lipopolysaccharide (LPS)- and tumor necrosis factor (TNF)-induced ABIN-3 mRNA expression in different cell lines. Total RNA was isolated from THP-1, U937, Jurkat, HeLa, HepG2, or A549 cells that were either left untreated or stimulated for 3 h with 100 ng/mL LPS or 1,000 IU/mL TNF as indicated. RT-PCR was performed using ABIN-3-specific primers to amplify a 700 bp fragment of the ABIN-3 open reading frame. RT-PCR for β-actin was used as a control.

and 1 min at 72°C; 1 cycle: 30 s at 94°C, 30 s at 56°C, and 1 min at 72°C; 1 cycle: 30 s at 94°C, 30 s at 53°C, and 1 min at 72°C; 25 cycles: 30 s at 94°C, 30 s at 51°C, and 1 min at 72°C; 7 min at 72°C and cool down to 20°C.

11. Analyze 7.5 μL of the amplified sample using agarose gel electrophoresis. An example of the results produced in a typical experiment is shown in **Fig. 3**.

4. Notes

1. Due to instability, this solution can only be used for 1 month.
2. Water-saturated isobutanol. Shake equal volumes of water and isobutanol in a bottle and allow to separate into upper (isobutanol) and lower (water) phases overnight. Use the upper layer. Store at room temperature.
3. TBS-T was found to be the optimal buffer for detection of A20, whereas PBS-T was found to be the optimal buffer for detection of ABIN-3.
4. Incubation buffer with primary antibody for A20 and ABIN-3 or secondary antibodies can be reused up to three times if stored at −20°C.
5. After homogenization and before addition of chloroform, samples can be stored at −70°C for at least 1 month.
6. The aqueous phase is about 60% of the volume of TRIzol Reagent used for homogenization.
7. The RNA precipitate can be stored in 75% ethanol at 4°C for at least 1 week, or at −70°C for at least 1 year.

Acknowledgements

This work was supported in part by grants from the "Interuniversitaire Attractiepolen" (IAP6/18), the "Fonds voor Wetenschappelijk Onderzoek-Vlaanderen" (FWO; grant 3G010505), and the "Geconcerteerde Onderzoeksacties" of the University of Ghent (GOA; grant 01G06B6). Kelly Verhelst is supported as a predoctoral research fellow with the IWT, Lynn Verstrepen is supported as a predoctoral fellow by the FWO and the Emmanuel van der Schueren Stichting.

References

1. Kawai, T. and Akira, S. (2006) TLR signaling. *Cell Death Diff.* 13, 816–825.
2. Miggin, S. M. and O'Neill, L.A. (2006) New insights into the regulation of TLR signaling. *J. Leukoc. Biol.* 80, 220–226.
3. Wullaert, A., Heyninck, K., Janssens, S. and Beyaert, R. (2006) Ubiquitin: tool and target for intracellular NF-κB inhibitors. *Trends Immunol.* 27, 533–540.
4. Beyaert, R., Heyninck, K. and Van Huffel, S. (2000) A20 and A20-binding proteins as cellular inhibitors of nuclear factor-kappa B-dependent gene expression and apoptosis. *Biochem. Pharmacol.* 60, 1143–11451.
5. Boone, D. L., Turer, E. E., Lee, E. G., Ahmad, R.C., Wheeler, M.T., Tsui, C., Hurley, P., Chien, M., Chai, S., Hitotsumatsu, O., McNally, E., Pickart, C. and Ma, A. (2004) The ubiquitin-modifying enzyme A20 is required for termination of Toll-like receptor responses. *Nat. Immunol.* 5, 1052–1060.
6. Wullaert, A., Verstrepen, L., Van Huffel, S., Adib-Conquy, M., Cornelis, S., Kreike, M., Haegman, M., El Bakkouri, K., Sanders, M., Verhelst, K., Carpentier, I., Cavaillon, J. M., Heyninck, K. and Beyaert, R. (2007) LIND/ABIN-3 is a novel lipopolysaccharide-inducible inhibitor of NF-kappaB activation. *J. Biol. Chem.* 282, 81–90.
7. Weaver, B. K., Bohn, E., Judd, B. A., Gil, M. P. and Schreiber, R. D. (2007) ABIN-3: a molecular basis for species divergence in IL-10 induced anti-inflammatory actions. *Mol. Cell Biol.* 27, 4603–4616.
8. Wertz, I. E., O'Rourke, K. M., Zhou, H., Eby, M., Aravind, L., Seshagiri, S., Wu, P., Wiesmann, C., Baker, R., Boone, D. L., Ma, A., Koonin, E. V. and Dixit, V. M. (2004) De-ubiquitination and ubiquitin ligase domains of A20 downregulate NF-kappaB signaling. *Nature* 430, 694–699.
9. Heyninck, K. and Beyaert, R. (2005) A20 inhibits NF-kappaB activation by dual ubiquitin-editing functions. *Trends Biochem. Sci.* 30, 1–4.
10. Mauro, C., Pacifico, F., Lavorgna, A., Mellone, S., Iannetti, A., Acquaviva, R., Formisano, S., Vito, P. and Leonardi, A. (2006) ABIN-1 binds to NEMO/IKKgamma and co-operates with A20 in inhibiting NF-kappaB. *J. Biol. Chem.* 281, 18482–18488.
11. Heyninck, K., De Valck, D., Vanden Berghe, W., Van Criekinge, W., Contreras, R., Fiers, W., Haegeman, G. and Beyaert, R. (1999) The zinc finger protein A20 inhibits TNF-induced NF-kappaB-dependent gene expression by interfering with an RIP- or TRAF2-mediated transactivation signal and directly binds to a novel NF-kappaB-inhibiting protein ABIN. *J. Cell Biol.* 145, 1471–1482.
12. Van Huffel, S., Delaei, F., Heyninck, K., De Valck, D. and Beyaert, R. (2001) Identification of a novel A20-binding inhibitor of nuclear factor-kappa B activation termed ABIN-2. *J. Biol. Chem.* 276, 30216–30223.
13. Heyninck, K., Kreike, M. M. and Beyaert, R. (2003) Structure–function analysis of the A20-binding inhibitor of NF-kappa B activation, ABIN-1. *FEBS Lett.* 536, 135–140.
14. Krikos, A., Laherty, C. D. and Dixit, V. M. (1992) Transcriptional activation of the tumor necrosis factor alpha-inducible zinc finger protein, A20, is mediated by kappa B elements. *J. Biol. Chem.* 267, 17971–17976.
15. Verstrepen, L., Adib-Conquy, M., Kreike, M., Carpentier, I., Adrie, C., Cavaillon, J.M. and Beyaert, R. (2008) Expression of the NF-kappa B inhibitor ABIN-3 in response to TNF and toll-like receptor 4 stimulation is itself regulated by NT-kappa B. *J. Cell. Mol. Med.* 12, 316–329.

Chapter 14

Characterisation of Viral Proteins that Inhibit Toll-Like Receptor Signal Transduction

Julianne Stack and Andrew G. Bowie

Summary

Toll-like receptor (TLR) signalling involves five TIR adapter proteins, which couple to downstream protein kinases that ultimately lead to the activation of transcription factors such as nuclear factor-κB (NF-κB) and members of the interferon regulatory factor (IRF) family. TLRs play a crucial role in host defence against invading microorganisms, and highlighting their importance in the immune system is the fact that TLRs are targeted by viral immune evasion strategies. Identifying the target host proteins of such viral inhibitors is very important because valuable insights into how host cells respond to infection may be obtained. Also, viral proteins may have potential as therapeutic agents. Luciferase reporter gene assays are a very useful tool for the analysis of TLR signalling pathways, as the effect of a putative viral inhibitor on a large amount of signals can be examined in one experiment. A basic reporter gene assay involves the transfection of cells with a luciferase reporter gene, along with an activating expression plasmid, with or without a plasmid expressing a viral inhibitor. Induction of a signalling pathway leads to luciferase protein expression, which is measured using a luminometer. Results from these assays can be informative for deciding which host proteins to test for interactions with a viral inhibitor. Successful assays for measuring protein–protein interactions include co-immunoprecipitations (Co-IPs) and glutathione-*S*-transferase (GST)-pulldowns. Co-IP experiments involve precipitating a protein out of a cell lysate using a specific antibody bound to Protein A/G sepharose. Additional molecules complexed to that protein are captured as well and can be detected by Western blot analysis. GST-pulldown experiments are similar in principle to Co-IPs, but a bait GST-fusion protein complexed to glutathione-sepharose (GSH) beads is used to pull down interaction partners instead of an antibody. Again, complexes recovered from the beads are analysed by Western blotting.

Key words: TLR, NF-κB, MAP kinases, Transfection, Reporter gene assay, Immunoprecipitation, Viral immune evasion, Signal transduction, Poxviruses.

Claire E. McCoy and Luke A.J. O'Neill (eds.), *Methods in Molecular Biology, Toll-Like Receptors, vol. 517*
DOI: 10.1007/978-1-59745-541-1_14

1. Introduction

Accumulating evidence shows that many Toll-like receptors (TLRs) are involved in the detection of viruses, and viral proteins and nucleic acids have been shown to activate antiviral signalling pathways via TLRs, leading to cytokine and interferon induction (reviewed in *ref.1*). Important examples of such antiviral signalling pathways include activation of MAP kinases such as ERK, p38, and JNK, and of transcription factors such as nuclear factor κB (NFκB) and interferon regulatory factors (IRFs). Highlighting the importance of TLRs in antiviral immunity, a number of proteins which viruses employ against TLR signalling has been identified, and these represent anti-immunity viral subversion and evasion strategies *(1)*. For example, the Toll-IL-1 receptor (TIR) domain-containing signalling adapter molecule, TRIF, is a target of the hepatitis C virus protease, NS3/4A. The NS3/4A serine protease causes specific proteolysis of TRIF, reducing its abundance and inhibiting dsRNA signalling through the TLR3-TRIF pathway to IRF3 and NFκB *(2)*. NS3/4A can also target IPS-1 *(3)*; thus HCV can disable both the TLR3 and RIG-I detection pathways with a single protein. This potentially limits the expression of multiple host defence genes, thereby promoting persistent infections with this medically important virus *(2)*. Likewise, vaccinia virus (VACV) encodes a number of proteins, which interfere with TLR signalling, for example, A46 and A52 *(4)*. In fact, A46 was the first viral protein identified to inhibit TLR signalling. It was initially identified because it contained a signature TIR domain. Upon infection of cells by VACV, A46 is expressed by the virus and can interact with the TIR signalling adaptor molecules Mal, MyD88, TRAM, and TRIF, via its TIR domain *(5)*. This allows A46 to block multiple signalling pathways emanating from TLR complexes *(5)*. A46 can also inhibit IL-1 signalling, since IL-1 also utilises MyD88. Further, deletion of the *A46R* gene led to an attenuated phenotype in a murine intranasal model of infection *(5)*. Thus, targeting TLR signalling pathways in infected cells confers an advantage on VACV in vivo.

As more viral inhibitors of TLR signalling are being identified (e.g. by bioinformatics or by using functional screens with cloned viral genes), it is important to be able to characterise their mechanism of action. There are a number of reasons for this. Studying the effects of viral proteins on TLR signalling will generate important insights into how host cells respond to infection. Novel signalling proteins (targeted by the virus) may be identified. Viral proteins can be used as tools to probe the signalling pathways and examine the activities of host proteins. Further, viral proteins, or derivatives from them, may have potential as therapeutic agents. For example, a peptide derived from A52 was found to inhibit cytokine secretion in response to TLR-dependent signalling and to reduce bacterial-induced inflammation in vivo *(6, 7)*. This peptide may have application in the treatment of chronic inflammation initiated by bacterial or viral infections.

Using A46 as an example, this chapter describes assays that can be used to analyse the effects of a cloned viral gene on TLR signalling pathways, and to identify the TLR signalling proteins, which the viral inhibitor targets. A46, termed SalF9R in the Western Reserve strain of VACV *(8)*, was cloned by PCR amplification from WR DNA and ligated into the multiple cloning site of the mammalian expression vector pRK5 *(4)*. Firstly, cloned A46 is transfected into mammalian cells to measure its ability to modulate IL-1 and TLR signalling pathways by reporter gene assays. These reporter assays are mostly carried out in human embryonic kidney (HEK) 293 cells. However, a murine macrophage cell line, RAW264.7 cells, is also used. These cells express most TLRs *(9)* and are very useful for testing the effect of putative inhibitors on a wide variety of TLR agonists in a single experiment. A basic reporter gene assay involves the transfection of cells with a luciferase reporter gene (e.g. nuclear factor-κB (NF-κB)), a thymidine kinase (TK) *Renilla* luciferase control plasmid along with other required activating expression plasmids (e.g. CD4-TLR4), with or without a plasmid expressing A46. Induction of a signalling pathway (i.e. NF-κB activation induced by the transfection and expression of CD4-TLR4) leads to the expression of the firefly luciferase protein due to transcription of the reporter gene. Luciferase is then assayed by using a luminometer.

The Stratagene PathDetect System™ is used for MAP kinase assays. Each PathDetect *trans*-reporting system includes a unique fusion *trans*-activator plasmid that expresses a fusion protein. The fusion *trans*-activator protein consists of the activation domain of either the c-Jun, Elk1, or CHOP transcription activator fused with the DNA-binding domain (DBD) of the yeast Gal4 (residues 1–147). The transcription activators c-Jun, Elk1, and CHOP are phosphorylated and activated by c-Jun N-terminal kinase (JNK), ERK1/2, or p38 kinase, respectively, and their activity reflects the in vivo activation of these kinases and the corresponding signal transduction pathways. The pFR-luciferase reporter plasmid contains a synthetic reporter with five tandem repeats of the yeast Gal4 binding sites that control expression of the firefly luciferase gene. The DBD of the fusion *trans*-activator protein binds to the reporter plasmid at the Gal4 binding sites. Phosphorylation of the transcription activation domain of the fusion *trans*-activator protein will activate transcription of the luciferase gene from the reporter plasmid. Similar to the NF-κB reporter assays, the amount of luciferase expressed reflects the activation status of the signalling events.

Trans-reporting systems for measuring IRF3, IRF5, IRF7, and p65 activation have also been developed *(10–12)*. Similar to the Stratagene PathDetect System™, IRF3, IRF5, IRF7, and p65 fusion *trans*-activator proteins are used in conjunction with the pFR-luciferase reporter plasmid. The activity of upstream signalling events which lead to the phosphorylation of the transcription activation domains can be quantified by measuring the level of luciferase gene transcription from the reporter plasmid. In this way, a wide variety of IL-1R and TLR signals can be tested using a single end-point (luciferase expression), both by overexpressing signalling molecules and by stimulation of cells. Results from these assays will then be informative for deciding which host proteins to test for interaction with the viral inhibitor. For example, the observation that A46 blocked all IL-1 (NF-κB and MAP kinase) and TLR4 signals (NF-κB, MAPK, and IRF3) tested suggested it was acting close to the receptor complex *(5)*. Co-immunoprecipitation (Co-IP) and glutathione-*S*-transferase (GST)-pulldown experiments then demonstrated interactions between A46 and Mal, MyD88, TLR4, TRAM, and TRIF *(5)*.

Immunoprecipitation is the technique of precipitating an antigen out of solution using an antibody specific to that antigen. Co-IP experiments are a standard method to assess protein–protein interaction. By precipitating one protein believed to be in a complex, additional members of the complex are captured as well. The protein complexes, once bound to the specific antibody, are removed from the whole cell lysate by capture with an antibody-binding protein attached to a solid support such as an agarose bead. These antibody-binding proteins (Protein A, Protein G) were initially isolated from bacteria and recognise a wide variety of antibodies. Following the initial capture of a protein or protein complex, the solid support is washed several times to remove any proteins not specifically and tightly bound through the antibody. After washing, the precipitated proteins are eluted and analysed by Western blotting.

GST-pulldown experiments are another common method of examining interactions between a probe protein and a putative interaction partner. The probe protein is a GST-fusion, whose coding sequence is cloned into an isopropyl-β-D-thiogalactoside (IPTG)-inducible expression vector. This protein is expressed in bacteria and purified by affinity chromatography on glutathione-sepharose (GSH) beads. Cell lysates expressing the target protein and the GST-fusion protein conjugated to the GSH beads are incubated together and then washed to eliminate non-specific pulldown. Complexes recovered from the beads are analysed by Western blotting.

2. Materials

2.1. Cell Culture

1. Human embryonic kidney (HEK) 293T cells.
2. RAW 264.7 cells, a murine macrophage cell line.
3. Dulbecco's Modified Eagle's Medium (DMEM) supplemented with 10% FCS and 2 mM glutamine with 50 μg/ml gentamicin.
4. Trypsin-EDTA solution: 0.5 g/ml Trypsin, 0.2 g/ml EDTA.
5. T75 cell culture flasks.
6. Sterile cell culture 10-cm dishes, 6 and 96 well plates.

2.2. Transfection

1. Expression plasmids (*see* **Subheading 2.3**).
2. Sterile water.
3. Serum-free DMEM.
4. GeneJuice™ transfection reagent (Merck), a proprietary non-toxic formulation of a cellular protein and a small amount of a novel polyamine.
5. Sterile 1.5 ml Eppendorfs.
6. V-bottomed 96 well plates.
7. Flat-bottomed 96 well plates.
8. TLR ligands (*see* **Subheading 3.3.2**).

2.3. Expression Vectors

1. Empty vector, pcDNA3.1 (Invitrogen).
2. pRL-TK vector (Promega). This vector is used as an internal control reporter and can be used in combination with any experimental reporter vector. It provides low to moderate levels of *Renilla* luciferase expression in co-transfected mammalian cells. This protein functions as a genetic reporter immediately following translation.
3. NF-κB-luciferase reporter gene construct *(4)*.
4. IFN-β-luciferase reporter gene construct *(13)*.
5. IL-8 promoter reporter gene luciferase construct *(14)*.
6. RANTES promoter reporter gene luciferase construct *(15)*.
7. ISRE-luciferase reporter gene construct (Stratagene).
8. PathDetect™ CHOP, Elk-1, and c-Jun *trans*-reporting system (pFA-CHOP, pFA-Elk-1, pFA-c-Jun, pFR-Luc, pFC-MEK3, pFC-MEKK, and pFC-MEK1; Stratagene).
9. Gal 4-IRF3 *(10)*.
10. Gal 4-IRF5 *(11)*.
11. Gal4-IRF7 *(10)*.

12. Gal4-p65 *(12)*.
13. Flag-tagged hTLR3 *(11)*.
14. CD4-TLR4 *(16)*.
15. HA-tagged full-length hMal *(17)*.
16. AU1-tagged full-length hMyD88 *(18)*.
17. Flag-tagged full-length hTRAM *(10)*.
18. Flag-tagged full-length hTRIF *(19)*.
19. Flag-tagged full-length hSARM *(20)*.

2.4. Luciferase Assays

1. Luminometer capable of reading 96 well plates (Reporter™ microplate luminometer, Turner designs).
2. A 5X passive lysis buffer (Promega). Store at -20°C. Dilute to 1X using distilled water immediately prior to use.
3. Luciferase assay mix: 20 mM tricine, 1.07 mM $(MgCO_3)_4Mg(OH)_2{\cdot}5H_2O$, 2.67 mM $MgSO_4$, 0.1 M EDTA, 33.3 mM DTT, 270 mM co-enzyme A, 470 mM luciferin, 530 mM ATP. One hundred to five hundred millilitres of this mix can be made at a time and aliquoted into smaller volumes (10–50 ml). Store in the dark at -20°C and thaw to room temperature before use.
4. One milligram per millilitre of Coelenterazine (Insight Bio) in 100% ethanol, stored at -20°C. Coelenterazine is the substrate for the *Renilla* enzyme.
5. White opaque 96 well plates.

2.5. Co-immunoprecipitations

1. NP-40 lysis buffer: 50 mM HEPES, 100 mM NaCl, 1 mM EDTA, pH 7.5, 10% (v/v) glycerol, 1% (v/v) NP-40 (IGEPAL) containing 10 µl/ml aprotinin, 1 mM PMSF, and 1 mM sodium orthovanadate. PMSF is unstable in aqueous solutions and will need to be refreshed every 25 min during the lysis step.
2. Fifty per cent slurry of either Protein A- or Protein G-conjugated sepharose beads (Sigma; *see* **Note 1**).
3. A 3X laemmli sample buffer (for 10 ml): 3 ml glycerol, 0.6 g SDS, 30 mg bromophenol blue, 1.875 ml 1 M Tris-HCl pH 6.8, 5.125 ml dH_2O. Add 150 µl 1 M DTT/ml just before use. Prior to addition of DTT, 3X sample buffer can be stored at room temperature.

2.6. SDS–Polyacrylamide Gel Electrophoresis

1. Resolving gel buffer: 1.5 M Tris-HCl pH 8.8.
2. Stacking gel buffer: 1 M Tris-HCl pH 6.8.
3. Thirty per cent (v/v) acrylamide/bis solution (Protogel®; National Diagnostics). As the unpolymerised solution is a neurotoxin, gloves should be worn at all times.

4. *N,N,N,N′*-Tetramethyl-ethylenediamine (TEMED) (Sigma).
5. Ammonuim persulphate (APS): Make a 10% solution in water. For reliable results, it is best to make up this solution fresh each time as APS is quite unstable.
6. Isopropanol.
7. Resolving gel: 6 ml 30% (v/v) bisacrylamide mix, 3.75 ml 1.5 M Tris-HCl pH 8.8, 150 µl 10% (w/v) SDS, 150 µl 10% (w/v) APS, 7.5 µl TEMED made up to 15 ml with dH_2O. This is the volume required to pour two 0.75 mm 12% mini gels, and can be altered as required.
8. Stacking gel: 1 ml 30% bisacrylamide mix, 0.75 ml 1 M Tris-HCl pH 6.8, 60 µl 10% (w/v) SDS, 60 µl 10% (w/v) APS, and 6 µl TEMED made up to 6 ml with dH_2O. This is enough to pour the stacking gel for two 0.75 mm mini gels.
9. A 10X running buffer pH 8.3: 250 mM Tris, 1.92 M glycine, 10% (w/v) SDS. Dissolve 30 g Tris, 144 g glycine, and 10 g SDS in 1 l of distilled water. Check pH but do not alter due to the presence of SDS.
10. Pre-stained molecular weight marker (e.g. NEB Broad Range molecular weight markers).

2.7. Western Blotting

1. A 10X transfer buffer: 250 mM Tris, 1.92 M glycine. Dissolve 144 g of glycine and 30.3 g Tris in 1 l of water. Store at room temperature. Prior to use, dilute to 1X with the addition of water and add methanol to 20%. At this stage, the buffer may be stored at 4°C.
2. Polyvinylidene fluoride (PVDF) membrane (Millipore).
3. Phosphate-buffered saline (PBS) – Tween. For convenience, prepare a 10X solution of PBS: 687 mM NaCl, 39 mM NaH_2PO_4, 226 mM Na_2HPO_4. Dissolve 85 g NaCl, 4.68 g anhydrous NaH_2PO_4, and 32.2 g anhydrous Na_2HPO_4 in 1 l of water. Prior to use, dilute in water to 1X and add Tween 20 to 0.1% (v/v).
4. Blocking buffer: 5% (w/v) non-fat dry milk in PBS-Tween.
5. Antibody diluent: 5% (w/v) non-fat dry milk in PBS-Tween.
6. Primary antibody raised against viral protein of interest. For example, here we use anti-A46 rabbit polyclonal antibody raised against a purified, bacterial-expressed A46-GST-fusion protein. Alternatively if the protein is epitope tagged, an antibody against the epitope can be used (e.g. Flag, Myc, or HA).
7. Anti-rabbit IgG-conjugated horseradish peroxidase secondary antibody (Sigma) for polyclonal primary antibodies or

anti-mouse IgG-conjugated horseradish peroxidase secondary antibody (Sigma) for monoclonal primary antibodies.

8. Enhanced chemiluminescent (ECL) substrate. Commercial (e.g. Pierce) or home-made ECL substrate can be used. Lyophilised luminol and *p*-coumaric acid can be purchased from Sigma and used to make your own ECL substrate; 250 mM luminol stock: 0.44 g luminol in 10 ml DMSO. Store at -20°C in 1 ml aliquots, wrapped in tinfoil. Ninety millimolar *p*-coumaric acid stock: 0.15 g *p*-coumaric acid in 10 ml DMSO. Store at -20°C in 0.44 ml aliquots, wrapped in tinfoil. Solution A: 1 ml luminol stock; 0.44 ml *p*-coumaric acid; 10 ml 1 M Tris-HCl pH 8.5; make up to 100 ml with dH_2O; store wrapped in tinfoil at 4°C. Solution B: 10 ml 1 M Tris-HCl pH 8.5; make up to 100 ml with dH_2O; store wrapped in tinfoil at 4°C. Thirty per cent H_2O_2 solution is also required.
9. Stripping buffer: We use Re-Blot Plus Strong 10X solution (Chemicon). Dilute to 1X with water (*see* **Note 2**).

2.8. GST-Pulldown Assays

1. *E. coli* BL21 competent cells transformed with either pGEX or pGEX-A46.
2. Luria-Bertani (LB) broth: To 900 ml dH_2O, add 10 g Tryptone, 5 g yeast extract, and 5 g NaCl. Shake until the solutes dissolve. Make up to 1 l with dH_2O. Sterilise by autoclaving.
3. Terrific broth (TB): To 900 ml dH_2O, add 12 g Tryptone, 24 g yeast extract, and 4 ml glycerol. Adjust the volume to 1 l. Shake until the solutes dissolve. In 90 ml dH_2O, dissolve 2.31 g KH_2PO_4 and 12.54 g K_2HPO_4. Adjust the volume to 100 ml with dH_2O. Sterilise both solutions by autoclaving. Allow the solution to cool and then mix.
4. Ampicillin: Make 100 mg/ml stock in dH_2O, filter sterilise and aliquot in the hood. Store at -20°C. Use at 100 μg/ml, i.e. 1/1,000 dilution.
5. NETN extraction buffer: 300 mM NaCl, 1 mM EDTA, 20 mM Tris pH 8.0, 0.5% NP-40 (v/v) plus 1 μl/ml leupeptin (stock at 10 mg/ml), 5 μl/ml aprotinin, 10 μl/ml PMSF (100 mM stock), and 1 mg/ml lysozyme.
6. GSH beads (Sigma; *see* **Note 1**).
7. Coomassie blue stain: 50% (v/v) methanol, 10% (v/v) glacial acetic acid, 2.5 g/l Coomassie blue dye. This can be reused many times, until the blue colour has faded and does not stain as efficiently or the solution gets a bit dirty.
8. Destain solution: 50% (v/v) methanol, 10% (v/v) glacial acetic acid. This can be regenerated by charcoal treatment. This

is done by adding 10 g charcoal (Sigma) to 1 l of used destain solution. Shake the bottle and then filter using a funnel and filter paper. The filtrate can be used again immediately.

3. Methods

3.1. Growth and Maintenance of HEK 293T and RAW 264.7 Cell Lines

1. HEK 293T and RAW 264.7 cell lines are cultured in DMEM and maintained at 37°C in a humidified atmosphere of 5% CO_2. For continuing cell culture, cells are seeded at 1×10^5 cells/ml (15 ml per T75 flask) and sub-cultured two or three times a week when cells were at 50–80% confluence.
2. HEK 293T cells are removed from the surface of the flask by incubation with 5 ml of Trypsin-EDTA (0.05 mg/ml) for 1 min. Complete medium (10 ml) is then added to the cells. The contents of the flask are then transferred to a 30 ml tube and centrifuged at 110 × *g* for 5 min.
3. RAW 264.7 cells are removed from the surface of the flask using a cell scraper. The entire contents of the flask are then transferred to a 30 ml tube and centrifuged at 110 × *g* for 5 min.
4. The supernatant is removed and the pellet resuspended in 1 ml of complete medium. Cells are counted using a haemocytometer and a bright light microscope.

3.2. Transient Transfection Using GeneJuice

This is the basic protocol on which the transfections for co-immunoprecipitation and GST-pulldown experiments are based. **Subheadings 3.4.2** and **3.6.2** contain details of how much of particular plasmids are to be used in these experiments.

1. For 96 well plate transfections, cells are seeded at 2×10^4 cells per well (HEK 293) or 4×10^4 cells per well (RAW 264.7) and grown overnight (*see* **Note 3**).
2. Plasmids, serum-free medium, and GeneJuice mixes are made up in sufficient quantities for 3.5 treatments, enough therefore to perform the experiment in triplicate.
3. Cells are transfected with 230 ng (HEK 293) or 200 ng (RAW 264.7) DNA per transfection (*see* **Note 3**). Reporter plasmids such as the NF-κB-luciferase reporter or the IFNβ-luciferase reporter, TK *Renilla*, and plasmids expressing genes of interest such as A46, TLRs, the TIR adapters, or other signalling molecules are mixed together in a V-bottomed 96 well plate according to the experimental design. In all cases, the amount of DNA used per transfection is kept

constant using the appropriate amount of relevant empty vector control, pcDNA3.1.

4. The appropriate amount of GeneJuice (0.8 μl per transfection for HEK 293 cells or 0.5 μl per transfection for RAW 264.7 cells) is mixed with either 9.2 μl (HEK 293) or 9.5 μl (RAW 264.7) serum-free DMEM per transfection and incubated at room temperature for 5 min (*see* **Note 3**).
5. Thirty-five microlitres of this mixture are added to the triplicate amounts of DNA and incubated for 15 min at room temperature. Ten microlitres per well are then added to the cells, which are allowed to recover for 16 h at 37°C prior to stimulation.
6. For transfection of 10-cm plates, 8–12 μg DNA is used in combination with 15 μl GeneJuice and 235 μl serum-free DMEM.
7. All experiments are harvested at least 24 h after transfection.

3.3. Luciferase Gene Reporter Assay

3.3.1. Reporter Gene Assays

1. For the NF-κB and ISRE reporter gene assays, 60 ng of either κB-luciferase or ISRE-luciferase reporter genes are used.
2. The Stratagene PathDetect System™ is used for MAP kinase reporter assays. CHOP (0.25 ng), c-jun (0.25 ng), or Elk1 (5 ng) Gal4 fusion vectors are used in combination with 60 ng pFR-luciferase reporter to measure p38, JNK, and ERK1/2 activation, respectively.
3. For the p65, IRF3, and IRF7 transactivation assays, 1 ng p65-Gal4, 3 ng IRF3-Gal4, or 3 ng IRF7-Gal4 fusion vectors are used in combination with 60 ng pFR-luciferase reporter.
4. The IL-8, IFN-β, and RANTES assays are carried out using 60 ng of the IL-8, IFN-β, or RANTES promoter luciferase reporter genes.
5. In all assays, 20 ng of *Renilla* luciferase internal control (Promega) is used.
6. Fifty nanogram CD4-TLR4-, 25 ng Mal-, 25 ng MyD88-, 25 ng TRAM-, or 25 ng TRIF-expressing vectors are used to drive a wide variety of signals in reporter gene assays.
7. A 0.5 ng TLR3-expressing vector, in conjunction with stimulation of cells with poly(I:C), is used to examine TLR3 signals in reporter gene assays.
8. Twenty-five nanogram, 50 ng or 100 ng A46- and 100 ng SARM-expressing vectors are used in many different reporter gene assays to examine the effect of inhibitor proteins on TLR signalling pathways.

3.3.2. Stimulation of Cells

HEK293s are responsive to IL-1, TNF-α, and Phorbyl 12-myristate 13-acetate (PMA; Sigma). RAW 264.7 cells express all murine TLRs, and thus respond to Pam3Cys (Invivogen), MALP2 (Alexis Corporation), poly(I:C) (Amersham Biosciences), LPS (Alexis Corporation), Flagellin, R-848, and phosphothioate CpG DNA (Sigma), as well as IL-1, TNF-α, and PMA:

1. Cells are stimulated 18 h post-transfection with the required TLR ligands according to your experimental plan.
2. The concentrations of ligands used are as follows: 100 ng/ml IL-1, 100 ng/ml TNF-α, 125 nM MALP2, 5 mg/ml Pam3Cys, 1,000 ng/ml LPS, 250 ng/ml Flagellin, 1 μM R-848, 5 mg/ml CpG, or 10 ng/ml PMA.

3.3.3. Preparation of Cellular Lysates

1. Cells are transfected in 96 well plates as described in **Subheading 3.3**.
2. Six to eight hours post-stimulation or 24 h post-transfection (if no stimulation), medium is removed from HEK 293T cells and the cells are lysed for 15 min on a rocking platform at room temperature with 50 μl passive lysis buffer.
3. The RAW 264.7 cell line, however, is centrifuged at 1,000 × *g* for 5 min before removal of the medium. Cells are then lysed as in **step 2**.
4. Twenty microlitres of aliquots of these lysates are transferred into two opaque white plastic 96 well plates.

3.3.4. Measurement of Luciferase Activity

1. Firefly luciferase activity is assayed by the addition of 40 μl of luciferase assay mix to the 20 μl aliquots in one of the white 96 well plates.
2. Coelenterazine is diluted in PBS to give a 2 μg/ml final concentration. *Renilla* luciferase is assayed by the addition of 40 μl of this solution to the 20 μl aliquots in the other white 96 well plate.
3. Luminescence is read using either the Reporter™ (Turner designs) or Luminoskan™ (ThermoElectron Corp) microplate luminometer. Expression (or activity) levels of the firefly luciferase gene reflect the activation status of the signalling events. Expression (or activity) levels of *Renilla* luciferase reflect the levels of transfection efficiency in the experiment.
4. Firefly luminescence readings are normalised for transfection efficiency by dividing the figures with the relevant *Renilla* values and are expressed as fold stimulation over unstimulated empty vector control.
5. The white opaque 96 well plates can be reused almost indefinitely if washed thoroughly after each use.

3.4. Immunoprecipitation and Immunoblottinga

3.4.1. Antibody Pre-coupling

1. The relevant antibodies for immunoprecipitation are first pre-coupled to either Protein A- (polyclonal antibodies and Flag monoclonal antibody) or Protein G sepharose (all other monoclonal antibodies). This is done by incubation with 30 μl of 50% Protein A/G slurry per immunoprecipitation sample on a roller, overnight at 4°C. This is done by placing the 50% Protein A/G slurry/antibody mixture-containing Eppendorfs in a 50 ml tube and then putting that tube on a roller mixer, which is located in either a cold room or a cold cabinet. Two microlitres of the A46 antibody were used for each immunoprecipitation sample. For all other antibodies, 2 μg of antibody per immunoprecipitation sample should be used.
2. Just before use, the beads are washed three times to remove uncoupled antibody. This is done by centrifuging the 50% Protein A/G slurry/antibody mixture at 3,000 × *g* for 3 min and removing the supernatant. One millilitre of PBS is then pipetted into the Eppendorf, and the tube is inverted a number of times to ensure thorough separation of beads. The slurry is again centrifuged at 3,000 × *g* for 3 min and the supernatant removed. After the final wash, the beads are resuspended in the appropriate volume of PBS to ensure a 50% slurry once more.

3.4.2. Immunoprecipitation and Immunoblotting

1. HEK 293T cells are seeded into 10-cm dishes (1.5 × 10^6 cells per dish, in 15 ml medium) 24 h prior to transfection.
2. Four micrograms of each construct are transfected using GeneJuice™ as previously described. Where only one construct is expressed, the total amount of DNA (8 μg) is kept constant by supplementation with empty vector, pcDNA3.1 (*see* **Note 4**).
3. Cells are harvested 24 h post-transfection (*see* **Note 5**). The cells are washed with 1 ml room temperature PBS to remove any medium and are then removed from the plate by scraping in 1 ml ice-cold PBS. The samples are centrifuged at 16,100 × *g* for 5 min and the supernatant discarded.
4. The pellet of cells is lysed in 850 μl NP-40 lysis buffer for 30 min (*see* **Note 5**). The samples are then centrifuged for 10 min at 16,100 × *g*.
5. Four hundred microlitres of lysate are added to the relevant pre-coupled antibodies and rolled either overnight or for 2 h at 4°C. This is done by placing the cell lysate-containing Eppendorfs in a 50 ml tube and then putting that tube on a roller mixer which is located in either a cold room or a cold cabinet.
6. Twenty-five microlitres of 3X laemlli sample buffer are added to the remaining 50 μl of lysate, which is then boiled for 4 min and frozen immediately.

7. After rolling, the immune complexes are washed three times with 1 ml lysis buffer. This is done by centrifuging the immune complexes at 3,000 × *g* for 3 min and removing the supernatant. One millilitre NP-40 lysis buffer is then pipetted into the Eppendorf, and the tube is inverted a number of times to ensure thorough separation of beads. The tubes are again centrifuged at 3,000 × *g* for 3 min and the supernatant removed. After the final wash, all supernatant is removed (*see* **Note 6**) and beads are resuspended in 30 µl 3X laemmli sample buffer. The samples are boiled for 4 min and either frozen immediately or analysed by SDS-PAGE and subsequent Western blot.
8. Eighteen microlitres of the lysates are blotted for control purposes, 18 µl of the immune complex are immunoblotted for co-precipitating protein, and 10 µl are blotted for the protein directly recognised by the immunoprecipitating antibody (*see* **Note 4**).

3.5. Western Blot Analysis

3.5.1. SDS–Polyacrylamide Gel Electrophoresis (SDS-PAGE)

1. Samples are resolved on a sodium dodecylsulphate (SDS) polyacrylamide gel using a constant voltage of 150 V.
2. Samples are first run through a stacking gel to concentrate protein and then resolved according to size using 10–15% polyacrylamide gels (*see* **Note 7**).
3. Samples are run with pre-stained protein markers as molecular weight standards.

3.5.2. Transfer of Proteins to Membrane

1. The resolved proteins are transferred to Immobilon-P polyvinylidene diflouride (PVDF) membrane (Sigma) using the mini *Trans*-blot, a wet transfer system from Bio-Rad.
2. The PVDF membrane is cut to the required size (i.e. the size of the gel) and activated by placing in 100% methanol for a few seconds. It is then briefly rinsed in dH_2O.
3. All components (sponges, filter paper, PVDF membranes, and gels) are soaked beforehand in transfer buffer.
4. The Western blot "sandwich" is prepared as follows. A piece of sponge is placed on the black portion of the cassette. A layer of filter paper is laid on the sponge. The gel is placed on the filter paper and overlaid with the PVDF membrane. A second piece of filter paper is placed on the membrane, followed by a second sponge.
5. The cassette is placed in the transfer rig in the correct orientation (black-to-black and red-to-red), the chamber filled with transfer buffer, and a constant voltage of 100 V was applied for 1 h (*see* **Note 8**).

3.5.3. Antibody Blotting

1. Membranes are blocked for non-specific binding by incubation in blocking buffer at 4°C overnight or at room temperature for 1 h.
2. The membrane is washed for 5 min in PBS-Tween.
3. The membrane is then incubated for 1 h at room temperature or at 4°C overnight with the primary antibody of interest at 1:100 to 1:1,000 dilutions depending on the antibody in question (*see* **Note 9**).
4. The membrane is washed five times for 5 min in PBS-Tween and incubated with the appropriate secondary horseradish peroxidase-linked enzyme for 1 h at room temperature.
5. Again the blots are washed five times. The blots are then washed in PBS without Tween for 2 min.
6. Blots are developed by enhanced chemiluminescence using home-made ECL substrate. Just before use, equal volumes of solutions A and B are mixed as required (1 ml of each per mini blot) and 0.61 μl 30% H_2O_2 per millilitre of solution B is added. The solution is mixed gently by pipetting and then added dropwise to the blots. The blots are incubated with the ECL substrate for a minute or so. The excess solution is removed before placing the blots in a cassette for developing in the darkroom.
7. If required, blots are stripped by shaking in 10 ml 1X Re-Blot Plus (Chemicon) for 15 min. The blots are then re-blocked and probed with the desired antibodies.

3.6. GST Fusion Protein Interaction Assay

3.6.1. Preparation of GST-Fusion Proteins

1. Twenty millilitres LB broth (containing ampicillin) are inoculated with a single colony of *E. coli* BL21 cells transformed with pGEX-A46R or pGEX and incubated overnight at 37° C, 200 cycles/min.
2. The next day, the overnight cultures are diluted 1:50 (i.e. 10 ml into 500 ml broth) in ampicillin-containing terrific broth (TB) (pGEX-A46R) or LB broth (pGEX).
3. The cultures are incubated at 37°C, 200 cycles/min for approximately 2 h or until $O.D._{600}$ of the culture is approximately 0.5.
4. The cultures are then moved to a 30°C incubator and allowed to acclimate for 20 min.
5. IPTG is added to a final concentration of 70 μM and the cultures are incubated for a further 5 h (*see* **Note 10**).
6. The cultures are centrifuged using at 7,750 × *g* for 10 min, 4°C in 250 ml tubes. The supernatant is decanted and the pellets frozen until required at -70°C.

7. The pellets are left to thaw on ice and then resuspended in 15 ml NETN.
8. The resuspended pellets are transferred to 30 ml centrifuge tubes and each pellet is sonicated on ice with 20 s pulses followed by 10 s pauses to prevent over heating of the preparation, 12 times.
9. The suspensions are cleared of insoluble material by centrifugation 14,500 × *g* for 30 min, 4°C, and the supernatants pooled in a 50 ml tube.
10. Five hundred microlitres of GSH beads (50% slurry), which had previously been washed three times in 1 ml NETN plus inhibitors but without lysozyme, are added and the mixture is incubated for 2 h at 4°C on a roller.
11. The mixture is then centrifuged at 1,000 × *g* for 10 min at 4°C and the supernatant discarded.
12. Twenty-five millilitres of ice-cold NETN (plus inhibitors) are added to the beads, and the tube is rolled on a roller mixer for 5 min at 4°C before being centrifuged at 1,000 × *g* for 5 min. This washing step is repeated a further four times (five washes in total).
13. After the final wash, the beads are resuspended in the appropriate volume of NETN (plus inhibitors) to make a 50% slurry, transferred to an Eppendorf, and stored overnight at 4°C.
14. The protein-containing slurry is then analysed by SDS-PAGE and Coomassie staining for the expression of the fusion protein (*see* **Note 11**). Resolved gels are soaked in Coomassie blue stain for at least 1 h and then washed several times with destain solution until bands appear.

3.6.2. Affinity Purification of Complexes Using GST-Fusion Proteins

1. HEK 293T cells are seeded into 10-cm dishes (1.5 × 10^6 cells) 24 h prior to transfection. For GST-pulldown experiments, 8 μg of expression vector is transfected using Gene-Juice as previously described. The cells are harvested 24 h post-transfection in an identical fashion as that described in **Subheading 3.4.2**.
2. Twenty-five microlitres of 3X sample buffer are added to 50 μl of lysate, which was then boiled for 4 min and frozen immediately. Eight hundred microlitres of lysate are added to 50 μl of a 50% slurry prewashed GST-fusion protein or GST alone as a control and incubated for 2 h on a roller at 4°C as previously described.
3. The samples are washed three times with 1 ml NP-40 lysis buffer.

4. All supernatant is removed and beads are resuspended in 50 µl 3X sample buffer.
5. The samples are boiled for 4 min and analysed by Western blot.

4. Notes

1. Protein A, Protein G, and GSH beads arrive from their supplier in an ethanol suspension. Before use, swirl or agitate the container to resuspend the beads in their liquid. It is very useful to have a stock of P200 tips with their pointed ends cut off for handling the sepharose beads because the beads will block uncut tips. Remove an amount approximate or slightly more than you require and place in an Eppendorf. Centrifuge the beads gently (1,000 × g) for 1 min and remove the supernantant. Resuspend the beads in 1 ml PBS and centrifuge again. Remove supernatant, thus removing the last of the ethanol. Using your own judgement, assess the volume of beads in the Eppendorf and add the same volume of PBS. You now have a 50% slurry of Protein A or Protein G sepharose, which is stable for storage at 4°C for a number of weeks.
2. Ten millilitres 1X solution of Re-Blot Plus are sufficient for the stripping of one blot. If stored at 4°C, the solution can be reused a number of times (until the solution takes on a cloudy appearance).
3. Before attempting transfections in different cells to those used in these experiments, it is important to optimise cell numbers, GeneJuice volumes, and DNA amounts. Too much GeneJuice or DNA can be toxic to the cells. However, too little of either can result in poor transfection efficiency. Different combinations of reporter plasmids may be worth trying in other cell lines also. Cell density is another factor to optimise, as it is critical to transfection efficiency.
4. For any interaction experiments, stringent controls are vital. It is very important to demonstrate that the interaction is real, and not just an artefact, for example, due to non-specific antibody binding. For an experiment where you wish to see if protein X interacts with protein Y, the following experimental set-up is recommended. Transfect three 10-cm dishes as follows: X alone (4 µg), Y alone (4 µg), and X and Y together (4 µg of each). Lyse in 850 µl NP-40 lysis buffer as previously described. Incubate 400 µl of each lysate with Protein A/G sepharose beads conjugated to an antibody for X

(i.e. IP: X). Incubate the other 400 μl of each lysate with Protein A/G sepharose beads conjugated to an antibody for Y (i.e. IP: Y). Then load two gels in the following manner and analyse by SDS-PAGE and subsequent Western Blot. Gel 1: 18 μl lysates 1–3, 18 μl IP: X 1–3, 10 μl IP: Y. Probe this blot with the antibody for Y. Gel 2: 18 μl lysates 1–3, 18 μl IP: Y 1–3, 10 μl IP: X. Probe this blot with the antibody for X. In each of these two gels, lanes 1, 2, and 3 show expression of either X or Y in the appropriate samples and lanes 7, 8, and 9 show specific pulldown of either X or Y by their relevant antibodies. Lane 6 in both gels is where you will see an interaction or not, and you should not see a band in lanes 4 or 5 if your antibody pulldowns are specific.

5. There are a number of variations that can be attempted when optimising your immunoprecipitation experiments. You can vary the length of time for which the cells are transfected from 24 to 48 or even 72 h. For membrane-bound proteins, which can be difficult to extract, cells can be lysed for longer times such as 1 h or perhaps more. Other lysis buffers such as RIPA or modified RIPA may also be tried. However, we would advise against using too stringent a lysis buffer as you may end up disrupting your interaction.
6. When removing the final supernatant from your immunoprecipitation samples, it is extremely convenient to use gel-loading tips to remove the final traces of liquid. These tips have finely pointed ends of over 3 cm in length and are too finely bored to suck up the sepharose beads.
7. The heavy chain, or indeed the light chain, of the immunoprecipitating antibody can be a major annoyance when doing immunoprecipitation experiments if your protein of interest is of a similar or identical size to either of these proteins, as the chains appear as very strong bands on your resultant immunoblots. This "masking" of interesting bands can be circumvented in a number of ways. Firstly, try and separate your band from the antibody band as much as possible. In the case of the heavy chain, this can be achieved by slowly running your samples on a low per cent gel (8%). If it is the light chain that is the problem, running your samples on a very high per cent gel (15% or 18%) may help. Also, there are antibodies which recognise either the heavy or light antibody chain only, and these can be very useful. For example, anti-mouse IgG, Fab specific (Sigma) or anti-mouse IgG, Fc specific (Sigma). It is also possible to covalently bind your antibody to the sepharose beads using commercially available reagents.
8. Depending on the size of your protein of interest, you may want to vary the blotting conditions. For large proteins of

over 80 kDa, increasing your blotting time over the recommended 60 min may be useful. When blotting for very long time periods, it is important to prevent the apparatus from overheating. Replacing your cooling ice-block midway through the process or carrying out the procedure in the cold room are ways of maintaining optimal temperatures. For very small proteins, blotting for 30 min may mean greater retention of your protein on the membrane and better detection when developing your blots.

9. The basic factor that can be altered when optimising antibody conditions is primary antibody concentration. Optimal antibody dilutions can vary from 1/100 to 1/10,000. A good starting point is usually 1/1,000. The concentration of secondary antibody is also very important. If too much secondary antibody is used, non-specific bands may mask your results. On the other hand, blank blots or very faint bands may result if too little secondary antibody is used. Other variables are the blocking agent (powdered milk or BSA), diluent (PBS-Tween or TBS-Tween), secondary antibody concentration, and temperature of use.
10. It is a good idea to optimise your fusion protein expression conditions before attempting large-scale protein purification for experiments. Simple changes to temperature (18°C, 25°C, 30°C, or 37°C) and IPTG concentration (0.2–1 mM) may make a huge difference to protein yield and quality.
11. When you have analysed your GST-fusion protein production by SDS-PAGE, it may not be as clean as you would have hoped, i.e. there are a lot of other bands visible on the gel. By increasing your NaCl concentration in the NETN extraction buffer to 500 mM or even 750 mM and washing for longer, you will find that your extract is much purer. For future experiments, pre-clearing the bacterial lysates with sepharose beads that are not conjugated to glutathione may also improve matters.

References

1. Bowie, A. G. (2007). Translational minireview series on Toll-like receptors: recent advances in understanding the role of Toll-like receptors in anti-viral immunity. *Clin Exp Immunol* 147, 217–26.
2. Li, K., Foy, E., Ferreon, J. C., Nakamura, M., Ferreon, A. C., Ikeda, M., Ray, S. C., Gale, M. and Lemon S.M. (2005). Immune evasion by hepatitis C virus NS3/4A protease-mediated cleavage of the Toll-like receptor adapter protein TRIF. *Proc Natl Acad Sci U S A* 102, 2992–7.
3. Li, X.-D., Sun, L., Seth, R. B., Pineda, G. and Chen, Z. J. From the cover: hepatitis C virus protease NS3/4A cleaves mitochondrial antiviral signalling protein off the mitochondria to evade innate immunity. *Proc Natl Acad Sci U S A*102, 17717–22.
4. Bowie, A., Kiss-Toth, E., Symons, J. A., Smith, G. L., Dower, S. K. and O'Neill, L. A. (2000). A46R and A52R from vaccinia virus are antagonists of host IL-1 and toll-like receptor signaling. *Proc Natl Acad Sci U S A* 97, 10162–7.

5. Stack, J., Haga, I. R., Schroder, M., Bartlett, N. W., Maloney, G., Reading, P. C., Fitzgerald, K. A., Smith, G. L. and Bowie, A. G. (2005). Vaccinia virus protein A46R targets multiple Toll-like-interleukin-1 receptor adaptors and contributes to virulence. *J Exp Med* 201, 1007–18.
6. McCoy, S. L., Kurtz, S. E., Macarthur, C. J., Trune, D. R. and Hefeneider, S. H. (2005). Identification of a peptide derived from vaccinia virus A52R protein that inhibits cytokine secretion in response to TLR-dependent signaling and reduces in vivo bacterial-induced inflammation. *J Immunol* 174, 3006–14.
7. Tsung, A., McCoy, S. L., Klune, J. R., Geller, D. A., Billiar, T. R. and Hefeneider, S. H. (2007). A novel inhibitory peptide of Toll-like receptor signalling limits lipopolysaccharide-induced production of inflammatory mediators and enhances survival in mice. *Shock* 27, 364–369.
8. Smith, G. L., Chan, Y. S. and Howard, S. T. (1991). *J Gen Virol* 72, 1349–1376.
9. Applequist, S. E., Wallin, R. P. and Ljunggren, H. G. (2002). Variable expression of Toll-like receptor in murine innate and adaptive immune cell lines. *Int Immunol* 14, 1065–74.
10. Fitzgerald, K. A., Rowe, D. C., Barnes, B. J., Caffrey, D. R., Visintin, A., Latz, E., Monks, B., Pitha, P. M. and Golenbock, D. T. (2003). LPS-TLR4 Signaling to IRF-3/7 and NF-κB involves the Toll adapters TRAM and TRIF. *J Exp Med* 198, 1043–55.
11. Schoenemeyer, A., Barnes, B. J., Mancl, M. E., Latz, E., Goutagny, N., Pitha, P. M., Fitzgerald, K. A. and Golenbock, D. T. (2005). The interferon regulatory factor, IRF5, is a central mediator of TLR7 signaling. *J Biol Chem* 280, 17005–12.
12. Jefferies, C., Bowie, A., Brady, G., Cooke, E. L., Li, X. and O'Neill, L. A. (2001). Transactivation by the p65 subunit of NF-kappaB in response to interleukin-1 (IL-1) involves MyD88, IL-1 receptor-associated kinase 1, TRAF-6, and Rac1. *Mol Cell Biol* 21, 4544–52.
13. Sato, M., Suemori, H., Hata, N., Asagiri, M., Ogasawara, K., Nakao, K., Nakaya, T., Katsuki, M., Noguchi, S., Tanaka, N. and Taniguchi, T. (2000). Distinct and essential roles of transcription factors IRF-3 and IRF-7 in response to viruses for IFN-alpha/beta gene induction. *Immunity* 13, 539–48.
14. Kiss-Toth, E., Guesdon, F. M., Wyllie, D. H., Qwarnstrom, E. E. and Dower, S. K. (2000). A novel mammalian expression screen exploiting green fluorescent protein-based transcription detection in single cells. *J Immunol Methods* 239, 125–35.
15. Lin, R., Heylbroeck, C., Genin, P., Pitha, P.M. and Hiscott, J. (1999). Essential role of interferon regulatory factor 3 in direct activation of RANTES chemokine transcription. *Mol Cell Biol* 19, 959–66.
16. Medzhitov, R., Preston-Hurlburt, P. and Janeway, C. A., Jr. (1997). A human homologue of the Drosophila Toll protein signals activation of adaptive immunity. *Nature* 388, 394–7.
17. Fitzgerald, K. A., Palsson-McDermott, E. M., Bowie, A. G., Jefferies, C. A., Mansell, A. S., Brady, G., Brint, E., Dunne, A., Gray, P., Harte, M. T., McMurray, D., Smith, D. E., Sims, J. E., Bird, T. A. and O'Neill, L. A. (2001). Mal (MyD88-adapter-like) is required for Toll-like receptor-4 signal transduction. *Nature* 413, 78–83.
18. Muzio, M., Ni, J., Feng, P. and Dixit, V. M. (1997). IRAK (Pelle) family member IRAK-2 and MyD88 as proximal mediators of IL-1 signaling. *Science* 278, 1612–5.
19. Sato, S., Sugiyama, M., Yamamoto, M., Watanabe, Y., Kawai, T., Takeda, K. and Akira, S. (2003). Toll/IL-1 receptor domain-containing adaptor inducing IFN-beta (TRIF) associates with TNF receptor-associated factor 6 and TANK-binding kinase 1, and activates two distinct transcription factors, NF-kappa B and IFN-regulatory factor-3, in the Toll-like receptor signaling. *J Immunol* 171, 4304–10.
20. Liberati, N. T., Fitzgerald, K. A., Kim, D. H., Feinbaum, R., Golenbock, D. T. and Ausubel, F. M. (2004). Requirement for a conserved Toll/interleukin-1 resistance domain protein in the Caenorhabditis elegans immune response. *Proc Natl Acad Sci U S A* 101, 6593–8.

Part III

Genetic Techniques in Toll-Like Receptor Analysis

Chapter 15

Genetic Dissection of Toll-Like Receptor Signaling Using ENU Mutagenesis

Kasper Hoebe

Summary

Forward genetics has led to many discoveries and particularly in the field of Toll-like receptors (TLRs), it has played an important role in identifying key components involved in the innate sensing of pathogens. With the mouse genome fully sequenced and the ability to generate many mutant phenotypes through random germline mutagenesis, forward genetics has become an efficient means by which to identify key components involved in our immune response. In this chapter I provide a practical guide for performing germline mutagenesis in mice. I focus on the application of this technology to the identification of genes involved in TLR signaling.

Key words: Mouse, ENU, Mutagenesis, Phenotype, Genotype, TLR, Mapping.

1. Introduction

The mammalian genome consists of more than 25,000 annotated genes, most of which have been poorly described. Therefore, a major challenge in all areas of biological inquiry is to uncover gene function. Both forward and reverse genetic approaches have been successfully applied to this end. Forward (classical) genetics proceeds from phenotype to the identification of a causal genetic change (mutation). Reverse genetics, on the other hand, begins with the creation of a genetic change and ends with the identification of a phenotype.

The key to a forward genetic approach is the existence of a heritable trait resulting in distinct phenotypes. These phenotypes may be visible, e.g. white versus black coat color, but may also be

Claire E. McCoy and Luke A.J. O'Neill (eds.), *Methods in Molecular Biology, Toll-Like Receptors, vol. 517*
DOI: 10.1007/978-1-59745-541-1_15

a characteristic induced by exposure to a stimulus or challenge, e.g. the ability to cope with a viral challenge. Since the 1980s, it has been possible to establish the ultimate genetic causes of such phenotypes through positional cloning, which has led to important discoveries in many different areas of research. For example, positional cloning has made possible the elucidation of complex immune mechanisms involved in the host response against microbial pathogens *(1–4)*. In addition, forward genetics has played an important role in the identification of genes involved in the TLR signaling pathways. Foremost, the identification via positional cloning of TLR4 as the sole LPS receptor was a "breakthrough" discovery that lead to the realization that the family of TLRs serves a major role in pathogen sensing *(5)*.

More recently, the complexity of TLR signaling was further dissected through a forward genetic approach using *N*-ethyl-*N*-nitrosourea (ENU) germline mutagenesis. This approach revealed additional important components of the TLR signaling pathways, including (1) Trif, an essential adaptor molecule downstream of TLR3 and TLR4; (2) CD36, a coreceptor for TLR2/6 signaling; and (3) UNC93B, an essential component of the TLR3/7/9 signaling pathways in mice and humans. The cases of CD36 and UNC93B provide key examples of the advantages one may expect to derive from such an approach. First, phenotype is put before genotype and the need for guesswork about gene function is minimized. The involvement of CD36 as a receptor for diacylglycerides in the activation of the TLR2/6 signaling pathway was a surprise, found only after the identification of $CD36^{oblivious}$ mutation. This phenotype had gone unnoticed in mice with targeted deletions of the CD36 locus, and only later was the observed phenotype confirmed for existing CD36-knockout mice *(6)*. Thus, although targeted knockouts may exist, not all phenotypic effects of knockout mutations are examined, since intuition alone does not always guide the investigator to the correct biological function. Second, forward genetics leads to unexpected findings that would otherwise have been hard to predict. For instance the finding that UNC93B is an essential component for TLR3/7/9 signaling in mammals revealed a function quite distinct from what was previously known from work performed on UNC93 function in *Caenorhabditis elegans*. UNC93 in worms was shown to be part of a tripartite potassium channel and targeted knockouts revealed an uncoordinated body movement phenotype. Based on this phenotype, one would probably not predict an immunological function for mammalian UNC93B, as was ultimately revealed by forward genetics.

Although positional cloning, or "cloning by phenotype," was once considered an arcane art, two recent developments have significantly advanced the field. First, the determination of the mouse genome sequence has eliminated the need for contig construction and trivialized the identification of informative markers for high-resolution mapping. Second, the current availability of low cost high-throughput DNA sequencing has tremendously accelerated the process of finding mutations. Thus, the current limiting factor in positional cloning is the limited availability of strong monogenic phenotypes, which has been referred to as the "phenotype gap" *(7)*.

In mice, ENU is a powerful mutagen that at optimal doses can result in more than 1 mutation per locus per 700 gametes *(8–13)*. ENU introduces point mutations in spermatogonial stem cells, predominantly affecting A/T base pairs (44% A/T → T/A transversions and 38% A/T → G/C transitions), whereas at the protein level, ENU primarily results in missense mutations (64% missense, 26% splicing errors, and 10% non-sense mutations) *(14)*. ENU is believed to induce about one base-pair change per million base pairs of genomic DNA *(15)*.

Here, we provide a practical guide to ENU mutagenesis in mice and describe how it can uncover genes involved in TLR signaling. This entails an ex vivo approach to assess the ability of macrophages to respond to various TLR ligands. I also detail the procedures required to identify the causal mutations.

2. Materials

2.1. ENU Mutagenesis and Breeding

ENU is used in animal studies to elicit a mutagenic and/or carcinogenic response. ENU is classified as a potential human mutagen, carcinogen, and teratogen. Therefore, researchers working with the material must be aware of potential hazards while handling and disposing of the material and/or treated animals/tissues.

1. C57BL/6 male mice, 5–6 weeks old (The Jackson Laboratories, Bar Harbor, ME, USA).
2. Isopak ENU (Sigma; N3385), stored at -20°C.
3. Ninety-five percent EtOH.
4. Phosphate-citrate buffer: 0.1 M NaH_2PO_4, 0.05 M sodium citrate, pH 5.0–6.0.
5. ENU-inactivation solution: 5% $Na_2S_2O_3$ in 0.1 M NaOH.
6. Eighteen-gauge needles, 10 ml and 60 ml syringes for solution preparation.
7. Water bath with warm water.
8. One milliliter insulin syringe for injections.
9. Double gloves, face shields for injections.

10. Anesthetics: 10 mg/ml ketamine HCl (Fort Dodge, IA, USA) and 2 mg/ml xylazine (Butler Animal Health, Dublin, OH, USA) in phosphate-buffered saline (PBS) solution.

2.2. Functional Analysis of Toll-Like Receptor Signaling in Macrophages

1. Three percent thioglycolate medium, Brewer modified (Becton Dickinson, Sparks, MD, USA), prepared, and autoclaved according to manufacturer's instructions.
2. Anesthetics: 10 mg/ml ketamine HCl (Fort Dodge Animal Health, Overland Park, KS, USA) and 2 mg/ml xylazine (Butler Animal Health, Dublin, OH, USA) in PBS solution.
3. Heating pad.
4. Five milliliter syringes with 18-gauge needles containing PBS.
5. Hank's Buffered Salt Solution (HBSS), supplemented with 10% foetal bovine serum (FBS) and 2% penicillin (10,000 U/ml)/streptomycin (10,000 g/ml) (P/S) solution.
6. Dulbecco's Modified Eagles's Medium (DMEM), 5% FBS, and 2% P/S.
7. TLR ligands:
   (a) Lipopolysaccharide from *Salmonella minnesota* R595 (Re) (TLRgrade™) (liquid; # ALX-581-008-L002, Alexis Biochemicals, Lausen, Switzerland).
   (b) Poly I: poly C (Amersham, Piscataway, NJ, USA).
   (c) $Pam_3Cys\text{-}SK_4$, $Pam_2Cys\text{-}SK_4$, and MALP-2 (EMC microcollections GmbH, Tübingen, Germany).
   (d) Phosphorothioate-stabilized CpG oligodeoxynucleotide (CpG ODN): 5′-TCC-ATG-ACG-TTC-CTG-ATG-CT-3′ (Integrated DNA Technologies, Coralville, IA, USA).

2.3. Genome-Wide Mapping and Establishing a Critical Region

1. Outcross strain: C3H/HeN mice (Taconic, Hudson, NY, USA).
2. Thermocycler (i.e. Perkin Elmer) and PCR reagents.
3. A panel of microsatellite or single nucleotide polymorphisms (SNPs) informative for background and outcross strain, covering the entire mouse genome with regular intervals (~60 for backcross or ~120 for intercross mapping).
4. Microsatellites and/or SNPs to further define the critical region.
5. Appliances and software for SNP/microsatellite detection (i.e. ABI PRISM 3100/3700, MegaBACE 1000, ABI PRISM 7900H, and/or Pyrosequencing).

3. Methods

3.1. Generation of G0 ENU-Mutagenized Mice

Six-week-old male C57BL/6 mice (*see* **Note 1**) are anesthetized to prevent sudden movements and then treated with ENU administered in three weekly doses (90 mg/kg body weight) via intraperitoneal injection. *Wear protective clothing, double gloves and face shield, and perform this procedure in a fume hood.*

3.1.1. Preparation of ENU Solution and ENU Injection

1. Before preparation, bring ENU Isopak to room temperature.
2. Inject 10 ml of 95% EtOH using an 18-gauge needle and 10 ml syringe into Isopak. Properly dispose of the needle and syringe immediately.
3. Swirl the Isopak in the warm water bath until ENU powder is entirely dissolved.
4. Add 90 ml phosphate-citrate buffer using 60 ml syringes and 18-gauge needles.
5. ENU is easily degraded and the exact concentration needs to be determined by measuring the absorbance at $\lambda = 398$ nm using a spectrophotometer, and calculate the concentration using the formula: $0.72OD_{398\ nm} = 1$ mg/ml of ENU.
6. After preparation of the ENU solution, anesthetize mice with xylazine/ketamine injection, 100 μl intramuscularly. While they are asleep, weigh them and inject 90 mg/kg of ENU intraperitoneally.
7. Leave treated mice overnight in disposable cages in the fume hood.

3.1.2. Breeding of G1 and G3 Mice

After the last dose, mice are placed in isolation (one mouse per cage) for 12 weeks to allow recovery of fertility. Fertility testing is performed at 8 weeks to assure that mutagenized mice are sterile. Fertility at this stage indicates that a sub-optimal dose of ENU was administered. After the recovery period, each G0 male is bred to untreated, wild-type C57BL/6 female mice to generate a maximum of 20 G1 offspring (*see* **Note 2**). These G1 animals are used either for phenotypic screens or to produce G2 mice, which in turn are backcrossed to the G1 to generate G3 offspring (*see* **Fig.1**). Usually G1 males are used to propagate mutations to homozygosity, and two daughters are backcrossed to each G1 founder.

3.2. Considerations for Phenotypic Screening of ENU-Mutagenized Mice

Phenotypic screening assays should have limited variance, should target a large genomic footprint if possible (i.e. many genes with non-redundant function), and should ideally probe a poorly understood phenomenon. Both G1 and G3 mice can be used in

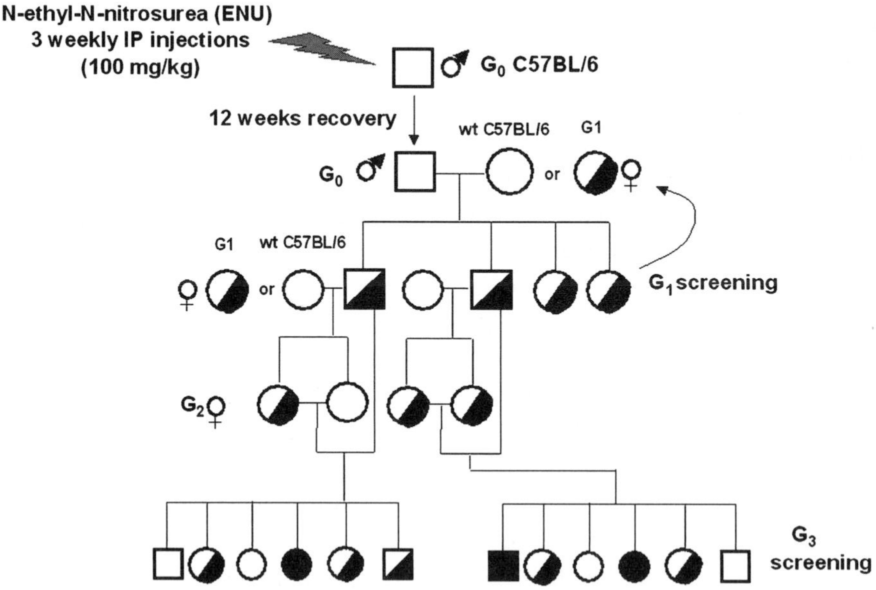

Fig. 1. ENU treatment of C57BL/6 mice and micropedigree construction for screening of dominant (G1) and recessive (G3) phenotypes. Each G1 male is bred with two G2 females and subsequently from each G2 female 3 G3 offspring are tested in recessive screens.

any screen to identify phenodeviants (*see* **Note 3**). A non-lethal screen is advantageous in that it allows for secondary confirmation and breeding of phenodeviants directly, but lethal screens of G3 mice are feasible if the G1 and G2 parents are retained to propagate mutations. Here, we provide a protocol for the screening of peritoneal macrophages isolated from mutagenized mice that has successfully been applied to identify genes involved in TLR signaling.

3.3. Functional Analysis of Toll-Like Receptor Signaling in Macrophages

3.3.1 Isolation and Plating of Peritoneal Macrophages

1. Inject 7- to 8-week-old mice with 1 ml of 3% thioglycolate solution intraperitoneally (IP).
2. After 3 days, anesthetize mice by injecting 100 µl of ketamine/xylazine solution intramuscularly (IM).
3. Harvest peritoneal macrophages through lavage of the peritoneum with 5 ml of sterile PBS using a syringe and 18-gauge needle, and allow the mice to recover in cages on heating pads when finished.
4. Collect cells in 15 ml conical polypropylene tubes containing 5 ml of HBSS medium and centrifuge at 500 × *g* for 10 min.
5. Resuspend cells in DMEM medium and bring to a final density of 1×10^6 cells/ml.
6. Add 100 µl of cell suspension per well to 96-well flat-bottom plates and after 1 h of incubation at 37°C (in 95% O_2: 5%

CO_2) expose cells to TLR ligands at a sub-optimal concentration (for dose–response curves see **ref.** ***16***). After an additional 4 h incubation, supernatants are collected and either stored at -80°C or used directly for measurement of secreted cytokines. For macrophage screening, measurements of TNFα and type I IFN are reliable endpoints covering most signaling pathways downstream of TLRs.

3.4. Confirmation of the Phenotype and Generation of a Homozygous Stock

The TLR screen allows for breeding of phenodeviants directly with C57BL/6 mice, after a phenodeviant has been identified. Subsequently, a backcross approach will be performed with the identified phenodeviant, after which 50% of the G3 offspring is expected to show the phenotype. This approach has two advantages: (1) an outcross with wild-type C57BL/6 mice will remove non-related mutations; (2) it allows for an assessment of the phenotype to be dominant or recessive by testing transmissibility in the F1. Ultimately, homozygous breeding pairs should be set up to fix the mutation. Fixation, verified by expanding an inbred line and testing for uniform susceptibility, is achieved as a prelude to mapping the mutation.

3.5. Genome-Wide Mapping and Defining the Critical Region

Once a mutation is fixed, it has to be first mapped to a chromosomal interval and then to a small region of the chromosome with a limited number of genes (the "critical region"). Direct sequencing of all the genes in the critical region is then performed to find the mutation that causes the phenotype. In the vast majority of cases, phenotypes emanate from changes in the amino acid sequence of proteins, resulting from missense, non-sense, or splicing mutations. Therefore, coding exons or the coding regions of cDNAs should be first examined to find the causative mutation. Since ENU creates mutations with low frequency (only one per million bp of DNA) and since coding regions comprise only about 1.5% of total genomic DNA, a small critical region (1–2 Mb in length) is very unlikely to have more than one coding change.

3.5.1. Choosing an Outcross Strain and Breeding Strategies

To map a mutation affecting TLR signaling, the mutant stock (C57BL/6 background) can be outcrossed to a second *Mus musculus* strain (e.g., C3H/HeN) and the F1 hybrids that result are backcrossed either to the mutant stock (for a recessive mutation) or to the outcross strain (for a dominant mutation). In choosing the strain used as a mapping partner, it is important to ensure that there is a strong phenotypic similarity between this partner and normal C57BL/6 mice and, moreover, that hybrid animals are similar to both parents. Outcrosses between C57BL/6J and C3H/HeN have proven to be effective for the mapping of TLR-related phenotypes in that limited differences in qualitative and quantitative ligand-induced TLR-responses are observed between these strains. Most *Mus musculus* strains have sufficient genetic variation (SNPs or microsatellite markers) to permit the mapping of a mutation produced on the C57BL/6 background, and even C57BL/10 mice, which are very similar to C57BL/6, can now be used for mapping at least to a chromosomal interval. Recessive mutations can also be mapped using an intercross strategy: i.e. the F1 hybrids are mated with each other rather than backcrossing them to the mutant stock. This will result in F2 mice that contain twice the number of meioses. Each F2 mouse is therefore more informative, but a higher density of markers throughout the genome is required in order to assure linkage between the phenotype and one or more markers in the panel. Once F2 offspring are produced they are tested for the phenotype of interest and genomic DNA is collected from each mouse.

3.5.2. Establishing Chromosomal Linkage for a Phenotype

The mutagenized background (i.e. C57BL/6) and outcross strains (i.e. C3H/HeN) are both inbred strains and thus homozygous at all loci (with the exception of the ENU-induced mutations on the C57BL/6 background). The C57BL/6 mice can be distinguished from C3H/HeN mice through germline differences at more than one million loci scattered along the chromosomes. These differences may be considered as potential markers and in an F2 mouse, can be used to trace the parental origin of a chromosomal interval. These markers can involve single base pair differences (single nucleotide polymorphisms; SNPs) or differences in the length of simple repetitive sequences or microsatellites (poly[CA], poly[AG], or poly[GATA], e.g.) (*see* **Note 4**). In the case of a backcross approach, F1 hybrids (heterozygous for all markers) are crossed to the homozygous C57BL/6 mutant stock. We can therefore expect each of the offspring to be homozygous C57BL/6 at about half of the marker loci and heterozygous at the remaining loci. To localize the mutation, all animals of the F2 generation are examined phenotypically to determine whether they are homozygous or heterozygous for the mutation of interest. In addition, they are examined genotypically (*see***Note 5**) to determine whether they are homozygous or heterozygous for each of the genome-wide markers. Concordance between presence of the phenotype and homozygosity for a marker, and absence of the phenotype and heterozygosity for that same marker, is evaluated for each of the markers in the mapping panel. An LOD (Log Odds Distance) score is calculated for each marker, and is used as an objective index of linkage. Usually, the analysis of ~20 mice (including those with phenotype and those without) is sufficient to obtain significant linkage to one or more of the chromosomal markers:

The LOD score (Z) developed by Newton E. Morton in 1955 is equivalent to (probability of a linkage)/(probability of non-linkage)

between the mutation and the marker. Given non-linkage, we expect 50% concordance between marker genotype and the phenotype of interest.

The LOD score can be calculated for each genome-wide marker using the formula

$$Z(\Phi)=\log[(1-\Phi)^{n-r}\Phi^{r}/(1/2)^{n}]$$

where Φ = fraction that is concordant, $1 - \Phi$ = fraction that is disconcordant, n = the total number of observations (mice), $(n - r)$ = number of discordant mice, and r = number of concordant mice. In principle, an LOD score of 3 or higher is considered as significant ($P < 0.01$) linkage, but ultimately one should aim for an LOD score of 4 or 5 to be absolutely confident. In a backcross, using a panel of 60 evenly spaced markers, 20–30 F2 mice are usually sufficient to achieve this.

3.5.3. Further Confinement of the Mutation to a "Critical Region"

Once significant linkage to one or more markers on a chromosome has been obtained, new markers within the chromosomal interval are examined. At this point, the original F2 as well as additional F2 animals should be included in the analysis of chromosomal markers so that a "critical region" can be established. This can be done by identifying one proximal and one distal marker that are each separated from the mutation by at least one crossover event. In a cross between two *Mus musculus* strains, one centiMorgan (cM; 1% crossover frequency) corresponds to approximately 1.6 Mb of DNA, the average critical region that is derived from 100 G2 mice would theoretically correspond to approximately 1.6 Mb in length, given a high density of informative markers. Many thousands of SNP and microsatellite markers have been published (*see* **Subheading 3.4.6**, **item 3**), but new informative microsatellite markers may often be found by randomly sequencing CA or GATA repeats as needed (*see* **Note 4**).

The mutation is confined to a region small enough to permit sequencing of all coding exons within the region, and for some laboratories, this will mean tighter confinement than it will for others. Candidate genes may be examined either at the cDNA level or the genomic level. The average Mb of genomic DNA contains 12 genes and the average gene contains 8.8 exons and, hence, approximately 100 exons with coding function. However, it is often found that many more genes reside within a small critical region, because genes tend to be clustered rather than randomly dispersed in the genome.

3.5.4. Identification of the Causative Mutation Within the Critical Region

In some instances, a "best" candidate gene presents itself for sequencing and the causal mutation is readily found. In other cases (perhaps the most interesting cases) there is a little basis for choice, and prioritization is difficult. Sequencing of candidate genes can be carried out at the cDNA level (preferable if a gene contains many exons and a small number of splice forms) or at the genomic level (if there are many exons and an unknown number of splice forms, or if there is uncertainty as to whether the mRNA is expressed). In addition to sequencing coding regions, the expression levels of the genes in the critical region can be assessed by Northern blot or quantitative PCR analysis. This may be helpful to identify mutations outside the coding region that affect the expression of a gene and that would otherwise be missed (fortunately a very rare circumstance).

3.5.5. Useful Tools and Web Sites to Efficiently Identify a Mutation in the Critical Region

To identify the mutation in a critical region, a variety of bioinformatic tools can help prioritize candidate genes and/or are essential for successful analysis of genomic information. Here, we describe some of the programs that are freely accessible and their specific applications to efficiently analyze any critical region:

1. *Ensembl* (http://www.ensembl.org/) Ensembl is a genome browser, jointly created by the EMBL-European Bioinformatics Institute (EBI) and the Sanger Institute, containing databases of selected eukaryotic genomes. The web site provides essential information on genomic sequences as well as available microsatellite markers and/or SNPs for any given region. The databases are regularly updated and as of this writing, annotation changes frequently. Other genome browsers include the Vertebrate Genome Annotation database (also known as VEGA, http://vega.sanger.ac.uk/index.html), which supplies manual annotation for specific regions of the mouse genome, and the NCBI Map Viewer (http://www.ncbi.nlm.nih.gov/mapview/map_search.cgi?taxid= 10090).
2. *National Center for Biotechnology Information (NCBI)* (http://www.ncbi.nlm.nih.gov/)This site contains genomic databases and software tools for analyzing genome data. In addition, NCBI contains a web-based catalog that contains thousands of entries for genes and genetic disorders and serves as a phenotypic companion to the Human Genome Project (OMIM).
3. *Repeatmasker (developed by A.F.A. Smit, R. Hubley & P. Green*; http://repeatmasker.org) This program is essential for processing sequencing data derived from the public databases (NCBI, Ensembl) or the Celera database. It screens DNA sequences for interspersed repeats and low-complexity DNA sequences. The output of the program is a detailed annotation of the repeats that are present in the query

sequence as well as a modified version of the query sequence in which all the annotated repeats have been masked (masked nucleotides are depicted as Ns). This program is useful for the design of unique amplification primers for genomic DNA.

4. Web sites for identification of SNPs/microsatellite markers:

   http://www.ncbi.nlm.nih.gov/projects/SNP/MouseSNP.cgi

   http://www.broad.mit.edu/snp/mouse/,and http://snp.gnf.org/,which also contains an extensive gene expression atlas

5. *The mouse phenome database* (http://phenome.jax.org/pub-cgi/phenome/): A broad collection of phenotypic and genotypic data for laboratory mice.

4. Notes

1. In principle any strain can be used for mutagenesis but one should consider the availability of genomic information as well as relevant background mutations that may affect the phenotype that will be tested. The sensitivity of the strain to ENU should be examined if it is not already known (considerable interstrain variation has been observed). The C57BL/6 strain has been a popular choice for many ENU laboratories.

2. An alternative breeding scheme to increase mutational rate is by using previously generated G1 females as breeders for G0 males or to breed G1 males with G1 females. These breeding schemes not only increase the mutation rate, but also allow for analysis of X-linked genes.

3. With one bp change per million bp and a total length of ~2,998 Mb for the mouse genome, one can calculate that each G1 male carries ~3,000 bp changes genome-wide. With the coding region being 1.3% of total genomic sequence and 76% of random bp changes creating a coding change, it follows that each G1 mouse carries about 30 coding changes genome-wide. These exist in a heterozygous form and do not necessarily lead to phenotypic changes. Many ENU-induced mutations are recessive or codominant at best, and phenotypic changes are then observed only in homozygotes. For any ENU mutagenesis project, it is important to decide whether one will generate and screen G1 mice only (dominant mutations) and/or whether one will generate and screen G3 mice (recessive mutations). In the G3 generation, each mouse will carry ~4 mutations in homozygous form. Although increasing the numbers of G3 mice for screening

of each pedigree will ultimately reveal all homozygous mutations, a point of diminishing return is reached, and many small pedigrees are more cost-effective than a few large ones for revealing phenotype. The probability of transmission of each G1 mutation to homozygosity can be calculated as $P = 1-[(0.5)+(0.5)*(3/4)^{g_3}]^{g_2}$, where g_2 is the number of G2 and g_3 is the number of G3 offspring per G2 mother. If pedigrees are constructed to produce two G2 daughters per G1 male, and three G3 pups per G2 daughter are directed to a given screen, ~50% capture of G1 mutations in homozygous form is achieved.

4. In some cases, no published markers reside within the critical region yet further meiotic mapping is desirable. New markers must then be found. First it is essential to determine whether the region is stringently conserved between the strains that were used in the mapping cross. An SNP database (e.g., http://snp.gnf.org/) can disclose whether coinheritance has occurred within the region. If coinheritance has not occurred, it may be most efficient to sequence parts of the region at random, whereon one can expect to find an SNP every 1–2 kb. If coinheritance has occurred, one can still look for simple sequence length polymorphisms by analyzing simple repeats such as CA (=TG), TA (=AT), and CT (=GA) or GATA (=TATC). GC repeats are exceedingly rare. All repeats can be polymorphic, but the frequency of polymorphisms is, in our experience, GATA > CA > CT > TA. On average, about 1 in 6 CA repeats is informative when two disparate *Mus musculus* strains are compared.
5. There are numerous DNA preparation/isolation kits commercially available. One important consideration may be the ability to apply the isolation method to high-throughput analysis of genomic DNA samples.

References

1. Brunkow, M. E., Jeffery, E. W., Hjerrild, K. A., Paeper, B., Clark, L. B., Yasayko, S. A., Wilkinson, J. E., Galas, D., Ziegler, S. F., and Ramsdell, F. (2001) Disruption of a new forkhead/winged-helix protein, scurfin, results in the fatal lymphoproliferative disorder of the scurfy mouse. *Nat. Genet.* 27, 68–73.
2. Lee, S. H., Webb, J. R., and Vidal, S. M. (2002) Innate immunity to cytomegalovirus: the Cmv1 locus and its role in natural killer cell function. *Microbes. Infect.* 4, 1491–1503.
3. Scalzo, A. A., Wheat, R., Dubbelde, C., Stone, L., Clark, P., Du, Y., Dong, N., Stoll, J., Yokoyama, W. M., and Brown, M. G. (2003) Molecular genetic characterization of the distal NKC recombination hotspot and putative murine CMV resistance control locus. *Immunogenetics* 55, 370–378.
4. Webb, J. R., Lee, S. H., and Vidal, S. M. (2002) Genetic control of innate immune responses against cytomegalovirus: MCMV meets its match. *Genes Immun.* 3, 250–262.
5. Poltorak, A., He, X., Smirnova, I., Liu, M. Y., Van Huffel, C., Du, X., Birdwell, D., Alejos, E., Silva, M., Galanos, C., Freudenberg, M., Ricciardi-Castagnoli, P., Layton, B., and

Beutler, B. (1998) Defective LPS signaling in C3H/HeJ and C57BL/10ScCr mice: mutations in Tlr4 gene. *Science* 282, 2085–2088.

6. Hoebe, K., Georgel, P., Rutschmann, S., Du, X., Mudd, S., Crozat, K., Sovath, S., Shamel, L., Hartung, T., Zahringer, U., and Beutler, B. (2005) CD36 is a sensor of diacylglycerides. *Nature* 433, 523–527.
7. Balling, R. (2001) ENU mutagenesis: analyzing gene function in mice. *Annu. Rev. Genomics Hum. Genet.* 2, 463–492.
8. Favor, J., Neuhauser-Klaus, A., and Ehling, U. H. (1998) The effect of dose fractionation on the frequency of ethylnitrosourea-induced dominant cataract and recessive specific locus mutations in germ cells of the mouse. *Mutat. Res.* 198, 269–275.
9. Favor, J., Neuhauser-Klaus, A., Ehling, U. H., Wulff, A., and van Zeeland, A. A. (1997) The effect of the interval between dose applications on the observed specific-locus mutation rate in the mouse following fractionated treatments of spermatogonia with ethylnitrosourea. *Mutat. Res.* 374, 193–199.
10. Rinchik, E. M., Carpenter, D. A., and Selby, P. B. (1990) A strategy for fine-structure functional analysis of a 6- to 11-centimorgan region of mouse chromosome 7 by high-efficiency mutagenesis. *Proc. Natl. Acad. Sci. U.S.A.* 87, 896–900.
11. Rinchik, E. M. and Carpenter, D. A. (1999) N-ethyl-N-nitrosourea mutagenesis of a 6- to 11-cM subregion of the Fah-Hbb interval of mouse chromosome 7: Completed testing of 4557 gametes and deletion mapping and complementation analysis of 31 mutations. *Genetics* 152, 373–383.
12. Russell, W. L., Hunsicker, P. R., Carpenter, D. A., Cornett, C. V., and Guinn, G. M. (1982) Effect of dose fractionation on the ethylnitrosourea induction of specific-locus mutations in mouse spermatogonia. *Proc. Natl. Acad. Sci. U.S.A.* 79, 3592–3593.
13. Russell, W. L., Hunsicker, P. R., Raymer, G. D., Steele, M. H., Stelzner, K. F., and Thompson, H. M. (1982) Dose–response curve for ethylnitrosourea-induced specific-locus mutations in mouse spermatogonia. *Proc. Natl. Acad. Sci. U.S.A.* 79, 3589–3591.
14. Justice, M. J., Noveroske, J. K., Weber, J. S., Zheng, B., and Bradley, A. (1999) Mouse ENU mutagenesis. *Hum. Mol. Genet.* 8, 1955–1963.
15. Concepcion, D., Seburn, K. L., Wen, G., Frankel, W. N., and Hamilton, B. A. (2004) Mutation rate and predicted phenotypic target sizes in ethylnitrosourea-treated mice. *Genetics* 168, 953–959.
16. Hoebe, K., Du, X., Goode, J., Mann, N., and Beutler, B. (2003) Lps2: a new locus required for responses to lipopolysaccharide, revealed by germline mutagenesis and phenotypic screening. *J. Endotoxin. Res.* 9, 250–255.

Chapter 16

Microarray Experiments to Uncover Toll-Like Receptor Function

Harry Björkbacka

Summary

This chapter is intended as a handbook for anyone interested in using microarrays to study Toll-like receptor (TLR) function or any other biological question. Although microarray technology has developed into a standard tool at many laboratories disposal, most of the actual microarray processing is done by core facilities using highly specialized equipment. This chapter only briefly describes these methods in principle and instead focus on the parts that investigators themselves can influence, such as the experimental design, RNA isolation, statistical analysis, cluster analysis, data visualization, and biological interpretation.

Key words: Microarray, RNA isolation, Clustering, Experimental design, Normalization, Statistical analysis.

1. Introduction

The knowledge of which genes are induced by pathogens and Toll-like receptor (TLR) ligands have been greatly expanded using microarray technology starting with the cataloguing of immune responses in *Drosophila* *(1, 2)* and later in immune cells such as macrophages *(3, 4)* as well as in whole animals infected with various pathogens *(5, 6)*. The role of TLRs in diseases other than infectious diseases, such as atherosclerosis *(7)* and rheumatoid arthritis *(8)*, has also been evaluated with the aid of microarray experiments. Microarrays can be used to

Claire E. McCoy and Luke A.J. O'Neill (eds.), *Methods in Molecular Biology, Toll-Like Receptors, vol. 517*
DOI: 10.1007/978-1-59745-541-1_16

map TLR-activated pathways by studying effects on the global gene expression pattern upon manipulation and modulation of TLR pathway components, such as has been done in cells lacking functional MyD88 *(9)*, TRIF *(10)*, and IFN-β *(11)*. Also, microarray analysis has uncovered clusters of TLR-regulated genes that have been shown in follow-up studies to be coregulated by a common transcription factor *(12)*. Microarray experiments are ideally suited to study TLR-activated pathways with profound effect on a cell expression pattern that is not simply characterized by a few marker cytokines. Also, microarray experiments can uncover novel genes, functions, and interactions not easily discovered by other methods.

1.1. Microarray Platforms

The first choice one faces as an investigator is to choose microarray platform. This choice dictates how the experiment will be designed. Essentially there are two choices: (1) one-color microarrays or high-density oligonucleotide microarrays (e.g. Affymetrix GeneChips) or (2) two-color microarrays or spotted oligonucleotide/cDNA microarrays. These and other available microarray platforms are discussed further elsewhere *(13, 14)*.

1.1.1. Two-Color Microarrays

In two-color microarrays, a test RNA sample is labeled with one fluorescent dye (e.g. Cy3) while a control RNA sample is labeled with another fluorescent dye (e.g. Cy5) and both samples are cohybridized to the microarray. The output data will be expression ratios of test over control sample mRNA expression. Thus two-color arrays do not measure levels of expression but rather a differential expression between two samples. The cohybridization of two samples effectively eliminates possible manufacturing differences between one microarray and the next and yields robust expression ratios. Two-color arrays are manufactured by depositing reporter nucleic acids in defined locations on a solid support using a robotic printer. The solid support is usually a glass microscope slide with a surface optimized for binding of the reporter nucleic acid and subsequent hybridization with the labeled target nucleic acid. The reporter nucleic acid can be a PCR-amplified product from a clone library or a synthesized nucleic acid. The reporter nucleic acid is robotically printed on the microarray with pins that upon contact release a certain volume of nucleic acid (contact printer) or by spraying the nucleic acid onto the microarray with a sort of ink-jet printer (noncontact printer). The first arrays produced were cDNA arrays, but nowadays, spotted oligonucleotide microarrays are most common and several companies sell oligonucleotide reporter libraries or already printed microarrays. A microscope slide can fit about 40,000–50,000 oligonucleotide reporters. Detailed protocols for producing spotted microarrays in an academic core laboratory can be found elsewhere *(15)*.

1.1.2. One-Color Microarrays

Affymetrix GeneChips are manufactured by a photolithographic technique to carry out parallel synthesis of a large number of oligonucleotides in situ on a solid surface *(16)*. In this technique, protective groups on growing oligonucleotide chains are removed by exposure to light through photolithographic masks allowing for site-specific reaction of the next nucleotide. Thus this technique, developed by computer chip manufacturers, can produce a large number of oligonucleotide reporters on a small surface ($>10^6$ sequences/cm^2). Currently Affymetrix Inc supplies GeneChips with about 29,000 human genes for expression analysis. Each gene is represented by multiple 25-mer oligonucleotide reporters targeting multiple regions of the target gene and they are grouped in pairs with either complete complementarity to their target sequences (perfect match (PM) probes) or with a single mismatch centered in the middle of the oligonucleotide (mismatch (MM) probe). This increases sensitivity and specificity of the relatively short oligonucleotide probes on a GeneChip.

1.1.3. Pros and Cons of the Platforms

The commercial Affymetrix platform is fixed and makes direct comparison of data obtained in different labs easier. Even so, studies have shown that lab-to-lab variability is an issue to consider *(17)*. Also, the relative ratios obtained by two-color microarrays may be more comparable then the absolute measures of one-color microarray *(17)*. Spotted microarrays can be more flexible as reporters of choice can be incorporated on the microarray. The Affymetrix GeneChip itself tends to be more expensive than spotted microarrays, but has the advantage that only one sample is hybridized to each microarray and thus the control samples are hybridized only once, while in two-color microarrays a control sample is hybridized together with the test sample on each microarray. This could be important if the control samples are limiting. Affymetrix requires less total RNA (about 0.1–1 µg) for a standard hybridization than two-color microarrays (about 5–20 µg). However, amplification methods can reduce the RNA requirement substantially for both methods (*see* **Subheading 3.2.2**).

1.2. Experimental Design

The experimental design needs to be considered carefully *(18)*. This point really needs to be emphasized. A microarray experiment does not differ from ordinary experiments in any major way other than that microarray experiments cost more than most experiments, and thus one cannot afford to not plan the experiment properly. In general, the same rules for biological and technical replication of your regular experiment also apply for a microarray experiment. If you want high confidence in your microarray data, you should spend time and thought on designing a well-planned experiment with adequate replication. If on the other hand you want a global overview or snap shot of the expression profile to be used more to generate new ideas and hypotheses, then fewer replicates may suffice, but you may have to accept wasting time on a few false positives in the list of interesting genes pursued with other experimental methods. There are many possible designs that can be employed in microarray analysis, such as direct comparison of samples, comparison to a reference sample or complex factorial designs *(18–20)*.

1.2.1. Simple Designs

In one-color experiments, each sample is hybridized to a single array. This gives the investigator freedom to compare the samples to each other in several different ways. In a two-color experiment, on the other hand, one has to decide which samples to hybridize together on the same array, and thus the experimental design becomes more important. The simplest design comparing two conditions is a design where the two conditions, such as untreated control cells (control) and TLR ligand-treated cells (treated/experimentally modulated), are compared directly to each other on the same array (**Fig.1A**). It is also possible that no logical control exists. For instance, it may not be ethical or valid scientifically to remove healthy tissue from subjects as control to gene expression in a diseased tissue or tumor. In such an experiment, it is possible to introduce a reference sample to which all diseased tissue expression profiles are compared allowing differences in expression among different patient groups to be uncovered (**Fig.1B**). Although the reference sample design can be used to compare any samples to each other by their relation to the reference sample, it may be better to still do a direct comparison if the experiment is aimed at comparing the two samples. Reference samples where RNA from multiple cell types and tissues has been pooled are commercially available for several species. It is also possible to combine the direct comparison and the reference design (**Fig.1B**).

1.2.2. Pooling and Biological Replicates

The subject of sample pooling is frequently discussed in experimental design. Generally, pooling should be avoided. Pooling may be tempting when the RNA source or the experimental budget is limiting. Comparing pooled samples hybridized to several arrays will reduce the array-to-array variability, but nothing is learned about the biological variability. For instance, a gene could be picked out as highly significant in four technical replicates of a pooled sample by the statistical analysis, but in actuality one of the pooled biological replicates could dominate the expression of the sample pool. If instead biological replicates are studied, this deviating biological sample could be identified and a better evaluation of the biological relevance of this gene made. A hybrid between pooling and separate biological replicates is to pool the control samples but to keep the treated samples as separate biological replicates. An often asked question is how many biological replicates one should perform as the cost of microarray

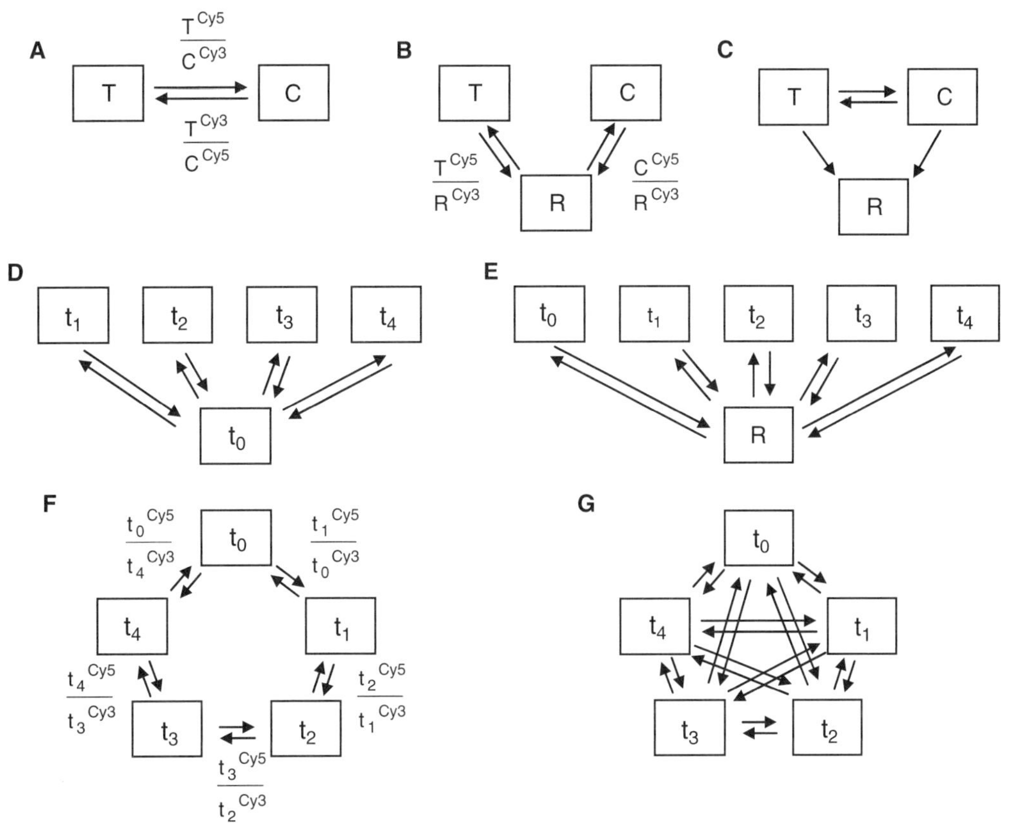

Fig. 1. Experimental designs in two-color microarray experiments. (**A**) The simplest design for an experiment comparing a treated sample, T, versus a control sample, C, is the direct comparison shown here. The treated sample could for instance be LPS-treated macrophages versus untreated control macrophages. Shown here is also a dye swap design (*double arrow*) where the treated and control samples are each labeled with the Cy3 and Cy5 dyes. (**B**) In the reference sample design, the treated and control sample is hybridized together with a reference sample, such as a commercial mixture of total RNA from several tissues thereby covering most transcripts. Since what ultimately will be compared is the treated versus the control sample the dye swap may be omitted. (**C**) A design combining the direct comparison and the reference sample design. (**D**) An experimental design for a time course experiment with time points, t_0–t_4, can be dealt with in several ways. Direct comparisons can be made of each time point versus the control time point, t_0, or (**E**) a reference design can be used with a common reference RNA. The nature of the control can also vary in a time course experiment affecting the design. Depending on what the biological question is the control sample could for instance be the zero time point at the start of the experiment or control cells incubated for equal time as the treated cells. (**F**) A loop design comparing each time point directly to the subsequent time point may be an efficient design to detect significant changes from one point to the next. This design does not require large quantities of control or reference samples. (**G**) A loop design with all possible pair wise comparisons that maximizes the ability to detect changes as all possible comparisons are made.

experiments may defer researchers from performing enough biological replicates. This is dangerous as a poorly replicated microarray experiment will have less power in the statistical analysis and considerable amounts of time and money will be spent on validating genes that could have been dismissed as not significant in an experiment with adequate biological replication. In my experience, at least four biological replicates should be performed.

1.2.3. Dye Swap Replicates and Technical Replicates

It is well known that the two dyes used in two-color array introduce a dye bias (*see* **Subheading 3.3.2.1**). The approach used to deal with dye bias is to perform dye swap replicates where the samples are reciprocally labeled and hybridized to a second array (**Fig.1A**). In a dye swap experiment, the pair of reciprocally labeled arrays also yields information about the technical variability. In my experience, the technical variation is always less than the biological variation (*see also* **Subheading 1.2.2**). Thus one can question the need for technical replicates in an experiment with adequate biological replicates. With many biological replicates, dye bias can be controlled for by swapping the dyes on every other array. In a reference sample design (**Fig.**1B), the dye swap replicates can be excluded as the interesting comparisons to be made are all labeled with the same dye.

1.2.4. Complex Designs

Experimental designs can be made increasingly complex. Let us consider a time course study to exemplify this. One could imagine combining each time point with a common zero time point control sample, or to use a reference sample, or making a loop design as illustrated in **Fig.1D–F**. An advantage of a loop design is the reduced amount of time zero or control samples required. Also, since this loop design compares subsequent time points to each other, it measures changes from one time point to the next more accurately than a common control design will. However, to maximize the ability to detect changes, all possible direct pair wise comparisons should be made (**Fig.1G**).

1.3. RNA Isolation and Quality Control

The purity and integrity of the RNA is crucial for a successful microarray experiment. To achieve a high purity RNA preparation, isolation methods are often combined such as in the protocol provided below. The provided protocol should not take more than 3–4 h to perform.

2. Materials

2.1. RNA Preparation and Isolation

1. TRIzol reagent (Invitrogen, Carlsbad CA, USA).
2. RNase-free disposable plastic (*see* **Note 1**): microcentrifuge screw-cap tubes with rubber seal, microcentrifuge tubes, wide-bore pipet tips, filter tips, and suitable tubes for tissue homogenizing such as 5 ml round-bottom tubes.
3. For tissue: RNA later solution (Ambion, Austin, TX, USA).
4. Omni TH tissue homogenizer and Omni Tip plastic probes (Omni International, Marietta GA, USA) or other equivalent tissue homogenizer.

5. RNase-free solutions: chloroform, isopropanol 100%, ethanol 70%, UltraPure DNase/RNase-free distilled water (Invitrogen).
6. Five milligram per milliliter linear polyacrylamide (LPA; Ambion, Austin, TX, USA; *see* **Note 2**).
7. Equipment: Microcentrifuge, pipettes.
8. RNeasy mini kit (Qiagen, Valencia CA, USA).
9. RNase-free solutions: UltraPure DNase/RNase-free distilled water (Invitrogen), β-mercaptoethanol, ethanol 100%, 1×TE buffer (optional).
10. Equipment: Microcentrifuge, pipettes.

3. Method

3.1. RNA Preparation and Isolation

1. Adherent cultured cells can be lysed directly in 1 ml TRIzol reagent. Transfer lysates to screw-cap tubes with a rubber seal to avoid leakage of the hazardous TRIzol reagent. Harvested tissue needs to be protected by RNA later if the RNA is to remain intact (*see* **Note 3**). Tissue can be homogenized in TRIzol with a tissue homogenizer keeping the samples on ice to avoid mechanical heating.
2. Add 0.2 ml chloroform per 1 ml of lysate/homogenate. Shake vigorously by hand for 15 s. Phase separation is achieved by centrifugation for 10 min at 10,000 × *g* at 4°C.
3. Prepare labeled, microcentrifuge tubes with 0.5 ml of RNase-free isopropanol and 1 μl of linear polyacrylamide (LPA) per 1 ml of TRIzol lysate (*see* **Note 2**).
4. Using a wide-bore pipette tip, remove the top, aqueous phase and transfer to the isopropanol and LPA-prepared tubes to precipitate the RNA, ensuring that none of the lower phase is transferred.
5. Mix tubes well by inversion. Pellet the RNA precipitate by centrifugation for 10 min at 10,000 × *g* at 4°C.
6. Discard the supernatant and wash the RNA pellet with 1 ml 70% ethanol per 1 ml of original TRIzol lysate. Vortex briefly.
7. Centrifuge for 5 min at 10,000 × *g*. Remove all of the supernatant.
8. Air-dry the RNA pellet with tubes open for a few minutes until the pellet is translucent. Do not over-dry pellets. Dissolve the RNA pellet in 100 μl of RNase-free water by gently pipetting over the pellet. You should be able to see the RNA

going into solution. The RNA can be incubated at 55°C to ensure that the pellets are completely dissolved. Samples may be frozen at this stage and the protocol continued later.

9. Add 350 µl buffer RLT to the 100 µl RNA sample (prepare with 10 µl β-mercaptoethanol per milliliter buffer RLT following Qiagen's instructions) (*see* **Note 4**).
10. Add 250 µl 100% ethanol and mix thoroughly by pipetting.
11. Apply the sample (700 µl) to an RNeasy mini column placed in the supplied 2 ml collection tube. Close the tube gently and centrifuge for 15 s at 10,000 × *g* and discard the flow through.
12. Transfer the RNeasy column to a new 2 ml collection tube and pipette 500 µl buffer RPE (with ethanol added as per Qiagen's protocol) onto the column. Centrifuge for 15 s at 10,000 × *g* and discard the flow through.
13. Pipette an additional 500 µl of buffer RPE to column. Centrifuge for 15 s at 10,000 × *g* and discard the flow through.
14. Centrifuge an additional 1 min to completely dry the membrane.
15. Transfer the column to a new, labeled, RNase-free microcentrifuge tube, and elute the RNA from the column by adding 40–100 µl RNase-free water. Let sit 1 min and then centrifuge for 1 min at 10,000 × *g*.
16. Optional: add an additional 40–100 µl RNase-free water to the column, let sit 1 min and then centrifuge for 1 min at 10,000 × *g* using the same collection tube. (Qiagen recommends using the initial eluate for second elution to obtain a higher concentration of RNA, but in my experience a second elution with water gives a higher total yield of RNA.)
17. Save a separate 5 µl aliquot for spectroscopy and integrity analysis before storing the purified total RNA samples at –80°C.

3.1.1. RNA Concentration and Purity

The concentration of the isolated total RNA can be determined by measuring absorbance at 260 nm. An absorbance of one unit at 260 nm corresponds to 40 µg of RNA per milliliter. Measuring the ratio of absorbance at 260 and 280 nm gives a measure of the purity as protein contaminants absorbing at 280 nm will result in a lower 260/280 ratio. Typically ratios >2.0 are expected for pure RNA dissolved in 10 mM Tris-HCl pH 7.5 buffer. This ratio can vary considerably if the RNA is dissolved in unbuffered water as pH influences the A260/A280 ratio greatly. Absorbance at >350 nm wavelengths indicates the presence of aggregates that scatter light. Residual phenol contaminations absorb light at around 270 and 230 nm and may interfere with RNA quantification.

The RNA concentration can be conveniently determined using a NanoDrop spectrophotometer (NanoDrop Technologies, Wilmington DE, USA) that requires only 1 μl sample.

3.1.2. RNA Integrity

Monitoring the RNA integrity is imperative as even a slightly degraded RNA will severely affect quantification (**Fig.2B**). If the RNA has started to degrade, perhaps due to exposure to RNases during the isolation procedure or lack of protection during tissue

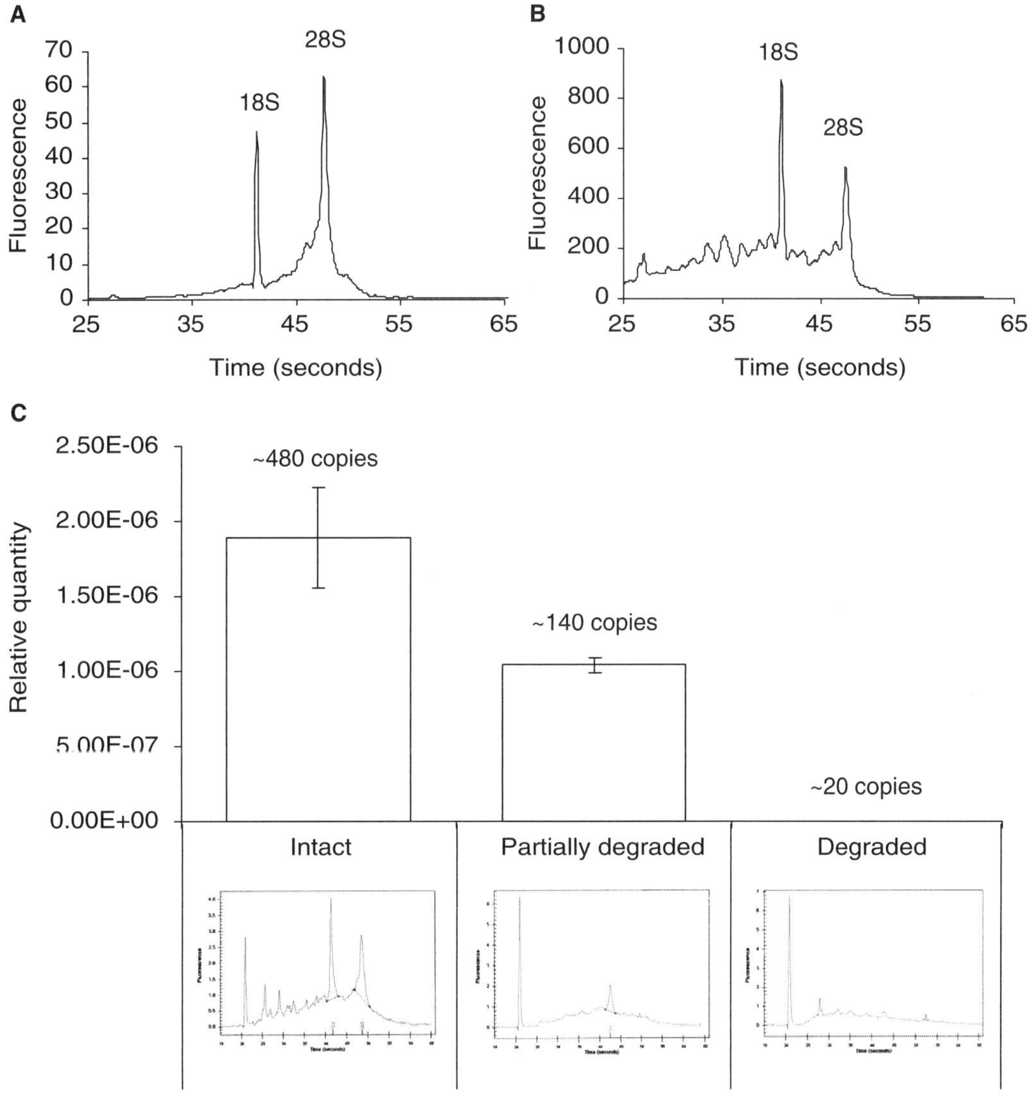

Fig. 2. RNA integrity analysis of total RNA isolated from spleen analyzed in an Experion Automated Electrophoresis System (Bio-Rad). (**A**) Nondegraded total RNA with a 28S–18S peak ratio greater than 1.5. (**B**) Partially degraded total RNA with diminished 28S peak and increase in shorter RNA fragments. (**C**) Quantitative real-time PCR analysis of mouse aorta β-actin expression in samples of varying total RNA integrity.

harvest, long RNA molecules such as the 28S ribosomal RNA will be cleaved first. The integrity of the RNA can thus be monitored by measuring the 28S to 18S ratio by electrophoresis. On an RNA gel, staining of the 28S rRNA band should appear twice as strong as the 18S band. However, the gold standard for RNA integrity analysis today is capillary electrophoresis using a Bioanalyzer 2100 (Agilent Technologies) or an Experion Automated Electrophoresis System from Bio-Rad (**Fig.2A**). Most microarray core facilities will require you to check the integrity of the RNA and most often they can perform this service using capillary electrophoresis.

3.2. Microarray Processing

This section briefly describes the methods involved in labeling the RNA for hybridization to the microarray and how the data are retrieved. There are several kits available for labeling the RNA and the microarray manufacturer will most likely recommend use of a certain kit. Often the processing of the microarrays is done by a core facility. The methodology is only briefly described in this section with the goal to inform about the multiple experimental factors that will affect the data quality and yield experimental variability. More detailed protocols can be found elsewhere and are referenced to when appropriate.

3.2.1. Labeling Methods

Several labeling methods are available such as direct chemical labeling of nucleotides with *cis*-platinum derivatives and direct incorporation of CyDye-coupled nucleotides into cDNA upon reverse transcription of RNA *(21)*. When choosing the labeling method, one needs to ensure that the labeled nucleic acid will be complementary to the reporter nucleic acid. For cDNA arrays, the reporter nucleic acid is a PCR product and thus the labeled nucleic acid can be complementary to either strand, while for oligonucleotide arrays the reporter nucleic acid is represented by the sense strand. Without amplification, the total RNA amount required ranges from a few hundred nanograms to 20 μg depending on labeling method. Thus the choice of labeling method could be influenced by the availability of sample. The range given above corresponds roughly to 100,000–5,000,000 mammalian cells *(21)*. Using more total RNA for labeling will improve the sensitivity of the microarray experiment *(18)*.

3.2.1.1. Labeling for One-Color Microarrays

Labeling for Affymetrix Genechips is based on in vitro transcription *(16)*. First, the RNA is reverse transcribed into cDNA using an oligo-dT primer with a T7-promotor. After second-strand synthesis, in vitro transcription from the T7 promotor will produce cRNA. Biotinylated nucleotides are incorporated in the in vitro transcription reaction to serve as tags for detection. The biotinylated cRNA is fragmented by hydrolysis to make transcripts of uniform length improving hybridization efficiency.

After hybridization the bound biotinylated cRNA fragments can be detected by staining with fluorescent phycoerythrin-labeled streptavidin. Additional signal amplification can be achieved by a second stain with biotinylated anti-steptavidin antibodies and phycoerythrin-labeled streptavidin. About 0.1–1 μg of total RNA is required per microarray.

3.2.1.2. Labeling for Two-Color Microarrays

Most commonly labeling is achieved by incorporating aminoallyl-modified nucleotides in a reverse transcription reaction producing cDNA from the template RNA *(15)*. The primary amines of the aminoallyl adduct is subsequently reacted with either Cy3 or Cy5 dye. Using aminoallyl-modified nucleic acids is preferred over using directly CyDye-conjugated nucleotides since the different chemistry of the dye molecules may give a bias in dye incorporation. Either oligo-dT, random primers or a mix of both can be used to prime the reverse transcription reaction. The choice of optimal priming may depend on the design of the reporter oligonucleotides, i.e., how close to the 3′ end, and subsequently the poly-A tail, of the gene the reporter sequence was selected. About 5–20 μg of total RNA is required.

3.2.2. Amplification of RNA for Spotted Oligonucleotide Arrays

If very small quantities of RNA are available, the messenger nucleic acid can be amplified by in vitro transcription *(22)*. In essence, labeling is performed similar to that for Affymetrix GeneChips, with reverse transcription into cDNA using an oligo-dT-T7 promotor primer. After second-strand synthesis, in vitro transcription produces cRNA that can be directly chemically labeled with Cy3 or Cy5. Multiple rounds of amplification can be performed, but should be avoided as each amplification step may skew the proportions of the original mRNA profile due to varying efficiency for individual mRNAs. Amplification tends to reduce the amplitude of differential expression ratios, but, on the other hand, amplification also tends to reduce variance. For a single round of amplification, about 50–100 ng of total RNA is required.

3.2.3. Hybridization

Hybridization is performed under a cover slip in a humidity chamber or in a special chamber limiting the hybridization volume. Today special automatic hybridization stations have been developed that can move the hybridization solution around on the microarray, thus greatly reducing background. Hybridization parameters include hybridization temperature, duration, buffer composition, additives (such as formamide), and stringency washes. Also the microarray slide chemistry may dictate choice of hybridization conditions. Competing DNA can be added to the hybridization solution to increase specificity. After complete hybridization, the microarrays are carefully washed to remove unspecifically bound nucleic acids, often in multiple steps with reduced ionic strength of the washing buffer.

3.2.4. Scanning

There are many commercially available microarray scanners used to read the fluorescence signal of the microarray at high resolution *(23)*, with the most common being the two-color microarray scanner utilizing photomultiplier tube (PMT) detection technology. Using this scanner, the red and green fluorescent molecules at each spatial pixel of the microarray are excited with independent lasers and discrete bands of the emitted red and green photons are passed through optical filters to independent PMTs for detection. The detectors and computer transform the photons received into a digital value and typically a 16-bit TIFF image corresponding to the intensity and location of each color of fluorescent molecule is created. Variation and errors in a microarray experiment can be introduced by the scanning process itself, either by the scanner instrumentation or the user, via the user-selected scanning parameters *(23)*.

3.2.5. Image Analysis

Image analysis and the detection of spots or features on the microarray are an important aspect of the microarray experiment commonly performed by a software supplied by the scanner manufacturer *(24)*. Automation of this process is required for high-throughput analysis. The analysis can be divided into three parts: (1) addressing or gridding, where coordinates for each spot are assigned by placing a grid that is automatically adjusted using an algorithm; (2) segmentation classifies pixels as either foreground, inside the spot boundary, or background just outside the spot boundary; and (3) intensity extraction, where the fluorescence intensities for each spot, both foreground and background for both fluorescence channels, are calculated. In Affymetrix GeneChips, there is not enough space to calculate local backgrounds *(25)*. Instead the lowest second percentile of the probe values are chosen to represent the background. The GeneChip is divided into areas with individual background correction to account for uneven backgrounds across the chip.

3.2.6. Raw Data Storage

Even a small microarray operation will generate vast amounts of data that preferentially are stored in a relational database *(26)*. Both commercial and free database solutions developed by academic laboratories exist, and there are public microarray data repositories (*see* **Table 1**).

3.3. Data Analysis

Microarray data present some new analysis challenges. Usually biological experiments are concerned with a few measured parameters measured on several subjects, but in a microarray experiment thousands of parameters (gene expression levels) can be measured on a few subjects. A basic assumption in any microarray experiment is that the expression of most genes in the sample under study and the control sample is unchanged. This assumption allows us to normalize microarrays to each other. Another technical issue with the large datasets generated by microarray analysis is that experimental

Table 1
Analysis software. Free for academic users

Software	Description	Web Site
Statistical analysis, clustering, and visualization		
BioConductor	A software package with many tools for analysis and comprehension of microarray and genomic data	http://www.bioconductor.org
Cluster and Tree View	An integrated pair of programs for analyzing and visualizing the results of complex microarray experiments	http://rana.lbl.gov/EisenSoftware.htm
dChip	A software package for analysis of Affymetrix gene expression microarray data	http://www.dchip.org
Gene Expression Dynamics Inspector, GEDI *(38)*	A self-organizing map algorithm that allows the visualization, inspection, and navigation through large microarray data sets	http://www.childrenshospital.org/research/ingber/GEDI/gedihome.htm
GenePattern	An analysis platform with clustering, prediction, marker selection and pathway analysis tools	http://www.broad.mit.edu/cancer/software/genepattern/index.html
Multi Experiment Viewer (MeV)	A microarray data analysis tool, incorporating sophisticated algorithms for clustering, visualization, classification, statistical analysis, and biological theme discovery	http://www.tm4.org/mev.html
R software package	A software environment for statistical computing and graphics with many microarray analysis packages	http://www.r-project.org
SAM (Significance Analysis of Microarrays)	A statistical technique for finding significant genes in a set of microarray experiments	http://www-stat.stanford.edu/~tibs/SAM/
TM4	A suite of tools consisting of four major applications – Microarray Data Manager (MADAM), TIGR_Spotfinder, Microarray Data Analysis System (MIDAS), and Multi Experiment Viewer (MeV), as well as a Minimal Information About a Microarray Experiment (MIAME) – compliant MySQL database	http://www.tm4.org/
Gene ontology tools		
FatiGO	Finds differential distributions of GO terms between two groups of genes	http://fatigo.bioinfo.cipf.es/
GoSurfer	Visualizes GO terms as a hierarchical tree and uses statistical tests to find over-represented GO terms	http://bioinformatics.bioen.uiuc.edu/gosurfer/

(continued)

Table 1 (continued)

Software	Description	Web Site
GOstat	Finds statistically overrepresented GO terms within a group of genes	http://gostat.wehi.edu.au
Pathway tools		
Cytoscape	A software platform for visualizing molecular interaction networks and integrating these interactions with gene expression profiles	http://www.cytoscape.org/
GenMAPP	Visualizes gene expression data on maps representing biological pathways and groupings of genes	http://www.genmapp.org/
Gene information sources		
GeneCards	A database of human genes that includes automatically-mined genomic, proteomic and transcriptomic information, as well as orthologies, disease relationships, SNPs, gene expression, and gene function	http://www.genecards.org
Mouse Genome Informatics (MGI)	Provides integrated access to data on the genetics, genomics, and biology of the laboratory mouse	http://www.informatics.jax.org/
PubGene	A text mining tool for PubMed articles looking for co-citation of multiple genes or proteins and displays them as "literature networks"	http://www.pubgene.org
SOURCE	Collects and compiles data from many scientific databases into easy to navigate GeneReports	http://source.stanford.edu
Databases		
BioArray Software Environment (BASE)	A MIAME compliant, open source and Web-based database server to manage the massive amounts of data generated by microarray analysis	http://base.thep.lu.se/
Public data repositories		
ArrayExpress	A MIAME-compliant public repository for microarray data	http://www.ebi.ac.uk/arrayexpress/
Gene expression omnibus (GEO)	A gene expression repository supporting MIAME compliant microarray data submissions	http://www.ncbi.nlm.nih.gov/geo/

researchers often find it difficult to manage these large datasets in their common spreadsheet software such as Microsoft Excel. It is often practical to manipulate the large datasets using some scripting programming language such as Perl or Python.

3.3.1. Filtering

Before normalization, the data are filtered to remove unreliable data. Many different filters can be applied and as many philosophies on filtering exist. If the microarray experiment was performed with many replicates, one may use only a few filters and rely on the statistical analysis to sort out genes that for one or the other reason varies across the experiment. On the other hand, extensive filtering of the data could remove problem spots on individual arrays and perhaps uncover differentially expressed genes that would have been deemed nonsignificant due to large variability. Common filters to use include signal-to-noise filters to remove weak spots or spots with unusually high background, low fluorescence filters to remove genes with low expression and high fluorescence filters to remove spots overexposed and thus not within the dynamic range of the scanner. One can also filter based on spot size or variability of the fluorescence of individual pixels within a spot. A common filter to use is to set the lowest allowable spot intensity. This filter could be important in two-color microarray datasets, since one channel could have a strong fluorescence, while the other channel has no fluorescence at all or even slightly negative after background subtraction. This is common for instance for cytokines that show no expression unless induced by a TLR agonist. For such a spot, an expression ratio cannot be calculated and the data are lost. By setting a lowest allowable intensity, a ratio can be calculated and the data point saved, although being an underestimate of the true ratio.

3.3.2. Normalization

Microarray data are normalized to minimize systematic and random experimental variation *(14, 27)*. Normalization can be applied both per gene and per array. Normalization methods assume that the majority of genes are not differentially expressed between samples and that the expression is influenced by random effects. One should keep in mind that this assumption may not hold true for all data.

3.3.2.1. Two-Color Microarray Normalization

A well-known source of variation in two-color microarray experiments arises from the use of two dyes *(28)*. The Cy3 and Cy5 dye biases are due to differences in photosensitivity and incorporation efficiency between the two dyes. Some of this difference can be accounted for by scanning the dyes at different detection sensitivities, although different scanner settings may also introduce an additional bias. Normalization is required to minimize systematic and random experimental variations such as the dye bias described above. Although many strategies for normalization can be applied, the most commonly used normalization method is a locally weighted regression method known as LOWESS (locally weighted scatterplot smoothing) normalization. LOWESS operates by dividing data in a MA-plot (the log2 of the channel 1 over channel 2 ratio vs. log10 of the square root of channel 1 multiplied by channel 2) into narrow intervals using a sliding window approach. The data in each window are fitted to a linear polynomial and adjusted based on the assumption that the majority of the genes are not differentially expressed.

3.3.2.2. One-Color Microarray Normalization

Before comparing two Affymetrix GeneChips, scaling or normalization methods must be applied *(29)*. The simplest approach is to re-scale each GeneChip in an experiment to equalize the average signal intensity across all chips by dividing all measured intensities on an array to the average or median intensity. The intensities of two arrays can also be normalized to each other by linear regression. This can be done by transforming the intensities so that linear regression line becomes the identity line ($y = x$) in a plot of the intensities of array 1 versus array 2. A popular normalization method is quantile normalization. In this method, the quantile of each value in the distribution of intensities is computed and these are then matched up across all the arrays and transformed so that the smallest value on each array is identical, the second smallest is identical, and so forth.

3.3.3. Statistical Analysis

As microarray data are composed of many measurements of expression on few samples, many new statistical analysis tools and theory are being developed. Essentially a regular Student's *t*-test for significance will not suffice as even with a $p < 0.01$ measuring 10,000 genes will give 100 genes that according to the test could be significant just by chance, which is often too many to be acceptable. Instead Bonferroni or Benjamini–Hochberg correction for multiple testing is applied so that within a gene list with a $p < 0.05$ comprising of 100 genes, five genes are expected to be false positives *(30)*. Bonferroni is often too stringent, but Benjamini –Hochberg correction usually gives a manageable number of significant genes. Other statistical analysis methods include rank and permutation tests *(31–33)*. An intuitive software for statistical analysis is SAM (Significance Analysis of Microarrays) (*see* **Table 1**). In this application, one can chose an acceptable level of the false discovery rate in a simple graphical representation of the data. Thus one can weigh the stringency of the test to the size of the gene list one wishes to continue working on. The R programming language is widely used for statistical software development and data analysis, and has quickly become a favorite among statisticians and is also frequently used in microarray data analysis. Bioconductor is a genomic data analysis software based on the R programming language and incorporates many tools for microarray data analysis.

3.4. To Make Biological Sense Out of Microarray Data

Once the microarray data have been filtered, normalized, and subjected to statistical analysis, it is time to make biological sense of the expression profile *(27, 34)*. Since co-expressing genes tend to fulfill common roles in the cell, a better understanding of the gene expression profile can be achieved by grouping genes with similar expression together. These groups of genes can then be analyzed for the significant enrichment of biological terms with functional meaning. The grouping or clustering of genes also helps to visualize the large sets of data.

3.4.1. Data Clustering and Visualization

There are many clustering methods using different algorithms to find genes with a similar expression pattern in the data *(14, 35, 36)*. The rational of clustering data is the underlying assumption that genes with similar expression patterns also have similar biological functions. Clustering data will also help to globally visualize the dataset in a way that columns of numbers in spread sheet never can. Widely used clustering methods include hierarchical clustering, *K*-means clustering, and self-organizing maps (SOMs).

3.4.2. Hierarchical Clustering

In hierarchical clustering each data object is initially placed in its own cluster. Next, the closest pair of clusters is merged forming a new cluster. This process is then repeated until all data are contained in a single cluster or a termination condition is satisfied. Based on the metrics used to calculate the similarity of the clusters, hierarchical clustering can be divided into single linkage, complete linkage, or average linkage clustering. In single linkage hierarchical clustering, the distance between two clusters is determined by the distance between their closest members, while in complete clustering the cluster distance is determined by the most distant members. In average linkage clustering, the cluster distance is measured as the average distance between two clusters. The output of the hierarchical clustering algorithm is a dendrogram where the branches record the distance between objects and the formation of groups (**Fig. 3A**). This graphic representation has become a favorite method of visualization of global patterns in expression data. Hierarchical clustering allows grouping of both genes and arrays. Thus the method will not only reveal common gene expression patterns across all arrays but also which experimental conditions are the most similar.

3.4.3. K-Means Clustering

In *K*-means clustering, the investigator selects the number of clusters (*K*) to partition the data into. The algorithm then positions *K* number of random objects representing an initial cluster mean and calculates for each gene which cluster mean is the closest. The genes are then assigned to the closest cluster and a new mean for that cluster is calculated from its members. This process of calculating the nearest cluster mean and cluster re-assignment is then repeated for all genes until the boundaries of the clusters stop changing. A drawback of *K*-means clustering is that the investigator has to specify the number of clusters. Thus

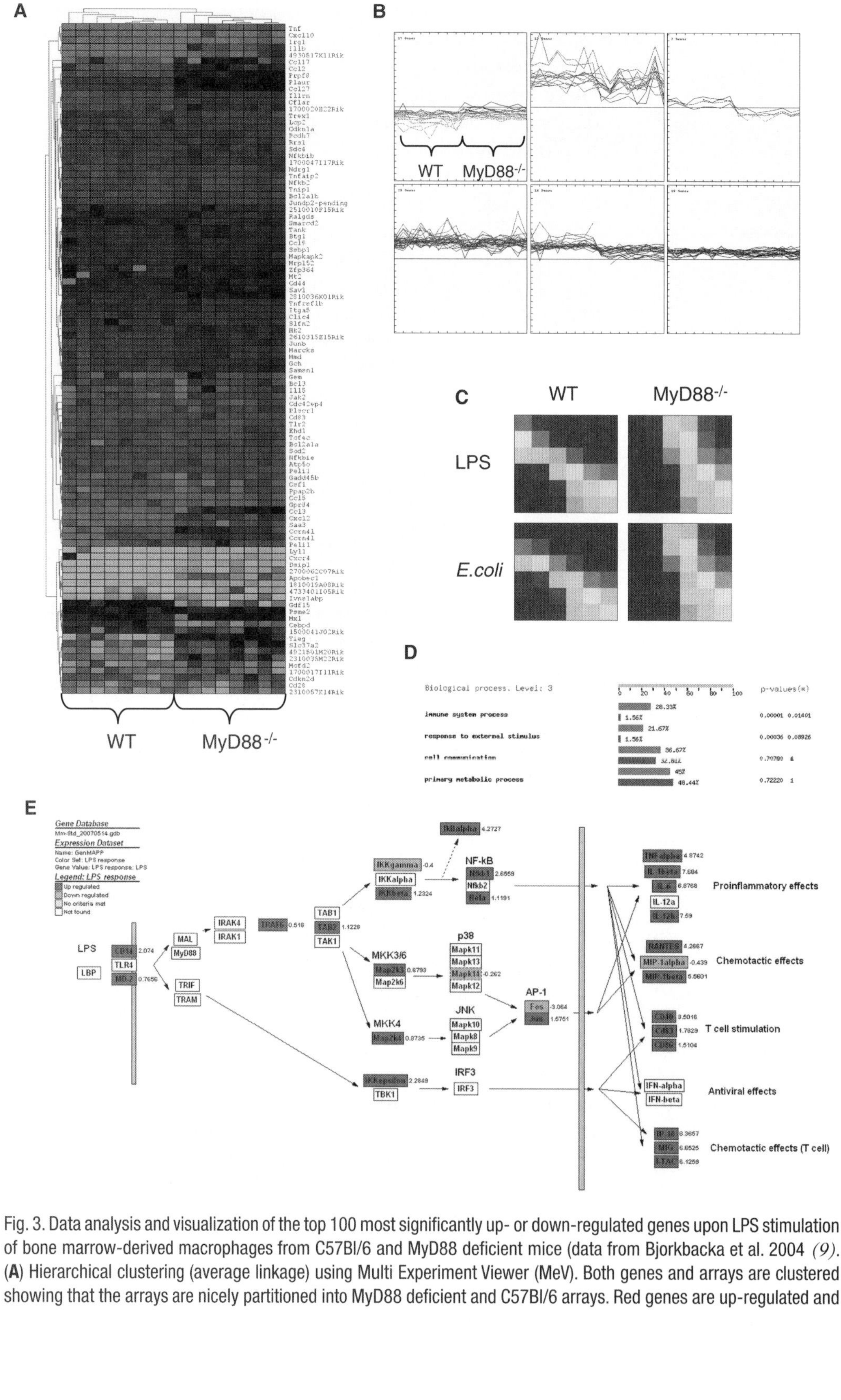

Fig. 3. Data analysis and visualization of the top 100 most significantly up- or down-regulated genes upon LPS stimulation of bone marrow-derived macrophages from C57Bl/6 and MyD88 deficient mice (data from Bjorkbacka et al. 2004 *(9)*. (**A**) Hierarchical clustering (average linkage) using Multi Experiment Viewer (MeV). Both genes and arrays are clustered showing that the arrays are nicely partitioned into MyD88 deficient and C57Bl/6 arrays. Red genes are up-regulated and

the algorithm needs to run repeatedly with different *K*-values to compare the clustering result before making a decision about the optimal number of clusters to partition the data into. The *K*-means clustering algorithm output is usually simple line graphs of the genes in each cluster (**Fig. 3B**), although each cluster can also be visualized in a similar way as hierarchical clustering.

3.4.4. Self-Organizing Maps

The self-organizing map (SOM) is a well-known un-supervised artificial neural network learning algorithm *(37)*. The artificial neural network is a collection of mathematical functions used to model input data to a defined output. The input data are gene expression vectors and the output are a number of reference vectors generated at random. Similar to *K*-means clustering, each gene is mapped to the closest reference vector, which is then modified to more closely resemble the gene, and then the process is repeated. This process where the reference vectors are adapting and competing for the genes is called training of the network. When the reference vectors have converged to fixed values and the training is complete, clusters are identified by mapping all data to the final output reference vectors. In the training process, the reference vectors are also adjusted so that they become more similar to neighboring reference vectors, and thus similarity between different clusters can be assessed. This is often visualized by placing similar clusters in close proximity to each other. Tools based on the SOM algorithm can be powerful visualization aids as is exemplified in **Fig. 3C** by the use of the Gene Expression Dynamics Inspector software *(38)*. Similar to *K*-means clustering, the investigator has to predefine the number of clusters of the SOM.

3.4.5. Gene Ontology Tools

Knowing how genes cluster is of little help unless the gene annotations in a particular cluster give the investigator some biological context. The Gene Ontology Consortium (http://www.geneontology.org) provides a controlled vocabulary to annotate biological knowledge. The gene ontology (GO) terms have been

green genes down-regulated. (**B**) *K*-means clustering into six clusters using MeV. Both clusters with MyD88-dependent and MyD88-independent genes can be identified. (**C**) Gene Expression Dynamics Inspector (GEDI) mosaic map visualization. Up-regulated genes are colored red and down-regulated blue. Each square in the mosaic consists of a cluster of genes that can be view by clicking that square. The software gives a nice global overview of the gene expression pattern. For instance, the top right corners seem to contain clusters with MyD88-independent up-regulated genes. Also, the LPS and *E. coli* induced gene expression profile is largely similar. (**D**) FatiGO analysis to find significantly over- or underrepresented gene ontology terms within the top 100 genes compared to a random sample of 100 genes also present on the microarray. The gene ontology term "immune system process" is significantly overrepresented among the top 100 genes ($p \sim 0.014$, adjusted for multiple testing), while for instance the term "primary metabolic process" is not. (**E**) GenMAPP visualization of expression data in the Toll-like receptor (TLR) signaling pathway (adapted from Kyoto Encyclopedia of Genes and Genomes (KEGG), http://www.genome.jp/kegg/). As can be seen, LPS treatment of C57Bl/6 macrophages regulates several genes in the TLR signaling pathway (*See Color Plates*).

organized into a hierarchical structure with three organizing principles: biological processes, cellular component, and molecular function. To aid interpretation of microarray data, statistical tools have been developed to find significantly over- or underrepresented GO terms within a list of genes *(39)*. Tools such as these may help to uncover a biological function that is overrepresented within a cluster of genes with similar gene expression (**Fig. 3D**).

3.4.6. Other Tools

Other tools that can be helpful in visualizing microarray data are pathway tools such as GenMAPP (http://www.genmapp.org). This software allows you to import microarray data to be viewed in your own constructed biological pathways or in already defined pathways and classifications, e.g. based on GO terms and KEGG pathways (**Fig. 3E**).

3.5. Independent Confirmation of Microarray Data and Publication

Microarray technology has developed into a standard tool at many laboratories disposal. Still, the expression of key genes needs to be confirmed by independent methods such as quantitative real-time PCR or by measuring protein expression by ELISA or immunoblotting. Prior to publication or planning of extensive follow-up experiments, the annotation of genes should be updated based on the reporter sequence as annotation information can be changed or updated in public databases *(17)*. Most scientific journals require that microarray data are made available to the public. There are several public microarray data repositories available (*see* **Table 1**). The scientific community has also worked out a standard for the minimum information about a microarray experiment (MIAME), which describes the minimum information required to ensure that microarray data can be easily interpreted and that the results can be independently verified *(40)*.

4. Notes

1. RNase-free disposable plastic can be purchased and is recommended over baked glassware. In my experience, even disposable plastic not specifically labeled "RNase free" can be used in most cases. Wear gloves at all times handling the RNA samples as it is very easy to contaminate your RNA preparation with the abundant RNases present on your skin. Using pipettes designated for RNA work and filter tips is also recommended.
2. Linear polyacrylamide (LPA) can be used as an efficient neutral carrier for precipitating nucleic acids with ethanol *(41)*. Although glycogen is often used as carrier in ethanol precipitation, it has been shown that glycogen inhibits the activity

of an ssDNA-binding protein on which linear polyacrylamide has no effect. A disadvantage of using linear polyacrylamide as carrier is that the pellet of polyacrylamide may not stick tightly on the bottom of microcentrifuge tubes. Be careful not to discard pellet when removing supernatants. Use of LPA makes precipitation steps in RNA isolation protocols much more efficient and effectively reduces precipitation time from 1 h at –20°C to a few minutes. Adding 10–20 μg of LPA into nucleic acid solutions before ethanol precipitation allows picogram amounts of nucleic acids longer than 20 bp to be precipitated without loss. LPA can be prepared by polymerizing a 5% acrylamide solution without bis-acrylamide in 40 mM Tris-HCl (pH 8), 20 mM sodium acetate, 1 mM EDTA by adding ammonium persulfate to a final concentration of 0.1% (w/v) and 1/1,000 volume of TEMED at room temperature for 30 min. When the solution has become viscous, the polymer can be precipitated with 2.5 volumes of ethanol, and then recovered by centrifugation. Dissolving the pellet in10 mM Tris-HCl (pH 8.0), 1 mM EDTA to a final polyacrylamide concentration of 5 mg/ml makes an LPA solution that can be stored refrigerated for years. If used in RNA applications prepare, make sure to use RNase-free LPA.

3. RNA is starting to degrade very fast in unprotected tissue. RNA later is crucial in preventing degradation if the tissue cannot be processed immediately. Tissue removed surgically needs to be placed in RNA later within minutes of the harvest. I have experienced degradation of RNA from mouse spleens collected and snap frozen in liquid nitrogen, but thawed on ice for 20 min without RNA later protection.
4. Most cell culture RNA samples do not depend on the RNeasy purification, but we still use them routinely as they get rid of any residual phenol contaminants from the Trizol reagent, thereby maintaining a consistent high quality of our isolated RNA. In my experience, the extra purification using the RNeasy kit is essential to get rid of proteoglycan contaminants.

Acknowledgments

The author is grateful for support from The Swedish Research Council, The Swedish Heart and Lung Foundation, The Crafoord Foundation, The Royal Physiographic Society, The Albert Påhlsson Foundation, Malmö University Hospital Foundation, and The Magnus Bergvall Foundation.

References

1. Irving, P., Troxler, L., Heuer, T. S., Belvin, M., Kopczynski, C., Reichhart, J. M., Hoffmann, J. A., and Hetru, C. (2001) A genome-wide analysis of immune responses in Drosophila. *Proc Natl Acad Sci U S A* 98, 15119–24.
2. De Gregorio, E., Spellman, P. T., Tzou, P., Rubin, G. M., and Lemaitre, B. (2002) The Toll and Imd pathways are the major regulators of the immune response in Drosophila. *EMBO J* 21, 2568–79.
3. Gao, J. J., Diesl, V., Wittmann, T., Morrison, D. C., Ryan, J. L., Vogel, S. N., and Follettie, M. T. (2003) Bacterial LPS and CpG DNA differentially induce gene expression profiles in mouse macrophages. *J Endotoxin Res* 9, 237–43.
4. Schmitz, F., Mages, J., Heit, A., Lang, R., and Wagner, H. (2004) Transcriptional activation induced in macrophages by Toll-like receptor (TLR) ligands: from expression profiling to a model of TLR signaling. *Eur J Immunol* 34, 2863–73.
5. Rodriguez, N., Mages, J., Dietrich, H., Wantia, N., Wagner, H., Lang, R., and Miethke, T. (2007) MyD88-dependent changes in the pulmonary transcriptome after infection with *Chlamydia pneumoniae*. *Physiol Genomics* 30, 134–145.
6. Burch, L. H., Yang, I. V., Whitehead, G. S., Chao, F. G., Berman, K. G., and Schwartz, D. A. (2006) The transcriptional response to lipopolysaccharide reveals a role for interferon-gamma in lung neutrophil recruitment. *Am J Physiol Lung Cell Mol Physiol* 291, L677–82.
7. Bjorkbacka, H., Kunjathoor, V. V., Moore, K. J., Koehn, S., Ordija, C. M., Lee, M. A., Means, T., Halmen, K., Luster, A. D., Golenbock, D. T., and Freeman, M. W. (2004) Reduced atherosclerosis in MyD88-null mice links elevated serum cholesterol levels to activation of innate immunity signaling pathways. *Nat Med* 10, 416–21.
8. Pierer, M., Rethage, J., Seibl, R., Lauener, R., Brentano, F., Wagner, U., Hantzschel, H., Michel, B. A., Gay, R. E., Gay, S., and Kyburz, D. (2004) Chemokine secretion of rheumatoid arthritis synovial fibroblasts stimulated by Toll-like receptor 2 ligands. *J Immunol* 172, 1256–65.
9. Bjorkbacka, H., Fitzgerald, K. A., Huet, F., Li, X., Gregory, J. A., Lee, M. A., Ordija, C. M., Dowley, N. E., Golenbock, D. T., and Freeman, M. W. (2004) The induction of macrophage gene expression by LPS predominantly utilizes Myd88-independent signaling cascades. *Physiol Genomics* 19, 319–30.
10. Weighardt, H., Jusek, G., Mages, J., Lang, R., Hoebe, K., Beutler, B., and Holzmann, B. (2004) Identification of a TLR4- and TRIF-dependent activation program of dendritic cells. *Eur J Immunol* 34, 558–64.
11. Thomas, K. E., Galligan, C. L., Newman, R. D., Fish, E. N., and Vogel, S. N. (2006) Contribution of interferon-beta to the murine macrophage response to the Toll-like receptor 4 agonist, lipopolysaccharide. *J Biol Chem* 281, 31119–30.
12. Gilchrist, M., Thorsson, V., Li, B., Rust, A. G., Korb, M., Kennedy, K., Hai, T., Bolouri, H., and Aderem, A. (2006) Systems biology approaches identify ATF3 as a negative regulator of Toll-like receptor 4. *Nature* 441, 173–8.
13. Heller, M. J. (2002) DNA microarray technology: devices, systems, and applications. *Annu Rev Biomed Eng* 4, 129–53.
14. Ehrenreich, A. (2006) DNA microarray technology for the microbiologist: an overview. *Appl Microbiol Biotechnol* 73, 255–73.
15. Hager, J. (2006) Making and using spotted DNA microarrays in an academic core laboratory. *Methods Enzymol* 410, 135–68.
16. Dalma-Weiszhausz, D. D., Warrington, J., Tanimoto, E. Y., and Miyada, C. G. (2006) The Affymetrix GeneChip platform: an overview. *Methods Enzymol* 410, 3–28.
17. Yauk, C. L., and Berndt, M. L. (2007) Review of the literature examining the correlation among DNA microarray technologies. *Environ Mol Mutagen* 45, 380–394.
18. Neal, S. J., and Westwood, J. T. (2006) Optimizing experiment and analysis parameters for spotted microarrays. *Methods Enzymol* 410, 203–21.
19. Churchill, G. A. (2002) Fundamentals of experimental design for cDNA microarrays. *Nat Genet* 32 Suppl, 490–5.
20. Townsend, J. P. (2003) Multifactorial experimental design and the transitivity of ratios with spotted DNA microarrays. *BMC Genomics* 4, 41.
21. Brownstein, M. (2006) Sample labeling: an overview. *Methods Enzymol* 410, 222–37.
22. Baugh, L. R., Hill, A. A., Brown, E. L., and Hunter, C. P. (2001) Quantitative analysis of mRNA amplification by in vitro transcription. *Nucleic Acids Res* 29, E29.
23. Timlin, J. A. (2006) Scanning microarrays: current methods and future directions. *Methods Enzymol* 411, 79–98.
24. Yang, Y. H., Buckley, M. J., and Speed, T. P. (2001) Analysis of cDNA microarray images. *Brief Bioinform* 2, 341–9.

25. Do, J. H., and Choi, D. K. (2006) Normalization of microarray data: single-labeled and dual-labeled arrays. *Mol Cells* 22, 254–61.
26. Hess, K. R., Zhang, W., Baggerly, K. A., Stivers, D. N., and Coombes, K. R. (2001) Microarrays: handling the deluge of data and extracting reliable information. *Trends Biotechnol* 19, 463–8.
27. Breitling, R. (2006) Biological microarray interpretation: the rules of engagement. *Biochim Biophys Acta* 1759, 319–27.
28. Dobbin, K. K., Kawasaki, E. S., Petersen, D. W., and Simon, R. M. (2005) Characterizing dye bias in microarray experiments. *Bioinformatics* 21, 2430–7.
29. GeneChip® Expression Analysis. Data Analysis Fundamentals. http://www.affymetrix.com/support/downloads/manuals/data_analysis_fundamentals_manual.pdf (Last accessed date, June 23, 2007)
30. Gusnanto, A., Calza, S., and Pawitan, Y. (2007) Identification of differentially expressed genes and false discovery rate in microarray studies. *Curr Opin Lipidol* 18, 187–93.
31. Lee, M. L., Gray, R. J., Bjorkbacka, H., and Freeman, M. W. (2005) Generalized rank tests for replicated microarray data. *Stat Appl Genet Mol Biol*4, Article3.
32. Yang, H., and Churchill, G. (2007) Estimating p-values in small microarray experiments. *Bioinformatics* 23, 38–43.
33. Tan, Y. D., Fornage, M., and Fu, Y. X. (2006) Ranking analysis of microarray data: a powerful method for identifying differentially expressed genes. *Genomics* 88, 846–54.
34. Dopazo, J. (2006) Functional interpretation of microarray experiments. *Omics* 10, 398–410.
35. Belacel, N., Wang, Q., and Cuperlovic-Culf, M. (2006) Clustering methods for microarray gene expression data. *Omics* 10, 507–31.
36. Quackenbush, J. (2001) Computational analysis of microarray data. *Nat Rev Genet* 2, 418–27.
37. Toronen, P., Kolehmainen, M., Wong, G., and Castren, E. (1999) Analysis of gene expression data using self-organizing maps. *FEBS Lett* 451, 142–6.
38. Eichler, G. S., Huang, S., and Ingber, D. E. (2003) Gene expression dynamics inspector (GEDI): for integrative analysis of expression profiles. *Bioinformatics* 19, 2321–2.
39. Beissbarth, T. (2006) Interpreting experimental results using gene ontologies. *Methods Enzymol* 411, 340–52.
40. Brazma, A., Hingamp, P., Quackenbush, J., Sherlock, G., Spellman, P., Stoeckert, C., Aach, J., Ansorge, W., Ball, C. A., Causton, H. C., Gaasterland, T., Glenisson, P., Holstege, F. C., Kim, I. F., Markowitz, V., Matese, J. C., Parkinson, H., Robinson, A., Sarkans, U., Schulze-Kremer, S., Stewart, J., Taylor, R., Vilo, J., and Vingron, M. (2001) Minimum information about a microarray experiment (MIAME)-toward standards for microarray data. *Nat Genet* 29, 365–71.
41. Gaillard, C., and Strauss, F. (1990) Ethanol precipitation of DNA with linear polyacrylamide as carrier. *Nucleic Acids Res* 18, 378.

Chapter 17

Uncovering Novel Gene Function in Toll-Like Receptor Signalling Using siRNA

Michael Carty, Sinéad Keating, and Andrew Bowie

Summary

In the last number of years siRNA has emerged as a key technique in understanding gene function. While siRNA has been used in lower organisms such as *Caenorhabditis elegans*, its use in mammalian cells, where gene manipulation is difficult, is where its greatest benefit has been realised. The advancements made in siRNA technology now provide us with an alternative approach to relying on "knockout mice" in uncovering mammalian gene function. In addition, siRNA provides us with a complementary approach to overexpression systems in cultured mammalian cells. siRNA is a superior method of post-transcriptional gene silencing (PTGS) compared to ribozyme and anti-sense technologies both in terms of potency and specificity.

Key words: siRNA, Gene function, Post-transcriptional gene silencing.

1. Introduction

siRNA makes use of the endogenous RNA interference (RNAi) pathway which is evolutionarily conserved among plants, invertebrates and mammals *(1)*. This cellular defence pathway is initiated by the presence of dsRNA molecules greater than 30 bp in length, which may be transcriptional products from randomly integrated transposons, transgenes or viral genes. These dsRNAs are processed by the RNase III like enzyme Dicer producing dsRNAs of 21–23 nucleotides in length which also possess dinucleotide 3′ overhangs. These small interfering or "siRNA" molecules are then incorporated into the multi-protein complex RISC (RNA-Induced Silencing Complex). The components of

Claire E. McCoy and Luke A.J. O'Neill (eds.), *Methods in Molecular Biology, Toll-Like Receptors, vol. 517*
DOI: 10.1007/978-1-59745-541-1_17

RISC are responsible for siRNA unwinding and strand separation. This results in either one of the two strands, known as the guide strand, to integrate and direct RISC to the target mRNA. As Dicer produces two RNA strands, two functionally distinct RISC complexes may emerge; however, this does not occur as the anti-guide strand is degraded during the activation of RISC *(2)*. The use of siRNA to reduce target mRNA relies upon the anti-sense strand entering RISC. The catalytic component of RISC, an endonuclease known as Argonaute, then mediates target mRNA degradation. This riboprotein-endonuclease complex is maintained to degrade other target mRNAs.

Recently, a new class of regulatory RNAs have been identified *(3)*. These micro-RNAs or miRNA are derived from a class of non-coding RNA genes or from intronic regions of transcribed genes. The primary transcripts or *pri-miRNAs* are processed in the nucleus by the nuclear enzyme Drosha and are then exported to the cytoplasm as pre-miRNAs of approximately 70 nucleotides. In the cytoplasm these molecules are processed by Dicer and subsequently loaded into RISC. Therefore, the miRNA and RNAi pathways converge at the point of RISC. However, while siRNA-mediated gene silencing relies on full complementarity with the target mRNA sequence, followed by target cleavage *(4, 5)* miRNAs are only partially complementary to the mRNA target and gene silencing is due to transcriptional repression *(6, 7)*. The importance of miRNA in cell homeostasis is underscored by the fact that up to 30% of mammalian genes are regulated by this mechanism *(8)*.

Early attempts by researchers to trigger the RNAi pathway in mammalian cells were hampered by the activation of the anti-viral interferon response by the introduction of >30 bp dsRNAs *(9)*. This interferon response results in PKR activation leading to phosphorylation and subsequent inactivation of the eIF-2α transcription initiation factor *(10)* along with concomitant induction of the 2′-5′-oligoadenylate synthase(OAS)/RNAse L pathway leading to non-specific RNA degradation *(11)*. However, this issue of inducing the anti-viral interferon response was circumvented by the discovery that 21–25 nucleotide dsRNAs evaded the interferon pathway and could still elicit effective and specific knock-down of target gene expression *(12)*. Thus, the current experimental method used to mimic the RNAi pathway centres on the introduction of 21 nucleotide siRNA duplexes which are complementary to the target mRNA of interest. There are two main delivery methods commonly used to introduce siRNA into cells. Cationic lipid-mediated transfection of siRNAs into cells is based on the electrostatic interactions between positively charged lipids and negatively charged siRNAs resulting in siRNA delivery by endocytosis.

An alternative method uses viral or non-viral expression systems to express siRNAs as small hairpin or shRNAs. Viral-based shRNA delivery systems are most useful in cells that are difficult to transfect such as terminally differentiated or non-dividing cells. The shRNAs are transcribed from dsDNA that encodes a sense loop anti-sense RNA molecule allowing it to form the characteristic hairpin structure. These shRNAs are then transported from the nucleus to the cytoplasm where they are processed by Dicer thus producing 21–23 nucleotide siRNAs. For the purpose of this chapter, however, we will focus solely on the use of cationic lipid-mediated transfection of siRNA.

1.1. Design of siRNA

A number of criteria have been identified for effective design of siRNA *(13, 14)*. These include: (1) Target sequence that is 75 to 100 bp downstream of the start codon of the open reading frame, and avoiding both 5′ and 3′ UTRs. This is to avoid mRNA sequence that may be occupied by regulatory proteins such as translation initiation complexes. (2) Identify a region of the sequence, 5′ AA(dN19)UU that has a GC content of approximately 50%. If there are no sequence stretch with 5′ AA(dN19)UU, widen the search to 5′ AA(dN21) or 5′ NA(N21). The AA dinucleotide defines the composition of the 3′ overhangs, which in this case will be dTT. (3) Perform database searches, such as BLASTn (Basic Local Alignment Search Tool nucleotide) to ensure the siRNA targets only the gene of interest. In general it is recommended that at least three candidate siRNAs are screened for functionality either by Western blotting or reverse transcription–polymerase chain reaction (RT-PCR). The guidelines described above are referred to as conventional siRNA design.

Additional studies by Reynolds et al. *(15)* and Jagla et al. *(16)* identified eight further features of the 19mer duplex which enhance the function of siRNA (3′ overhangs excluded).

1. G/C content of between 36% and 52%
2. The presence of one or more A/U in positions 15–19 (sense strand)
3. Absence of internal repeats
4. An A base at position 19 (sense strand)
5. An A base at position 3 (sense strand)
6. A U base at position 10 (sense strand)
7. A base other than G or C at position 19 (sense strand)
8. A base other than G at position 13 (sense strand)

The well-defined phenomenon of non-specific or "off target" siRNA effects *(17)* may be minimised by the careful siRNA design and by the chemical modifications of the siRNA duplex which limits the sense strand entering RISC and contributing to off target effects. RISC guide strand formation relies on the phosphorylation of the 5′ hydroxyl residues and this can be prevented by replacing the 5′ hydroxyl group by aminopropyl phosphate residues *(18)*. A recent review by Patzel *(19)* described an additional number of siRNA features that enhance functionality.

These include avoidance of stretches of "G" bases; lipid conjugation of siRNA to enhance uptake; siRNA backbone consisting of phosphorothioate linkages to enhance resistance to cellular nucleases. Another consideration of relevance to innate immunity is the avoidance of motifs that are immunostimulatory such as 5′-GUCCUUCAA-3′, which are recognised by TLR7; Also the selection of siRNA that will have "unstructured" guide or anti-sense strands should be avoided. The avoidance of off target effects can be further accomplished by the use of the Smith–Watermann alignment tool which is more sensitive than BLASTn to recognise short stretches of complementary sequence.

The criteria described above allow for computational or *In Silico* selection of highly potent siRNAs and are incorporated into a number of siRNA design tools available on the Web from a variety of sources. It is also worth mentioning that a number of companies such as Dharmacon and Invitrogen have pre-designed siRNAs to an increasing number of mouse and human genes. To ensure the specificity of the siRNA it is necessary to include a control or negative siRNA. This may be a scrambled version of the selected active siRNA or alternatively it may be directed to a gene absent in the cells such as GFP or luciferase.

Here we will describe in detail the siRNA methods we have used to evaluate the contribution of our target genes of interest in TLR signalling, namely SARM and IRAK2 *(20, 21)*. In our experiments we have used unmodified siRNAs both designed and supplied by Qiagen and have used Lipofectamine 2000 (Invitrogen) as our transfection reagent. We will describe a number of methods both in HEK293 cells and human peripheral blood mononuclear cells (PBMCs) to determine the efficacy of the siRNA and the functional consequence of "knock-down" using both reporter gene assays and ELISA in 96-well format.

2. Materials

2.1. Cell Culture

1. Dulbecco's Modified Eagle's Medium (DMEM, Sigma) supplemented with 10% fetal bovine serum (FBS) and 2 mM glutamine (Sigma) with or without 50 μg/ml gentamycin (Sigma).
2. PBMC medium: consists of RPMI (Sigma) supplemented with 10% FBS and 2 mM glutamine (Sigma) with or without 50 μg/ml gentamycin (Sigma).
3. Trypsin-EDTA solution (0.5 g/ml Trypsin, 0.2 g/ml EDTA, Sigma).
4. Sterile 96-well cell culture plates (Corning).
5. Cell culture flasks (Corning).

6. Lipopolysaccharide (Alexis).
7. Poly(I:C) (GE Healthcare).
8. Lymphoprep (Axis-Shield).

2.2. siRNA Transfection

1. Functional siRNA (20 μM stock, Qiagen) (*see* **Note 1**).
2. Control non-silencing siRNA (20 μM stock, Qiagen).
3. Serum-free DMEM.
4. Lipofectamine 2000 (Invitrogen).
5. Sterile V-bottomed 96-well plates (Bibby Sterilin).

2.3. RNA Isolation

1. Tri® Reagent (Sigma).
2. Molecular biology grade water (RNase and DNase free, Sigma).
3. Chloroform (Sigma).
4. Isopropanol (Sigma).

2.4. Reverse Transcription–Polymerase Chain Reaction

1. Random primers (Invitrogen).
2. Nuclease-free water (Promega).
3. $MgCl_2$ (25 mM stock) (Novagen).
4. dNTPs (2 mM stock) (Novagen).
5. RNase OUT (Invitrogen).
6. ImProm-II Reverse Transcriptase (Promega).
7. ImProm-II 5X reaction buffer (Promega).
8. SARM forward primer 5′-TGCATGGCTTCAGTGTCTTC-3′ (100 pmol/μl stock).
9. SARM reverse primer 5′-ACCCTCCAAACTGGTGTCAG-3′ (100 pmol/μl stock).
10. Actin forward primer: 5′-CGCGAGAAGATGACCCAGATC-3′ (100 pmol/μl stock).
11. Actin reverse primer: 5′-TTGCTGATCCACATCTGCTGG-3′ (100 pmol/μl stock).
12. *Taq* DNA polymerase (Invitrogen).
13. *Taq* DNA polymerase reaction buffer (10X stock solution, Invitrogen).
14. $MgCl_2$ (50 mM stock) (Invitrogen).
15. DNA loading dye (10X stock solution): 0.125% (w/v) xylene cyanol, 0.125% bromophenol blue, 0.625% SDS, 62.5% glycerol.

2.5. SDS–Polyacrylamide Gel Electrophoresis (PAGE)

1. Resolving gel buffer: 1.5 M Tris-HCl (pH 8.8).
2. Stacking gel buffer: 1 M Tris-HCl (pH 6.8).

3. Thirty per cent acrylamide/bis solution (37.5:1): Gloves should be worn at all times while handling this solution as the unpolymerised solution is a neurotoxin.
4. *N,N,N,N'*-tetramethyl-ethylenediamine (TEMED, Sigma).
5. Ammonuim persulphate: Make a 10% solution in water and store in single use aliquots at –20°C.
6. Water-saturated butanol: Mix equal volumes of water and isobutanol and shake vigorously in a glass container. Allow to separate and isolate the top layer. Store at room temperature.
7. Running buffer (10X stock soultion): 250 mM Tris, 1.92 M glycine, 10% SDS. Running buffer should be at pH 8.3. This should be confirmed but do not alter the pH due to the presence of SDS.
8. Pre-stained molecular weight markers (broad range molecular weight markers, New England Biolabs).
9. Laemmli sample buffer: 62 mM Tris-HCl, pH 6.8, 2% (w/v) SDS, 10% glycerol, 0.1% (w/v) bromophenol blue, 50 mM dithiothreitol (DTT). Prepare the solution without DTT and store in aliquots at –20°C. Add DTT to a final concentration of 50 mM immediately prior to use.

2.6. Western Blotting

1. Transfer buffer (10X stock solution): 250 mM Tris, 1.92 M glycine. Prior to use, dilute to 1X with distilled water and supplement with 20% methanol.
2. Polyvinylidene fluoride (PVDF) membrane (Millipore).
3. Phosphate-buffered saline containing Tween-20 (PBS-T). For convenience, prepare a 10X solution of PBS (687 mM NaCl, 39 mM NaH_2PO_4, 226 mM Na_2HPO_4) and dilute to 1X with distilled water. Add Tween-20 to a final concentration of 0.1% (v/v).
4. Blocking buffer: 5% (w/v) non-fat dry milk in PBS-T.
5. Antibody diluent: 1% (w/v) bovine serum albumin (BSA, Sigma) in PBS-T.
6. Anti-SARM rabbit polyclonal antibody (Prosci).
7. Anti-β-actin monoclonal antibody (Sigma).
8. Horseradish peroxidase conjugated anti-rabbit IgG secondary antibody (Sigma).
9. Horseradish peroxidase conjugated anti-mouse IgG secondary antibody (Sigma).
10. Enhanced chemiluminescent substrate (solution A: 2.5 mM Luminol (Sigma), 0.396 mM *p*-coumaric acid (Sigma), 0.1 M Tris-HCl pH 8.5; solution B: 0.1 M Tris-HCl pH 8.5; 30% H_2O_2). A 250 mM stock solution of Luminol in DMSO

can be prepared and stored as 1 ml aliquots at –20°C. Similarly a 90 mM solution of *p*-coumaric acid in DMSO can be stored in 0.44 ml aliquots at –20°C. This is convenient for preparation of a 100 ml working stock of solution A which is stored at 4°C. In addition, solution B and the 30% H_2O_2 are also stored at 4°C. To prepare a working stock immediately before use combine 5 ml of solution A and solution B and 3.05 µl of 30% H_2O_2.

11. Stripping buffer: Re-Blot Plus Strong 10X solution (Chemicon). Dilute to 1X with distilled water.
12. Kodak X-Omat LS film (Sigma).

2.7. Measurement of Relative Luciferase Activity

1. Luminometer (Luminoskan Ascent, Thermo Electron Corporation).
2. Passive lysis buffer (5X stock solution, Promega). Store at –20°C and dilute to 1X using distilled water immediately prior to use.
3. 2x Firefly luciferase substrate, made using the following recipe: Chemicals listed in order of addition, for 228 ml final volume:
   (1) 20 mM Tricine (0.817 g).
   (2) 2.67 mM $MgSO_4 \cdot 7H_2O$ (1.21 ml of 500 mM).
   (3) 0.1 mM EDTA (45.6 µl of 500 mM EDTA, pH 8.0).
   (4) 33.3 mM DTT.
   (5) 530 µM ATP (12 ml of fresh 10 mM solution).
   (6) 270 µM Acetyl CoEnzyme A (Lithium salt, Sigma) 1.89 ml of 25 mg/ml.
   (7) 30 mg Luciferin (Biosynth).

   From now on keep the solution in the dark.

   (8) Add H_2O to bring volume up to 222.6 ml.
   (9) Add 570 µl of 2 M NaOH (solution should now turn yellow).
   (10) Add 1.21 ml of 50 mM magnesium carbonate hydroxide ($(MgCO_3)_4Mg(OH)_2 \cdot 5H_2O$).
   (11) Store wrapped in foil at –20°C in aliquots of 10–30 ml.

   Note: This solution is stable to multiple freeze-thaws. Thaw to RT before use. This solution can be diluted 50:50 in water before use.
4. Renilla luciferase enzyme substrate: 1 mg/ml coelenterazine (Biotium Inc.) reconstituted in 100% ethanol and stored at –20°C. Prepare 2 µg/ml working solution with 1X PBS directly prior to use.
5. COSTAR white polystyrene non-treated 96-well assay plates (Corning).

2.8. Elisa

1. Human RANTES, IL-8 and TNF-α ELISA kits (R&D systems).

3. Methods

3.1. Culture of HEK 293 Cells and PBMCs

3.1.1. Culture of HEK 293 Cells

HEK 293 cells are grown in T75 flasks and are passaged twice weekly using the following method:

1. Remove the medium from the cell monolayer and rinse once in PBS.
2. To the cell monolayer add 3 ml of trypsin-EDTA and place the flask in the 37°C incubator. Incubate for 2–5 min.
3. To the trypsinised cells add 10 ml of medium containing FCS to inactivate the trypsin.
4. Centrifuge the cells at 120 × *g* for 5 min at room temperature.
5. Remove the supernatant and resuspend the cell pellet in 2–3 ml of fresh medium.
6. Count the cells and seed them in T75 vented flasks at a density of 1×10^5 cells per ml in a total volume of 15 ml.

3.1.2. Isolation of PBMCs

1. If isolating the PBMCs from buffy coats dilute the blood at a ratio of 1:7 using sterile PBS. However, if you are using fresh blood, dilute the blood 1:2 again using PBS.
2. Add 15 ml of lymphoprep reagent to 50 ml tubes for the number of samples required.
3. Slowly layer 35 ml of the diluted blood onto the lymphoprep. This must be done very slowly so that the diluted blood remains on top of the lymphoprep.
4. Centrifuge at 1,400 × *g* for 30 min. For this step it is crucial that the centrifuge break be switched off, otherwise the gradient will be disrupted and isolation of the PBMCs will be impossible.
5. Using a sterile transfer pipette remove the cells in the buffy coat layer to 50 ml tubes. The buffy coat layer is located at the interface of the lymphoprep layer at the bottom and the serum layer at the top.
6. Dilute the cells 1:2 in PBS and centrifuge at 120 × *g* for 8 min. At this point the break on the centrifuge may be left on.
7. Wash the cell pellet in PBS and centrifuge as in **step 6**.
8. Resuspend the pellet in an appropriate volume of medium (4–6 ml).
9. Count the cells and seed plates for experiments.

3.2. siRNA Transfection of HEK 293 Cells and PBMCs

3.2.1. siRNA Transfection of HEK 293 Cells in 6-Well Plates

1. Seed HEK293T cells at a cell density of 2×10^5 cells per well in 6-well plates (2 ml of medium per well) 24 h prior to siRNA transfection (*see* **Note 2**).
2. Add 40 pmol of siRNA to 100 μl of serum-free DMEM medium (SFM).
3. Equilibrate 2 μl of Lipofectamine 2000 in 98 μl of SFM for each transfection required for 5 min at room temperature.
4. Combine the 100 μl SFM/Lipofectamine 2000 mix and the diluted siRNA and incubate at room temperature for 20 min paying attention not to exceed this 20 min incubation *(14)*.
5. Add the 200 μl siRNA/Lipofectamine 2000/SFM transfection mix to one well of cells in a 6-well plate.
6. Repeat the siRNA transfection 24 h later and isolate RNA 24 to 48 h following the second siRNA transfection. These samples can now be analysed by RT-PCR as described in **Subheading 3.3**.

3.2.2. siRNA Transfection of PBMCs in 10 cm Dishes

1. Seed PBMCs into 10 cm² dishes at a cell density of 1×10^6 cells per ml in a total volume of 15 ml and incubate cells at 37°C for at least 1 h before the first siRNA transfection.
2. For each transfection required add 6.25 μl of Lipofectamine 2000 to 243.75 μl of SFM and incubate at room temperature for 5 min.
3. Add 250 μl of the Lipofectamine 2000/SFM mix to 300 pmol of siRNA diluted to a final volume of 250 μl in SFM and incubate for 20 min at room temperature. Pipette the entire 500 μl transfection mix onto one 10 cm² dish of PBMCs.
4. Repeat the siRNA transfection 24 h later and harvest the cells 24 to 48 h after the second siRNA transfection by scraping the cells into the medium, transferring to a suitable tube and centrifuging at 120 × *g* for 8 min at 4°C.
5. Discard the supernatant and resuspend the cell pellet in 100 μl of 1X Laemmli sample buffer. Sonicate the samples, boil for 5 min at 100°C, and either resolve directly by SDS-PAGE (*see* **Subheading 3.4**) or store at –20°C.

3.3. To Determine Efficacy of siRNA Using RT-PCR

RT-PCR is performed to determine that messenger RNA is being suppressed following the transfection of cells with gene specific siRNA (**Fig. 1**). Alternatively or in addition, proteins may be assayed by Western blot to demonstrate reduced protein expression if an antibody against the protein is available (*see* **Subheading 3.4** and **Fig. 2**). In either case it should be possible to visualise both reduced messenger RNA and protein levels.

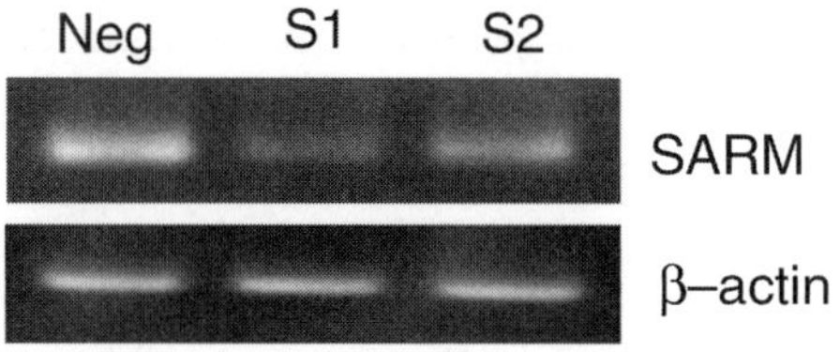

Fig. 1. Knock-down of SARM mRNA following transfection with SARM targeting siRNA in HEK293T cells. HEK293T cells were transfected with either the non-targeting siRNA, Neg or one of the SARM targeting siRNAs, S1 or S2. Twenty-four hours later the siRNA transfection was repeated. RNA was isolated, reverse transcribed, and used as template to PCR amplify SARM (*top panel*) or β-actin (*lower panel*).

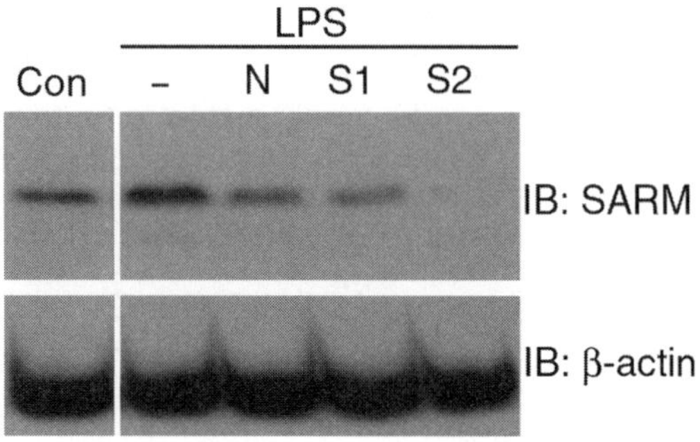

Fig. 2. Knock-down of SARM protein expression following transfection with SARM targeting siRNA in human PBMCs. Human PBMCs were seeded at a density of 1 $\times 10^6$/ml and were allowed to equilibrate for 1 h. The cells were then treated with either the non-targeting siRNA, N or one of the SARM targeting siRNAs, S1 or S2. Twenty-four hours later the siRNA transfection was repeated and the cells were then treated with LPS for 1 h. Immunoblot of cell lysate for SARM using an anti-SARM antibody (*upper panel*). As a loading control the blot was stripped and re-probed for β-actin (*lower panel*).

3.3.1. RNA Isolation from HEK293T Cells

1. Following transfection of siRNAs into HEK293T cells (*see* **Subheading 3.2.1**), discard the cell supernatants, add 200 μl of TRI® Reagent to each well of the 6-well plate and resuspend cells gently using a pipette tip. Transfer the resulting lysate to a sterile Eppendorf and incubate for 5 min at room temperature.
2. Add 40 μl of chloroform to each sample, vortex thoroughly until the mixture turns pink and incubate for 15 min at room temperature. Centrifuge samples at 16,100 × *g* for a further 15 min at 4°C. Samples must be kept at 4°C from now on to minimise the degradation of RNA.
3. Transfer the upper aqueous layer containing the RNA to a fresh sterile Eppendorf, taking care not to remove any of the interface between the two layers as this contains genomic DNA.

4. Add an equal volume of isopropanol to the aqueous layer and incubate on ice for at least 20 min. Alternatively, the samples may be stored overnight at −20°C as this may improve RNA precipitation.
5. Centrifuge the precipitated RNA at 16,100 × *g* for 15 min at 4°C. Discard the supernatant and add 500 μl of 75% ethanol (diluted in diethylpyrocarbonate [DEPC] water).
6. Centrifuge in 75% ethanol at 16,100 × *g* for 5 min at 4°C to wash the RNA pellet. Once again, discard the supernatant and allow the pellet to air-dry ensuring the removal of all traces of ethanol but taking care not to over-dry it as this will make it more difficult to dissolve. Once dry, dissolve the pellet in 20 μl of molecular biology grade water (DNase and RNase free) and store at −70°C.

3.3.2. Reverse Transcriptase Production of cDNA

1. To generate double stranded cDNA add 1 μg of isolated RNA sample along with 0.5 ng of random primers to a suitable sterile RNase-free Eppendorf and make up to a final volume of 5 μl in nuclease free water.
2. Heat to 70°C for 5 min and then place on ice for a further 5 min.
3. In the given order, add the indicated volumes of the following reagents into an Eppendorf, (*see* **Table 1**) scaling up the volumes for the number of samples required.
4. Add 15 μl of this master-mix to each RNA/random primer sample and incubate at 25°C for 5 min followed by an extension step at 42°C for 1 h.
5. Heat the samples to 70°C for 15 min to inactivate the reverse transcriptase enzyme and store the cDNA at −20°C. Avoid freeze/thawing the cDNA samples.

3.3.3. PCR to Detect SARM

1. On ice, add the indicated volumes of the reagents listed below to an Eppendorf, scaling up the volumes for the number of samples required (*see* **Table 2**).
2. Add 23 μl of this PCR master-mix to 2 μl of each cDNA sample (*see* **Subheading 3.3.2**) and carry out a PCR using the following conditions:
   - 95°C for 5 min (1 cycle)
   - 95°C for 30 s, 56°C for 30 s, 72°C for 1 min (30 cycles)
   - 72°C for 10 min (1 cycle)

   Store the samples temporarily at 4°C or at −20°C for longer periods.
3. Add 1.25 μl of 10X DNA loading dye to 12.5 μl of each PCR sample and resolve the entire volume by submerged horizontal agarose gel electrophoresis using a 1% agarose gel.

Table 1
Generation of cDNA by Reverse Transcription

Reagent	End concentration	Volume required (μl)
Nuclease-free water	n/a	1.3
5X ImProm-II reaction buffer[a]	1X	4
25 mM $MgCl_2$[a]	4 mM	3.2
2 mM dNTPs (Novagen)	0.5 mM	5
RNase OUT (Invitrogen)	0.5 μl per reaction	0.5
ImProm-II™ reverse transcriptase[a]	1 μl per reaction	1

[a] From Promega.

Table 2
PCR to detect SARM in cDNA

Reagent	End concentration	Volume required (μl)
H_2O	n/a	12.25
10X reaction buffer (Invitrogen)	1X	2.5
100 pmol/μl SARM forward primer[a] 5′-TGCATGGCTTCAGTGTCTTC-3′	4 pmol/μl	1
100 pmol/μl SARM reverse primer[a] 5′-ACCCTCCAAACTGGTGTCAG-3′	4 pmol/μl	1
50 mM $MgCl_2$ (Invitrogen)	2 mM	1
2 mM dNTPs (Novagen)	0.4 mM	5
Taq DNA polymerase (Invitrogen)	0.25 μl per reaction	0.25

[a] Using these primers will produce a PCR product of 395 bp.

3.3.4. PCR to Detect β-Actin

1. Make up a PCR master-mix containing the reagents listed in **Table 2** that is sufficient for the number of samples required. However, the primers in this instance are

   β-actin forward primer: 5′-CGCGAGAAGATGACCCAGATC-3′

   β-actin reverse primer: 5′-TTGCTGATCCACATCTGCTGG-3′

Using these primers will produce a PCR product of 731 bp.

2. Again, add 23 μl of the PCR mix to 2 μl of each cDNA sample and carry out a PCR reaction under the following conditions:

- 95°C for 5 min (1 cycle)
- 95°C for 20 s, 60°C for 40 s, 72°C for 1 min (30 cycles)
- 72°C for 10 min (1 cycle)

The samples can be kept temporarily at 4°C but for longer term storage the samples should be kept at –20°C.

3. Add 1.25 μl of 10X DNA loading dye to 12.5 μl of each PCR sample and resolve the entire volume by submerged horizontal agarose gel electrophoresis using a 1% agarose gel.

3.4. To Determine Efficacy of siRNA Using SDS-PAGE to Detect Endogenous Protein

3.4.1. SDS-PAGE

1. The following protocol is for use with the Mini-PROTEAN® 3 electrophoresis system from Bio-Rad which accommodates the use of 0.75 or 1.5 mm SDS-PAGE gels. Using the Bio-Rad gel casting apparatus assemble a 0.75 mm spacer plate together with a short plate, ensuring that the plates have been thoroughly cleaned up with detergent and then rinsed with distilled water followed by 70% ethanol.
2. To prepare a single 0.75 mm 10% polyacrylamide resolving gel combine and mix the following reagents: 1.33 ml water, 1.19 ml 30% acrylamide/bis solution, 910 μl 1.5 M Tris-HCl (pH 8.8), 35 μl 10% SDS, 35 μl 10% ammonium persulphate, and 1.4 μl TEMED. These volumes can be scaled up as required if more than one gel is needed and similarly if 1.5 mm spacer plates are used. Immediately pour this gel mix into the pre-assembled gel casting apparatus and overlay with water-saturated butanol. The gel should polymerise within approximately 30 min.
3. Pour off the water-saturated butanol overlay and rinse the gel with distilled water.
4. Prepare the stacking gel by combining and mixing the following ingredients: 1.4 ml water, 0.33 ml 30% acrylamide/bis solution, 0.25 ml 1 M Tris-HCl (pH 6.8), 20 μl 10% SDS, 20 μl 10% ammonium persulphate, and 2 μl TEMED. Immediately layer the stacking gel mix over the polymerised resolving gel and insert a comb with the correct number of wells. The stacking gel should polymerise more rapidly than the resolving gel, taking approximately 20 min.
5. Once the stacking gel has polymerised, remove the comb, rinse the wells with distilled water, and assemble in the electrophoresis apparatus.
6. Fill the inner chamber of the electrophoresis apparatus with 1X running buffer and pour a sufficient amount into the outer chamber to cover the electrode.
7. Load 10 μl of prepared sample per well and reserve one well for 8 μl of pre-stained molecular weight markers.

8. Connect the apparatus to a power supply and initially run the gel at 100 V until the proteins have migrated through the stacking gel whereupon the voltage may be increased to 150 V. The gel should be run until the dye fronts have run off to ensure good separation of proteins.

3.4.2. Western Blot Analysis of SARM Expression

1. Once the proteins have been resolved by SDS-PAGE they are transferred to PVDF membrane by either semi-dry or wet transfer. Instructions for wet transfer using the Bio-Rad mini Trans-Blot tank system are given below.
2. A single gel holder cassette is required for each SDS-PAGE gel. Place a fibre pad which has been soaked in 1X transfer buffer onto the black surface of the cassette. Layer three pieces of 3 mm Whatman filter paper, which has been similarly equilibrated in 1X transfer buffer, onto the fibre pad.
3. Cut a piece of PVDF membrane slightly larger than gel size, immerse it in methanol to activate it, and then equilibrate in 1X transfer buffer.
4. Remove the glass plate sandwich from the electrophoresis unit and separate the short plate from the spacer plate to release the gel. The stacking gel may be cut from the resolving gel and discarded. Carefully place the gel onto the filter paper and place the PVDF membrane gently over it.
5. Soak a further three pieces of filter paper in 1X transfer buffer, set down over the membrane, and cover the gel/blot sandwich with a second equilibrated fibre pad. Take care to remove all air bubbles by rolling out the gel/blot sandwich with a tube or pipette and then carefully close the gel holder cassette.
6. Insert the cassette into the electrode assembly with the black surface of the cassette beside the black surface of the electrode assembly. This orientation is critical to ensure transfer of proteins onto the PVDF membrane as opposed to in the incorrect direction into the buffer.
7. Place the electrode assembly, along with a Bio-Ice cooling unit, into the buffer chamber and fill with 1X transfer buffer. Then, connect the lid of the apparatus to a power supply and transfer at 100 V for 1 h.
8. On completion of the transfer step, disassemble the apparatus and incubate the membrane in blocking buffer (5% milk in PBS-T) for 1 h at room temperature with shaking (*see* **Note 3**).
9. Rinse the blot briefly in PBS-T and incubate with anti-SARM antibody diluted 1 in 1,000 in 1% BSA in PBS-T for 1 h at room temperature with shaking. Alternatively, the blot may be incubated in primary antibody at 4°C overnight.

10. Wash the blot three times in PBS-T for 5 min each and subsequently incubate the blot with HRP-conjugated anti-rabbit secondary antibody, diluted 1 in 10,000 in 1% BSA in PBS-T, for 1 h at room temperature with shaking.
11. Again, wash the blot three times with PBS-T for 5 min each, followed by a further wash in PBS to remove tween which may interfere with ECL detection.
12. For each blot, add 1 ml of ECL solution A to 1 ml of solution B and add 0.61 µl of 30% H_2O_2. Discard the last wash, dab off any excess wash buffer from the membrane onto paper towels, and place the membrane onto a clean surface. Add 2 ml of this freshly prepared ECL reagent onto each blot and incubate at room temperature for 5 min.
13. Remove any excess ECL reagent from the membrane, again by dabbing onto some tissue, and place the blot between two sheets of acetate contained in an exposure cassette. Expose the blot to x-ray film for 30 s or longer if required for bands of interest to appear.

3.4.3. Stripping and Re-Probing Blots for β-Actin

1. Dilute the Re-Blot Plus Strong 10X solution to 1X with distilled water and incubate the blot for 20 min completely immersed in this solution with shaking.
2. Discard the Re-Blot solution and wash the blot twice in PBS-T for 5 min each (*see* **Note 9**).
3. Re-block the membrane in 5% milk in PBS-T and probe with anti-β-actin monoclonal primary antibody diluted 1 in 4,000 in 5% milk in PBS-T and proceed as in **Subheading 3.4.2. 9**.

3.5. Use of siRNA in Reporter Gene Assays to Assess a Role for Gene of Interest in Specific Signalling Cascades

3.5.1. Luciferase Reporter Assays in HEK293 Cells

1. Seed HEK293 cells stably expressing a TLR of interest, e.g. HEK293TLR3 cells or HEK293TLR4 cells, at a cell density of 3 × 10^4 cells per well in a 96-well plate (in 200 µl of medium) 24 h prior to transfection with siRNA (*see* **Note 4**).
2. All reporter gene assay transfections are done in triplicate. Therefore, sufficient plasmid DNA and siRNAs for 3.5 transfections are aliquoted as required into a well of a sterile V-bottomed 96-well plate. For each transfection, 60 ng of either the NFκB- or IFNβ-luciferase reporter gene plasmid are included. Similarly, 20 ng of the phRL-TK reporter plasmid (Promega) are transfected to normalise data for transfection efficiency. The total plasmid DNA concentration is kept constant at 230 ng per transfection by the addition of pcDNA 3.1 empty vector DNA (Invitrogen).
3. Add the required amount of siRNA to the plasmid DNA in each well of the V-bottomed 96-well plate (*see* **Note 5**) and dilute the siRNA and plasmid DNA in 35 µl of SFM.

4. Add 0.2 μl of Lipofectamine 2000 to 9.8 μl of SFM for each single transfection required (*see* **Note 6**). As sufficient plasmid DNA and siRNA for 3.5 transfections are included for each triplicate, similarly, sufficient Lipofectamine 2000 for 3.5 transfections must be allowed for each triplicate. Incubate this Lipofectamine 2000/SFM mix for 5 min at room temperature.
5. Add 35 μl of this mix to each well of the V-bottomed 96-well plate containing the diluted plasmids and siRNAs and equilibrate for 20 min at room temperature.
6. Add 20 μl of the transfection mix onto each of the triplicate wells of the 96-well plate of cells.
7. The cells are transfected with siRNA only 24 h following the first transfection exactly as described above but without any plasmid DNA (*see* **Note 7**).
8. Cells are stimulated with the required TLR agonist, e.g. 25 μg/ml poly(I:C) (for TLR3) or 100 ng/ml LPS (for TLR4), 6 h prior to harvesting cells and 48 h following the initial transfection (*see* **Note 8**).
9. To harvest the cells for measurement of luciferase activity, the cell supernatant is discarded simply by flicking off and each well of adherent cells is lysed in 50 μl of 1X passive lysis buffer (diluted in distilled water) for 15 min at room temperature with vigorous shaking.
10. The resulting lysate is then divided between two 96-well assay plates to determine firefly luciferase and Renilla luciferase activity. Thus, 40 μl of 1X luciferase assay mix is added to 20 μl of cell lysate to determine firefly luciferase activity and 40 μg/ml coelenterazine is similarly added to 20 μl of cell lysate to determine Renilla luciferase. Luminescence is measured using a luminometer. Data are normalised for transfection efficiency using the values obtained for the constitutively active Renilla luciferase and all data are expressed as the mean of triplicate values (±SD) relative to control levels.

3.5.2. ELISA Assays in HEK293 and PBMCs

The contribution of a gene of interest to TLR signalling may also be assessed by measuring cytokine release induced by treatment with a particular TLR agonist, such as poly(I:C) for TLR3 or LPS for TLR4, in cells treated with control non-silencing siRNAs compared to specific target gene siRNAs. Here, the experiment is carried out exactly as described (*see* **Subheading 3.5.1**) except plasmid DNA transfection is not required. Supernatants are harvested as opposed to preparing cell lysates and cytokine release is determined by ELISA according to the manufacturer's instruction.

To confirm the contribution of a particular gene of interest to a specific TLR signalling pathway in a primary human cell model, PBMCs may be treated with siRNA and cytokine release used as a measurement of TLR activity in an experimental set-up very similar to that outlined above except:

1. PBMCs isolated either from buffy coats or fresh blood samples from donors are seeded at a cell density of 2×10^5 cells per well in 200 µl of PBMC medium (*see* **Subheading 2.1**) and allowed to equilibrate at 37°C for at least 1 h prior to siRNA transfection.
2. Slightly more Lipofectamine 2000 is used for each transfection when transfecting PBMCs in a 96-well plate. Thus, 0.25 µl of Lipofectamine 2000 is diluted in 9.75 µl of SFM for each single transfection required (*see* **Note 6**).
3. In the experiments using PBMCs 4 pmol of siRNA per well proved to be most effective (*see* **Note 5**).

4. Notes

1. Tips for the use of siRNA:
   (a) On receipt of the siRNA, reconstitute in the supplied siRNA resuspension buffer. Heat to 90°C for 1 min and incubate at 37°C for 1 h to remove higher order aggregates that may have formed in the lyophilisation process. The siRNA can then be stored at –20°C.
   (b) As far as possible keep the siRNA on ice to prevent hydrolysis. However, the siRNA may be freeze thawed multiple times without affecting stability as long as RNase free conditions are maintained at all times *(14)*.
2. For siRNA it is recommended that antibiotics be left out of the medium as the presence of antibiotics decreases cell viability, especially during transfection. However, in our experiments, we have found this largely not to be the case.
3. After completion of the transfer, the PVDF membranes can be allowed to air-dry and later activated with methanol.
4. Finding the optimal cell density for each cell type used greatly affected the results obtained. Ideally the cells should be 30–50% confluent upon the first siRNA transfection.
5. In many of the experiments we have found that the effective siRNA concentrations varied depending on a number of parameters, such as cell type, the target gene, and the siRNA used. In addition to this we have found that the transfection efficiencies varied slightly, thus the effective siRNA dose

altered between experiments. For this reason we recommend that a range of siRNA concentrations be used. Therefore for these experiments we have used 2, 4, 10, and 20 pmol of siRNA per well.

6. We have found that the amount of Lipofectamine 2000 was a critical parameter in the experiments. While 0.2 μl of Lipofectamine 2000 per well of 96-well plate was effective in the HEK293 cells, 0.25 μl was optimal for transfection of the PBMCs. Thus for optimising the transfection of different cell types we recommend trying a dose range of 0.2–0.5 μl.
7. A variation of this method may also be employed where the siRNA alone is transfected on the first day and the plasmids are introduced with the siRNA in the second transfection.
8. Sonicate the LPS for 2 min before use. Discard working stocks of LPS contained in plastic Eppendorfs as LPS sticks to plastic.
9. The 1x Re-Blot solution may be re-used up to 3 times or until it appears cloudy. The 1x Re-Blot solution can be stored at 4°C.

References

1. Tomari, Y. & Zamore, P. D. (2005). Perspective: machines for RNAi. *Genes Dev* 19, 517–529.
2. Gregory, R. I., Chendrimada, T. P., Cooch, N. & Shiekhattar, R. (2005). Human RISC couples microRNA biogenesis and posttranscriptional gene silencing. *Cell* 123, 631–640.
3. Bartel, D. P. (2004). MicroRNAs: genomics, biogenesis, mechanism, and function. *Cell* 116, 281–297.
4. Elbashir, S. M., Lendeckel, W. & Tuschl, T. (2001). RNA interference is mediated by 21- and 22-nucleotide RNAs. *Genes Dev* 15, 188–200.
5. Caudy, A. A., Ketting, R. F., Hammond, S. M., Denli, A. M., Bathoorn, A. M., Tops, B. B., Silva, J. M., Myers, M. M., Hannon, G. J. & Plasterk, R. H. (2003). A micrococcal nuclease homologue in RNAi effector complexes. *Nature* 425, 411–414.
6. Hutvagner, G. & Zamore, P. D. (2002). A microRNA in a multiple-turnover RNAi enzyme complex. *Science* 297, 2056–2060.
7. Doench, J. G., Petersen, C. P. & Sharp, P. A. (2003). siRNAs can function as miRNAs. *Genes Dev* 17, 438–442.
8. Lewis, B. P., Burge, C. B. & Bartel, D. P. (2005). Conserved seed pairing, often flanked by adenosines, indicates that thousands of human genes are microRNA targets. *Cell* 120, 15–20.
9. Stark, G. R., Kerr, I. M., Williams, B. R., Silverman, R. H. & Schreiber, R. D. (1998). How cells respond to interferons. *Annu Rev Biochem* 67, 227–264.
10. Manche, L., Green, S. R., Schmedt, C. & Mathews, M. B. (1992). Interactions between double-stranded RNA regulators and the protein kinase DAI. *Mol Cell Biol* 12 5238–5248.
11. Minks, M. A., West, D. K., Benvin, S. & Baglioni, C. (1979). Structural requirements of double-stranded RNA for the activation of 29,59-oligo(A) polymerase and protein kinase of interferon-treated HeLa cells. *J Biol Chem* 254, 10180–10183.
12. Caplen, N. J., Parrish, S., Imani, F., Fire, A. & Morgan, R. A. (2001). Specific inhibition of gene expression by small double-stranded RNAs in invertebrate and vertebrate systems. *Proc Natl Acad Sci U S A* 98, 9742–9747.
13. Elbashir, S. M., Martinez, J., Patkaniowska, A., Lendeckel, W. & Tuschl, T. (2001). Functional anatomy of siRNAs for mediating efficient RNAi in *Drosophila melanogaster* embryo lysate. *EMBO J* 20, 6877–6888.
14. Elbashir, S. M., Harborth, J., Weber, K. & Tuschl, T. (2002). Analysis of gene function in somatic mammalian cells using small interfering RNAs. *Methods* 26, 199–213.
15. Reynolds, A., Leake, D., Boese, Q., Scaringe, S., Marshall, W. S. & Khvorova, A. (2004).

Rational siRNA design for RNA interference. *Nat Biotechnol* 22, 326–330.

16. Jagla, B., Aulner, N., Kelly, P. D., Song, D., Volchuk, A., Zatorski, A., Shum, D., Mayer, T., De Angelis, D. A., Ouerfelli, O., Rutishauser, U. & Rothman, J. E. (2005). Sequence characteristics of functional siRNAs. *RNA* 11, 864–872.
17. Jackson, A. L., Bartz, S. R., Schelter, J., Kobayashi, S. V., Burchard, J., Mao, M., Li, B., Cavet, G. & Linsley, P. S. (2003). Expression profiling reveals off-target gene regulation by RNAi. *Nat Biotechnol* 21, 635–637.
18. Chiu, Y. L. & Rana, T. M. (2002). RNAi in human cells: basic structural and functional features of small interfering RNA. *Mol Cell* 10, 549–561.
19. Patzel, V. (2007). In silico selection of active siRNA. *Drug Discov Today* 12, 139–148.
20. Carty, M., Goodbody, R., Schroder, M., Stack, J., Moynagh, P. N. & Bowie, A. G. (2006). The human adaptor SARM negatively regulates adaptor protein TRIF-dependent Toll-like receptor signaling. *Nat Immunol* 7, 1074–1081.
21. Keating, S. E., Maloney, G. M., Moran, E. M. & Bowie, A. G. (2007). IRAK-2 participates in multiple toll-like receptor signaling pathways to NFkappaB via activation of TRAF6 ubiquitination. *J Biol Chem* 282, 33435–33443.

Chapter 18

Genotyping Methods to Analyse Polymorphisms in Toll-Like Receptors and Disease

Chiea-Chuen Khor

Summary

It is now well accepted that a significant genetic component governs host susceptibility to different infectious diseases. As the Toll-like receptors (TLRs), together with their co-receptors and their downstream signalling partners, play such a crucial role in pathogen recognition and subsequent activation of the host immune response, any genetic mutation (polymorphism) that alters the protein structure and in so doing affects the ability of the TLRs or their co-receptors to bind to their associated pathogen-associated molecular patterns (PAMPs) will likely affect host susceptibility towards infection. Examination of the TLR signalling cascade suggests the existence of several bottlenecks or rate-limiting steps, obvious ones being at the level of the receptors, the adaptor proteins, TNF receptor-associated factor 6 (TRAF6), as well as at the IκB/NF-κB interaction point. Mutations to these downstream members might confer either resistance or increased susceptibility, depending on their nature. Indeed, it has been demonstrated time and again that natural variation in some of the molecules mentioned above does affect differential susceptibility to infectious diseases (e.g. invasive bacterial infections, tuberculosis, and malaria) specific to the binding spectrum of the TLRs involved.

Key words: Toll-like receptors, Adaptor protein, Genetic, Polymorphism, Infectious disease.

1. Introduction

The members of the Toll-like receptor (TLR) signalling cascade, being a cornerstone of the innate immune response, are likely to play an important role in the host recognition and host immune response process against invading microbes *(1, 2)*. Polymorphisms which impair the ability of cascade members to signal are likely to be involved in differential susceptibility to infectious pathogens, and recent data from large-scale studies employing candidate gene approaches have shown this to be true

Claire E. McCoy and Luke A.J. O'Neill (eds.), *Methods in Molecular Biology, Toll-Like Receptors, vol. 517*
DOI: 10.1007/978-1-59745-541-1_18

(4–13). Novel polymorphisms in candidate genes can be identified by direct sequencing of DNA. These polymorphisms can then be genotyped by a number of techniques, which will be described below, together with post-experimental statistical analysis. Here, several genotyping methodologies are discussed: the classical restriction fragment length polymorphism (RFLP), genotyping via the Sequenom and Illumina platforms, as well as direct sequencing. The RFLP method is cost effective, but only suited for one marker at a time. The Sequenom platform is excellent for mid-level genotyping throughput (up to 2,500 samples and 50–1,000 markers). Genotyping using the Illumina system usually entails whole-genome screens of 10,000 patients and up to a million markers.

2. Materials

2.1. Patient Samples and DNA Extraction

1. Buccal swabs or heparinised blood tubes for patient sample collection.
2. DNA extraction kit (e.g. Qiagen PAXgene Blood DNA validation kit).

2.2. The Polymerase Chain Reaction (PCR)

1. Five microlitres of genomic DNA at approximately 3 ng/μl.
2. 1.5 μl of 10X PCR buffer (ABI).
3. 0.6–1.5 μl of 25 mM $MgCl_2$.
4. 0.47 μl 8 mM dNTP mix.
5. 0.4 μl of each forward and reverse PCR primers (stock at 10 μM).
6. 0.04 μl of AmpliTaq Gold (ABI).
7. Distilled water to make up a final reaction volume of 15 μl.
8. 96 well/384 well microtitre plate for the PCR.
9. Mineral oil or adhesive sealing lids to prevent evaporation.

2.3. Agarose Gel Electrophoresis

1. Agarose powder (Sigma Aldrich).
2. Tris-acetate (TAE) buffer (Sigma Aldrich). A 10X concentration, but diluted to 1X for use.
3. Ethidium bromide (Generic).
4. Loading dye: 0.025 g xylenol orange is added to equal volumes of glycerol and sterile water.

2.4. Direct DNA Sequencing

2.4.1. Sequencing Reaction

1. Qiagen PCR purification kit.
2. $MultiScreen_{96}$ PCR plates, cat. no.: MANU 030 (Millipore).
3. BigDye vs 3.1 terminator mix (ABI).

4. BigDye buffer (ABI).
5. PCR primers diluted to 10 μM.

2.4.2. Ethanol Precipitation

1. Sodium acetate (3 M, pH 5.5).
2. One hundred per cent (v/v) ethanol.
3. Hi-Di formamide (ABI).

2.5. The Sequenom Mass-Array Platform

1. For each polymorphism to be genotyped, a set of two PCR primers and one extension primer (a total of three primers per assay) should be ordered.
2. Ten nanograms of genomic DNA per reaction.
3. 2.5 mM $MgCl_2$.
4. Two hundred micromolar of dNTPs.
5. Fifty nanomolar of primers per primer pair.
6. Taq polymerase (Hot Star Taq, Qiagen, or alternatively Titanium Taq, BD) for a final reaction volume of 7.5 μl.

2.6. Restriction Fragment Length Polymorphism

1. PCR products for the reaction.
2. One to two units of restriction enzyme per reaction.
3. Two to three per cent (w/v) agarose gel for resolving the digested PCR products.

2.7. The Illumina Platform

The Illumina platform is an ultra-high throughput system more suited for very large scale, whole-genome association studies (where upwards of 100,000 polymorphisms are genotyped simultaneously) compared to the more focused and targeted candidate gene approach. For this assay to be economically feasible, study populations upwards of 10,000 samples are recommended. In summary, this technique involves selecting the polymorphisms one wishes to genotype in a given sample set and sending this requisition to Illumina. The customised chips will be built for the genotyping exercise. Illumina also supplies ready-made chips (e.g. the HumanHap 550 chip, which can genotype up to 500,000 polymorphisms at one go). These details of this technique are too complicated for the purpose of this book.

3. Methods

3.1. Extracting DNA from Patient Samples

Nowadays, DNA extraction from clinical samples (whether blood or buccal swabs) is very straightforward. Many excellent extraction kits (e.g. Nucleon II kits from Scotlab Bioscience, Buckingham, UK, or alternatively Qiagen kits) can be found in the market, and they all revolve around the same basic principles:

(a) Breaking down the cells with a lysis buffer to release cellular contents
(b) Washing to remove cellular debris resulting from the lysis step
(c) Digestion with protease to remove proteins interacting with the DNA
(d) Precipitating with isopropanol and ethanol
(e) Drying of the DNA pellet (usually by air) and re-suspension in distilled water (*see* **Note 1**)

As kits produced by different companies have slightly differing protocols and reagent additives, please refer to the kit you are using for the explicit protocol.

3.2. Basic PCR

As the PCR is the heart of all genotyping techniques, the technique in setting up a basic PCR should be mastered before any large scale genotyping. The PCR is performed before the actual genotyping step, be it the RFLP assay, the Sequenom-MassArray assay, or direct DNA sequencing. It is essential that the PCR step be optimised to yield clean and specific PCR products for the actual genotyping step, as non-specific PCR products will interfere with the genotyping step and affect the genotyping calls. The specificity of the PCR product can be visualised via agarose gel electrophoresis (*see* **Subheading 3.3**, **Fig. 1**).

3.2.1. Setting up the PCR

1. Place 5 μl (approximately 15 ng) of genomic DNA (working stocks at a concentration of 3 ng/μl is suggested) in an Abgene 96 well PCR plate.

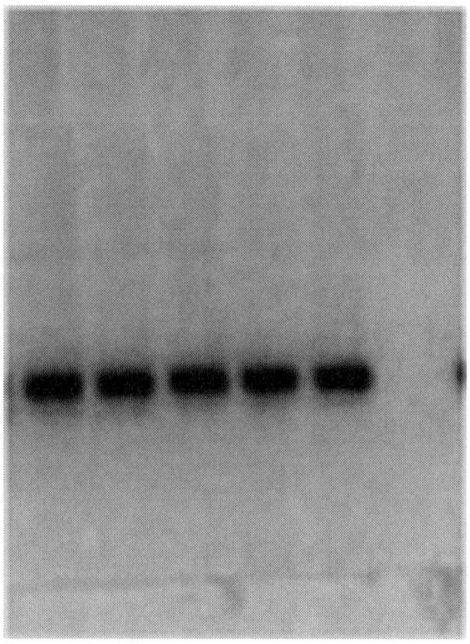

Fig. 1. Specific polymerase chain reaction (PCR) products at 190 bp in size for the region around *Mal/TIRAP* S180L. Note that the PCR product is clean, with no non-specific bands.

2. Prepare the PCR cocktail as appended in **Subheading 2.2, items** 2–7. Add the distilled water, followed by the Taq polymerase. Great care should be taken with the Taq polymerase, as it is apt to go off; if it has not been stored at cold enough temperatures (–20°C being ideal), or if has been left sitting on the bench.
3. Ten microlitres of the prepared PCR cocktail are then added to each sample of genomic DNA.
4. Centrifuge the plate at 250–350 × *g* for 15 s.
5. Add 15 μl of mineral oil to the PCR mix to avoid evaporation during thermal cycling.
6. The reaction mix is then cycled using a thermal cycler (highly recommended are MJ Research tetrads) using the following conditions:
   (a) Ninety-five degrees Celsius for 15 min
   (b) Ninety-five degrees Celsius for 30 s
   (c) Fifty-five to sixty-five degrees Celsius for 30 s
   (d) Seventy-two degrees Celsius for 1–2 min
   (e) Seventy-two degrees Celsius for 5 min

Steps (b)–(d) should be repeated for 37 additional cycles.

7. The amplified products will then be separated by being subjected to agarose gel electrophoresis. For difficult PCRs, suggestions for troubleshooting are included at the end of this chapter (*see* **Note 2**).

3.3. Agarose Gel Electrophoresis

The concentration of the agarose gels that is generally used ranges from 2% to 5% (w/v).

1. Agarose powder is added to 600 ml 1X Tris-acetate (TAE) buffer in a 1.5 L conical flask.
2. Dissolve the agarose powder by heating the mixture for 5–8 min on high power in a microwave oven.
3. Remove the mixture from the oven and swirl it gently around every 2–3 min to ensure even heating.
4. Cool the molten agarose mixture under running water till just above the setting point; the setting point will be achieved if the flask is cool to the touch.
5. Add in 5 μl of ethidium bromide (EtBr) per 100 ml of molten agarose.
6. Quickly swirl the agarose-EtBr mixture and pour off into a suitable gel tray.
7. Add well-forming gel combs so as to leave wells in the agarose gel for sample loading.
8. Leave the gel to set completely.

9. Add 5 μl of loading dye to 5 μl of PCR product in a separate, fresh plate.
10. Load the samples into the wells and electrophorese the PCR samples for 30–40 min at 200 V.
11. The electrophoresed samples can now be visualised via ultraviolet (UV) light transillumination (*see* **Fig 1**. as an example).

3.4. Direct DNA Sequencing

For direct sequencing, the region of interest (which should be between 300 and 1,000 bp) must first be amplified by PCR (*see* **Subheading 3.2**). After ensuring that the PCR yields a clean and specific product (*see* **Subheading 3.3**), the amplified product containing the region of interest can be purified via two ways: By using the

(a) Qiagen PCR purification kit
(b) Vacuum suction method

If the number of samples to be purified is small, then the Qiagen PCR purification kit could be used. If there were many samples to be purified (e.g. 50 or more), then the vacuum suction method is preferable.

The purification protocol using the Qiagen kit is based on the principles of DNA extraction and purification (*see* **Subheading 3.1**). Explicit instructions should be sought in the protocol provided. Again, the final elution step should be with distilled water rather than TE buffer (*see* **Note 1**).

For larger-scale sample purification, the samples are transferred onto Millipore plates (MultiScreen$_{96}$ PCR plates, cat. no.: MANU 030) and the liquid phase suctioned through the membrane using a vacuum manifold, thus leaving the amplified product bound to the membrane of the Millipore plates. After elution with distilled water, the samples are then ready for the sequencing reaction.

3.4.1. Sequencing Reaction

Ideally, this reaction should be performed in a 96 well plate.

1. Prepare the Sequencing cocktail by adding the following per reaction:
   (a) One microlitre of BigDye terminator mix (ABI)
   (b) Three microlitres of BigDye Buffer
   (c) 0.16 μl of primer (diluted to 10 μM)
   (d) Distilled water to add up to 8 μl per reaction
2. Add 2 μl of the purified PCR product to 8 μl of sequencing cocktail for a final reaction volume of 10 μl.
3. The reaction mix is then cycled on a thermocycler (MJ Research) at the following conditions:
   (a) Ramp of one degrees Celsius per second to 96°C
   (b) Ninety-six degrees Celsius for 10 s

(c) Ramp of one degrees Celsius per second to 50°C
(d) Fifty degrees Celsius for 10 s
(e) Ramp of one degrees Celsius per second to 60°C
(f) Sixty degrees Celsius for 4 min

The above six steps should be repeated for 25 cycles.

3.4.2. Ethanol Precipitation

After the sequencing reaction has been performed, the DNA strands are precipitated as follows:

1. Add 2 μl of sodium acetate (3 M, pH 5.5) and 50 μl of 100% (v/v) ethanol per reaction.
2. Centrifuge at 400–500 × *g* for 30 min.
3. Remove the supernatant carefully by pipette without disturbing the DNA pellet at the bottom of the well.
4. Add 100 μl of 70% (v/v) ethanol per sample to wash the DNA pellet.
5. Centrifuge at 400–500 × *g* for 15 min.
6. Again, remove the supernatant carefully without disturbing the pellet.
7. Remove the residual supernatant by inverting the plate and centrifuging at 9 × *g* for 10 s.
8. Leave the samples to air-dry on the bench.
9. When dried, add 10 μl of Hi-Di formamide (ABI) and 10 μl of distilled water to each sample.
10. The sequenced samples can now be run on an ABI PRISM 3700 capillary sequencer (Applied Biosystems). Nowadays, many institutions provide a centralised core support for the running of this step. Please check with your relevant department for administrative details regarding this step.
11. The output sequences can now be aligned and visualised using many different computational softwares, including but not limited to the polyphred Phrap (written by Brent Ewing and Phil Green of the University of Washington Genome Centre, Seattle) and consed (Gordon, Abajian, and Green, 1998) software. Commercially available softwares such as Sequencher (Gene Codes) can also be used.

3.5. The Sequenom MassArray Platform

The Sequenom system is arguably the first of many serious attempts to create a platform for high-throughput genotyping *(3)*. To date, apart from the Sequenom system, there exist other commercially available systems capable of ultra-high throughput genotyping (e.g. Illumina). However, for most genotyping studies using the candidate gene approach, the Sequenom system is much preferred for its flexibility, capability for focused, rapid-response genotyping.

Many core-genomics facilities have optimised in-house protocols in running the Sequenom platform. Reference laboratories include the W.M. Keck Facility at Yale, the Whitehead Institute at MIT, and the Wellcome Trust Center at Oxford.

The Sequenom assay is performed as follows:

1. Amplify the genomic DNA of interest using a PCR cocktail optimised for the Sequenom platform; final reaction volume of 7.5 μl.
   (a) Ten nanograms of genomic DNA
   (b) 2.5 mM $MgCl_2$
   (c) Two hundred micromolar of dNTPs
   (d) Fifty nanomolar per primer pair
   (e) Hot Star Taq (Qiagen), or Titanium Taq (BD)

N.B. The amount of each reagent will vary depending on the size of the multiplex (number of assays being run in a single well)

2. Cycle the PCR cocktail under the following conditions:
   (a) Ninety-five degrees Celsius for 15 min
   (b) Ninety-five degrees Celsius for 20 s
   (c) Fifty-six degrees Celsius for 30 s
   (d) Seventy-two degrees Celsius for 1 min

Repeat steps (b)–(d) for 45 cycles.
   (e) Final extension of 72°C for 3 min

A notable exception here is that mineral oil is not used for the Sequenom system. Instead, plastic seals are applied to the PCR plates prior to thermo-cycling.

After the initial PCR step detailed above, the subsequent steps in the assay (e.g. the clean-up step with artic shrimp alkaline phosphatase, the extension reaction using the extension primer, and spotting onto chips prior to analysis by mass spectrometry) are usually performed by institutional core-genomics facilities using the standard conditions prescribed by Sequenom.

3.5.1. Optimising the Sequenom Platform

The Sequenom system performs extremely well when markers were run singly (uniplex) or in pairs (duplex). Higher levels of multiplexing using the standard protocol generally resulted in inconsistent genotyping success rates, ranging from 40% to 75%, and also frequently results in the failure of the assay for one or more markers. As such, the assay was tested using several different conditions and reagents, and the optimisation points are described below:

1. Emphasis was placed on the usage of concentrated genomic DNA for the initial PCR, or alternatively preamplified DNA using the GenomiPhi protocol advocated by GE Healthcare. At least 10 ng of template DNA was required for multiplex

levels of three or higher. The usage of 15 ng per reaction was often ideal. The success rate was also found to decrease sharply when too much template DNA was used (quantities in excess of 20 ng per reaction often adversely affects the PCR step).

2. For higher multiplexing levels (four or more primer pairs), it had been found that lowering the final concentration of each primer pair to 40 nM or so enables each primer pair to bind successfully to the template DNA, and thus increasing the genotyping success rate. For heptaplex assays, a final concentration of 30 nM per primer pair was often optimal.
3. Each primer pair should be tested separately prior to multiplexing. Each of them should yield specific PCR products. Failure to do so should result in the assay being redesigned.

This optimised method yields genotype success rates of ~90% with all three genotypes (wild-type, heterozygote, and homozygous variant) clearly distinguishable on the scatter-plots with no ambiguity (**Fig 2**.).

3.6. Restriction Fragment Length Polymorphism

The classical RFLP is based on natural variation in the genome, which creates or destroys sites recognisable by restriction enzymes. For our purposes, the genomic location surrounding the mutation of interest is amplified via PCR, followed by restriction enzyme digestion of this amplified product.

1. Plate out 5–10 µl of the amplified fragment of DNA (*see* **Subheading 3.2**) containing the genomic region of interest into a 96 well plate.
2. To each DNA sample, add 1–2 U of restriction enzyme, together with the accompanying buffer supplied with the enzyme to make up a final reaction volume of 10–15 µl (*see* **Note 3**).
3. Incubate the digest reaction at the temperature specified in the manufacturer's instructions (usually between 37°C and 65°C overnight).
4. Prepare a 2–4% (w/v) agarose gel.
5. Perform agarose gel electrophoresis on the digested PCR products (*see* **Subheading 3.3**).
6. Visualise the disgested fragments via UV transillumination (sample output in **Fig 3**.).
7. When no suitable restriction enzyme could be found, genotyping by this method can nevertheless be performed by using a mismatched oligonucleotide to generate an appropriate restriction site. This is known as "site-directed mutagenesis-assisted RFLP". Please refer to **Note 4** for the details in designing this assay using the *TIRAP* S180L mutation as an example.

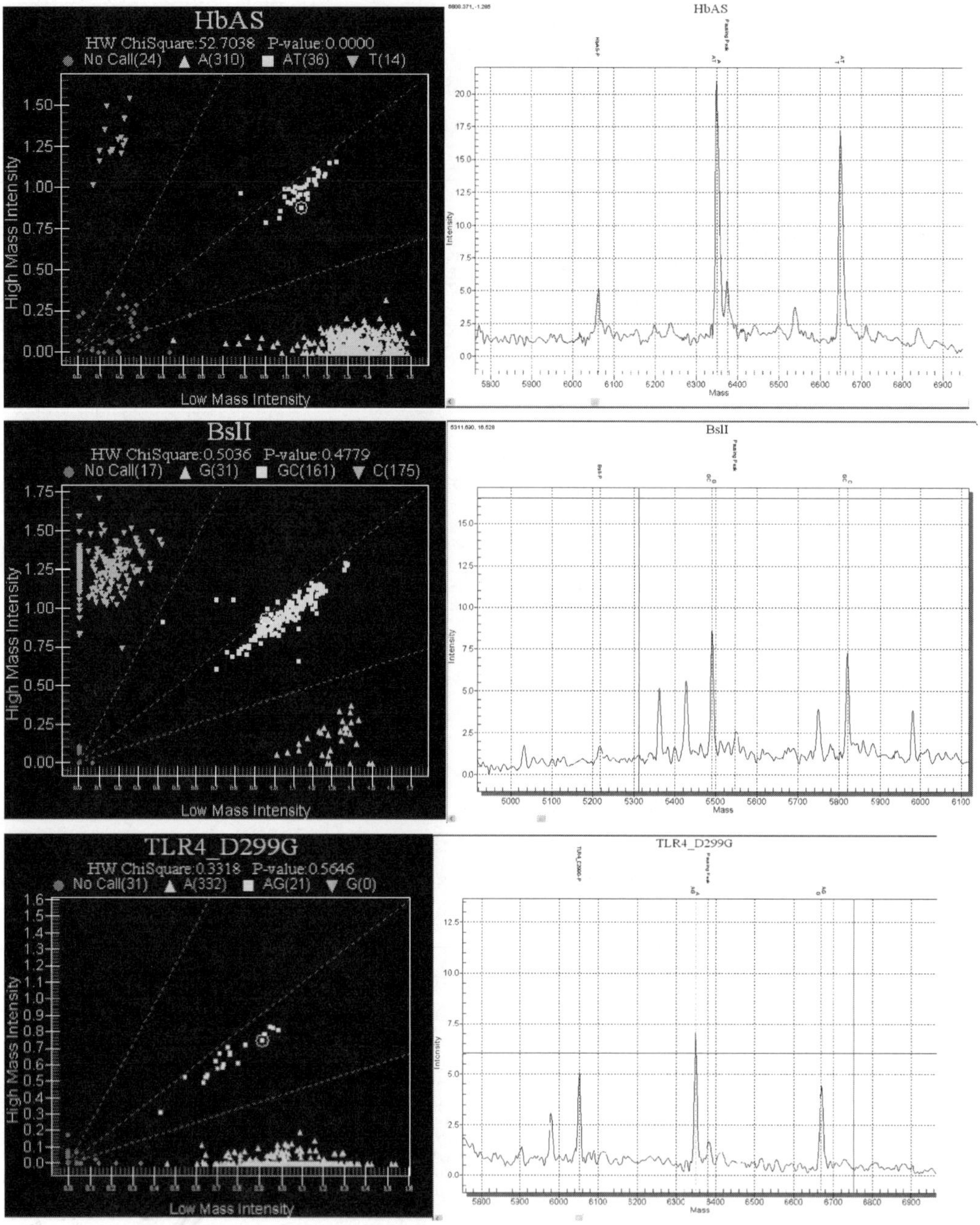

Fig. 2. Scatter-plots and trace-files of three markers genotyped using the Sequenom platform. All three trace-files on the right were from heterozygous individuals, where the peaks denoting the two alleles are clearly distinguishable.

3.7. Quality Control Measures

Ideally, two independent techniques (alternative chemistry) should be used to verify the accuracy of genotyping. For example, Sequenom calls being counter checked via RFLP or direct sequencing. Other quality control checks were that the genotyping success rate had to be above 90%, and the genotype distribution in the

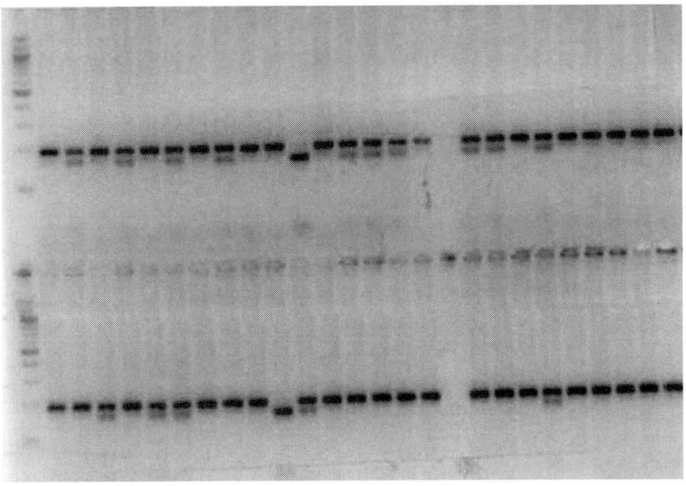

Fig. 3. Restriction fragment length polymorphism (RFLP) assay for *TIRAP* 180Ser Leu with *BstX*I restriction enzyme. The wild-type 180Ser is uncut, whereas the mutant 180Leu, which creates a restriction site for *BstX*I, is denoted by the cut, lower fragment. Heterozygote carriers display both cut and uncut bands.

control group was checked for significant deviation from Hardy–Weinberg equilibrium. Markers whose genotype distribution in the controls significantly deviated from Hardy–Weinberg equilibrium ($P < 0.05$) may be analysed, but the results should be interpreted with extreme caution.

3.8. Statistical Analysis

The Pearson's chi-squared (χ^2) test is used to compare the differences in genotype and allele frequencies differences in genotype distribution between the cases and controls. If the polymorphism typed is rare (e.g. less than ten cell counts being present), then the more conservative Fisher's exact test should be used. The conventional threshold for declaring statistical significance in single marker analysis is $P < 0.05$. However, in today's world of large-scale genome-wide association studies utilising hundreds of thousands of markers, a conservative threshold of declaring statistical significance has been proposed to be in the neighbourhood of $P = 10^{-7}$. Replication in an independent study population would also lend credence to the initial observation.

4. Notes

1. Try to avoid re-suspending the DNA pellet in TE buffer, as TE buffer contains EDTA, which will chelate the magnesium ions from the PCR mix. Magnesium is a crucial co-enzyme for Taq polymerase to work.

2. There are bound to be regions in the human genome with multiple repeating sequences, or strings, or A/T repeats. For these regions, where multiple non-specific products are often amplified, thus making clean PCR products difficult to obtain, dimethyl sulphoxide (DMSO) can be added to the PCR mix to make up 5% of the final reaction mix volume (0.75 μl). DMSO inhibits the binding of the PCR primers to genomic regions, and thus could help reduce the incidence of non-specific priming.
3. It is sometimes necessary to optimise the RFLP protocol by adding more (up to 5 U) of restriction enzyme per reaction to obtain reasonable resolution between fragments. An example of a restriction enzyme which has this property is MfeI.
4. The genomic sequence of the region of interest in the seventh exon of *TIRAP* is shown below. The 180Ser → Leu mutation is highlighted. It represents a C → T point mutation. As there were no naturally occurring restriction enzyme recognition sites occurring in the genomic region, a single point mutation has been introduced into the forward PCR primer to artificially create a site. The region where the forward primer is designed is underlined. A C → A point mutation introduces a *BstXI* restriction enzyme site (recognition site: CCANNNNNNTGG, highlighted in grey) for the 180Leu mutation only. The wild-type 180Ser remains uncut. Heterozygote carriers display both cut and uncut bands. No changes were made to the reverse primer.

CTCATCACGCCGGGCTTCCTTCAGGACCCCT-
GGTGCAAGTACCAGATGCTGCAGGCCCTGAC-
CGAGGCTCCAGGGGCCGAGGGCTGCACCATCC
CCCTGCTGT**[C/T]**GG GCCTCAGCAGAGCTGCYTAC-
CCACCTGAGCTCCGATTCATGTACTACCTCGAT-
GGCAGGGGCCCTGATGGTGGCTTTCGTCAAGTCAAA-
GAAGCTGTCATGCGTTGTAAGCTACTACAGGAGGGA-
GAAGGGGAACGGGATTCAGCTACAGTATCTGATC-
TACTTTGACTTTTAGGAGACAGCCCTGTAGCCTAG-
TAGTTCAAAGCGCAGCTTCTGGAARAGGCTGTCG-
GGGTTTGTATCCTGGCTCCTGCACTT
Forward primer: CTCCAGGGGCCGAGGGCTGCACCATCC
CC[C → A]TGCTG
Reverse primer: TACTGTAGCTGAATCCCGTTCC

Acknowledgements

The author would like to thank Adrian Hill, Luke O'Neill, Stephen Chapman, and Fredrik Vannberg for their helpful suggestions, encouragement, and support.

References

1. Hill, A. V. (2006) Aspect of genetic susceptibility to human diseases. *Annu. Rev. Genet.* 40: 469–86
2. Takeda, K., Kaisho, T., Akira, S. (2003) Toll-like receptors. *Annu. Rev. Immunol.* 21: 335–76.
3. Jurinke, C., van den Boom, D., Cantor, C. R., Koster, H. (2002) Automated genotyping using the DNA MassArray technology. *Methods Mol. Biol.* 187: 179–92.
4. Poltorak, A., He, X., Smirnova, I., Liu, M.Y., Van Huffel, C., Du, X., Birdwell, D., Alejos, E., Silva, M., Galanos, C., Freudenberg, M., Ricciardi-Castagnoli, P., Layton, B., Beutler, B. (1998) Defective LPS signaling in C3H/HeJ and C57BL/10ScCr mice: mutations in Tlr4 gene. *Science.* 282: 2085–8.
5. Hoshino, K., Takeuchi, O., Kawai, T., Sanjo, H., Ogawa, T.,Takeda, Y., Takeda, K., Akira, S. (1999) Cutting edge: Toll-like receptor 4 (TLR4)-deficient mice are hyporesponsive to lipopolysaccharide: evidence for TLR4 as the Lps gene product. *J. Immunol.* 162: 3749–52
6. Smirnova, I., Mann, N., Dols, A., Derkx, H.H, Hibberd, M.L., Levin, M., Beutler, B. (2003) Assay of locus-specific genetic load implicates rare Toll-like receptor 4 mutations in meningococcal susceptibility. *Proc. Natl. Acad. Sci. U. S. A.* 100: 6075–80.
7. Hawn, T.R., Verbon, A., Lettinga, K.D., Zhao, L.P., Li, S.S., Laws, R.J., Skerrett, S. J., Beutler, B., Schroeder, L., Nachman, A., Ozinsky, A., Smith, K.D., Aderem, A. (2003) A common dominant TLR5 stop codon polymorphism abolishes flagellin signaling and is associated with susceptibility to legionnaires' disease. *J. Exp. Med.* 198:1563–72.
8. Hawn, T.R., Verbon, A., Janer, M., Zhao, L.P., Beutler, B., Aderem, A. (2005) Toll-like receptor 4 polymorphisms are associated with resistance to Legionnaires' disease. *Proc. Natl. Acad. Sci. U. S. A.* 102:2487–9.
9. Hoebe, K., Georgel, P., Rutschmann, S., Du, X., Mudd, S., Crozat, K., Sovath, S., Shamel, L., Hartung, T., Zahringer, U., Beutler, B. (2005) CD36 is a sensor of diacylglycerides. *Nature* 433: 523–7.
10. Khor, C.C., Chapman, S.J., Vannberg, F.O., Dunne, A., Murphy, C., Ling, E.Y., Frodsham, A.J., Walley, A.J., Kyrieleis, O., Khan, A., Aucan, C., Segal, S., Moore, C.E., Knox, K., Campbell, S.J., Lienhardt, C., Scott, A., Aaby, P., Sow, O.Y., Grignani, R.T., Sillah, J., Sirugo, G., Peshu, N., Williams, T.N., Maitland, K., Davies, R.J., Kwiatkowski, D.P., Day, N.P., Yala, D., Crook, D.W., Marsh, K., Berkley, J.A., O'Neill, L.A., Hill, A.V. (2007) A Mal functional variant is associated with protection against invasive pneumococcal disease, bacteremia, malaria and tuberculosis. *Nat. Genet.* 39: 523–38
11. Krishnegowda, G., Hajjar, A.M., Zhu, J., Douglass, E.J., Uematsu, S., Akira, S., Woods, A.S., Gowda, D.C. (2005) Induction of proinflammatory responses in macrophages by the glycosylphosphatidylinositols of *Plasmodium falciparum*: cell signalling receptors, glycosylphosphatidylinositol (GPI) structural requirement, and regulation of GPI activity. *J. Biol. Chem.* 280: 8606–16.
12. Dunne, A., Ejdeback, M., Ludidi, P.L., O'Neill, L.A., Gay, N.J. (2003) Structural complementarity of Toll/interleukin-1 receptor domains in Toll-like receptors and the adaptors TIRAP and MyD88. *J. Biol. Chem.* 278: 41443–51
13. Thoma-Uszynski, S., Stenger, S., Takeuchi, O., Ochoa, M.T., Engele, M., Sieling, P.A., Barnes, P.F., Rollinghoff, M., Bolcskei, P.L., Wagner, M., Akira, S., Norgard, M.V., Belisle, J.T., Godowski, P.J., Bloom, B.R., Modlin, R.L. (2001) Induction of direct antimicrobial activity through mammalian toll-like receptors. *Science* 291: 1544–7.

Part IV

Tool-Like Receptors and Disease

Chapter 19

Experimental Models of Acute Infection and Toll-Like Receptor Driven Septic Shock

Ruth Ferstl, Stephan Spiller, Sylvia Fichte, Stefan Dreher, and Carsten J. Kirschning

Summary

Mainly Gram-negative and Gram-positive bacterial infections, but also other infections such as with fungal or viral pathogens, can cause the life-threatening clinical condition of septic shock. Transgression of the host immune response from a local level limited to the pathogen's place of entry to the systemic level is recognised as a major mode of action leading to sepsis. This view has been established upon demonstration of the capacity of specific pathogen-associated molecular patterns (PAMPs) to elicit symptoms of septic shock upon systemic administration. Immune stimulatory PAMPs are agonists of soluble, cytoplasmic, as well as/or cell membrane-anchored and/or -spanning pattern recognition receptors (PRRs) such as Toll-like receptors (TLRs). However, reflection of pathogen–host crosstalk triggering sepsis pathogenesis upon an infection by a host response to challenge with an isolated PAMP is incomplete. Therefore, an experimental model more reflective of pathogen–host interaction requires experimental host confrontation with a specific pathogen in its viable form resulting in a collective stimulation of a variety of specific PRRs. This chapter describes methods to analyse innate pathogen sensing by the host on both a cellular and systemic level.

Key words: PAMP, PRR, TLR, NF-κB, Luciferase activity, IL-6, TNFα, NO, LPS, Shock.

1. Introduction

LPS is an eminently stable pathogen-associated molecular pattern (PAMP) whose structure has been known for some time *(1, 2)*. Like most PAMPs, LPS lacks intrinsic toxicity since it has no decomposing activity or acrid properties. This statement was supported by the unresponsiveness of specific biological species to LPS challenge while the immune stimulatory capacity of LPS and analysis of LPS-resistant strains of mice led to the proposal of its role as an agonist to a receptor encoded by the so-called

Claire E. McCoy and Luke A.J. O'Neill (eds.), *Methods in Molecular Biology, Toll-Like Receptors, vol. 517*
DOI: 10.1007/978-1-59745-541-1_19

LPS-gene (*LPS*) *(3, 4)*. Cloning of LPS binding protein (LBP, an acute phase serum protein), as well as identification of LBP and CD14 (a glycosyl phosphatidyl inositol-anchored and leucine-rich domain (LRR)-rich monocyte differentiation marker) as LPS co-receptors *(5, 6)*, spurred the search for other pattern recognition receptors (PRRs) mediating LPS-induced cell activation *(7)*. Identification of *Drosophila* Toll carrying a CD14-like (LRR-rich) extracellular domain and an IL-1 receptor-like intracellular domain (TIR) as a developmental and antimicrobial signal-transducing receptor was the basis for the identification of mammalian Toll-like receptors (TLRs) and the intracellular signalling capacity of TLR4 *(8–15)*. Subsequently, TLR2 was identified as an "outside–inside" cellular LPS signal transducer *(16, 17)*. At the same time, genomic analysis of LPS-resistant mouse strains identified TLR4 as the *LPS* product evidencing the dominant LPS sensor function of TLR4 *(18, 19)*. The LPS signal transducing function of TLR4 relies on TLR4-MD-2 complex formation *(20, 21)*.

Immune stimulatory microbial or viral products such as specific lipoproteins have been assigned as agonists to TLR 1-2, 2-6, or 2-2 dimers *(22–29)*, while DNA, flagellin, polyinosinic–polycytidylic acid (poly I:C) (double-stranded RNA mimic), single-stranded RNA, and protozoan profilin-like protein together with uropathogenic bacteria are assigned as agonists to TLR9, TLR5, TLR3, TLR7, and TLR8, as well as murine TLR11, respectively *(30–38)*. The TLR–agonist pairs listed above are representative only since TLR specificity is more comprehensive *(39, 40)*. Agonists of human TLR10, as well as murine TLR12 and TLR13, have not yet been described. However, a high importance for all TLRs residing at the cell surface or/and in the endosome to sense invading pathogens is implied *(41, 42)*. Specific plasma membrane traversing or cytoplasmic signal-transducing PRRs other than TLRs have substantial impacts on pathology accompanying infections or chronic inflammations as well *(43)*. Possibly all TLR ligands have cytoplasmic receptors additionally *(44)*.

Conditional activation of the host immune system upon sensing of PAMPs like LPS mediates immediate fighting off microbial invaders. PAMP challenge via PRRs such as TLRs is also required to activate the adaptive immune system, which is a requisite for the clearance of infection *(39)*. This feature is exploited for vaccination through usage of immune stimulatory PAMPs as adjuvant *(45)*. Otherwise, immune over-activation seen in sepsis damages the host organism itself, which leads to septic shock eventually due to high PAMP dose-dependent PRR activity *(46–48)*. Understanding the mechanisms underlying pattern recognition as the initial step of beneficial or adverse immune activation *(42, 49)* might improve purposive modulation of responsiveness to infection or other immune stimulatory challenges such as those operative in vaccination, autoimmunity, transplantation, or trauma.

The human embryonic kidney (HEK) 293 cell line is applied most widely for specific PRR function analysis for two reasons mainly. First, HEK293 cells are largely non-responsive to PAMP challenge yet are responsive to cytokines such as IL-1 and TNFα. This feature implies functionality of cytoplasmic signalling molecules and transcription factors such as TNF receptor-associated factors (TRAFs), receptor interacting proteins (RIPs), MyD88, and further Toll/IL-1 receptor (TIR) domain-containing adapter molecules such as Toll/IL-1 receptor domain-containing adaptor-inducing IFNβ (TRIF), Toll/IL-1 receptor domain-containing adaptor-inducing IFNβ-related adaptor molecule (TRAM), and/or Toll/IL-1 receptor domain-containing adaptor protein (TIRAP)/MyD88 adaptor-like (Mal) in HEK293 cells *(50)*. Furthermore, mitogen-activated protein kinases (MAPKs), activator protein (AP) 1, specific interferon-regulated factors (IRFs), nuclear factor (NF)-κB family members, as well as inhibitors of NF-κB and their kinases, are expressed by HEK293 cells *(39)*. Therefore, HEK293 cells normally gain responsiveness to specific agonists upon overexpression of merely the cellular PRR(s) they interact with. Secondly, HEK293 cell transfection is simple and protein overexpression levels are often high. In contrast to HEK293 cells, immune cells such as macrophages or dendritic cells (DC) typically express specific arrays of different PRRs endogenously due to their surveillance function at the first line of the immune system. Their challenge with an agent such as a PAMP and subsequent analysis of the resulting activation status are generally informative on the immune stimulatory capacity of the agent. Addressing the question for a specific PRR involved in cellular recognition of an immune stimulatory agent using primary immune cells sometimes requires inactivation of the candidate PRR. This can be accomplished by pre-treatment with a specific receptor antagonist (e.g. neutralising ligand or ligand derivatives, intracellular signalling inhibitor, or neutralising antibody), or inhibition or prevention of mRNA expression (knock down or targeted gene knock out, respectively). If analysis for involvement of all known PRRs in recognition of an agent reveals negative results, identification of a yet unknown PRR might rely on consideration of a protein not yet implied in pattern recognition or screening of a cDNA library for expression cloning using ligand-specific cell activation as read-out.

Primary immune cells are prepared for analysis in vitro typically by isolation from compartments and tissues such as blood, spleen, lymph nodes, peyer's patches, peritoneum, or lung or interstitial space, or by systematic differentiation of bone marrow stem cells *(51)*. Analysis of effects of systemic application for instance by intravenous (i.v.) or intraperitoneal (i.p.) injection, ingestion, or inhalation of PAMPs, or infection via one of the routes named above requires availability of an approved experimental model system. The three different methods listed below describe methods to characterise a pattern recognition-specific aspect of a pathogen–host interrelation involving PRR-specific gain of function or loss of function both at the cellular and systemic level.

2. Materials

2.1. Cell Culture and PAMP Ligands

1. Dulbecco's Modified Eagle's Medium (DMEM) for culture of human embryonic kidney (HEK) 293 cells.
2. RPMI1640 containing 50 μM mercapto-ethanol for primary peritoneal macrophage culture.
3. Media for both cell types should contain 10% fetal calf serum (FCS) (optional: reduction to 2% in order to minimise potential pattern recognition inhibitory serum effects) and antibiotics such as penicillin and streptomycin to exclude interference by culture infection. Cells used should be free of persisting infections such as with mycoplasma.
4. Autoclaved thioglycolate (Sigma) solubilised in filter-sterilised phosphate-buffered saline (PBS) *(52)*.
5. Syringe (2 ml or larger) coupled to a 21G or similar injection needle.
6. Inbred mice strains of potentially relevant phenotypes housed under specific pathogen-free (SPF) conditions, regular night–day cycling, 21°C, air conditioning, weekly cage exchange, and access to standard nutrition and drinking water ad libitum.
7. Suspensions of viable, inactivated, or lysed microbial organisms or viruses if not purified or synthetic products such as
   - Synthetic bacterial lipopeptide analogues (e.g. Pam_3CSK_4, an N-terminally tripalmitoylated hexapeptide carrying N-terminal cysteine, e.g. EMC microcollections)
   - Polyinosinic–polycytidylic acid (poly I:C) (Sigma, Calbiochem, or Fluka)
   - LPS (rough LPS lacking a long saccharide chain such as that of *Salmonella enteridis* serovar Minnesota substrain Re595 or smooth LPS carrying a long saccharide chain such as that of *Escherichia coli* O111:B4) (Sigma or List Biological laboratories)
   - R848 (Alexis)
   - CpG DNA oligonucleotides (e.g. TIB molbiol) are to be analysed for their PRR recruitment. Preparations of isolated compounds (but not necessarily of viable or inactivated organisms) should be solubilised and made ready for application to cells by sonication or dissolving otherwise as appropriate.

8. Pro-inflammatory cytokines such as IL-1β or TNFα, as well as/or phorbol 12 myristate 13 acetate (PMA, Sigma), are candidates for positive control stimulants.

2.2. Reporter Gene Activation Assay

1. DNA plasmids driving
   (a) Constitutive PRR such as membrane bound TLR or cytoplasmic nuclear oligomerisation domain-containing receptor (NOD) expression
   (b) Co-receptor such as MD-2 or/and CD14 expression
   (c) And/or signalling molecule (for instance MyD88 or specific interferon responsive factor) expression under control of promoters such as that of cytomegalovirus
   (d) As well as first reporter protein (such as renilla luciferase whose expression might be controlled by constitutively active promoter) expression
   (e) And second/inducible (such as NF-κB- or other transcription factor-driven) reporter protein such as firefly luciferase expression for transfection by a method such as calcium phosphate precipitation (see below).
2. Two molar Ca_3Cl_2.
3. A 2x HBS: 5 g HEPES, 0.37 g KCl, 0.105 g Na_2HPO_4, 8 g NaCl, adjust to 500 ml (aqua bidest), and pH 7.1 *(53)*.
4. Lysis buffer (Promega).
5. Renilla luciferin substrate solution: 64.28 g NaCl (1.1 M), 6.42 g Na_2EDTA (2.2 M), 5.44 g KH_2PO_4 (0.22 M), 0.44 g BSA, 0.084 g NaN_3 (1.3 M), 0.6 mg coelenterazine (1.43 μM), adjust to 1 l (aqua bidest), and pH 5 *(54)*.
6. Firefly luciferin substrate solution quenching renilla luciferase activity: 292 mg ATP (530 μM), 0.52 g $(MgCO_3)_4 Mg(OH)_2{\cdot}5H_2O$ (1.07 mM), 322 mg $MgSO_4{\cdot}7H_2O$ (2.67 mM), 3.584 g tricine (20 mM), 37.2 mg EDTA (0.1 M), 5.14 g DTT (33.3 mM), 210 mg coenzyme A (270 μM), 132 mg D-luciferin (470 μM), adjust to pH 7.8, and 1 l (aqua bidest) *(55)*.
7. Luciferase activity reader such as for single tube or 96 well plate reading with/without (injection by pipetting) integrated injection unit (e.g. Berthold, for instance injection of 50 μl luciferin solution and immediate measurement of light flashes for 10 s).

2.3. Assay of Primary Macrophage Activation

1. Cultured cells (*see* **Subheading 3.1**), primed overnight optionally.
2. Griess reagent for measurement of nitrite concentration in cellular supernatants *(56)*:

(a) 0.4 g *N*-(1-napthyl) ethylene diamine dihydrochloride in 200 ml aqua bidest.

(b) Four grams of sulphanylamide and 10 g H_3PO_4 in 200 ml aqua bidest.Both solutions are stable for several weeks in the dark at room temperature if stored separately. One molar $NaNO_2$ solution is used for preparation of standard and can be stored for 2 weeks at 4°C (standard concentrations such as 300, 100, 30, 10, 3, 1, 0.3 μM). Reader for absorption of light at 450 nm (and 570 nm for normalisation of plate absorption) is required.

3. Enzyme-linked immunosorbent assay (ELISA) kits for analysis of concentrations of TNFα, IL-6, of other cytokines, chemokines, or of other proteinaceous inflammatory mediators present in cell culture supernatants potentially. Self-plating and purchasing separate coating and detection antibodies is economical (Biosource or duo kit from R&D).

2.4. Analysis of Mouse Responsiveness to Systemic PAMP Challenge

1. Inbred mice stains (*see* **Subheading 2.1, item 6**).
2. Syringes (such as standard insulin syringes) for injection of agents.
3. PAMP or whole pathogen preparation to be applied (lethal doses of LPS in mice are 48 mg/kg body weight if applied alone or 4 mg/kg if applied together with D-galactosamine, D-galN).
4. D-galN (Sigma, 0.5 mg/ml sterile PBS, 800 mg/kg body weight).
5. Recombinant murine IFNγ (1.25 μg/300 μl sterile PBS).

3. Methods

The immune stimulatory potency of different PAMPs varies. Therefore, the doses of specific compounds or microbial preparations to be applied might have to be titrated to find a concentration at which cell or systemic activation is effective. LPS might serve as a positive control in both cellular and systemic analysis. In exceptional cases, other PAMPs such as lipopeptide or DNA oligonucleotides can be used depending on the expression of the respective PRRs in the experimental system applied. Instead of application of pure PAMPs, whole pathogens such as bacteria may also be applied. However, the activity of pathogens applied may lead to disintegration of mammalian cells ablating analysis of cell or immune activation. For analysis of pattern recognition specifically, application of inactivated pathogens might be required. Inactivation can be performed for example by heat treatment, sonification, irradiation, solubilisation in solvent, or through the use of antibiotics.

Specific PRRs might not function immediately upon infection, instead priming is required. For example, full sensitiveness of macrophages for TLR2 ligands and NO release from macrophages depends on activation of first line sensors such as TLR4 leading to release of IFNγ by other immune cells such as natural killer cells retroacting on macrophages. Accordingly, maximal TLR2 activity and NO release of macrophages in vitro rely on priming such as with IFNγ (recombinant, 20 ng/ml) (*see* **Subheading 3.3**). Systemic IFNγ priming mimicking an underlying infection can be achieved by injection of 50 μg/kg body weight murine IFNγ into the tail vein 45 min prior to systemic PAMP challenge of mice (*see* **Subheading 3.4**). This consideration is not an issue to be considered upon ectopic PRR overexpression necessarily.

3.1. Cell Culture

1. Adherent HEK293 cells are maintained in DMEM and should be grown below confluence. For experimental procedures, cells are seeded upon trypsinisation at a low density such as 7.5×10^3 cells in a single well of a 96 well plate to enable separation of individual cells from other cells on the plate (*see* **Note 1**).
2. For preparation of primary peritoneal macrophages, wild-type and another genotype such as TLR knockout are given an intraperitoneal injection of 2–4 ml of autoclaved 4% thioglycolate solution in PBS (*see* **Notes 2** and **3**). This elicits macrophage invasion into the peritoneum.
3. Five days later, mice are sacrificed by incubation in a CO_2-saturated atmosphere followed by opening of ventral fur by careful longitudinal incision while maintaining the integrity of the peritoneum. Subsequently, intraperitoneal washout is initiated by injection of 10 ml cold and sterile PBS containing 2% FCS using a sterile syringe. Positioning of the syringe tip has to be controlled by visual inspection to prevent inner organ damage and clotting of the needle during fivefold alternate injection and aspiration of the 10 ml 2% FCS solution by pushing and pulling the piston to collect peritoneal content of dissolute cells (*see* **Note 4**).
4. Transfer cell solutions into sterile 15 ml plastic tubes followed by centrifugation for 5 min at $600 \times g$ (*see* **Note 5**).
5. The cell pellet is carefully solubilised in 10 ml of full cell culture medium. Centrifugation and solubilisation are repeated.
6. Cells are counted and seeded in a 96 well plate where $1.5–2 \times 10^5$ cells can be transferred into each well.
7. Two hours upon incubation at standard cell culture conditions, the supernatant is replaced with fresh medium to remove any non-adherent cells (*see* **Note 6**).

3.2. PAMP Challenge of Transfected Cells and Luciferase Assay

1. Freshly split HEK293 cells are seeded at a density at which each cell adheres to the bottom in the absence of direct contact with neighbouring cells. The 7.5×10^3 cells or up to 3×10^5 cells could be seeded in 200 μl medium per well of a 96 well plate (*see* **Note 7**).
2. After a minimum of 2 h, cells can be transfected using calcium phosphate (such as $Ca(H_2PO_4)_2$, $CaHPO_4$, and/or $Ca_3(PO_4)_2$) DNA precipitation *(52)*. If NF-κB-dependent reporter gene activation is to be analysed, an expression plasmid for ELAM promoter-driven luciferase expression that is used widely might be applied *(57)*. Also, constitutively active reporter plasmid for normalisation, as well as CMV promoter-driven effector molecule such as TLR and co-receptor expression plasmids, can be transfected into cells. One well of a 96 well plate might be transfected exemplarily for analysis of TLR4 activity with a solution containing:
   - 32 ng ELAM-luc plasmid
   - 64.5 ng renilla-luc plasmid
   - 0.15 ng empty vector
   - 2.5 ng MD-2 plasmid
   - 1 ng TLR4 plasmid (*see* **Note 8**)
   - 0.98 μl $CaCl_2$ (2 M)
   - 7.8 μl H_2O

This mixture is then pipetted into an opened tube positioned on a vortexer containing 7.8 μl 2x HBS. The resulting mixture is then pipetted into the cell culture supernatant of one well of a 96 well plate. In practice, a master-mix containing both of the two solutions (DNA-$CaCl_2$ and 2x HBS) at volumes such as 500 μl each can be used to transfect multiple wells. Rather than incubation of the mix for a longer time, it should be added to cells rapidly to prevent formation of large crystals. Particles lying on cells can be visible under the microscope 15 min after addition of transfection solution and incubation.

3. After 16 h incubation, medium should be removed by careful aspiration and new medium added to remove remaining crystals.
4. After an additional incubation for 8 h, cells can be challenged with PAMP or whole pathogen preparations, by pipetting into wells without touching the bottom of the well with the pipette tip (*see* **Note 9**). In order to determine cell responsiveness, media can be removed quantitatively between 6 and 16 h, and assayed for cytokine content such as by human IL-8 ELISA.
5. Thirty microlitres of 1X lysis buffer should be added to the adherent cells in each well of a 96 well plate, the plate incubated between 10 min and 30 min at room temperature on a Western blot swing, and 20 μl transferred into a well of a non-transparent black or white 96 well plate.

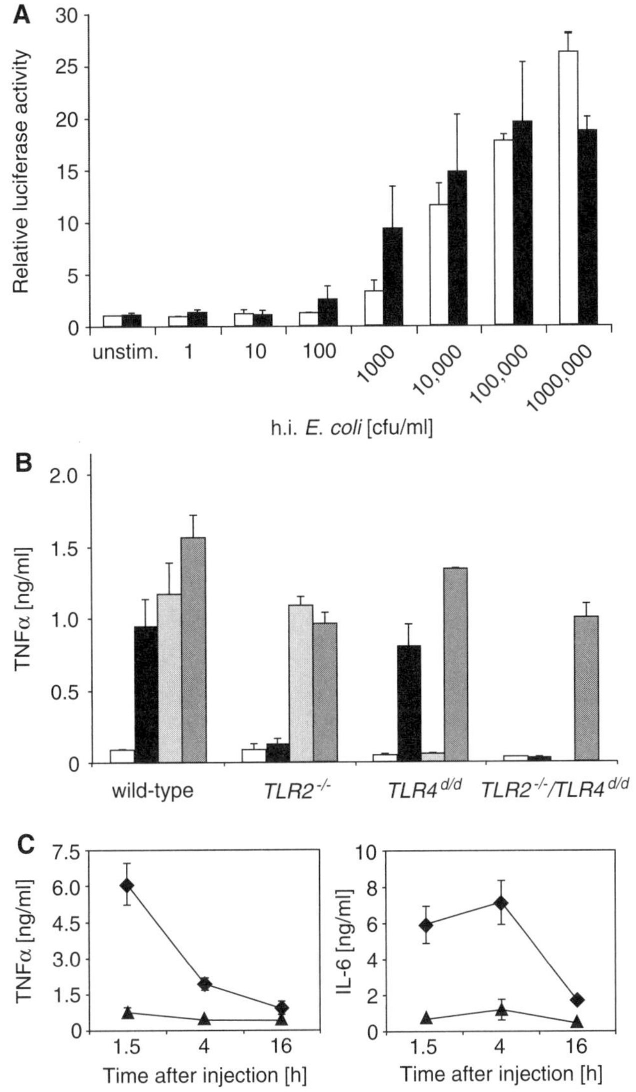

Fig. 1. TLR-mediated responsiveness to pathogen-associated molecular pattern (PAMP) challenge. (**A**) Human embryonic kidney (HEK) 293 cells were transfected with DNA plasmids driving nuclear factor (NF)-κB-dependent firefly luciferase expression and constitutive expression of renilla luciferase, as well as of TLR2 and CD14 (*white columns*) or of TLR4 and MD-2 (*black columns*). Transfected cells were challenged with washed and heat inactivated (h.i., 20 min at 100°C) *Escherichia coli* that had been cultured for 16 h before. Reporter gene activity was monitored upon cell lysis 16 h after challenge start. (**B**) Thioglycolate elicited peritoneal washout macrophages from C3H mice of the genotypes indicated were left untreated (*white columns*) or challenged with 1 mg/ml triacylated hexapeptide (Pam_3CSK_4, *black columns*), 100 ng/ml LPS (*light grey columns*), or 25 μg/ml poly I:C (*dark grey columns*). Supernatants were sampled after 8 h and subjected to enzyme-linked immunosorbent assay (ELISA). (**C**) Wild-type (♦) and *TLR4*$^{-/-}$ (▲) mice were challenged by intraperitoneal injection of 48 mg/kg of LPS. Mice were anaesthetised and approximately 100 μl of serum was drawn at the time points indicated. Serum samples were analysed by ELISA ($n = 2$ for each of the two experimental groups).

6. The 96 well plate containing lysates can be inserted into a 96 well plate light emission reader to let inject luciferin and read the numbers of resulting light emissions for 10 s (**Fig. 1A**).

3.3. PAMP Challenge of Primary Macrophages and ELISA or/and NO Assay

1. Cultured primary macrophages prepared as shown in **Subheading 3.1** are incubated overnight in the presence of 20 ng/ml IFNγ for priming or left untreated. IFNγ priming is obligatory if NO release is to be measured.
2. Cells might be pretreated with PRR-specific antagonists such as neutralising antibody or inhibitory nucleic acid oligo nucleotides.
3. Cells are then challenged for 30–60 min with PRR agonists such as purified PAMPs, synthetic analogues, or with whole pathogens in a viable or inactivated state. If cells are infected with viable pathogens such as bacteria which might be subjected to antibiotics in cell culture supernatants later, the laboratory safety standard has to accord with the bacteria's endangering potential.
4. Four to forty-eight hours of PAMP challenge supernatants can be analysed.
5. Nitrite content of 50 or 100 μl supernatant should be analysed immediately by mixing with 50 or 100 μl of Griess reagent consisting of equal parts of solution A and B (*see* **Subheading 2.3**, mixed shortly prior to usage) in a transparent 96 well flat bottom plate. Light absorptive capacity upon resulting colour reaction (10 min at room temperature) is to be analysed in a standard ELISA reader. Supernatant can be stored upon freezing (–20°C or –80°C) for later application to ELISA to determine concentration of a specific cytokine, chemokine, or related inflammatory mediator (Fig. **1B**). Alternatively, multi-array beads for flow cytometric analysis or protein array membranes can be used for synchronous detection of several different proteins present in the supernatant potentially.
6. Normalisation by relating to an intrinsic standard and statistic analysis shall follow raw data acquisition for result illustration.

3.4. PAMP Challenge and Analysis of Systemic Immune Activation in Mice

1. In order to increase sensitivity to PAMP challenge, mice can be primed by IFNγ administration intravenously (i.v.) or intraperitonealy (i.p.). Specifically, 50 μg/kg recombinant IFNγ might be injected i.v. 45 min prior to PAMP or microbial challenge such as infection (*see* **Note 3**).
2. PAMP preparations should be treated by sonification for solubilisation prior to application (*see* **Note 9**). If bacterial infection is to be performed, quantification of microbial load by plating and culturing of dilutions has to be included in the procedure as control. In order to prevent interference of medium components or released microbial products with recognition of the pathogen, the pathogen cell culture might be centrifuged (typically 15 min at 6,000 × *g* for bacteria) to suspend the pellet containing cells in sterile PBS. Again, the route of PAMP challenge or infection depends on the

clinically relevant infection whose experimental modelling is aimed at. Increased sensitiveness can be achieved by systemic co-application of 800 mg/kg D-galN *(58)*.

3. Depending on the parameters to be analysed, serum might be drawn from challenged mice at specific time points upon anaesthetisation from the retrobulbar plexus using capillaries to collect approximately 100 μl of blood from the retrobulbar plexus. Mouse resuscitation should follow serum removal. For instance upon LPS challenge, TNFα concentration peaks at the 90 min time point while maximal serum accumulation of other cytokines such as IL-6 or IL-12 occurs later. Seventy-five microlitres of fentanyl (0.05 mg/ml) might be applied per mouse (25 mg) for anaesthetisation upon which administration of 240 μl naloxone (0.4 mg/ml) resuscitates each mouse.
4. Serum might be heparinised or incubated at ice for 30 min and centrifuged at 2,500 rpm in a table-top centrifuge. Supernatant/serum can be analysed directly such as by plating to analyse microbial load or Griess assay, or subjected to ELISA either immediately or upon freezing for prior storage (**Fig. 1C**).

4. Notes

1. Use materials including containers, pipettes, solutions, and water that are pyrogen free to prevent interferences with specific pattern recognition.
2. Peptone induced instead of thioglycolate is sometimes used to induce peritoneal macrophages. Bone marrow derived macrophages are also being widely used.
3. This procedure does encompass injection of an agent and subsequent organ/tissue/cell removal and thus constitutes an animal experiment. An animal experiment has to be planned, an application has to be filed, it has to be approved by the responsive institutional and/or governmental committee(s), and an animal experiment is to be performed according to legal regulation. Degree of medical and scientific impact and novelty (lack of redundancy with published reports), as well as animal experiment expertise of the performer, choice of animal species, access to an approved animal facility, and reason of numbers of animals to be applied are important parameters to be considered. Alternatively other cells such as alveolar macrophages, splenocytes, or bone marrow derived macrophages or DC of different kinds such as myeloid (m)DC or plasmacytoid

(p)DC the generation of the latter of which depends on application of macrophage colony stimulating factor (MCSF), granulocyte (G)MCSF, and IL-4, or Flt3 ligand to differentiation medium can be prepared and applied *(51)*.

4. Pressure exerted on the syringe plunger should be limited to minimise shearing stress operative within and at the edges of the needle and thus optimise survival rate of taken-up cells. Positioning of the needle opening should be followed by spatial fixation throughout the injection-soaking procedure and varied only for maintenance of free access of fluid to the needle opening to prevent perforation of tissue.
5. If sample number is above 4, cells can be stored on wet ice for 1 h without detectable impairment of function in subsequent assaying. Erythrocyte lysis *(51)* is necessary only in the exceptional case of blood content of the washout.
6. Adherent cells are macrophages (85–95% $CD14^+$ $CD11b^+$). While peritonea of naïve mice contain approximately 1×10^6 peritoneal macrophages, their number increases to up to 1×10^7 per mouse into which thioglycolate had been injected 5 days earlier. Negative and positive controls, as well as triplicate plating of individual samples, are parameters to be considered for experimental design.
7. Use cells for limited time periods such as 6 weeks and thaw aliquots stored in liquid nitrogen to ensure optimal performance and absence of infection.
8. Ligand-dependent cell activation does not depend on maximal TLR expression level necessarily. For instance, TLR4 whose activity depends on coexpression of MD-2 can become active constitutively upon too strong overexpression. An optimal expression plasmid DNA concentration used for transfection at which sensor function is detectable but constitutive activity is not yet operative might need to be determined by titration.
9. PAMP solutions can be stored frozen at –20°C for 3 months. Prior to application to cells, PAMP solutions should be thawed and sonicated in a standard laboratory sonication water bath for 20 s for PAMP solubilisation.

Acknowledgement

This work was supported by German research foundation (DFG/SFB/TR22-A5).

References

1. Raetz, C. R., Ulevitch, R. J., Wright, S. D., Sibley, C. H., Ding, A., and Nathan, C. F. (1991) Gram-negative endotoxin: an extraordinary lipid with profound effects on eukaryotic signal transduction. *FASEB J* 5, 2652–60.
2. Rietschel, E. T., Brade, H., Holst, O., Brade, L., Muller-Loennies, S., Mamat, U., Zahringer, U., Beckmann, F., Seydel, U., Brandenburg, K., Ulmer, A. J., Mattern, T., Heine, H., Schletter, J., Loppnow, H., Schonbeck, U., Flad, H. D., Hauschildt, S., Schade, U. F., Di Padova, F., Kusumoto, S., and Schumann, R. R. (1996) Bacterial endotoxin: chemical constitution. biological recognition, host response, and immunological detoxification. *In* Current Topics in Microbiology and Immunology (Wagner, H., and Rietschel, E. T., Eds.), Vol. 216, pp. 39–81, Springer Berlin Heidelberg, New York.
3. Iliev, D. B., Roach, J. C., Mackenzie, S., Planas, J. V., and Goetz, F. W. (2005) Endotoxin recognition: in fish or not in fish? *FEBS Lett* 579, 6519–28.
4. Watson, J., and Riblet, R. (1974) Genetic control of responses to bacterial lipopolysaccharides in mice. I. Evidence for a single gene that influences mitogenic and immunogenic respones to lipopolysaccharides. *J Exp Med* 140, 1147–61.
5. Schumann, R. R., Leong, S. R., Flaggs, G. W., Gray, P. W., Wright, S. D., Mathison, J. C., Tobias, P. S., and Ulevitch, R. J. (1990) Structure and function of lipopolysaccharide binding protein. *Science* 249, 1429–31.
6. Wright, S. D., Ramos, R. A., Tobias, P. S., Ulevitch, R. J., and Mathison, J. C. (1990) CD14, a receptor for complexes of lipopolysaccharide (LPS) and LPS binding protein. *Science* 249, 1431–3.
7. Ulevitch, R. J., and Tobias, P. S. (1995) Receptor-dependent mechanisms of cell stimulation by bacterial endotoxin. *Annu Rev Immunol* 13, 437–57.
8. Anderson, K. V., Bokla, L., and Nusslein-Volhard, C. (1985) Establishment of dorsal-ventral polarity in the Drosophila embryo: the induction of polarity by the Toll gene product. *Cell* 42, 791–8.
9. Gay, N. J., and Keith, F. J. (1991) Drosophila Toll and IL-1 receptor. *Nature* 351, 355–6.
10. Lemaitre, B., Nicolas, E., Michaut, L., Reichhart, J. M., and Hoffmann, J. A. (1996) The dorsoventral regulatory gene cassette spatzle/Toll/cactus controls the potent antifungal response in Drosophila adults. *Cell* 86, 973–83.
11. Taguchi, T., Mitcham, J. L., Dower, S. K., Sims, J. E., and Testa, J. R. (1996) Chromosomal localization of TIL, a gene encoding a protein related to the Drosophila transmembrane receptor Toll, to human chromosome 4p14. *Genomics* 32, 486–8.
12. Medzhitov, R., Preston-Hurlburt, P., and Janeway, C. A., Jr. (1997) A human homologue of the Drosophila Toll protein signals activation of adaptive immunity. *Nature* 388, 394–7.
13. Rock, F. L., Hardiman, G., Timans, J. C., Kastelein, R. A., and Bazan, J. F. (1998) A family of human receptors structurally related to Drosophila Toll. *Proc Natl Acad Sci U S A* 95, 588–93.
14. Chaudhary, P. M., Ferguson, C., Nguyen, V., Nguyen, O., Massa, H. F., Eby, M., Jasmin, A., Trask, B. J., Hood, L., and Nelson, P. S. (1998) Cloning and characterization of two Toll/interleukin-1 receptor-like genes TIL3 and TIL4: evidence for a multi-gene receptor family in humans. *Blood* 91, 4020–7.
15. Muzio, M., Natoli, G., Saccani, S., Levrero, M., and Mantovani, A. (1998) The human toll signaling pathway: divergence of nuclear factor kappaB and JNK/SAPK activation upstream of tumor necrosis factor receptor-associated factor 6 (TRAF6). *J Exp Med* 187, 2097–101.
16. Yang, R. B., Mark, M. R., Gray, A., Huang, A., Xie, M. H., Zhang, M., Goddard, A., Wood, W. I., Gurney, A. L., and Godowski, P. J. (1998) Toll-like receptor-2 mediates lipopolysaccharide-induced cellular signalling. *Nature* 395, 284–8.
17. Kirschning, C. J., Wesche, H., Ayres, T. M., and Rothe, M. (1998) Human toll-like receptor 2 confers responsiveness to bacterial lipopolysaccharide. *J Exp Med* 188, 2091–7.
18. Poltorak, A., He, X., Smirnova, I., Liu, M. Y., Huffel, C. V., Du, X., Birdwell, D., Alejos, E., Silva, M., Galanos, C., Freudenberg, M., Ricciardi-Castagnoli, P., Layton, B., and Beutler, B. (1998) Defective LPS signaling in C3H/HeJ and C57BL/10ScCr mice: mutations in Tlr4 gene. *Science* 282, 2085–8.
19. Qureshi, S. T., Lariviere, L., Leveque, G., Clermont, S., Moore, K. J., Gros, P., and Malo, D. (1999) Endotoxin-tolerant mice have mutations in Toll-like receptor 4 (Tlr4). *J Exp Med* 189, 615–25.
20. Shimazu, R., Akashi, S., Ogata, H., Nagai, Y., Fukudome, K., Miyake, K., and Kimoto, M. (1999) MD-2, a molecule that confers lipopolysaccharide responsiveness on Toll-like receptor 4. *J Exp Med* 189, 1777–82.

21. Nagai, Y., Akashi, S., Nagafuku, M., Ogata, M., Iwakura, Y., Akira, S., Kitamura, T., Kosugi, A., Kimoto, M., and Miyake, K. (2002) Essential role of MD-2 in LPS responsiveness and TLR4 distribution. *Nat Immunol* 3, 667–72.
22. Brightbill, H. D., Libraty, D. H., Krutzik, S. R., Yang, R. B., Belisle, J. T., Bleharski, J. R., Maitland, M., Norgard, M. V., Plevy, S. E., Smale, S. T., Brennan, P. J., Bloom, B. R., Godowski, P. J., and Modlin, R. L. (1999) Host defense mechanisms triggered by microbial lipoproteins through toll-like receptors. *Science* 285, 732–6.
23. Aliprantis, A. O., Yang, R. B., Mark, M. R., Suggett, S., Devaux, B., Radolf, J. D., Klimpel, G. R., Godowski, P., and Zychlinsky, A. (1999) Cell activation and apoptosis by bacterial lipoproteins through toll-like receptor-2. *Science* 285, 736–9.
24. Hirschfeld, M., Kirschning, C. J., Schwandner, R., Wesche, H., Weis, J. H., Wooten, R. M., and Weis, J. J. (1999) Cutting edge: inflammatory signaling by Borrelia burgdorferi lipoproteins is mediated by toll-like receptor 2. *J Immunol* 163, 2382–6.
25. Ozinsky, A., Underhill, D. M., Fontenot, J. D., Hajjar, A. M., Smith, K. D., Wilson, C. B., Schroeder, L., and Aderem, A. (2000) The repertoire for pattern recognition of pathogens by the innate immune system is defined by cooperation between toll-like receptors. *Proc Natl Acad Sci U S A* 97, 13766–71.
26. Takeuchi, O., Kawai, T., Muhlradt, P. F., Morr, M., Radolf, J. D., Zychlinsky, A., Takeda, K., and Akira, S. (2001) Discrimination of bacterial lipoproteins by Toll-like receptor 6. *Int Immunol* 13, 933–40.
27. Bulut, Y., Faure, E., Thomas, L., Equils, O., and Arditi, M. (2001) Cooperation of Toll-like receptor 2 and 6 for cellular activation by soluble tuberculosis factor and Borrelia burgdorferi outer surface protein A lipoprotein: role of Toll-interacting protein and IL-1 receptor signaling molecules in Toll-like receptor 2 signaling. *J Immunol* 167, 987–94.
28. Alexopoulou, L., Thomas, V., Schnare, M., Lobet, Y., Anguita, J., Schoen, R. T., Medzhitov, R., Fikrig, E., and Flavell, R. A. (2002) Hyporesponsiveness to vaccination with Borrelia burgdorferi OspA in humans and in TLR1- and TLR2-deficient mice. *Nat Med* 8, 878–84.
29. Takeuchi, O., Sato, S., Horiuchi, T., Hoshino, K., Takeda, K., Dong, Z., Modlin, R. L., and Akira, S. (2002) Cutting edge: role of Toll-like receptor 1 in mediating immune response to microbial lipoproteins. *J Immunol* 169, 10–4.
30. Hemmi, H., Takeuchi, O., Kawai, T., Kaisho, T., Sato, S., Sanjo, H., Matsumoto, M., Hoshino, K., Wagner, H., Takeda, K., and Akira, S. (2000) A Toll-like receptor recognizes bacterial DNA. *Nature* 408, 740–5.
31. Hayashi, F., Smith, K. D., Ozinsky, A., Hawn, T. R., Yi, E. C., Goodlett, D. R., Eng, J. K., Akira, S., Underhill, D. M., and Aderem, A. (2001) The innate immune response to bacterial flagellin is mediated by Toll-like receptor 5. *Nature* 410, 1099–103.
32. Gewirtz, A. T., Navas, T. A., Lyons, S., Godowski, P. J., and Madara, J. L. (2001) Cutting edge: bacterial flagellin activates basolaterally expressed TLR5 to induce epithelial proinflammatory gene expression. *J Immunol* 167, 1882–5.
33. Alexopoulou, L., Holt, A. C., Medzhitov, R., and Flavell, R. A. (2001) Recognition of double-stranded RNA and activation of NF-kappaB by Toll-like receptor 3. *Nature* 413, 732–8.
34. Diebold, S. S., Kaisho, T., Hemmi, H., Akira, S., and Reis e Sousa, C. (2004) Innate antiviral responses by means of TLR7-mediated recognition of single-stranded RNA. *Science* 303, 1529–31.
35. Heil, F., Hemmi, H., Hochrein, H., Ampenberger, F., Kirschning, C., Akira, S., Lipford, G., Wagner, H., and Bauer, S. (2004) Species-specific recognition of single-stranded RNA via toll-like receptor 7 and 8. *Science* 303, 1526–9.
36. Lund, J. M., Alexopoulou, L., Sato, A., Karow, M., Adams, N. C., Gale, N. W., Iwasaki, A., and Flavell, R. A. (2004) Recognition of single-stranded RNA viruses by Toll-like receptor 7. *Proc Natl Acad Sci U S A* 101, 5598–603.
37. Zhang, D., Zhang, G., Hayden, M. S., Greenblatt, M. B., Bussey, C., Flavell, R. A., and Ghosh, S. (2004) A toll-like receptor that prevents infection by uropathogenic bacteria. *Science* 303, 1522–6.
38. Yarovinsky, F., Zhang, D., Andersen, J. F., Bannenberg, G. L., Serhan, C. N., Hayden, M. S., Hieny, S., Sutterwala, F. S., Flavell, R. A., Ghosh, S., and Sher, A. (2005) TLR11 activation of dendritic cells by a protozoan profilin-like protein. *Science* 308, 1626–9.
39. Akira, S., and Takeda, K. (2004) Toll-like receptor signalling. *Nat Rev Immunol* 4, 499–511.
40. Wagner, H. (2006) Endogenous TLR ligands and autoimmunity. *Adv Immunol* 91, 159–73.
41. O'Neill, L. A. (2004) Immunology. *After the toll rush. Science* 303, 1481–2.
42. Latz, E., Verma, A., Visintin, A., Gong, M., Sirois, C. M., Klein, D. C., Monks, B. G., McKnight, C. J., Lamphier, M. S., Duprex, W. P., Espevik, T., and Golenbock, D. T. (2007)

Ligand-induced conformational changes allosterically activate Toll-like receptor 9. *Nat Immunol* 8, 772–9.

43. Meylan, E., Tschopp, J., and Karin, M. (2006) Intracellular pattern recognition receptors in the host response. *Nature* 442, 39–44.
44. Kanneganti, T. D., Lamkanfi, M., Kim, Y. G., Chen, G., Park, J. H., Franchi, L., Vandenabeele, P., and Nunez, G. (2007) Pannexin-1-mediated recognition of bacterial molecules activates the cryopyrin inflammasome independent of Toll-like receptor signaling. *Immunity* 26, 433–43.
45. van Duin, D., Medzhitov, R., and Shaw, A. C. (2006) Triggering TLR signaling in vaccination. *Trends Immunol* 27, 49–55.
46. Cohen, J. (2002) The immunopathogenesis of sepsis. *Nature* 420, 885–91.
47. Annane, D., Bellissant, E., and Cavaillon, J. M. (2005) Septic shock. *Lancet* 365, 63–78.
48. Hotchkiss, R. S., and Nicholson, D. W. (2006) Apoptosis and caspases regulate death and inflammation in sepsis. *Nat Rev Immunol* 6, 813–22.
49. Jin, M. S., Kim, S. E., Heo, J. Y., Lee, M. E., Kim, H. M., Paik, S. G., Lee, H., and Lee, J. O. (2007) Crystal structure of the TLR1-TLR2 heterodimer induced by binding of a tri-acylated lipopeptide. *Cell* 130, 1071–82.
50. O'Neill, L. A., and Bowie, A. G. (2007) The family of five: TIR-domain-containing adaptors in Toll-like receptor signalling. *Nat Rev Immunol* 7, 353–64.
51. Coligan, J. E., Kruisbeek, A. M., Margulies, D. H., Shevach, E. M., and Strobe, W. (1990) Current Protocols in Immunology, Wiley, New York.
52. Sambrock, J., Fritsch, E. F., Maniatis, T., and Irwin, N. (1989) Molecular Cloning, Cold Spring Harbor Laboratory Press, New York.
53. Graham, F. L., and van der Eb, A. J. (1973) Transformation of rat cells by DNA of human adenovirus 5. *Virology* 54, 536–9.
54. Dyer, B. W., Ferrer, F. A., Klinedinst, D. K., and Rodriguez, R. (2000) A noncommercial dual luciferase enzyme assay system for reporter gene analysis. *Anal Biochem* 282, 158–61.
55. Alam, J., and Cook, J. L. (1990) Reporter genes: application to the study of mammalian gene transcription. *Anal Biochem* 188, 245–54.
56. Green, L. C., Wagner, D. A., Glogowski, J., Skipper, P. L., Wishnok, J. S., and Tannenbaum, S. R. (1982) Analysis of nitrate, nitrite, and [15 N]nitrate in biological fluids. *Anal Biochem* 126, 131–8.
57. Schindler, U., and Baichwal, V. R. (1994) Three NF-kappa B binding sites in the human E-selectin gene required for maximal tumor necrosis factor alpha-induced expression. *Mol Cell Biol* 14, 5820–31.
58. Galanos, C., Freudenberg, M. A., and Reutter, W. (1979) Galactosamine-induced sensitization to the lethal effects of endotoxin. *Proc Natl Acad Sci U S A* 76, 5939–43.

Chapter 20

Toll-Like Receptors and Rheumatoid Arthritis

Fabia Brentano, Diego Kyburz, and Steffen Gay

Summary

Rheumatoid arthritis (RA) is a chronic inflammatory disease that ultimately leads to the progressive destruction of cartilage and bone in numerous joints. There is mounting evidence for an important function of innate immunity in the pathogenesis of RA. Activation of cells by microbial components and also by endogenous molecules via Toll-like receptor (TLR) results in the production of a variety of proinflammatory cytokines, chemokines, and destructive enzymes, some of which can characteristically be found in RA.

By immunohistochemistry we found elevated TLR2, 3, and 4 expressions in the rheumatoid synovium. In the synovial lining layer and at sites of invasion into cartilage, RA synovial fibroblasts (RASF) are the major cells expressing TLR2, 3, and 4. Stimulation of cultured RASF in vitro with the TLR2 ligand bacterial lipoprotein (bLP), the TLR3 ligand poly(I-C), and the TLR4 ligand LPS was shown to upregulate IL-6 as well as matrix metalloproteinases (MMPs) 1 and 3. These results suggest an important role for TLR2, 3, and 4 in the activation of synovial fibroblasts in RA leading to chronic inflammation and joint destruction.

Key words: TLR, Rheumatoid arthritis, Synovial tissues, Synovial fibroblasts, Bacterial lipoprotein, Poly(I-C), Lipopolysaccharide , IL-6, MMP1.

1. Introduction

Rheumatoid arthritis (RA) is a chronic inflammatory disease with progressive articular damage caused by inflammatory cells and synoviocytes. Cartilage RA synovial fibroblasts (RASF) must be considered key cells in mediating the destruction of cartilage and bone in the affected joints. They are mainly found in the synovial lining layer, which is thickened and hyperplastic in RA *(1)*. RASF are part of the innate immune system and they express pattern-recognition receptors, such as the Toll-like receptors (TLRs), which sense certain highly conserved structures that are found on

Claire E. McCoy and Luke A.J. O'Neill (eds.), *Methods in Molecular Biology, Toll-Like Receptors, vol. 517*
DOI: 10.1007/978-1-59745-541-1_20

many different bacterial and viral products. The recognition of specific microbial/viral structures, such as bacterial lipoproteins (bLP, TLR2 ligand), double-stranded RNA (dsRNA, TLR3 ligand), or lipopolysaccharides (LPS, TLR4 ligand), by TLRs results in the upregulation of costimulatory molecules as well as in the induction of proinflammatory and destructive mediators. In the inflamed joints of patients with RA, endogenous TLR ligands are present. Endogenous TLR2 and 4 ligands include several heat shock proteins while RNA released from necrotic synovial fluid cells acts as an endogenous ligand for TLR3 *(2, 3)*. Considering the important role of TLR signaling as a critical link between innate and adaptive immunity, it has been proposed that a dysregulation in TLR signaling might be associated with autoimmune diseases, such as RA *(4, 5)*.

To investigate whether TLR2, 3, and 4 are expressed in RA synovial tissues, we performed immunohistochemistry on snap frozen RA synovial tissues with commercially available polyclonal antibodies. As noninflammatory controls, we also stained synovial tissues derived from patients with osteoarthritis (OA). By this technique we found TLR2, 3, and 4 proteins to be broadly expressed in synovial tissues derived from patients with RA *(6, 7)*. In particular, there was a pronounced expression in the synovial lining layer. In comparison to RA tissues, TLR2, 3, and 4 protein expression was markedly lower in OA synovial tissues.

The RA synovial lining layer is composed of activated synovial fibroblasts and macrophages. To characterize the TLR3 expressing cells in the synovial lining layer, tissue sections were double stained for TLR3 and the macrophage marker CD68. The majority of cells expressing TLR3 did not stain for CD68, indicating that most synoviocytes expressing TLR3 were not macrophages but synovial fibroblasts *(2)*.

It is known that IL-6 exerts stimulatory effects on T- and B-cells, thus favoring chronic inflammatory responses, whereas matrix metalloproteinases (MMPs) have been closely linked to the progressive destruction of articular cartilage in rheumatoid joints. To analyze whether TLR signaling might contribute to the elevated IL-6, MMP1, and MMP3 levels in the rheumatoid joint, RASF were isolated from RA synovial tissues and stimulated with the TLR2, 3, and 4 ligands. Twenty-four hours after stimulation, cell culture supernatants were collected and the cells were lysed to isolate total RNA. In the cell culture supernatants, IL-6 levels were measured by a commercially available enzyme-linked immunosorbent assay (ELISA) set. Furthermore upregulation of MMP1 and MMP3 mRNA was determined by TaqMan real-time polymerase chain reaction (PCR). We found a strong upregulation of IL-6, MMP1, and MMP3 by all three TLR ligands, with a most marked upregulation reached by the TLR3 ligand poly(I-C) *(2, 8)*. These results document a strong proinflammatory and joint destructing role of TLR signaling pathways in the pathogenesis of RA and identify TLRs as potential therapeutic targets.

2. Materials

2.1. Synovial Tissue Preparation for Immunohistochemistry

1. Synovial tissue specimens were obtained during synovectomy or joint replacement surgery from patients with RA and OA, after informed consent has been obtained (Department of Orthopedic Surgery, Schulthess Clinic, Zurich, Switzerland). Embedded synovial tissues were maintained at –80°C until cryosectioned.
2. Tissue-Tek® OCT™ compound (Tissue-Tek TT 4583; Sakura Finetech, Torrance, CA). Store at room temperature (RT).
3. Freezing mold: Simport Biotubes (T100-20, Omnilab, Switzerland). Autoclave tubes before use.
4. Superfrost™ Plus Slides (Menzel-Gläser, Braunschweig, Germany).

2.2. Immunohistochemistry

1. Fixation solution for frozen synovial tissue sections: Precooled acetone (histological grade –99.5%). Precool the amount of acetone needed for the fixation (around 100 ml of acetone) at –20°C (*see* **Note 1**).
2. Blocking of endogenous peroxidase activity: 0.1% H_2O_2 (*see* **Note 2**).
3. Washing buffer (Tris-buffered saline (TBS)): 100 mM Tris-HCl, pH 7.6, 150 mM NaCl.
4. Blocking buffer: 2% fetal bovine serum (FBS) (Omnilab, Switzerland) in TBS.
5. Primary antibodies. Affinity-purified goat antihuman TLR-3 polyclonal IgGs (Cat. No. JM-3445-100, MBL International, Woburn, MA), store at 4°C.
6. Negative control antibodies: ChromPure Goat IgG, whole molecule (Cat. No. 005-000-003, Jackson ImmunoResearch Europe Ltd., Suffolk, UK), store at 4°C.
7. Dilution buffer for secondary antibodies: 2% bovine serum albumin (BSA, Sigma-Aldrich, Basel, Switzerland), 5% human serum, in TBS (*see* **Note 3**).
8. Secondary antibodies: Polyclonal rabbit antigoat IgGs, biotinylated (Cat. No. E0466, DakoCytomation, Glostrup, Denmark), store at 4°C.
9. Vectastain Elite ABC Kit for horseradish peroxidase (HRP) (PK6100, Vector Laboratories, UK), store at 4°C.

10. HRP substrate chromogen: Aminoethylcarbazole (AEC) chromogen substrate (K3464, DakoCytomation), store at 4°C.
11. Macrophage-specific primary antibodies: Monoclonal mouse antihuman CD68 (Cat. No. M0718, DakoCytomation), store at 4°C.
12. Negative control antibodies: Mouse IgG1 (Cat. No. X0931, DakoCytomation), store at 4°C.
13. Secondary antibodies: Polyclonal rabbit antimouse IgG, alkaline phosphatase (AP) conjugated (DakoCytomation), store at 4°C.
14. AP substrate chromogen: Fast Blue BB reagent: mix naphthol-AS-MX phosphate dissolved in *N,N*-dimethylformamide immediately before use with Fast Blue BB dissolved in TBS, pH 8.5 (Sigma-Aldrich, Switzerland).
15. Nuclear counterstain: Hematoxylin solution according to Mayer (Cat. No. 51275, Sigma-Aldrich).
16. Mounting medium: Glycine gelatine.

2.3. Cell Culture

1. Protease for the dissociation of synovial tissues to release individual cells: Dispase II (Cat. No. 17105-041, Gibco, Basel, Switzerland).
2. Fetal bovine serum (FBS) (Omnilab, Switzerland): Heat inactivate FBS for 30 min at 56°C. Store heat inactivated FBS at –20°C.
3. Cell culture medium: Dulbecco's Modified Eagle's Medium/F-12 supplemented with 10% heat inactivated FBS, 50 IU/ml penicillin–streptomycin, 2 mM L-glutamin, 10 mM HEPES and 2% Fungizone® Antimycotic (all from Gibco). Sterile filtration through 0.22-µm syringe filter (TPP, St. Louis, MO) before storage at 4°C (*see* **Note 4**).
4. Trypsin/EDTA –0.05% Trypsin, 0.53 mM EDTA·4Na (Cat. No. 25300-054, Gibco), store at –20°C.
5. Phosphate-buffered saline (PBS): 136.9 mM NaCl, 2.7 mM KCl, 10 mM Na_2HPO_4, 1.8 KH_2PO_4 (adjust to pH 7.4 with HCl if necessary). PBS is autoclaved before storage at room temperature.

2.4. TLR Ligands for In Vitro Stimulation of RASF

1. Synthetic bLP as TLR2 ligand: *N*-palmitoyl-*S*-[2,3-bis(palmitoyloxy)-(2*RS*)-propyl]-[*R*]-Cys-[*S*]-Serl-[*S*]-Lys *(4)* trihydrochloride (Pam_3CSK_4, Invivogen, San Diego, CA). Dissolve bLP to a concentration of 1 mg/ml in endotoxin-free deionized water (dH_2O) and store in single use aliquots at –20°C.

2. Synthetic dsRNA as TLR3 ligand: Polyriboinosinic polyribocytidylic acid (poly(I-C), Invivogen, San Diego, CA). Dissolve poly(I-C) to a concentration of 2.5 mg/ml in endotoxin-free dH_2O and store in single use aliquots at –20°C.
3. Lipopolysaccharide (LPS) as TLR 4 ligand: *Escherichia coli* derived (List Biological Laboratories, Campbell, CA). Dissolve LPS to a concentration of 100 μg/ml in dH_2O and store aliquots at –20°C.

2.5. TaqMan Real-Time PCR

1. RNA isolation: Total RNA from cultured RASF is isolated with the RNeasy Mini kit (Qiagen, Basel, Switzerland), including treatment with RNase-free DNase (Qiagen).
2. Reverse transcription reagents (Applied Biosystems, Rotkreuz, Switzerland): MultiScribe™ reverse transcriptase (RT) (50 U/μl), random hexamers (50 μM), dNTP (10 mM solution of 2.5 mM each of dATP, dCTP, dGTP, and dTTP), 10x PCR buffer II (contains no $MgCl_2$), RNase inhibitor (20 U/μl). Store all reagents at –20°C.
3. Primer/probe sequences used for real-time PCR (Microsynth, Balgach, Switzerland): MMP1 forward primer: 5′-tgt-gga-cca-tgc-cat-tga-ga-3′; MMP1 reverse primer: 5′-tct-gct-tga-ccc-tca-gag-acc-3′; FAM/TAMRA labeled MMP1-specific probe: 5′-agc-ctt-caa-act-ctg-gag-taa-tgt-cac-acc-3′. MMP3 forward primer: 5′-ggg-cca-tca-gag-gaa-atg-ag-3′; MMP3 reverse primer: 5′-cac-ggt-tgg-agg-gaa-acc-ta-3′; FAM/TAMRA labeled MMP3-specific probe: 5′-agc-tgg-ata-ccc-aag-agg-cat-cca-cac-3′. Dissolve stock primer solutions to a working concentration of 11.25 μM and the probe sequences to a working concentration of 5 nM in dH_2O before storage at –20°C.
4. Eukaryotic 18S rRNA endogenous control: 18S VIC/TAMRA labeled Probe, Primer Ltd. (Applied Biosystems), store at –20°C.
5. qPCR Master Mix: 2x reaction buffer containing dNTP/dUTP, HotGoldStar, $MgCl_2$ (5 mM final concentration), UNG, stabilizers, and ROX passive reference (Eurogentech, Genève, Switzerland), store at –20°C.
6. MicroAmp Optical 96-well reaction plate and corresponding MicroAmp optical caps (Applied Biosystems).

2.6. ELISA for IL-6

BD OptEIA™ Human IL-6 ELISA set: Capture antibody: antihuman IL-6 (store at 4°C); detection antibody: biotinylated antihuman IL-6; enzyme reagent (store at 4°C): avidin–horseradish peroxidase conjugate (store at 4°C); standard: recombinant human IL-6 (store in single use aliquots at –80°C).

3. Methods

To analyze whether TLRs might play a role in the pathogenesis of RA the first step was to investigate whether TLRs are present in the rheumatoid synovium. Furthermore, TLR expression was compared between patients with RA and noninflammatory OA, to study the pathogenic relevance of TLR expression in RA. To obtain reliable results, it is essential that tissue specimens obtained after surgery are immediately embedded in OCT compound and snap frozen in liquid nitrogen.

3.1. Preparation of Fresh Frozen Tissue Sections

1. OCT embedding: Label plastic tissue mold with proper tissue identification and prepare the mold with a small layer of OCT. Immediately after surgery place the tissue in desired orientation and completely fill the mold with OCT. Note that the cutting begins at the bottom of the mold. The OCT compound begins to turn white as it freezes. Freeze tissue by placing mold on the surface of liquid nitrogen until 70–80% of the block turns white. Then, immerse the embedded tissue into liquid nitrogen for at least 5 min. Retrieve the mold from liquid nitrogen and store the embedded tissue at –80°C or move directly to cryostat.
2. Sectioning: Transfer the embedded tissue on dry ice to cryostat and let it equilibrate 5 min to cryostat temperature (–20°C). Sections are usually obtained when the temperature is from –18°C to –20°C. The OCT-embedded tissue is fixed to chuck with OCT media. Cut 8-μm sections and mount sections on positively charged slides. Let sections air-dry for 30 min at room temperature. After drying, slides are stored at –20°C (*see* **Note 5**).

3.2. Immunohistochemical Staining on Frozen Synovial Tissue Sections

3.2.1. Immunohistochemical Staining for TLR3

1. Fixation: Let the tissue sections air-dry for 30 min. Subsequently place the sections in a Coplin jar (*see* **Note 6**) with –20°C precooled 100% acetone for 15 min. After fixation, the tissue sections have to be air-dried for at least 60 min.
2. Rehydrate tissue sections with two washes of dH_2O (*see* **Note 7**).
3. Block endogenous peroxidase activity with 0.1% H_2O_2 for 10 min. Rinse the sections once with dH_2O and once for 5 min with TBS (washing buffer).
4. Incubate sections in blocking buffer (TBS, 2% FBS) for 30 min at room temperature to reduce nonspecific binding of antibodies. Perform the incubation in a sealed humidity chamber to prevent air-drying of the tissue sections.
5. Gently shake off excess blocking buffer and cover sections with primary antibodies or negative control antibodies diluted

in blocking buffer (5 μg/ml antihuman TLR3 antibodies or goat IgGs, respectively, diluted in TBS, 2% FBS). Place the sections in the humidity chamber and incubate either at room temperature for 1 h or overnight at 4°C (*see* **Note 8**).

6. Rinse the sections three times for 5 min in TBS with gentle shaking.
7. Gently remove excess TBS and cover sections with 1:2,000 diluted biotinylated rabbit antigoat antibodies in dilution buffer (TBS, 2% BSA, 5% human serum) for 30 min at room temperature in the humidity chamber. The addition of human serum decreases unspecific background on human tissue sections
8. During the incubation period, prepare Vectastain ABC reagent, according to the manufacturer's instruction (*see* **Note 9**).
9. After incubation, rinse tissue sections three times for 5 min in TBS with gentle shaking.
10. Gently remove excess TBS and cover sections with Vectastain ABC reagent (*see* **Step 8**). Incubate tissue sections for 30 min in the humidity chamber.
11. Rinse sections three times for 5 min in TBS with gentle shaking.
12. Apply AEC substrate-chromogen solution and incubate 2–5 min or until desired color intensity has developed.
13. Rinse the sections two times for 5 min with dH_2O with gentle shaking.
14. Counterstain the sections for 5–10 s with the hematoxylin counterstaining solution.
15. Rinse the slides under gently running tap water for 5 min (avoid a direct jet which may wash off or loosen the section).
16. Mount the sections using aqueous mounting medium such as glycerol gelatine. Coverslip may be sealed with clear nail polish (*see* **Note 10**).

An example of the results produced is shown in **Fig. 1**.

3.2.2. Double Staining for TLR3 and CD68

Because TLR3- and CD68-specific antibodies are derived from different host species, the simultaneous staining method can be applied. Simultaneous staining is less time consuming than the sequential method, since primary and secondary antibodies can be mixed together in two incubation steps.

1. Follow **steps 1–4** according to the **Subheading 3.2.1**: fixation *(1)* and rehydration *(2)* of the tissue sections, followed by blocking the peroxidase activity *(3)* and blocking of unspecific binding *(4)*.

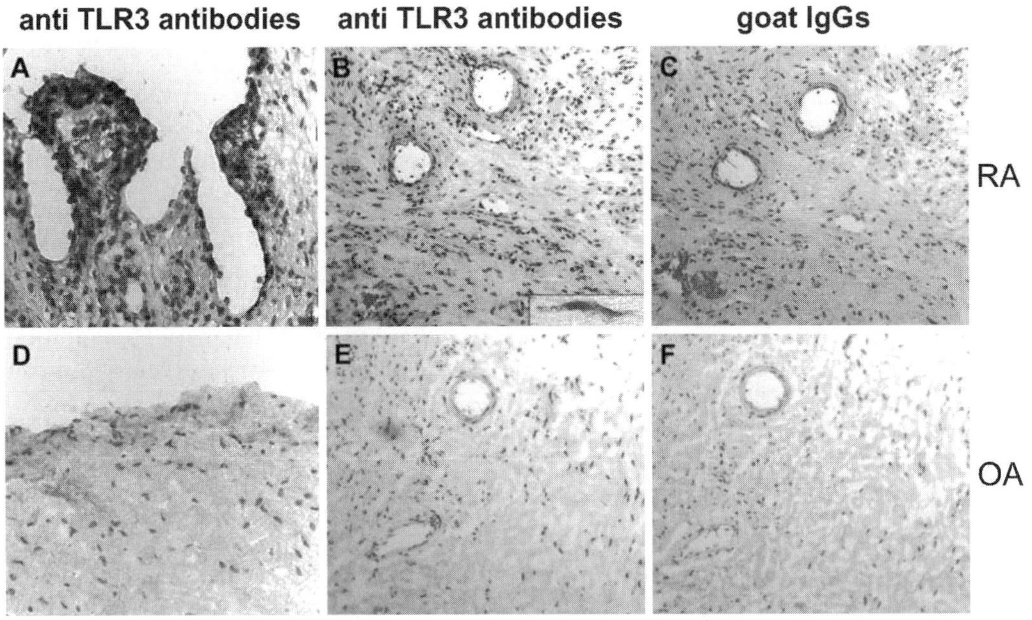

Fig. 1. Detection of Toll-like receptor-3 (TLR3) protein in synovial tissues from patients with rheumatoid arthritis (RA) and osteoarthritis (OA). Tissue sections of patients with RA (A–C) and OA (D–F) were stained with anti-TLR3 antibodies (A, B, D, E) or with the respective isotype control antibodies (C, F). TLR3 protein cells appear in red. In A–F, nuclei were stained with hematoxylin. In RA synovial tissues, TLR3 is abundantly present with a pronounced expression in the synovial lining layer. In OA synovial tissue, TLR3 expression is markedly lower in comparison to RA (from Brentano et al. 2005) *Arthritis Rheum* 52, 2656–2665, Fig. 1).

2. Gently shake off excess blocking buffer and cover sections with TLR3 and CD68 specific antibodies (5 µg/ml antihuman TLR3 antibodies/5 µg/ml mouse antihuman CD68 antibodies) or corresponding negative control antibodies (5 µg/ml goat IgGs/5 µg/ml mouse IgG1) diluted in blocking buffer (TBS, 2% FBS). Replace the sections in the humidity chamber and incubate either at room temperature for 1 h or overnight at 4°C.
3. Rinse sections three times for 5 min in TBS with gentle shaking.
4. Gently remove excess TBS and cover sections with biotinylated rabbit antigoat (dilution 1:4,000) and AP labeled rabbit antimouse (dilution 1:500) antibodies in the dilution buffer (TBS, 2% BSA, 5% human serum). Incubate tissue sections for 30 min at room temperature in the humidity chamber (*see* **Note 11**).
5. During incubation period, prepare Vectastain ABC reagent (as per kit instruction). After incubation rinse tissue sections three times for 5 min in TBS with gentle shaking.

6. Gently remove excess TBS and cover sections with Vectastain ABC reagent (*see* **Step 5**). Incubate tissue sections for 30 min in the humidity chamber.
7. Rinse sections three times for 5 min in TBS with gentle shaking.
8. Apply AEC substrate-chromogen solution and incubate 2–5 min or until desired color intensity has developed (TLR3-positive cells appear in red).
9. Rinse sections two times for 5 min with dH_2O with gentle shaking
10. Apply AP substrate-chromogen solution and incubate for 1–5 min or until desired color intensity is developed (CD68-positive cells appear in blue).
11. Rinse sections two times for 5 min with dH_2O with gentle shaking.
12. Coverslip the sections using aqueous mounting medium such as glycerol gelatine (*see* **Note 12**). An example of the results produced is shown in **Fig. 2**.

3.3. Cell Culture: Isolation of Primary Synovial Fibroblasts

1. Mince fresh RA synovial tissue in a Petri dish. During mincing, add 2–5 ml cell culture medium so that the synovial tissue does not dry out.
2. Transfer minced tissue into a 100 ml glass bottle and add 50 ml of Dispase solution (1.5 mg/ml Dispase II in cell culture

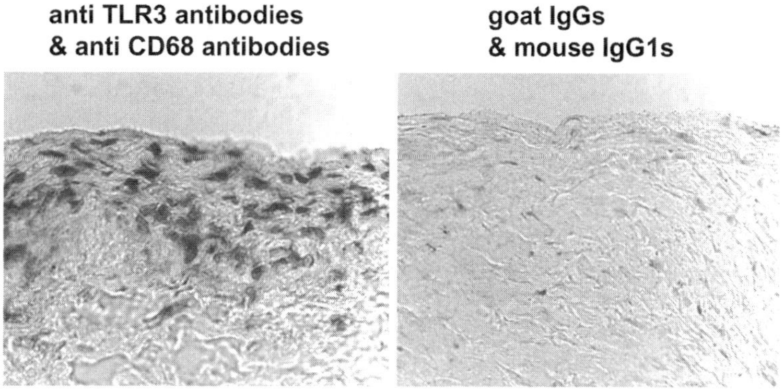

Fig. 2. Detection of Toll-like receptor-3 (TLR3) and CD68 double-positive cells. Rheumatoid arthritis (RA) synovial tissue sections were double stained for TLR3 and CD68 or with corresponding isotype antibodies, respectively. TLR3-positive cells appear in red, CD68-positive cells in deep blue. Most TLR3-positive cells are not positive for CD68. Since the synovial lining layer consists of macrophages and activated synovial fibroblasts the results indicate that rheumatoid arthritis synovial fibroblasts (RASF) are the major cells in the synovial lining expressing TLR3 (from Brentano et al. 2005) *Arthritis Rheum* 52, 2656–2665, Fig. 1).

medium). Stir minced synovial tissue with a magnetic stirrer for 90 min at 37°C.

3. Transfer Dispase solution containing the detached cells into 50 ml centrifuge tubes. After centrifugation at 400 × *g* for 10 min, discard the supernatant and resuspend the cells in 5 ml cell culture medium.
4. Incubate the synovial cells in cell culture medium (total volume of 15 ml) in a 75-cm^2 cell culture flask for 24 h. Then, change cell culture medium to remove nonadherent cells. When the adherent cells reach a confluence of approximately 80%, usually after 3–7 days of incubation, cells can be further passaged.
5. Passage of adherent RASF: Aspirate off the cell culture medium and discard. Add PBS carefully to the flask (10-ml/-5-cm^2 flask), without disturbing the monolayer. Rinse the monolayer by gently rocking the flask. Remove the PBS and discard (*see* **Note 13**). Add trypsin/EDTA (3-ml/-75-cm^2 flask) and rock the flask to ensure that the entire monolayer is covered with the trypsin solution. Return the culture flask to the incubator and leave for 5–10 min until the cells are roundshaped and begin to detach. The side of the flasks may be gently tapped to release any remaining attached cells. Then inactivate the trypsin by adding cell culture medium containing FBS (10-ml/75-cm^2 flask) and pipette the cells up and down until the cells are dispersed into a single cell suspension. Collect the cell suspension in a 15 ml centrifuge tube and pellet the cells by centrifugation at 400 × *g* for 10 min. Aspirate off the supernatant and resuspend the pellet in 10 ml cell culture medium. To split the cells in a ratio of 1:2, transfer 5 ml of the cell suspension to a new labeled flask containing 5 ml prewarmed medium. Incubate the cultures until the cells are approaching confluence (7–14 days) and then repeat the process. After four passages the primary cell cultures contain fibroblast-like cells (RASF) only.

3.4. Stimulation of RASF with TLR Ligands

1. Detach RASF with trypsin as described above (*see* **Subheading 3.3.5**), and pellet cells at 400 × *g* for 10 min. Subsequently resuspend the cell pellet in 1 ml cell culture medium and count cells with the use of the Neubauer-counting chamber. Calculate the total cell number and dilute the cells to a final concentration of 6×10^4 RASF/ml (*see* **Note 14**).
2. Place 2 ml of cell suspension per well into 4 wells of a 6-well plate (1.2×10^5 RASF per well). Make sure that RASF are evenly distributed over the well and after that let the cells settle overnight in the incubator.

3. The next day, dilute TLR ligands in prewarmed cell culture medium to obtain the appropriate working concentrations. For 300 ng/ml bLP working solution, dilute 3 µl bLP stock solution in 10 ml culture medium; for 10 µg/ml poly(I-C) working solution, dilute 40 µl poly(I-C) stock solution in 10 ml medium; and for 100 ng/ml LPS working solution, dilute 10 µl LPS stock solution in 10 ml medium.
4. Subsequently stimulate RASF with TLR ligands: Remove 6-well plate from the incubator, aspirate off culture medium, and wash the RASF monolayer gently with PBS. Pipette off PBS and add 2 ml cell culture medium as nonstimulated control to 1 well of the 6-well plate. For TLR2 stimulation, add 2 ml bLP working solution; for TLR3 stimulation, add 2 ml poly(I-C) working solution; and for TLR4 stimulation, add 2 ml LPS working solution.
5. Incubate nonstimulated and TLR ligand-stimulated RASF for 24 h.

3.5. Sample Preparation: Collection of Culture Supernatants and Cell Lysis for RNA Isolation

1. Twenty-four hours after TLR ligand stimulation, remove the 6-well plate from the incubator, and transfer cell culture supernatants of TLR ligand and unstimulated RASF in four labeled Eppendorf tubes correspondingly. Remove cell debris in cell culture supernatants by centrifugation at 1,000 × *g* for 2 min and transfer supernatants in fresh tubes correspondingly. Store cell culture supernatants at –20°C (*see* **Note 15**).
2. After rinsing the RASF gently with 2 ml PBS per well, lyse the cells by adding 350 µl RLT buffer per well for subsequent total RNA isolation. Mix thoroughly and transfer the cell lysates of the test samples to four labeled RNase-free Eppendorf tubes, correspondingly. Cell lysates can be stored at –20°C until proceeding to RNA isolation.

3.6. TaqMan Real-Time PCR

1. Isolate total RNA according to the manufacturer's instructions (RNeasy Mini Kit) including treatment with RNase-free DNase (*see* **Note 16**).
2. Reverse transcriptase (RT): Total RNA is reverse transcribed into cDNA with random hexamer primers to transcribe all RNA (mRNA, rRNA, tRNA). Prepare RT reaction mix for the four total RNA test samples by pipetting 8 µl 10X RT buffer, 17.6 µl $MgCl_2$, 16 µl dNTP, 4 µl random hexamer, 1.6 µl RNase inhibitor, and 2 µl MultiScribe RT in a RNase-free Eppendorf tube (20 µl reaction per sample). Label four 0.2 ml MicroAmp reaction tubes for the four total RNA test samples. Pipette 12.3 µl of the RT reaction mix (1X PCR buffer, 5.5 mM $MgCl_2$, 2 mM dNTP, 0.4 U/ml RNase inhibitor, 1.25 U/ml MultiScribe RT) to each MicroAmp reaction tube. Transfer 300–500 ng total RNA to the corresponding

reaction tube in a volume of 7.7 µl (fill up with DEPC water, if necessary, total volume 20 µl per reaction). Cap the reaction tubes and tap gently to mix the reactions. Centrifuge the tubes briefly to force the solution to the bottom and to eliminate air bubbles from the mixture. Transfer the tubes to the thermal cycler block and perform RT: 10 min at 25°C, 30 min at 48°C, 5 min at 95°C, hold at 4°C.

3. TaqMan real-time PCR: Prepare 18S, MMP1, and MMP3 TaqMan reaction mix for the four cDNA test samples (25 µl reaction per sample per well, in duplicates). For the 18S TaqMan reaction mix, pipette 100 µl qPCR Master Mix, 10 µl 18S primer/probe, and 82 µl dH_2O in an Eppendorf tube. For the MMP1 and MMP3 TaqMan reaction mix, pipette 100 µl qPCR Master Mix, 8 µl probe, 16 µl forward primer, 16 µl reverse primer, and 52 µl dH_2O in an Eppendorf tube. Vortex the TaqMan reaction mix briefly and pipette 24 µl of the reaction mix per well of a MicroAmp 96-well reaction plate (eight wells per TaqMan reaction mix). Take the four cDNA test samples and pipette 1 µl cDNA per well to the according reaction mix (in duplicates). Cap and centrifuge the reaction plate briefly to force the solution to the bottom and to eliminate air bubbles from the mixture. Transfer the reaction plate to the ABI Prism 7700 Sequence Detection system (Applied Biosystems) (*see* **Note 17**).
4. Use the endogenous control 18S cDNA to normalize the results, according to the comparative threshold cycle (C_t) method for relative quantification, as described by the manufacturer. Calculate the differences in C_t values (ΔC_t) between the sample and the 18S cDNA. Then calculate relative expression levels according to the following formula: $\Delta\Delta C_t = \Delta C_t$ (stimulated sample) – ΔC_t (unstimulated sample). The value to plot relative expression is calculated according to the expression: An example of the results produced is shown in **Fig. 3**.

3.7. Il-6 ELISA

To determine the levels of IL-6 protein in the cell supernatants by ELISA, use the OptEIA human IL-6 set according to the manufacturer's instructions. Briefly, coat IL-6 capture antibodies overnight on 96-well immunosorbant plates. Block plates with PBS/10% FBS. Add diluted cell supernatants (1:50–1:500) and incubate for 1 h. Detect IL-6 protein with a complex of biotinylated anti-IL-6 and avidin–horseradish peroxidase conjugate and use tetramethylbenzidine hydrogen peroxide as substrate. Measure absorption at 450 nm. An example of the results produced by analysis of the data using Revelation software, version 4.22 (Dynex Technologies, Denkendorf, Germany), is shown in **Fig. 4**.

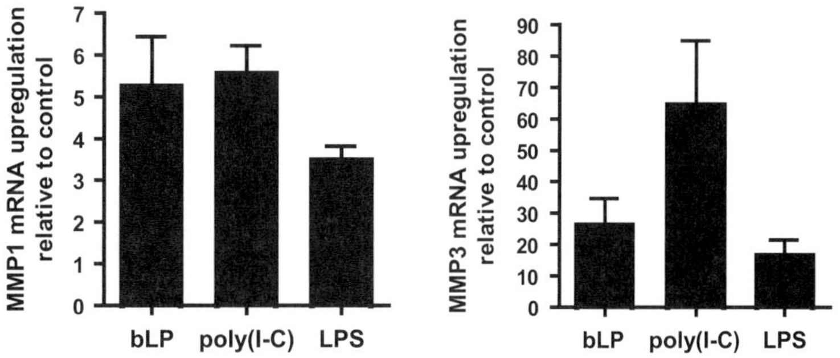

Fig. 3. Metalloproteinases (MMP) 1 and MMP3 mRNA upregulation in rheumatoid arthritis synovial fibroblasts (RASF) following stimulation with Toll-like receptor (TLR) 2, 3, and 4 ligands. RASF derived from five individual patients with RA were stimulated with the TLR2, 3, and 4 ligands bacterial lipoproteins (bLP), poly(I-C), and LPS for 24 h. MMP1 and MMP3 mRNA expression was significantly upregulated by the three TLR ligands compared to nonstimulated RASF cultures.

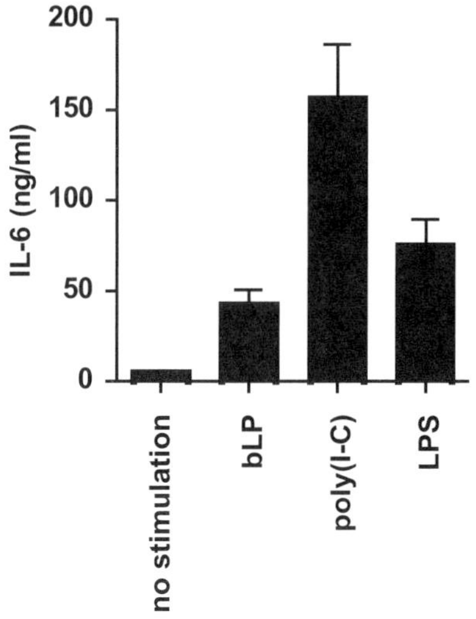

Fig. 4. IL-6 production by RASF after Toll-like receptor (TLR) ligand stimulation. Rheumatoid arthritis synovial fibroblasts (RASF) derived from individual patients with RA ($n = 7$) were untreated or stimulated with the TLR2 ligand bacterial lipoproteins (bLP), the TLR3 ligand poly(I-C) or with the TLR4 ligand LPS. RASF treated with TLR ligands upregulated the IL-6 production significantly compared to untreated cultures. The most marked upregulation was found after stimulation with poly(I-C).

4. Notes

1. Precooled acetone can be reused several times for the fixation of tissue sections.
2. Unless stated otherwise, all solutions should be prepared in deionized water (dH_2O).

3. Human serum was collected following Ficoll gradient centrifugation of heparinized blood.
4. Cell culture medium has to be prewarmed at 37°C before use.
5. Alternatively, tissue sections may be fixed in –20°C precooled acetone for 10 min, dried, and then stored at room temperature.
6. A Coplin jar is a moulded glass jar with drop-on lid and integral large diameter base. The jar has grooves to accept 76 ×26 mm^2 microscope slides vertically.
7. All rinsing steps are performed in Coplin jars. Always use large amounts of wash buffer, especially after antibody incubations, to reduce background.
8. For economy, use only 100–150 µl of diluted antibodies per sample.
9. Streptavidin–avidin complex formation needs at least 30 min.
10. When using AEC substrate, do not use alcohol-containing solutions for counterstaining (e.g., Harris' hematoxylin, acid alcohol), since the AEC stain formed by this method is soluble in organic solvents. The slide must not be dehydrated, brought back to toluene (or xylene), or mounted in toluene-containing mountants. AEC is also susceptible to further oxidation when exposed to light and thus it will fade over time. Dark storage and brief light viewing are recommended.
11. Generally, for the simultaneous staining method, it is advantageous to use secondary antibodies raised in the same host to prevent any unexpected interspecies cross reactivity.
12. Fast Blue substrate, similar to AEC substrate, is soluble in alcoholic and other organic solvents, so aqueous mounting media must be used.
13. Serum contains trypsin inhibitors. Thus it is important to remove traces of serum in this step.
14. To work with a homogenous synovial fibroblast cell culture, use RASF that were passaged 4–6 times. It has been shown that over six passages, significant changes in gene expression might occur.
15. When multiple experiments are performed with the same cell culture supernatants, aliquot the samples to avoid freeze/thaw cycles.
16. TaqMan real-time PCR analysis is sensitive to very small amounts of DNA; therefore, on-column DNase digestion is necessary to obtain reliable results.

17. For economy, only 12.5 µl instead of 25 µl TaqMan reactions per sample can be used. In case of using 12.5 µl TaqMan reactions, transfer 0.5 µl cDNA per well to the reaction mix.

References

1. Ospelt, C., Neidhart, M., Gay, R. E. & Gay, S. (2004). Synovial activation in rheumatoid arthritis. *Front Biosci.* 9, 2323–2334.
2. Brentano, F., Schorr, O., Gay, R. E., Gay, S. & Kyburz, D. (2005). RNA released from necrotic synovial fluid cells activates rheumatoid arthritis synovial fibroblasts via Toll-like receptor 3. *Arthritis Rheum.* 52, 2656–2665.
3. Roelofs, M. F., Boelens, W. C., Joosten, L. A., bdollahi-Roodsaz, S., Geurts, J., Wunderink, L. U., Schreurs, B. W., van den Berg, W. B. & Radstake, T. R. (2006). Identification of small heat shock protein B8 (HSP22) as a novel TLR4 ligand and potential involvement in the pathogenesis of rheumatoid arthritis. *J. Immunol.* 176, 7021–7027.
4. Brentano, F., Kyburz, D., Schorr, O., Gay, R. & Gay, S. (2005). The role of Toll-like receptor signalling in the pathogenesis of arthritis. *Cell Immunol.* 233, 90–96.
5. Sacre, S. M., Andreakos, E., Kiriakidis, S., Amjadi, P., Lundberg, A., Giddins, G., Feldmann, M., Brennan, F. & Foxwell, B. M. (2007). The Toll-like receptor adaptor proteins MyD88 and Mal/TIRAP contribute to the inflammatory and destructive processes in a human model of rheumatoid arthritis. *Am. J. Pathol.* 170, 518–525.
6. Pierer, M., Rethage, J., Seibl, R., Lauener, R., Brentano, F., Wagner, U., Hantzschel, H., Michel, B. A., Gay, R. E., Gay, S. & Kyburz, D. (2004). Chemokine secretion of rheumatoid arthritis synovial fibroblasts stimulated by Toll-like receptor 2 ligands. *J. Immunol.* 172, 1256–1265.
7. Seibl, R., Birchler, T., Loeliger, S., Hossle, J. P., Gay, R. E., Saurenmann, T., Michel, B. A., Seger, R. A., Gay, S. & Lauener, R. P. (2003). Expression and regulation of Toll-like receptor 2 in rheumatoid arthritis synovium. *Am. J. Pathol.* 162, 1221–1227.
8. Kyburz, D., Rethage, J., Seibl, R., Lauener, R., Gay, R. E., Carson, D. A. & Gay, S. (2003). Bacterial peptidoglycans but not CpG oligodeoxynucleotides activate synovial fibroblasts by toll-like receptor signaling. *Arthritis Rheum.* 48, 642–650.

Chapter 21

Practical Techniques for Detection of Toll-Like Receptor-4 in the Human Intestine

Ryan Ungaro, Maria T. Abreu, and Masayuki Fukata

Summary

The human intestine has evolved in the presence of a diverse array of luminal microorganisms. In order to maintain intestinal homeostasis, mucosal immune responses to theses microorganisms must be tightly regulated. The intestine needs to be able to respond to pathogenic organisms while at the same time maintain tolerance to normal commensal flora. Toll-like receptors (TLRs) play an important role in this delicate balance. TLRs are transmembrane noncatalytic receptor proteins that induce activation of innate and adaptive immune responses to microorganisms by recognizing structurally conserved molecular patterns of microbes. Expression of TLRs by intestinal epithelial cell is normally down-regulated to maintain immune tolerance to the luminal microorganisms.

One of the challenges of TLR research in the human intestine is that it is difficult for many experimental methods to detect very low expression of TLRs within the intestinal mucosa. Quantitative methods such as PCR are limited in their ability to detect TLR expression by specific cell types within a tissue sample, which can be important when studying the contribution of TLR signaling to pathological conditions. In this regard, immunohistochemistry (IHC) is advantageous in that one can visualize the distribution and localization of target proteins within both normal and pathologic parts of a given tissue sample. We found that a subset of human colorectal cancers over-express TLR4 by means of immunofluorescence (IF) and IHC methods. Localization of TLR4 within cancer tissue often appears to be patchy, making IHC an appropriate way to examine these changes. We will describe our current techniques to detect TLR4 in paraffin-embedded human large intestine sections. Establishing a practical IHC technique that may provide consistent results between laboratories will significantly enhance understanding of the role of TLRs in human intestinal health and disease.

Key words: Toll-like receptor (TLR), Intestine, Colon, Cancer, Inflammation, Immunofluorescence, Immunohistochemistry, Western blot analysis.

Claire E. McCoy and Luke A.J. O'Neill (eds.), *Methods in Molecular Biology, Toll-Like Receptors, vol. 517*
DOI: 10.1007/978-1-59745-541-1_21

1. Introduction

Despite extensive technological advances in research methods, there is always the need to improve and refine established qualitative and quantitative research techniques employed in the detection and measurement of molecular and pathological phenomena. Immunohistochemistry (IHC) is a widely used general research tool that allows for detection of specific targets within tissue samples through the use of labeled antibodies and reagents that can then be visualized with enzymatic reactions. IHC has the advantage of allowing one to visualize the distribution and localization of specific targets within a specimen. The location of an antigen within a cell (organelle, nucleus, membrane, etc.) can be determined while simultaneously appreciating anatomical and biological changes within the tissue such as neoplasia. In fact, IHC is widely used in the diagnosis of cancer. Specific markers detectable with IHC may further characterize types and/or stages of cancer.

The gastrointestinal tract is a unique organ in which the intestinal epithelium is in direct contact with a diverse array of microorganisms. The human intestine has developed to have a dual relationship with the commensal flora. The epithelium serves as a barrier against bacterial invasion while at the same time taking advantage of these bacteria to assist in the maintenance and function of the intestine. In order to maintain intestinal homeostasis, the mucosal immune response to the microorganisms must be tightly controlled. TLRs play important immune and nonimmune functions in establishing this delicate balance in the intestinal mucosa. TLRs are members of a conserved interleukin-1 (IL-1) superfamily of transmembrane receptors that recognize pathogen-associated molecular patterns (PAMPs) *(1)*. The interaction of TLR4 with its ligand lipopolysaccharide results in NF-κB activation and multiple gene expression causing rapid immune responses. The mitogen activated protein kinase (MAPK) pathway can also be activated by TLR4 mediated signaling *(2)*.

Although the expression of TLRs in the gastrointestinal tract has been examined, the expression, localization, and function of individual TLRs are still unclear. Intestinal epithelial cells (IECs) in the human large intestine have been thought to express most of the TLRs. However, in vitro data have shown that TLR expression and signaling in IECs appear to be down-regulated *(3–5)*. In contrast, up-regulation of TLR expression in IECs has been seen in chronic inflammatory conditions. In the intestinal mucosa of patients with inflammatory bowel disease (IBD), one of the typical chronic inflammatory conditions of the intestine, increased expression of TLR4 has been reported by immunofluorescent methods *(6)*. Others, however, have reported that expression of TLR4 and TLR2 is only increased in lamina propria macrophages in IBD by IHC detection *(7)*. Although the reason of this discrepancy is still unclear, both results imply TLR signaling in the intestinal mucosa, especially by TLR4, may be increased in certain situations.

Since expression of TLR4 is normally very low, assaying for changes in its expression is sometimes difficult because of the limitations of detection methods. Although polymerase chain reaction (PCR) based method has an advantage to detect such fine biological changes, results of intestinal expression of TLR4 by PCR method usually vary within the whole mucosa. Both IEC and infiltrating immune cells potentially express TLR4 making the expression level vary according to how many TLR4 expressing cells exist in each sample. IHC may be a good tool to overcome this problem since we may look at expression of TLR4 in various cell types in a larger tissue sample. We have attempted to detect expression changes of TLR4 by IHC in the human intestine, specifically in colorectal cancer samples. In this chapter, we describe our current techniques used to enhance detection of the extremely low level of TLR4 protein within tissue samples while attempting to reduce nonspecific antibody reactions.

2. Materials

2.1. Immunofluorescence (IF) for TLR4

1. Xylenes, certified ACS (Fisher Scientific, Fair Lawn, NJ). Xylene is flammable and should be stored in an appropriate location at room temperature.
2. Graded series of ethanol 100%, 95%, and 75%. Store with flammables at room temperature.
3. A 10 mM citrate buffer (pH 6.0): We make a stock solution of 5X 10 mM citrate buffer pH 6.0 by mixing 19.2 g of anhydrous citric acid in 2 L of deionized water adjusted to pH 6.0 and then prepare working concentrations as needed by diluting 1:5 with deionized water. Both stock and working solutions can be stored at room temperature. Working solution should be replaced once a significant amount of specimen tissue is seen floating, usually after 5–10 uses.
4. Phosphate buffered saline 10X (PBS) (Fisher Scientific): Create 1X (0.1 M) working concentration by diluting in deionized water and store at room temperature.
5. $NaBH_4$ (Sigma-Aldrich) mixed at a concentration of 1 mg/ml in PBS buffered to pH 8.0. This solution must be made fresh.

6. Avidin/Biotin blocking kit (Zymed, Carlsbad, CA) stored at 4°C (*see* **Note 1**).
7. Skim milk diluted 1:20 (5%) in PBS: This can be stored as working concentration stock at 4°C for 2 weeks.
8. Primary antibody: Biotin-conjugated mouse antihuman TLR4 monoclonal immunoglobulin (HTA125) (eBioscience, San Diego, CA). Stored at 4°C.
9. Fluorescein (FITC) conjugated Streptavidin (eBioscience). Stored at 4°C protected from light.
10. A 0.02% Triton X-100 (working concentration): We create a stock of 10% Triton X-100 (Fisher Scientific) diluted in deionized water and then add the appropriate amount depending on the volume of the wash buffer (1X PBS). For example, we add 460 μl of 10% Trition X-100 to 230 ml of wash buffer. Store stock at room temperature.
11. SlowFade Gold antifade reagent (Invitrogen) stored at 4°C protected from light. Place at room temperature 30 min prior to use.
12. Cover slips 20 × 40.

2.2. IHC for TLR4

1. Xylenes, certified ACS (*see* **Subheading 2.1**, **item 1**).
2. Ethanol (*see* **Subheading 2.1**, **item 2**).
3. A 10 mM citrate buffer (pH 6.0) (*see* **Subheading 2.1**, **item 3**).
4. Three percent hydrogen peroxide diluted in methanol stored at 4°C. Replace after ten uses. We create this working concentration by diluting 30% hydrogen peroxide 1:10 in methanol. Total amount of solution should be adjusted to size of wash basin.
5. PBS 1X (*see* **Subheading 2.1**, **item 4**).
6. Avidin/Biotin blocking kit (Zymed) stored at 4°C (*see* **Note 1**).
7. Tyramide Signal Amplification (TSA) Biotin System (Perkin Elmer/NEN Life Science, Boston, MA). The entire kit is stored at 4°C and is stable for 6 months according to the manufacturer. Multiple components of this kit are utilized in this staining methodology (*see* **Note 2**):
   (a) Blocking buffer is made, as directed by manufacturer, by adding the blocking reagent supplied in the TSA kit to 0.1 M Tris-HCl (VWR International, West Chester, PA), pH 7.5 with 0.15 M NaCl (Fisher Scientific) while heating gradually to 60°C. The blocking buffer contains caffeine and high purity casein. The process of making this buffer can take a few hours. Blocking

buffer can be stored as single use aliquots at –20°C. Thaw prior to use.

(b) Streptavidin–horseradish peroxidase (Streptavidin-HRP) is provided with kit but alternative streptavidin-HRP may be substituted.

(c) Biotinyl Tyramide is provided as a powder that is reconstituted with dimethyl sulfoxide (DMSO) molecular biology grade as specified by the particular TSA kit, which come in different sizes. Thaw prior to use.

8. Primary antibody: Rabbit antihuman TLR4 polyclonal immunoglobulin (Imgenex, San Diego, CA) stored at 4°C.
9. One percent bovine serum albumin (BSA) (Sigma-Aldrich) in PBS. Store at 4°C and discard if white flakes appear in solution or if solution becomes cloudy.
10. Wash buffer: 0.5 M NaCl in PBS. We make 1X PBS (0.1 M) first and then add 290 g of NaCl to 2 L of 1X PBS and mix for 20 min or until dissolved and then add to rest of stock. Store at room temperature.
11. A 0.02% Triton X-100 (working concentration): We create a stock of 10% Triton X-100 (Fisher Scientific) diluted in deionized water and then add the appropriate amount depending on the volume of the wash buffer. For example, we add 460 μl of 10% Trition X-100 to 230 ml of wash buffer (1X PBS). Store stock at room temperature.
12. Secondary antibody: Biotin-conjugated goat $F(ab')_2$ antirabbit monoclonal immunoglobulin (Invitrogen) stored at 4°C (*see* **Note 3**).
13. A 3,3′-diaminobenzidine tetrahydrochloride (DAB) (Sigma-Aldrich): DAB comes as a powder packaged in a 100 ml bottle and should be reconstituted by injecting 10 ml of deionized water through rubber bottle cap. The solution can then be aliquot and stored as single use aliquots at –20°C. DAB is a suspected carcinogen and during handling direct contact should be avoided.
14. A 0.01 M sodium azide (NaN_3) in 0.05 M Tris-HCl buffer pH 7.6: Dissolve NaN_3 (Sigma-Aldrich) and Tris-HCl (VWR International) in deionized water and store at room temperature (*see* **Note 4**).
15. DAB mixture: Mix 2 ml 0.01 M sodium azide (NaN_3) in 0.05 M Tris-HCl buffer pH 7.6, 2 μl 30% hydrogen peroxide, and 100 μl DAB. Vortex. Make fresh immediately prior to visualization reaction.
16. Hematoxylin (Sigma-Aldrich). Store at room temperature.

17. A 0.5% acid alcohol: Combine 995 ml of 70% ethanol with 5 ml of concentrated HCl solution (1 N).
18. Ammonia water: Add eight drops of concentrated NH_4OH (1 N) to 230 ml of tap water (enough to fill wash basin).
19. Cytoseal XYL (Richard-Allan Scientific, Kalamazoo, MI) stored at room temperature.
20. Cover slips 20 × 40.

3. Methods

TLR4 is a key regulator of both inflammation and epithelial homeostasis in the human intestine. TLR4 is normally down-regulated in the gastrointestinal tract in order to prevent continuous activation of the innate immune system. However, under certain pathologic conditions, TLR4 expression appears to be increased. Therefore, the ability to assay for this receptor may impact understanding of an important mechanism in the pathogenesis of gastrointestinal diseases.

The intestine is a complex organ with a diverse array of cell types in addition to the normally high concentration of commensal microorganisms. This presents a unique challenge when assaying for TLR4 in that its expression is typically low and there is the potential for nonspecific staining in an environment so rich in antigens. We will describe our current methods to detect TLR4, which usually has a low level of expression, in human intestinal tissues. We have developed both IF and IHC methods. An overview of the major steps in each of these methods is provided in **Table 1**. IF and IHC have similar applications but can be used to complement one another (**Table 2a, b**). Our technique includes antigen retrieval, enhancement of positive signals, inhibition of nonspecific binding of antibodies, and reduction of general background staining.

3.1. IF for TLR4

1. Fill three wash basins with xylene. Place paraffin-embedded human tissue slides in slide holder and immerse in xylene for 5 min a total of three times, once in each basin. This step should be done under a hood since xylene gives off a strong odor. Insufficient removal of paraffin may lead to an increase in background staining.
2. Fill two wash basins with 100% ethanol, two wash basins with 95% ethanol, and one with 75% ethanol. Transfer slides first to 100% ethanol wash basin. Immerse slides in 100% ethanol for 5 min twice, once in each basin, followed by

Table 1
Comparison of IHC and IF major steps

IF Method	IHC Method
1. Deparaffinize and rehydrate slides	1. Deparaffinize and Rehydrate Slides
2. Antigen retrieval with microwave boiling in 10 mM citrate buffer (pH 6)	2. Antigen retrieval with microwave boiling in 10 mM citrate buffer (pH 6)
3. Wash slides in $NaBH_4$ for 15 min	3. Block endogenous peroxide with 3% hydrogen peroxide for 15 min
4. Apply Avidin and Biotin Blockers for 10 min each	4. Apply Avidin and Biotin Blockers for 10 min each
5. Apply 5% milk solution for 30 min	5. Apply blocking buffer for 30 min
6. Apply primary antibody 1:500 and incubate overnight at 4°C	6. Apply primary antibody 1:2,000 and incubate overnight at 4°C and let sit at room temperature for 30 min prior to washing
7. Apply streptavidin-FITC 1:200 for 1 h	7. Apply secondary antibody 1:1,000 and incubate for 30 min
8. Apply antifade reagent with DAPI and mount each slide with cover slip	8. Apply streptavidin-HRP 1:100 for 30 min
9. Visualize with IF microscope	9. Apply tyramide 1:50 for 10 min
	10. Apply streptavidin-HRP 1:100 for 30 min
	11. Develop slides with DAB
	12. Counterstain with Hematoxylin
	13. Rehydrate slides and mount with Cytoseal

Note: Perform appropriate rinses and washes in between these main steps.

two washes in 95% ethanol, once in each basin for 5 min each, and finally one wash in 75% ethanol for 5 min. We recommend transferring the slides from xylene to ethanol in the hood. After the final ethanol wash, place slides in a basin filled with deionized water for 5 min to wash away the ethanol.

3. Fill a 1,000 ml beaker with 500 ml of 10 mM citrate buffer (pH 6) and immerse slides in slide rack in solution. Place beaker in a standard household microwave and bring to a boil at high power. This usually takes 5 min. Once solution is brought to a boil, stop the microwave and open door until bubbles disappear. Then restart microwave on high power and bring to a boil again (usually another 15–20 s). Stop

Table 2
Advantages and disadvantages of IHC (a) and IF (b)

Advantages	Disadvantages
Immunohistochemistry (a)	
Allows for better cellular localization of staining	Formalin fixation can increase background
Able to more easily archive stained tissue	Difficult to quantify differences in staining intensity between samples
Immunofluorescence (b)	
Easier to quantify differences in staining intensity between samples	Slides fade
	More difficult to archive

microwave and open door. Repeat this step for a total of five additional boils. Remove beaker from microwave and let cool at room temperature for at least 20 min. Slides may cool in citrate buffer for as long as necessary. However, after 20 min the beaker may be placed in cool tap water to speed up the cooling process. Once beaker is cool to touch, remove slides and place in wash basin with deionized water for 3 min (*see* **Note 5**).

4. Place slides in $NaBH_4$ for 15 min and then wash for 3 min in PBS. Small bubbles of hydrogen gas should be seen. This step can inhibit autofluorescence caused by formaldehyde.
5. Apply avidin blocker to each slide. Ensure that entire tissue is covered by gently spreading the solution with a pipette. Place slides in a humidified chamber for 10 min at room temperature. Then gently rinse slides with deionized water and place in PBS wash twice for 3 min each.
6. Apply biotin blocker to each slide. Ensure that entire tissue is covered by gently spreading the solution with a pipette. Place slides in a humidified chamber for 10 min at room temperature. Then gently rinse slides with deionized water and place in PBS wash twice for 3 min each.
7. Apply 5% milk solution to each slide. Ensure that entire tissue is covered by gently spreading the solution with a pipette. Place slides in a humidified chamber for 30 min at room temperature. Following incubation, gently tap side of slide on paper towel to remove excess blocking buffer prior to applying primary antibody. Do not wash slides. To block nonspecific binding at this step, 5–10% normal serum

corresponding to the animal from which the secondary antibody was made is typically used. However, since we do not use a secondary antibody we utilized milk instead.

8. Apply primary antibody at a 1:500 dilution in 1% BSA in PBS. One hundred μl per slide is usually sufficient. Ensure that entire tissue is covered by gently spreading the solution with a pipette. Place slides in a humidified chamber and incubate overnight at 4°C.
9. The next morning, gently rinse slides with deionized water and wash in PBS three times for 5 min each.
10. Apply streptavidin-FITC diluted 1:200 in PBS. One hundred μl per slide is usually sufficient. Ensure that entire tissue is covered by gently spreading the solution with a pipette. Place slides in a humidified chamber and incubate for 1.5 h at room temperature. From this step on, make sure slides are protected from light (incubation chamber should be opaque or wrapped in tinfoil as should wash basins).
11. Wash slides in PBS three times for 5 min each. Then wash slides once in PBS with 0.02% Triton X-100 for 5 min. Finally, wash again in PBS three times for 5 min each.
12. Remove slides and wipe away excess PBS. Apply SlowFade Gold antifade reagent onto each tissue sample and cover with cover slip. Seal sides of cover slip with clear nail polish and protect from light. Visualize with IF microscope. We recommend viewing slides the day after mounting with antifade reagent to allow sufficient time for tissue to absorb this reagent (**Fig. 1**).

3.2. IHC for TLR4

1. Deparaffinize and rehydrate slides as described in **Subheading** 3.1, **steps 1–2**.
2. Boil slides in 10 mM citrate buffer (pH 6) as described in **Subheading** 3.1, **step 3**.
3. Place slides in 3% hydrogen peroxide diluted in methanol for 15 min then rinse away hydrogen peroxide from slides by dipping in basin filled with deionized water and wash in PBS three times for 3 min each (*see* **Note 6**).
4. Apply avidin and biotin blockers as described (*see* **Subheading** 3.1, **steps 5–6**).
5. Apply blocking buffer (supplied in TSA Biotin System kit) to each slide, 100 μl per slide is usually sufficient. Ensure that entire tissue is covered by gently spreading the solution with a pipette. Place slides in a humidified chamber for 30 min at room temperature. Following incubation, gently tap side of slide on paper towel to remove excess blocking buffer prior to applying primary antibody. Do not wash slides.

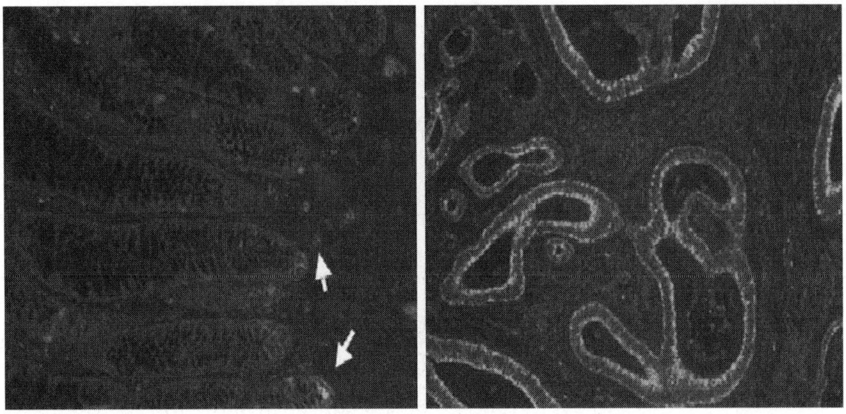

Fig. 1. Normal intestinal epithelial cell (IEC) (*left*) are weakly positive for Toll-like receptor-4 (TLR4) at the base of crypts (*arrow*). Several scattered lamina propria cells are also positive for TLR4 in the normal large intestine. Several colon cancer specimens showed strong positivity of TLR4 in IEC (*right*).

6. Apply primary antibody at a 1:2,000 dilution in 1% BSA in PBS. One hundred μl per slide is usually sufficient. Ensure that entire tissue is covered by gently spreading the solution with a pipette. Place slides in a humidified chamber and incubate overnight at 4°C. The next morning, prior to rinsing and washing slides, let stand at room temperature for 30 min. If slides appear dry, increase amount of primary antibody applied in subsequent experiments. Gently rinse slides with PBS and place in wash solution three times for 5 min each. The second of the three washes should include 0.02% Triton X-100 (*see* **Notes** 7 and **8**).
7. Apply biotin-conjugated secondary antibody at a 1:1,000 dilution in 1% BSA in PBS. One hundred μl per slide is usually sufficient. Ensure that entire tissue is covered by gently spreading the solution with a pipette. Place slides in a humidified chamber and incubate at room temperature for 30 min. Gently rinse slides with PBS and place in wash solution three times for 5 min each. The second of the three washes should include 0.02% Triton X-100.
8. Apply streptavidin-HRP at a 1:100 dilution in 1% BSA in PBS. One hundred μl per slide is usually sufficient. Ensure that entire tissue is covered by gently spreading the solution with a pipette. Place slides in a humidified chamber and incubate at room temperature for 30 min. Gently rinse slides with deionized water and place in wash solution three times for 5 min each. The second of the three washes should include 0.02% Triton X-100.

9. Apply tyramide at a 1:50 dilution in amplification diluent supplied in TSA biotin system kit. One hundred μl per slide is usually sufficient. Ensure that entire tissue is covered by gently spreading the solution with a pipette. Place slides in a humidified chamber and incubate at room temperature for 10 min. Gently rinse slides with deionized water and place in wash solution three times for 5 min each. The second of the three washes should include 0.02% Triton X-100.
10. Apply streptavidin-HRP again at a 1:100 dilution in 1% BSA in PBS. One hundred μl per slide is usually sufficient. Ensure that entire tissue is covered by gently spreading the solution with a pipette. Place slides in a humidified chamber and incubate at room temperature for 30 min. Gently rinse slides with deionized water and place in wash solution three times for 5 min each. The second of the three washes should include 0.02% Triton X-100.
11. Freshly prepare DAB mixture (while slides are undergoing final wash). Remove slides one by one from wash basin, dry back of slide and apply DAB. One hundred μl per slide is usually sufficient. Ensure that entire tissue is covered by gently spreading the solution with a pipette. Place slide under a light microscope to monitor for development of dark brown reaction product. Once tissue has developed to desired intensity, interrupt staining by gently rinsing with deionized water over waste cup. Place in a separate wash basin with deionized water. Repeat for all slides. Make sure to dispose of remaining DAB mixture and contents of waste cup in an appropriate organic waste container (*see* **Note 9**).
12. Place slide rack with developed slides in a larger basin and place under running tap water for 1–2 min. The basin should be large enough so that the stream of water does not directly touch the slides.
13. Place slides in hematoxylin for 40 s and rinse under tap water until water is running clear of hematoxylin (*see* **Note 10**).
14. Dip slides 2–3 times in 0.5% acid alcohol to remove excess hematoxylin and rinse under tap water again for 15–30 s.
15. Dip slides 8–10 times in ammonia water to set hematoxylin blue and rinse under tap water again for 15–30 s.
16. Transfer slides to wash basin filled with 95% ethanol twice, 1 min each. Then place slides in 100% ethanol twice, 1 min each. Lastly, place slides in xylene basin three times, 1 min each.
17. Remove slides individually from xylene and wipe away xylene on front and back of slide (do not touch tissue sample). Place two drops of cytoseal at base of slide, apply cover slip at an

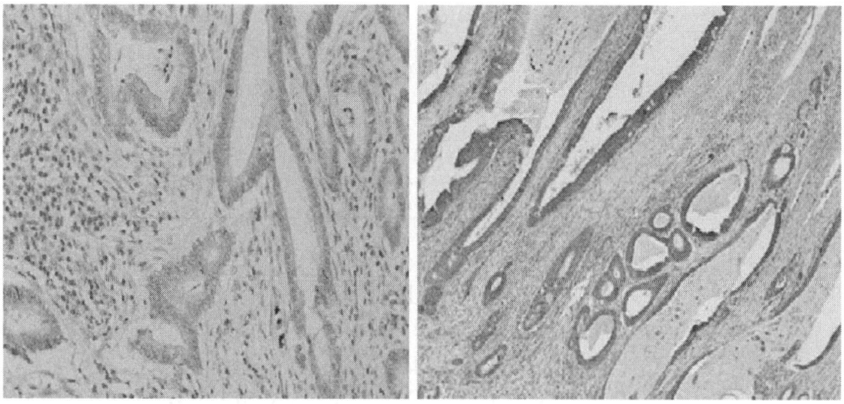

Fig. 2. Comparison between streptavidin–horseradish peroxidase method (*left*) with tyramide amplification method (*right*) in a colon cancer specimen. Notice the marked difference in intensity of staining (*See Color Plates*).

angle so as to begin to spread the cytoseal, and slowly lower the cover slip onto the tissue while cytoseal spreads underneath. Repeat for all slides (*see* **Note 11**) (**Fig. 2**).

4. Notes

1. We have found it necessary to use avidin and biotin blockers because endogenous biotin can cause false positives in IHC methods that utilize streptavidin to bind biotin-conjugated antibodies. As detailed by Mount and Cooper endogenous biotin is present in many human tissues with abundant mitochondria *(8)*. This was of importance in our colon cancer studies because many cancers have high levels of endogenous biotin (**Fig. 3**).
2. We have found the TSA Biotin System to significantly increase the sensitivity of detecting antigen. Multiple tyramide molecules deposit at sites where HRP is present, such as where streptavidin-HRP is bound to biotin-conjugated secondary antibodies. Since the tyramide is biotinylated, this results in greater amounts of biotin at sites of immunoreactivity *(9)* (**Figs. 4A, B**). A study by Toda et al. reported a 10-fold signal enhancement compared to unamplified slides without an increase in background *(10)*. However, we have found that tyramide can increase background when using higher concentrations of antibodies. This effect may also be due to insufficient blockade of endogenous peroxidase in the tissue.

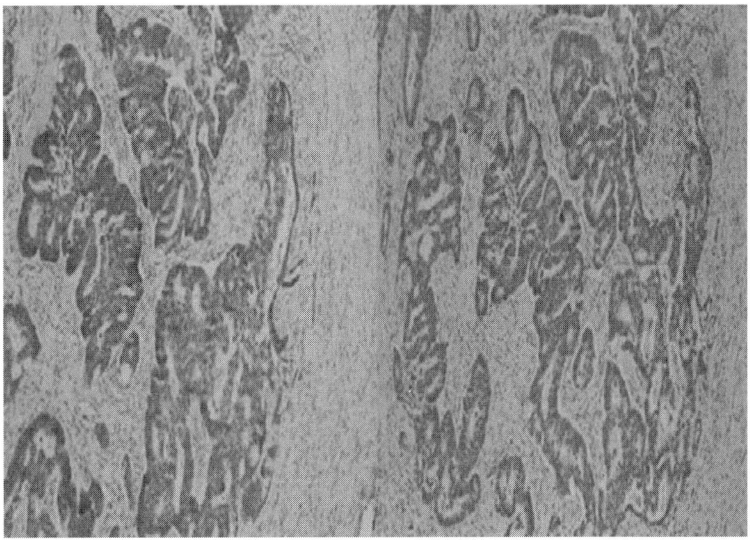

Fig. 3. Comparison between slides omitting primary antibody without avidin and biotin blockers (*left*) and with avidin and biotin blockers (*right*) (*See Color Plates*).

In fact, one of the advantages of tyramide is that it allows for conservation of expensive antibody since lower concentrations can provide a strong signal. Because the tyramide is subject to repeated freeze–thaw cycles, we have suspected that it may become ineffective after many uses. If staining suddenly stops working, one potential problem may be the activity of the tyramide stock.

3. Using an $F(ab')_2$ antibody fragment instead of a full antibody reduces the chance of nonspecific binding and in our experience results in less background staining. The Fc portion of antibodies may react nonspecifically and thus increase the chance of increased background and false positives (**Fig. 5**).
4. Using 0.01 M sodium azide (NaN_3) in 0.05 M Tris-HCl buffer pH 7.6 as the diluent for the DAB mixture used in the visualization step results in reduced background staining by inhibiting development of excess HRP. Adding a very small amount of sodium azide may decrease background created by endogenous peroxidase *(11)*.
5. This step is used for antigen retrieval. During the fixation process, target proteins or antigens may "shrivel" and become inaccessible to detection antibodies. Antigen retrieval denatures the protein enough to expose enough of the molecule to allow binding. We prefer microwave antigen retrieval to using enzymes such as 0.1% trypsin, 0.05% pronase, 0.4% pepsin because it allows for greater preservation of tissue integrity.

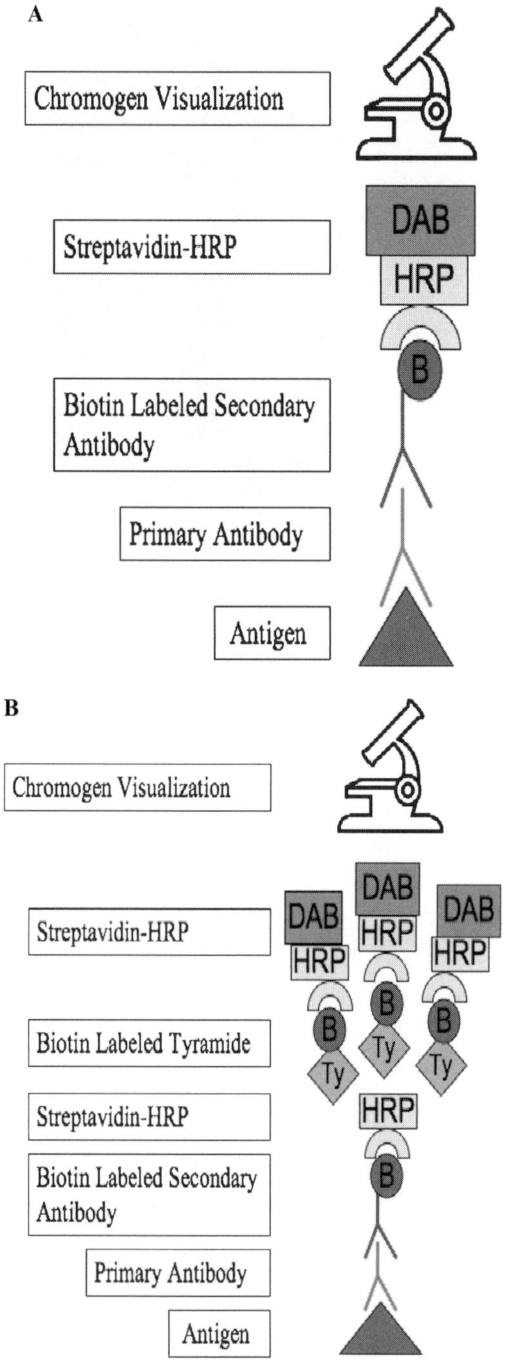

Fig. 4. (**A**) Standard streptavidin–horseradish peroxidase immunohistochemistry (IHC) method. (**B**) Tyramide amplification IHC method.

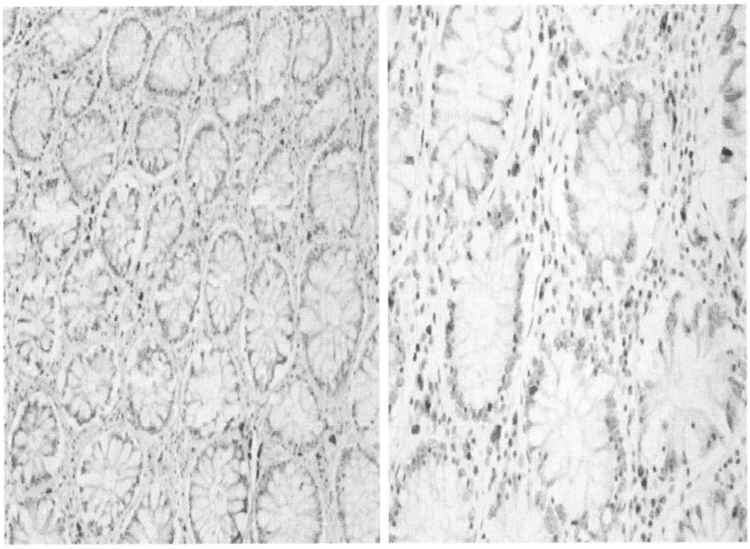

Fig. 5. Comparison of full IgG secondary antibody (*left*) with an F(ab′)$_2$ secondary antibody (*right*) with reduction in background staining (*See Color Plates*).

6. Placing the slides in hydrogen peroxide serves to block endogenous peroxidase activity, which is important in any system that utilizes horseradish peroxidase conjugated reagents to avoid false positives. Furthermore, this step is of particular importance in the intestine where there are high levels of endogenous peroxidase. In our experience with human intestine tissue, 15 min is ample time to block endogenous peroxidase activity while avoiding tissue damage.

7. We recommend titrating the dilution of primary and secondary antibodies to optimize balance between strength of primary signal and background staining. As an alternative to this polyclonal primary antibody we have tried the monoclonal antibody utilized for the IF method described above at a dilution of 1:50 (manufacturer recommends titrating antibody in range of 1–10 μg/ml). Instead of incubating overnight at 4°C we have found that the signal is stronger when incubating overnight at 20°C with minimal change in background. In this case, it is important to ensure that the slides do not dry out and it is often necessary to apply extra antibody. However, we have had inconsistent results with the monoclonal antibody in the IHC method with greater batch-to-batch variability. We suspect that this antibody is more labile when incubated at 20°C. To confirm specificity of primary antibodies, we recommend using primary antibody in Western blot analysis. If unexpected bands are seen by this method, consider possibility of false positive reaction in the staining (**Fig. 6**).

8. We have found that using the 0.5 NaCl in PBS wash in combination with 0.02% Triton X-100 leads to a much cleaner stain by decreasing the amount of dust and other contaminating particles adhering to the tissue. Hypertonic saline may block nonspecific binding of antibodies by electronic charge within the tissue. This technique appears to be more effective at washing away reagents in between steps leading to reduced background and edge staining.
9. During this step, each slide must be carefully monitored so that once the intensity of staining is sufficient the DAB mixture is rinsed off. Leaving DAB on the slide for too long may create extra background. We have generally found that strongly positive slides will develop within 5 min. However, there can be much tissue-to-tissue variability. One factor to be aware of is that formalin-fixed tissue that was left in formalin for an extended period of time will be more prone to increased nonspecific staining and these slides will need to have DAB removed more quickly.
10. Counterstaining can be adjusted to create desired contrast between staining and nonstaining tissue as well as for cellular localization.
11. Let slides sit to dry under a hood. Moving slides prior to drying of cytoseal may cause cover slips to become displaced and may damage underlying tissue. For this IHC method, we recommend using it as a qualitative measure of TLR4 expression instead of as a quantitative measure. This is because certain steps in the staining process can lead to batch-to-batch variability in staining intensity. For example, leaving DAB on longer in one batch may make the same slide appear to be more intensely positive than when stained in a previous batch. Therefore, this is a good method of establishing whether a tissue is positive or negative for TLR4 but it is difficult to subsequently quantify the staining intensity.

Acknowledgment

We would like to thank Anli Chen for her technical assistance.

References

1. Medzhitov, R., P. Preston-Hurlburt, E. Kopp, A. Stadlen, C. Chen, S. Ghosh, and C. A. Janeway, Jr. (1998) MyD88 is an adaptor protein in the hToll/IL-1 receptor family signaling pathways. *Mol Cell* 2, 253–258.
2. O'Neill, L. A., A. Dunne, M. Edjeback, P. Gray, C. Jefferies, and C. Wietek (2003) Mal and MyD88: adapter proteins involved in signal transduction by Toll-like receptors. *J Endotoxin Res* 9, 55–59.
3. Melmed, G., L. S. Thomas, N. Lee, S. Y. Tesfay, K. Lukasek, K. S. Michelsen, Y. Zhou, B. Hu, M. Arditi, and M. T. Abreu (2003) Human intestinal epithelial cells are broadly unresponsive to Toll-like receptor 2-dependent bacterial ligands: implications for host-microbial interactions in the gut. *J Immunol* 170, 1406–1415.
4. Otte, J. M., E. Cario, and D. K. Podolsky (2004) Mechanisms of cross hyporesponsiveness to Toll-like receptor bacterial ligands in intestinal epithelial cells. *Gastroenterology* 126, 1054–1070.
5. Abreu, M. T., P. Vora, E. Faure, L. S. Thomas, E. T. Arnold, and M. Arditi (2001) Decreased expression of Toll-like receptor-4 and MD-2 correlates with intestinal epithelial cell protection against dysregulated proinflammatory gene expression in response to bacterial lipopolysaccharide. *J Immunol* 167, 1609–1616.
6. Cario, E., and D. K. Podolsky (2000) Differential alteration in intestinal epithelial cell expression of toll-like receptor 3 (TLR3) and TLR4 in inflammatory bowel disease. *Infect Immun* 68, 7010–7017.
7. Hausmann, M., S. Kiessling, S. Mestermann, G. Webb, T. Spottl, T. Andus, J. Scholmerich, H. Herfarth, K. Ray, W. Falk, and G. Rogler (2002) Toll-like receptors 2 and 4 are up-regulated during intestinal inflammation. *Gastroenterology* 122, 1987–2000.
8. Mount, S. L., and K. Cooper (2001) Beware of biotin: a source of false-positive immunohistochemistry. *Current Diagnostic Pathol* 7, 161–167.
9. King, G., S. Payne, F. Walker, and G. I. Murray (1997) A highly sensitive detection method for immunohistochemistry using biotinylated tyramine. *J Pathol* 183, 237–241.
10. Toda, Y., K. Kono, H. Abiru, K. Kokuryo, M. Endo, H. Yaegashi, and M. Fukumoto (1999) Application of tyramide signal amplification system to immunohistochemistry: a potent method to localize antigens that are not detectable by ordinary method. *Pathol Int* 49, 479–483.
11. Tabata, H., A. Yamakage, and S. Yamazaki (1997) Cutaneous localization of endothelin-1 in patients with systemic sclerosis: immunoelectron microscopic study. *Int J Dermatol* 36, 272–275.

Chapter 22

Toll-Like Receptor-Dependent Immune Complex Activation of B Cells and Dendritic Cells

Melissa B. Uccellini, Ana M. Avalos, Ann Marshak-Rothstein, and Gregory A. Viglianti

Summary

High titers of autoantibodies reactive with DNA/RNA molecular complexes are characteristic of autoimmune disorders such as systemic lupus erythematosus (SLE). In vitro and in vivo studies have implicated Toll-like receptor 9 (TLR9) and Toll-like receptor 7 (TLR7) in the activation of the corresponding autoantibody producing B cells. Importantly, TLR9/TLR7-deficiency results in the inability of autoreactive B cells to proliferate in response to DNA/RNA-associated autoantigens in vitro, and in marked changes in the autoantibody repertoire of autoimmune-prone mice. Uptake of DNA/RNA-associated autoantigen immune complexes (ICs) also leads to activation of dendritic cells (DCs) through TLR9 and TLR7.

The initial studies from our lab involved ICs formed by a mixture of autoantibodies and cell debris released from dying cells in culture. To better understand the nature of the mammalian ligands that can effectively activate TLR7 and TLR9, we have developed a methodology for preparing ICs containing defined DNA fragments that recapitulate the immunostimulatory activity of the previous "black box" ICs. These reagents reveal an important role for nucleic acid sequence, even when the ligand is mammalian DNA.

Key words: AM14 transgenic BCR, Rheumatoid factor B cell, Immune complex, Autoantibodies, B cells, IFNα, Flt3L-DCs, TLR7, TLR9, Endogenous ligands, Biotinylated DNA.

1. Introduction

Autoimmune diseases such as systemic lupus erythematosus (SLE) are characterized by the presence of autoantibodies directed against endogenous DNA and RNA ligands. Immune system activation and immune complex (IC) deposition in vital organs

Claire E. McCoy and Luke A.J. O'Neill (eds.), *Methods in Molecular Biology, Toll-Like Receptors, vol. 517*
DOI: 10.1007/978-1-59745-541-1_22

lead to tissue destruction, with concomitant release of self-DNA and RNA that results in sustained tissue damage and inflammation *(1)*. Toll-like receptor 9 (TLR9) and Toll-like receptor 7(TLR7) are innate immune receptors located in the endosomal compartments, originally shown to be responsible for immune responses to viral ssRNA and hypomethylated CpG DNA, respectively *(2–5)*. While their role in response to infection is firmly established, studies performed in vivo and in vitro have also shown that aberrant expression of TLR9 and TLR7 has a profound effect on the autoimmune phenotypes found in diverse mouse models of systemic autoimmune disease *(6, 7)*. These observations have led to the hypothesis that under the appropriate conditions, the prevalence of certain RNA and DNA autoantigens could lead to activation of an immune response through engagement of TLR9 and/or TLR7. In this case, autoantigens would serve as autoadjuvants to the immune system *(8)*.

Use of the AM14 rheumatoid factor (RF) B cell receptor (BCR) transgenic model has been instrumental in demonstrating that activation of autoreactive B cells by endogenous DNA is BCR and TLR9-dependent. AM14 transgenic B cells recognize IgG2a of the a or j allotype with low affinity, and therefore serve as prototypic autoreactive RF B cells. In vitro, AM14 B cells provide an excellent system to test the immunostimulatory capacity of specific autoantigens by simply adding, as ligands, ICs consisting of autoantigen-specific IgG2a and the candidate autoantigen. Using this system, we have shown that ICs containing mammalian chromatin activate AM14 B cells in a TLR9-dependent manner *(9, 10)*. However, at the time, these results were somewhat surprising given that mammalian DNA is generally considered a poor ligand for TLR9. To better define the mammalian DNA ligand, we have produced defined ICs incorporating dsDNA fragments of known sequence composition. To avoid the high background stimulation associated with DNA-reactive antibodies, which invariably bind to cell debris, we have developed a simple method for labeling DNA fragments with biotin, or the hapten trinitrophenol (TNP). These DNA fragments can then be delivered to the AM14 BCR with antibodies specific for biotin or TNP. To demonstrate that the dsDNA fragments have been appropriately modified, the derivatized fragments can be mixed with the anti-biotin or anti-TNP antibodies, and formation of ICs can be assessed by a modified electrophoresis mobility shift assay. This technique allows us to directly compare the immunostimulatory capacity of various sequences. This kind of analysis has shown that ICs containing CpG-rich dsDNA fragments can activate AM14 B cells, while fragments lacking CpG motifs cannot. We have also demonstrated that IgG2a antibodies specific for RNA-associated autoantigens stimulate AM14 B cells in a TLR7-dependent manner *(11)*.

The adjuvant activity of defined DNA ICs, or RNA-containing ICs, can also be tested on cells that express an activating Fcγ receptor. In this case, the Fcγ receptor binds the IC and delivers the ligand to an intracellular compartment containing the TLR. Activation of both conventional DC (cDCs) and plasmacytoid (pDCs) can be assessed by measuring cytokine production *(12–14)*. The contribution of pDCs, a lymphoid DC subset responsible for interferon-α (IFN-α) production, is of particular interest since this cytokine is present at high levels in many patients with SLE. In addition, elevated amounts of IFN-α induce global changes in gene expression designated the "IFN-α signature" *(15)*. Interestingly, B cell responses to RNA-associated IC are significantly enhanced by IFN-α, suggesting that cross talk between these two cell types appears to be important in disease progression.

2. Materials

2.1. Antibody Preparation

1. IgG depleted medium: growth medium passed over a protein G column (*see* **Note 1**).
2. Disposable polypropylene columns (Pierce, Rockford, IL).
3. Protein G sepharose™4 Fast Flow (Amersham, Piscataway, NJ).
4. Phosphate-buffered saline (PBS, Invitrogen, Carlsbad, CA): 1 mM KH_2PO_4, 155 mM NaCl, 3 mM Na_2HPO_4.
5. Spectrophotometer.
6. Elution buffer: 0.1 M glycine-HCl pH 2.7.
7. Neutralization buffer: 1 M Tris-HCl pH 9.
8. pH paper (Fisher, Hampton, NH).
9. Storage buffer: 0.02% (w/v) Sodium Azide in PBS.
10. Slide-A-Lyzer dialysis cassettes, 10,000 molecular weight cut-off (Pierce).

2.2. DNA Fragment Preparation

1. Dam-/dcm-competent *Escherichia coli* (e.g. GM2163).
2. Carbenicillin (1000X, American Bioanalytical, Natick, MA) is dissolved in dH_2O at 50 mg/mL and stored in aliquots at −20°C for up to 1 year.
3. Luria-Bertani carbenicillin medium (LB carbenicillin): 1% (w/v) tryptone, 0.5% (w/v) NaCl, 0.5% (w/v) yeast extract are dissolved in water. Bring to pH 7.4 with NaOH and autoclave. Store at 4°C and add carbenicillin to 50 µg/mL just before use.

4. Luria-Bertani carbenicillin plates (LB carbenicillin plates): LB carbenicillin medium is made with 1.2% (w/v) agar, autoclaved, and cooled to 50°C prior to addition of carbenicillin to 50 μg/mL. Pour plates and store at 4°C for up to 2 months.
5. EndoFree® Plasmid Maxi Kit (Qiagen, Valencia, CA).
6. EndoFree® TE buffer (Qiagen): 10 mM Tris-HCl, 1 mM EDTA, pH 8.0.
7. TAE running buffer (50X): 2 M Tris base, 2 M acetate, 50 mM EDTA, pH 8.0.
8. Agarose gel: 1% (w/v) SeaKem® GTG® agarose (Cambrex, East Rutherford, NJ) is dissolved in 1X TAE buffer containing 0.5 μg/mL ethidium bromide (Sigma, St. Louis, MO).
9. EcoRI buffer (10X, New England BioLabs, Ipswich, MA). Store at –20°C.
10. Bovine serum albumin (BSA, 10X, New England BioLabs). Store at –20°C.
11. EcoRI and BamHI (New England BioLabs). Store in enzyme block at –20°C.
12. Distilled water DNAse, RNAse free (dH_2O, Invitrogen).
13. DNA Clean & Concentrator-25™ kit (Zymoresearch, Orange, CA).
14. Loading dye (6X): 0.2% Orange G, 50% (v/v) glycerol in dH_2O.
15. Zymoclean Gel DNA Recovery Kit™ (Zymoresearch).
16. CGneg primers: H931 (5′-AACTGGATCCCCTGGCCTTTTAGAGACATCAGAAGG-3′) H1560 (5′-GGCAGAATTCGGGATAGGTGGATTATGTGTCATCCATCC-3′).
17. GoTaq® Flexi Buffer (5X, Promega Madison, WI).
18. GoTaq® Flexi DNA Polymerase (Promega).
19. Deoxyribonucleotide triphosphates (dNTPs, Strategene, La Jolla, CA) are dissolved in dH_2O at 2.5 mM and aliquots are stored at –20°C. Avoid multiple freeze–thaws.
20. Biotin-16-deoxyuridine triphosphate (biotin-16-dUTP, 1 mM, Roche, Indianapolis, IN). Store at –20°C, loses activity after 1 year.
21. Klenow Fragment (3′ → 5′ exo-) (New England BioLabs). Store in enzyme block at –20°C.
22. PBS (*see* **Subheading 2.1, item 4**).
23. 5-(3-Aminoallyl)-2′-deoxy-uridine 5′-triphosphate, trisodium salt (Aminoallyl dUTP, Molecular Probes, Carlsbad, CA). Store at –20°C.

24. Sodium bicarbonate buffer: sodium bicarbonate (Sigma) is dissolved in dH_2O at 25 mg/mL and aliquots are stored at −20°C for up to 1 year.
25. TNP-e-Aminocaproyl-OSu (reactive TNP, Biosearch Technologies, Novato, CA) is dissolved in dimethylformamide at 20 mg/mL and stored in aliquots at −20°C for up to 1 year.

2.3. RNA Particle Preparation

1. RNP/Sm antigen (Arotec Diagnostics Limited, New Zealand).
2. RPMI 1640 (Invitrogen).
3. Amicon Ultra-4 filter units (centricon) 10,000 molecular weight cut-off (Millipore, Bedford, MA).

2.4. Preparation of ICs

1. RPMI medium (RPMI, Invitrogen): RPMI 1640 is supplemented with 10% (v/v) fetal bovine serum (Hyclone, Ogden, UT), Penicillin/Streptomycin/Glutamine solution: 100 units/mL penicillin G sodium, 100 μg/mL streptomycin sulfate, 0.29 mg/mL l-glutamine (Invitrogen), 10 mM HEPES pH 7.5, 22 mM β-mercaptoethanol (Sigma). Filter-sterilize and keep at 4°C for up to 1 month.
2. Serum-free RPMI: RPMI, without addition of fetal bovine serum.
3. DNA fragments, RNP/Sm particles, antibodies (*see* **Note 2**).

2.5. B Cell Proliferation Assay

1. Cell culture plasticware: 60 × 15 mm Petri dishes, 5 mL syringes, 96-well flat bottom plates, 15 mL conical tubes (Fisher).
2. A 25 gauge needle (Fisher).
3. Frosted slides (Fisher).
4. Cotton-plugged and unplugged Pasteur pipettes (Fisher).
5. Brandel Harvester (Brandel, Gaithersburg, MD).
6. Glass fiber filter (90 × 120 mm), printed filtermat A (Perk inElmer, Waltham, MA).
7. Filtermat sample bag (PerkinElmer).
8. Betaplate scintillation cocktail (PerkinElmer).
9. Microbeta Trilux counter (PerkinElmer).
10. Hanks balanced salt solution (HBSS, Invitrogen): HBSS medium is supplemented with 10 mM sodium phosphate (3.2 mM Na_2HPO_4; 7.2 mM $NaH_2PO_4{\cdot}2H_2O$) pH 7.2, and 5% (v/v) fetal bovine serum. Filter-sterilize and keep at 4°C.
11. IMag buffer: PBS (*see* **Subheading 2.1**, **item 4**) is supplemented with 0.5% (w/v) bovine serum albumin fraction V (Roche) and 2 mM EDTA. Store at 4°C.

12. Anti-Mouse CD45R/B220 magnetic particles (BD Biosciences, San Jose, CA).
13. IMagnet (BD Biosciences).
14. RPMI medium (*see* **Subheading 2.4, item 1**).
15. Mouse interferon alpha-A (PBL, Piscataway, NJ).
16. ^{3}H-thymidine medium: RPMI medium is supplemented with 25 μCi/mL [methyl-^{3}H]thymidine (Amersham). Store at 4°C.

2.6. Dendritic Cell Cytokine Assay

1. Cell culture plasticware: 60 × 15 mm Petri dishes, 20 mL syringes, 15 mL conical tubes, 70 μm cell strainer (Fisher).
2. A 25 gauge needle (Fisher).
3. PBS (*see* **Subheading 2.1, item 4**).
4. RBC lysis buffer (Sigma).
5. RPMI 1640 (Invitrogen).
6. RPMI medium (*see* **Subheading 2.4, item 1**).
7. FL B16 cells: Fms-like tyrosine kinase ligand (Flt3L)-transfected B16 melanoma cell line.
8. ELISA plates (Fisher).
9. Mouse interferon alpha-A standard (PBL).
10. Rat monoclonal antibody against mouse interferon alpha, 2 mg/mL (PBL).
11. ELISA wash buffer: 0.05% (v/v) Tween-20 (Sigma) in PBS.
12. ELISA blocking buffer: 1% (w/v) bovine serum albumin fraction V (Roche) in PBS.
13. Rabbit polyclonal antibody against mouse interferon alpha, 0.94 mg/mL (PBL).
14. $F(ab')_2$ donkey antirabbit IgG (H + L)-horseradish peroxidase, 0.8 mg/mL (Jackson Immunoresearch, West Grove, PA).
15. 3,3′,5,5′-Tetramethyl benzidine (TMB) Liquid Substrate (Sigma).
16. ELISA stop solution: 1 M H_3PO_4.
17. Spectrophotometer.

3. Methods

3.1. Antibody Preparation

AM14 RF B cells bind to IgG2a$^{a/j}$; therefore IgG2a$^{a/j}$ antibodies of the correct specificity can be used to deliver ligands to

TLR9 and TLR7. By an analogous mechanism DCs can take up antibodies through Fcγ receptors and deliver ligands to TLR9 and TLR7. Monoclonal antibodies of interest are obtained from hybridomas and antibody is purified from culture supernatants using protein G sepharose.

1. Grow hybridoma of interest to high density in appropriate IgG depleted medium, and harvest 500 mL of supernatant.
2. Prepare protein G column by pipetting enough of the protein G slurry into the column for about 1 mL of packed sepharose. Rinse column with 20 mL of PBS (*see* **Note 3**).
3. Apply hybridoma supernatant to column and save flow-through fraction.
4. Wash column with 30 mL of PBS. Read A_{280} of wash on spectrophotometer and continue rinsing with PBS until reading is below 0.01 absorbance unit.
5. Elute column with 10 mL of elution buffer. Collect 900 μL fractions (approximately 10) into 100 μL of neutralization buffer (*see* **Note 4**).
6. Read A_{280} of fractions and pool the most concentrated 3–4 fractions. Apply antibody to dialysis cassette and dialyze against PBS with at least three changes of PBS.
7. Read A_{280} of the antibody to determine concentration (A_{280} of a 1 mg/mL solution of IgG is approximately 1.5 absorbance units). Antibody should also be tested for endotoxin contamination (*see* **Note 5**). Aliquots of antibody are stored at −80°C. Use a fresh aliquot for each assay, and avoid freeze-thawing.
8. Rinse column with PBS and check that the column has returned to pH 7.6 using pH paper. Store column in storage buffer with both ends capped.

3.2. DNA Fragment Preparation

AM14 RF B cells can be activated by ICs composed of anti-nucleosome or anti-DNA antibodies in association with cell debris present in the culture supernatant. This activation was found to be DNase-sensitive and TLR9-dependent, supporting a model in which the B cell receptor binds to ICs containing chromatin/DNA and delivers them to an intracellular compartment where they are able to engage TLR9 *(9, 10)*. DCs can also take up ICs through Fcγ receptors and deliver ligands to TLR9 *(12, 13)*. We have used the antibodies PL2-3 (anti-nucleosome) *(16)*, and PA4 (anti-DNA) *(17)* to assess the role of TLR9 in autoreactive B cell and DC activation. In addition, we have used the defined DNA fragments CGneg (629 bp, no CpG) and CG50 (607 bp, 50 optimal CpG) *(18, 19)* to assess the role of CpG motifs in the activation of TLR9.

3.2.1. DNA Preparation from Plasmid

1. CGneg and CG50 are cloned into the EcoRI and BamHI restriction sites of pUC19 and LITMUS 29, respectively. Plasmids are transformed into dam/dcm-deficient *E. coli* (*see* **Note 6**) and streaked onto LB carbenicillin plates. Single colonies are used to inoculate a 3 mL starter culture of LB carbenicillin, and 50 μL of the starter culture is used to inoculate 100 mL of LB carbenicillin (*see* **Note 7**). Plasmid DNA is prepared using the using the EndoFree Plasmid Maxi Kit according to the manufacturer's instructions and resuspended in 500 μL of TE buffer. DNA concentration is determined by running a 1 μL sample on a 1% agarose gel and comparing to a known concentration of DNA ladder. Alternatively, DNA concentration can be determined by reading A_{260} (A_{260} of a 50 μg/mL solution is 1 absorbance unit).
2. DNA fragments are prepared by digesting plasmid DNA with EcoRI and BamHI to separate the CGneg and CG50 fragments from the plasmid backbone. Incubate 16 h at 37°C: 200 μg of plasmid DNA, 30 μL of EcoRI buffer, 30 μL BSA, 400 U EcoRI, 400 U BamHI, and dH_2O to 300 μL (*see* **Note 8**). Run a 100 ng sample on a 1% agarose gel to check for complete digestion, indicated by the presence of only two bands on the gel.
3. CGneg and CG50 fragments are agarose gel isolated. Add 60 μL loading dye to restriction digest and mix. Pour a 1% agarose gel with a comb large enough to hold 360 μL (*see* **Note 9**). Load the DNA sample and run at 120 V until the two bands are separated by at least 3 cm (*see* **Note 10**) and cut out CGneg or CG50 band with a new razor blade, removing any excess agarose. Use Zymoclean Gel DNA Recovery Kit™ according to the manufacturer's instructions to extract DNA from agarose (*see* **Note 11**). The maximum theoretical yield is 40 μg; actual yields are usually 70–80% of this.

3.2.2. DNA Preparation by PCR

1. DNA sequences of interest may also be prepared by PCR. As an example, to prepare CGneg by PCR, primers that include the EcoRI and BamHI restriction sites are used. Make PCR master mix by combining to a final concentration: GoTaq Flexi buffer to 1X, $MgCl_2$ to 1.5 mM, H931, and H1560 primers to 0.8 μM each, dNTP mix to 250 μM, 2.5 ng CGneg fragment, and 0.5 U Promega GoTaq to a total volume of 100 μL with dH_2O. Amplify using cycling conditions: 94°C for 5 min, 35 cycles of 95°C for 30 s, 65°C for 40 s, and 72°C for 45 s, followed by 72°C for 5 min. Estimate DNA concentration by running a 5 μL sample on a 1% agarose gel and comparing to a known concentration of DNA ladder.

2. Run DNA through DNA Clean & Concentrator-25™ kit according to the manufacturer's instructions and elute in TE buffer (*see* **Note 12**).
3. Digest DNA with EcoRI and BamHI as above.

3.2.3. End-Labeling with Biotin

1. To deliver DNA to TLR9, we have used ICs composed of an anti-biotin antibody and biotinylated DNA fragments. Biotinylated DNA is made by filling in 5′ overhangs left by digestion with EcoRI and BamHI, with biotin-16-dUTP using the Klenow Fragment of DNA polymerase I. DNA is prepared and digested with EcoRI and BamHI (*see* **Subheading 3.2.1, item 2**), and an agarose gel is run to check that the digestion is complete.
2. DNA is end-labeled by adding dATP, dCTP, dGTP, and biotin-16-dUTP to the digestion reaction to a final concentration of 25 μM each in the presence of 0.25 U/μg of Klenow Fragment (3′ → 5′ exo-), and incubating for 30 min at 37°C (*see* **Notes 13** and **14**).
3. DNA prepared from plasmid is then gel purified or DNA prepared by PCR is run through DNA Clean & Concentrator-25™ kit (*see* **Subheading 3.2.2, item 2**).

3.2.4. Internal Labeling with Biotin

1. DNA can alternatively be labeled internally with biotin by PCR. CGneg PCR is performed as above, except that individual nucleotides are used with biotin-16-dUTP substituted for a portion of the dTTP.
2. dATP, dCTP, and dGTP are added to a final concentration of 62.5 μM each. To label 10% of the dTTP residues, biotin-16-dUTP is added to a final concentration of 6.3 μM and dTTP is added to a final concentration of 56.3 μM. PCR is performed as above.
3. Free biotin and primers are removed by running DNA through the DNA Clean & Concentrator-25™ kit.

3.2.5. Gel Shift to Confirm Labeling

1. A DNA gel shift is used to confirm that DNA is labeled with biotin. Fifty nanograms of biotin-labeled DNA are combined with 2 μg of anti-body 1D4 (anti-biotin) or irrelevant control antibody Hy1.2 (anti-TNP) *(20)* in 10 μL PBS. No incubation is necessary.
2. Separate sample on an agarose gel. 2 μL of loading dye is added and samples are loaded on a 1% agarose gel. Biotin labeling is confirmed by complete shift of the free DNA band (*see* **Note 15**). An example of the results is shown in **Fig. 1**.

3.3. RNA Particle Preparation

AM14 RF B cells can be activated by ICs composed of anti-RNA antibodies in association with cell debris present in the culture

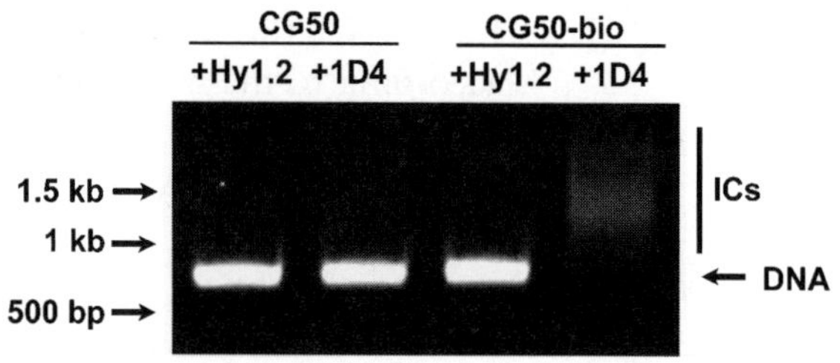

Fig. 1. Gel shift has to confirm biotin labeling of DNA. 50 ng of DNA was mixed with 2 μg of antibody and run on a 1% agarose gel. Left two lanes contain unmodified DNA combined with irrelevant control anti-TNP antibody (Hy1.2) or anti-biotin antibody (1D4). Right two lanes contain biotin labeled DNA combined with irrelevant control anti-TNP antibody (Hy1.2) or anti-biotin antibody (1D4). Biotin labeling is indicated by depletion of free DNA and presence of shifted DNA.

supernatant. Additionally, AM14 B cells are activated by sn/RNP particles in complex with anti-Sm/RNP antibodies. This activation was found to be RNAse-sensitive and TLR7-dependent, supporting a model in which the B cell receptor binds to ICs containing RNA/RNPs and delivers them to an intracellular compartment where they are able to engage TLR7 *(11)*. DCs can also take up ICs through Fcγ receptors and deliver ligands to TLR7 *(12, 13)*. We have used the antibodies BWR4 (anti-RNA) *(21)* and Y2 (anti-SmD) *(22)* in combination with snRNP particles to assess the role of TLR7 in autoreactive B cell activation. AM14 B cell stimulation by BWR4 depends on interaction with RNA present in cell debris in the culture, and hence addition of BWR4 alone induces activation. Conversely, sm/RNP particles are not present in high enough concentrations in cell debris in the culture, and hence addition of sm/RNP particles to Y2 is necessary to observe stimulation.

1. RNP/Sm particles are supplied in glycerol, which must be removed by filtration prior to use in tissue culture. Wash centricon 2X with 4 mL of RPMI 1640 by centrifuging 20 min at 1455 × *g* at 4°C. Discard flow through.
2. Combine 25 μg of RNP/Sm antigen with 4 mL RPMI 1640, add to centricon, and spin at 1455 × *g* at 4°C for 20 min. Discard flow through.
3. Wash centricon 2X with 4 mL RPMI 1640.
4. After the last centrifugation, use a Pipetman to measure the remaining volume of sm/RNP particles in the top of the centricon. Complete final volume to 1.25 mL by adding RPMI 1640. This will yield a stock solution of 20 μg/mL. Store in 100 μL aliquots at –80°C for up to 1 month.

3.4. Preparation of ICs

3.4.1. Preparation of DNA ICsa

1. DNA ICs are formed by combining DNA, antibody, and RPMI.
2. Calculate volumes for preparation of 4X stocks of DNA and antibody in duplicates per stimulation condition (*see* **Note 16**).
3. Mix antibody and DNA and incubate complexes for 1–2 h at 37°C. If using biotin-labeled DNA, no incubation is necessary.

3.4.2. Preparation of RNA ICs

1. RNA-containing ICs are formed by combining anti-RNA antibody BWR4 and RPMI.
2. RNP/Sm IC are formed by combining RNP/Sm antigen and anti-SmD antibody Y2 in serum-free RPMI and incubating complexes for 1–2 h at 37°C (*see* **Note 17**).

3.5. B Cell Proliferation Assay

AM14 B cell activation by ICs is measured by incorporation of ^{3}H-thymidine into DNA of dividing cells. B cells are hence stimulated for 24 h after which they are pulsed with ^{3}H-thymidine, harvested, and incorporated radioactivity is measured in a scintillation counter.

3.5.1. B Cell Preparation

1. Pipette 10 mL HBSS into one 15 mL conical tube per spleen.
2. Sacrifice mouse, harvest spleen using aseptic technique, and transfer to tube containing HBSS.
3. Inside the hood, pour contents of tube into a Petri dish. Trim any excess fat from the tissue and perfuse spleen with 5 mL of HBSS using a 5 mL syringe and a 25 gauge needle.
4. Unwrap frosted slides and wet with HBSS from the cell suspension. Crush spleen gently between the frosted side of the slides, making sure the spleen and the slides are wet with media, until only white matrix is left on the slide.
5. Rinse the slides with 1–2 mL of cell suspension using a Pasteur pipette. Discard slides and matrix.
6. Transfer the cell suspension to a new 15 mL conical tube using a Pasteur pipette.
7. Rinse the Petri dish with 3 mL of fresh HBSS and add to cell suspension. Pipette up and down to break up cell clumps. Leave on ice for 5 min. Remove supernatant from cell suspension leaving debris at the bottom of the tube, and transfer to a new 15 mL conical tube.
8. Centrifuge tubes at 300 × g at 4°C for 5 min.
9. Aspirate supernatant using a Pasteur pipette attached to a vacuum flask.

10. Flick the tube to resuspend the cells, add 10 mL of IMag buffer and pipette up and down (*see* **Note 18**). Count cells.
11. Centrifuge tubes at 300 × *g* at 4°C for 5 min.
12. Aspirate supernatant using a Pasteur pipette attached to a vacuum flask. Flick to resuspend cells.
13. Add anti-mouse CD45R/B220 Magnetic Particles, and proceed to B cell purification following the manufacturer's specifications.
14. After the final wash, resuspend cells in 3 mL RPMI and count.

3.5.2. B Cell Stimulation

1. Plate 100 μL of purified B cells at a density of 4×10^6 cells/mL in a flat bottom, 96-well plate. Seed cells in duplicate per stimulation condition (*see* **Note 19**).
2. If testing the effect of inhibitors, preincubate cells with inhibitor for 1–2 h at 37°C before addition of ICs. If testing RNA IC, preincubate cells with IFN-α for 1–2 h at 37°C before addition of ICs (*see* **Note 20**).
3. Add ICs and controls for B cell stimulation, complete volume to 200 μL with RPMI if necessary.
4. Incubate plates for 24 h in 37°C incubator, 5% CO_2.
5. Pulse plates by adding 50 μL of ^{3}H-thymidine medium per well and incubating for 6 h in 37°C incubator, 5% CO_2.
6. Harvest cells using one filtermat per plate and rinsing 10X with water and once with methanol. Dry filtermat for at least 2 h at RT (*see* **Note 21**).
7. Insert filtermat into sample bag, add 4 mL of scintillation cocktail, and completely wet filtermat. Remove excess scintillation cocktail and seal bag.
8. Count filtermat in Microbeta Trilux counter. An example of the results is shown in **Fig. 2**.

3.6. Dendritic Cell Cytokine Assay

Activation of DCs by ICs is measured by cytokine production. Hematopoietic cells extracted from mouse bone marrow are stimulated with Flt3L, which induces differentiation of stem cells into pDC and cDCs. The resulting population, designated "FLt3L-DCs," consists of a mix of pDC and cDC. IFN-α is secreted exclusively by pDCs, and is used as indicator of pDC activation by ICs. IL-6 is secreted by both pDCs and cDC, but predominately by cDCs, and can be used as an indicator of cDC activation by ICs.

3.6.1. DC Preparation

1. Dissect femur and tibia and place in 10 mL of PBS in a 15 mL conical tube. Keep on ice.
2. Transfer to Petri dish and trim off extra tissue.

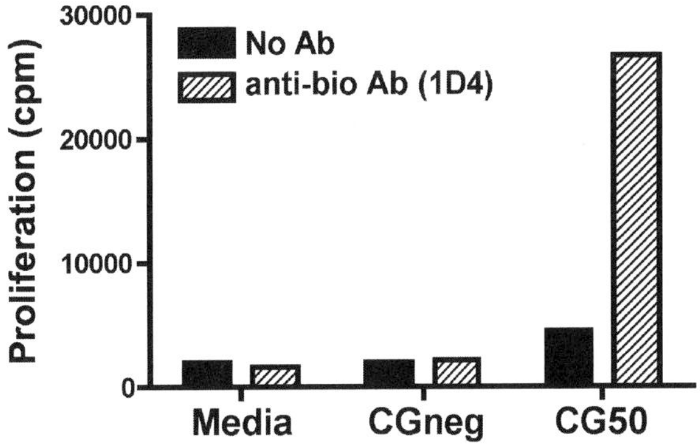

Fig. 2. AM14 B cell proliferation to ICs containing anti-biotin antibody and biotin-labeled DNA. AM14 B cells were purified with B220 magnetic beads and stimulated with ICs composed of 5 μg/mL anti-biotin antibody (1D4) and 100 ng/mL biotin-labeled CGneg or CG50 fragments. Proliferation was measured by incorporation of ^{3}H-thymidine.

3. Flush out bone marrow cells inside a 50 mL conical tube by applying 20 mL of RPMI 1640 with a 20 mL syringe and a 25 gauge needle.
4. Spin cells 300 × *g* for 5 min at 4°C.
5. Aspirate supernatant using a Pasteur pipette attached to a vacuum flask.
6. Flick cells to resuspend, add 500 μL of RBC lysis buffer, and incubate 1 min at room temperature.
7. Immediately add RPMI 1640 to bring to a total volume of 15 mL.
8. Spin cells at 300 × *g* for 5 min at 4°C, aspirate supernatant using a Pasteur pipette attached to a vacuum flask, and flick cells to resuspend.
9. Add 10 mL RPMI medium, run cells through a 70 μm cell strainer, and rinse strainer with 5 mL RPMI.
10. Spin cells at 300 × *g* for 5 min at 4°C, aspirate supernatant using a Pasteur pipette attached to a vacuum flask, and flick cells to resuspend.
11. Add 5 mL RPMI medium and count.
12. Seed cells at a 1.5 × 10^6 cells/mL in RPMI and add supernatant from FL-B16 cells at a final concentration of 7.5% (v/v) (*see* **Note 22**).
13. Incubate plates for 8 days in 37°C incubator, 5% CO_2.

3.6.2. DC Stimulation

1. Count cells and seed at a density of 3 × 10^6 cells/mL in 100 μL RPMI medium in a flat-bottom, 96-well plate (*see* **Note 23**).

2. Stimulate Flt3L-DC by addition of ICs (2X in 100 μL or 4X in 50 μL, complete to final volume of 200 μL).
3. Incubate plates for 24 h in 37°C incubator, 5% CO_2.
4. Harvest supernatants for measurement of cytokines.

3.6.3. Measurement of DC Cytokines by ELISA

1. Coat ELISA plates with 50 μL/well of rat monoclonal anti-mouse IFN-α at 2 μg/mL in PBS. Incubate O/N at 4°C.
2. Wash 3X with ELISA wash buffer.
3. Add 100 μL of ELISA blocking buffer and incubate for 2 h at room temperature.
4. Wash 3X with ELISA wash buffer.
5. Add 50 μL of supernatant, or mouse interferon alpha standard, at a range of 5,000–78 pg/mL in ELISA blocking buffer. Incubate O/N at 4°C.
6. Dilute rabbit polyclonal anti-mouse IFN-α to a final concentration of a 0.2 μg/mL in ELISA blocking buffer, add a 50 μL/well, and incubate 4 h at room temperature.
7. Wash 3X with ELISA wash buffer.
8. Dilute peroxidase-conjugated donkey anti-rabbit IgG $F(ab')_2$ fragment, to a final concentration of 80 ng/mL in ELISA blocking buffer, and add 50 μL per well. Incubate 1.5 h at room temperature.
9. Wash 5X with ELISA wash buffer.
10. Develop using 50 μL of TMB detection substrate. Stop reaction by adding 50 μL ELISA stop solution and measure absorbance at 450 nm.

4. Notes

1. Growth medium will vary depending on the hybridoma to be grown. Fetal calf serum contains IgG that is capable of binding to protein G during antibody purification, therefore medium must be depleted of IgG prior to use for growing hybridomas. IgG is removed by running medium over a protein G column as described for antibody purification, except that IgG is discarded.
2. All reagents, glassware, and plasticware used for cell culture should be sterile and endotoxin-free to avoid any confounding effects of TLR4 stimulation.
3. It is important to never let the protein G sepharose dry out. Column should be capped at both ends with liquid remaining above the sepharose whenever the column is not in use.

4. Before eluting, check that pH of elution buffer and neutralization buffer mix is around 7.6 using pH paper. The time that the antibody is in elution buffer should be minimized to prevent denaturing of antibody at low pH. If antibody is denatured using this method of elution, alternative elutions at higher pH can be used. See the manufacturer's information regarding alternative elution methods.
5. Antibodies should be tested for endotoxin contamination at 3X the concentration used in cell culture using the Limulus Amoebocyte Lysate Assay (Cambrex) according to the manufacturer's instructions. Endotoxin levels used in assay are below 0.03 EU/mL. Endotoxin can be removed from samples testing positive using a Triton® X-114 extraction *(23)*.
6. Dam/dcm-deficient *E. coli* are used to prevent methylation at adenosine and cytosine residues.
7. LITMUS 29 and pUC19 are high-copy plasmids, so do not exceed 100 mL of culture or grow for more than 16 h at 37°C at 250 rpm in a rotary shaker. Poor lysis and low DNA yield result from using too much bacteria, so we recommend weighing the bacterial pellet and using no more than 300 mg/maxiprep column. Including the LyseBlue reagent provided with the kit helps to monitor correct lysis.
8. This reaction can be scaled up, 2–4 U of each restriction enzyme/µg of DNA should be used, and enzyme should remain less than 10% of the total reaction volume. DNAse/RNAse free water should be used for all steps of DNA preparation.
9. If you do not have a comb this large, wells can be taped together using thick packing tape, or DNA can be loaded into multiple wells.
10. Running the gel this long is necessary to ensure complete separation of the plasmid backbone from the DNA fragment insert. We have also found that using high quality agarose such as SeaKem® GTG® agarose is critical for high fragment yields.
11. The CGneg and CG50 fragments make up approximately 20% of the plasmid, so 40 µg of fragment is expected from digesting 200 µg of plasmid. Each column binds a maximum of 25 µg of DNA, so two columns should be used. We have also used Qiagen's Gel Extraction kits and have found that yields are much better with the Zymoclean Gel DNA Recovery Kit™. DNAs are stored at 100–200 ng/µL in TE buffer. EDTA in TE buffer inhibits DNAses, but EDTA can be toxic to cells, so storing DNA at this concentration allows for dilution of the EDTA when adding to cell culture. All DNA preparations are tested for endotoxin using the Limulus Amoebocyte Lysate

Assay (Cambrex) according to the manufacturer's instructions, at 3X the concentration to be used in cell culture assays. Endotoxin levels used in assay are below 0.03 EU/mL. Most DNA made by these methods is endotoxin-free, but if DNA is contaminated, running DNA through an additional DNA Clean & Concentrator-25™ kit and/or extracting with Triton-X-114 *(23)* usually removes endotoxin.

12. Yields are usually about 5 μg/100 μL of PCR reaction, so 4–5 PCR reactions can be sequentially loaded onto one column.
13. The Klenow enzyme works most efficiently in the presence of all four dNTPs. Single ends may be labeled by leaving out selected dNTPs. Klenow Fragment has $3' \rightarrow 5'$ exonuclease activity in the absence of dNTPs, so ($3' \rightarrow 5'$ exo-) Klenow Fragment must be used if dNTPs are left out. As with restriction digests, enzyme should remain less than 10% of the total reaction volume.
14. Alternatively, ICs composed of antibody Hy1.2 (anti-TNP) and TNP-labeled DNA may be used. TNP-labeled DNA is made by substituting aminoallyl-dUTP for biotin-16-dUTP in the labeling reaction. DNA prepared from plasmid is then gel purified, or DNA prepared by PCR is run through DNA Clean & Concentrator-25™ kit. DNA is eluted in dH_2O (do not use TE buffer, as Tris base contains amines that interfere with labeling). TNP label is added by combining up to 10 μg of DNA in 50 μL of dH_2O, 30 μL of sodium bicarbonate buffer, and 20 μL of reactive TNP, vortexing, and incubating 60 min at room temperature in the dark. DNA is then run through a DNA Clean & Concentrator-25™ kit to remove excess TNP. Whenever possible we use biotin-labeling, as the DNA preparation involves less steps, yields are better, and measuring the extent of labeling is easier.
15. Alternatively, a gel shift can be used to confirm that DNA was labeled with TNP. The affinity of the anti-TNP antibody is lower than the anti-biotin antibody, making TNP labeling more difficult to measure. Because of this, it is critical that all steps are performed at 4°C. 100 ng of TNP-labeled DNA is combined with 2 μg of Hy1.2 (anti-TNP) or irrelevant control antibody 1D4 (anti-biotin) in 40 μL of PBS. 8 μL of loading dye is added and samples are incubated O/N at 4°C. A 2.5% agarose gel is run at 4°C with pre-chilled running buffer until the dye is about 0.5 in. below the wells (running further will dissociate complex).
16. We have routinely used antibodies at a final concentration of 0.1–10 μg/mL, and DNA fragments at a final concentration of 10 ng/mL–1 μg/mL. Antibodies and DNA fragments should be titrated to determine optimal concentrations.

17. We have used BWR4 at a range of 0.5–10 μg/mL and Y2 at 1–20 μg/mL. RNP/Sm particles are added at a final concentration of 1–2 μg/mL.
18. When working with primary B cell suspensions, avoid any bubbles and pipette up and down gently.
19. Have all the reagents ready for B cell stimulation after B cell preparation and seed cells as soon as possible. B cells die quickly and leaving them on ice while preparing other reagents will decrease your final yield.
20. AM14 B cells stimulated by RNA IC or snRNP IC need to be primed for 1–2 h with 1,000 U/mL of IFN-α. Prepare 4X stock solution of IFN-α and add 50 μL per well.
21. It is important to completely dry the filtermats, as remaining water in the filter may quench scintillation fluid with significant reduction of cpms. Drying for at least 2 h or O/N is recommended.
22. We have used conditioned medium from FL-B16 cells as a source of Flt3L. Alternatively, Flt3L is commercially available through R&D Systems, Minneapolis, MN.
23. Profiling of Flt3L-DCs by flow cytometry is recommended to determine the relative percentage of pDCs ($CD11b^{-}B220^{+}$) and cDCs ($CD11b^{+}B220^{-}$) after culture.

References

1. Chan, O. T., Madaio, M. P., and Shlomchik, M. J. (1999) The central and multiple roles of B cells in lupus pathogenesis. *Immunol Rev* 169, 107–21.
2. Diebold, S. S., Kaisho, T., Hemmi, H., Akira, S., and Reis e Sousa, C. (2004) Innate antiviral responses by means of TLR7-mediated recognition of single-stranded RNA. *Science* 303, 1529–31.
3. Hemmi, H., Takeuchi, O., Kawai, T., Kaisho, T., Sato, S., Sanjo, H., Matsumoto, M., Hoshino, K., Wagner, H., Takeda, K., and Akira, S. (2000) A Toll-like receptor recognizes bacterial DNA. *Nature* 408, 740–5.
4. Heil, F., Hemmi, H., Hochrein, H., Ampenberger, F., Kirschning, C., Akira, S., Lipford, G., Wagner, H., and Bauer, S. (2004) Species-specific recognition of single-stranded RNA via toll-like receptor 7 and 8. *Science* 303, 1526–9.
5. Lund, J. M., Alexopoulou, L., Sato, A., Karow, M., Adams, N. C., Gale, N. W., Iwasaki, A., and Flavell, R. A. (2004) Recognition of single-stranded RNA viruses by Toll-like receptor 7. *Proc Natl Acad Sci U S A* 101, 5598–603.
6. Christensen, S. R., Shupe, J., Nickerson, K., Kashgarian, M., Flavell, R. A., and Shlomchik, M. J. (2006) Toll-like receptor 7 and TLR9 dictate autoantibody specificity and have opposing inflammatory and regulatory roles in a murine model of lupus. *Immunity* 25, 417–28.
7. Pisitkun, P., Deane, J. A., Difilippantonio, M. J., Tarasenko, T., Satterthwaite, A. B., and Bolland, S. (2006) Autoreactive B cell responses to RNA-related antigens due to TLR7 gene duplication. *Science* 312, 1669–72.
8. Marshak-Rothstein, A. (2006) Toll-like receptors in systemic autoimmune disease. *Nat Rev Immunol* 6, 823–35.
9. Leadbetter, E. A., Rifkin, I. R., Hohlbaum, A. M., Beaudette, B. C., Shlomchik, M. J., and Marshak-Rothstein, A. (2002) Chromatin-IgG complexes activate B cells by dual engagement of IgM and Toll-like receptors. *Nature* 416, 603–7.
10. Marshak-Rothstein, A., Busconi, L., Lau, C. M., Tabor, A. S., Leadbetter, E. A., Akira, S., Krieg, A. M., Lipford, G. B., Viglianti, G. A., and Rifkin, I. R. (2004) Comparison of CpG

s-ODNs, chromatin immune complexes, and dsDNA fragment immune complexes in the TLR9-dependent activation of rheumatoid factor B cells. *J Endotoxin Res* 10, 247–51.

11. Lau, C. M., Broughton, C., Tabor, A. S., Akira, S., Flavell, R. A., Mamula, M. J., Christensen, S. R., Shlomchik, M. J., Viglianti, G. A., Rifkin, I. R., and Marshak-Rothstein, A. (2005) RNA-associated autoantigens activate B cells by combined B cell antigen receptor/Toll-like receptor 7 engagement. *J Exp Med* 202, 1171–7.
12. Boule, M. W., Broughton, C., Mackay, F., Akira, S., Marshak-Rothstein, A., and Rifkin, I. R. (2004) Toll-like receptor 9-dependent and -independent dendritic cell activation by chromatin–immunoglobulin G complexes. *J Exp Med* 199, 1631–40.
13. Means, T. K., Latz, E., Hayashi, F., Murali, M. R., Golenbock, D. T., and Luster, A. D. (2005) Human lupus autoantibody–DNA complexes activate DCs through cooperation of CD32 and TLR9. *J Clin Invest* 115, 407–17.
14. Yasuda, K., Richez, C., Maciaszek, J. W., Agrawal, N., Akira, S., Marshak-Rothstein, A., and Rifkin, I. R. (2007) Murine dendritic cell type I IFN production induced by human IgG-RNA immune complexes is IFN regulatory factor (IRF)5 and IRF7 dependent and is required for IL-6 production. *J Immunol* 178, 6876–85.
15. Baechler, E. C., Batliwalla, F. M., Karypis, G., Gaffney, P. M., Ortmann, W. A., Espe, K. J., Shark, K. B., Grande, W. J., Hughes, K. M., Kapur, V., Gregersen, P. K., and Behrens, T. W. (2003) Interferon-inducible gene expression signature in peripheral blood cells of patients with severe lupus. *Proc Natl Acad Sci U S A* 100, 2610–5.
16. Monestier, M., and Novick, K. E. (1996) Specificities and genetic characteristics of nucleosome-reactive antibodies from autoimmune mice. *Mol Immunol* 33, 89–99.
17. Monestier, M., Novick, K. E., and Losman, M. J. (1994) D-penicillamine- and quinidine-induced antinuclear antibodies in A.SW (H-2s) mice: similarities with autoantibodies in spontaneous and heavy metal-induced autoimmunity. *Eur J Immunol* 24, 723–30.
18. Viglianti, G. A., Lau, C. M., Hanley, T. M., Miko, B. A., Shlomchik, M. J., and Marshak-Rothstein, A. (2003) Activation of autoreactive B cells by CpG dsDNA. *Immunity* 19, 837–47.
19. Krieg, A. M., Wu, T., Weeratna, R., Efler, S. M., Love-Homan, L., Yang, L., Yi, A. K., Short, D., and Davis, H. L. (1998) Sequence motifs in adenoviral DNA block immune activation by stimulatory CpG motifs. *Proc Natl Acad Sci U S A* 95, 12631–6.
20. Shlomchik, M. J., Zharhary, D., Saunders, T., Camper, S. A., and Weigert, M. G. (1993) A rheumatoid factor transgenic mouse model of autoantibody regulation. *Int Immunol* 5, 1329–41.
21. Eilat, D., and Fischel, R. (1991) Recurrent utilization of genetic elements in V regions of antinucleic acid antibodies from autoimmune mice. *J Immunol* 147, 361–8.
22. Bloom, D. D., Davignon, J. L., Retter, M. W., Shlomchik, M. J., Pisetsky, D. S., Cohen, P. L., Eisenberg, R. A., and Clarke, S. H. (1993) V region gene analysis of anti-Sm hybridomas from MRL/Mp-lpr/lpr mice. *J Immunol* 150, 1591–610.
23. Aida, Y., and Pabst, M. J. (1990) Removal of endotoxin from protein solutions by phase separation using Triton X-114. *J Immunol Methods* 132, 191–5.

Chapter 23

Innate Immunity, Toll-Like Receptors, and Atherosclerosis: Mouse Models and Methods

Rosalinda Sorrentino and Moshe Arditi

Summary

Chronic inflammation and aberrant lipid metabolism represent hallmarks of atherosclerosis. Innate immunity critically depends upon Toll-like receptor (TLR) signalling. Recent data directly implicate signalling by TLR4 and TLR2 in the pathogenesis of atherosclerosis. The role that TLRs play in the pathogenesis of atherosclerosis can be assessed by using several animal models, which provide a double genetic deficiency in TLRs and molecules implicated in the lipid metabolism, such as ApoE or LDL receptor. Furthermore, a more recent technique, such as the bone marrow transplantation (BMT), can be a useful and straightforward method to elucidate the role of stromal versus hematopoietic cells in the acceleration of the atheroma.

Key words: Atherosclerosis, Double knockout mice, Lipid staining, Macrophage immunostaining, Bone marrow transplant.

1. Introduction

Atherosclerosis, formerly considered a lipid accumulation vascular disease, is actually classified as a focal and chronic ongoing inflammatory response. Reports dating back over 100 years have suggested that infectious agents could contribute to cardiovascular disease. However, only since the late 1980s evidence has emerged to highlight the link between the inflammatory response and the immune system regarding the induction of atherosclerosis *(1,2)*.

The modification of lipoproteins, especially in the subendothelial matrix, attracts diverse and multifactorial leukocytes, including monocytes, T cells, B cells, mast cells, dendritic cells (DCs), and neutrophils in the vessel wall. The recognition of

Claire E. McCoy and Luke A.J. O'Neill (eds.), *Methods in Molecular Biology, Toll-Like Receptors, vol. 517*
DOI: 10.1007/978-1-59745-541-1_23

excessive lipids in the intimal layer induces the production of endothelial-derived adhesion molecules, such as ICAM, VCAM, E-selectin, and the resulting interaction with which leads the circulating monocytes to bind, "roll", and get included into the vessel wall. If the monocytes become activated, they mature into macrophages, ingest modified lipids, and become foam cells, thereby promoting inflammation with the overcoming release of a variety of inflammatory mediators, as well as oxidants and proteases. Smooth muscle cells also participate in this process by expanding the extracellular matrix and releasing proteases. This tends to exacerbate inflammation and weaken the structure of the plaques, eventually causing a thrombogenic sub-endothelial process resulting in an arterial lumen occlusion followed by myocardial infarction, stroke, or sudden death. Hence, many observations have also revealed the accumulation of DCs in atherosclerosis-prone areas of the normal aorta and carotid arteries *(3)*, supporting the concept that immune mechanisms are involved in the formation of atherosclerotic lesions from the very early stages of the disease *(4, 5)*. After engulfing antigen in the arterial wall, vascular DCs likely migrate as veiled cells via the afferent lymph into regional lymph nodes where they activate T cells. The presence of DCs populating atherosclerotic lesions, which are key components in the regulation of innate immunity *(6, 7)*, highlights the contribution of the innate immune system, especially as the presence of *Chlamydia pneumoniae* has been identified in the cytoplasm of vascular DCs in the atherosclerotic lesions *(7)*. A number of other infectious agents have now been similarly associated with atherosclerotic cardiovascular disorders, including *Helicobacter pylori (8)*, CMV *(9)*, EBV *(10)*, HIV *(11)*, HSV1 *(12)*, HSV2 *(13)*, hepatitis B *(14)* and C *(15)*, and *Porphyromonas gingivalis (16)*.

One potentially important source of inflammatory vascular injury is LPS. Low amounts of LPS can be released into the circulation from Gram-negative bacteria that either colonize or indolently infect the human gastrointestinal, genitourinary, and respiratory tracts. Weekly injections of LPS accelerated the development of atherosclerotic lesions in rabbits on hypercholesterolemic diets *(17)*. These observations suggest that systemic pro-inflammatory mediators such as LPS may be pathogenically linked to the development and progression of atherosclerosis, and that pathogens, and hence Toll-like receptor (TLRs)-mediated signalling, could play a mechanistic role in atherosclerosis.

1.1. TLR Implication in Atherosclerosis: In Vivo Evidence

Innate immunity orchestrates the adaptive response through TLR signalling. Several reports have documented the expression of TLR4, TLR1, TLR2, and to a lesser extent TLR5 in both human plaques and murine animal models *(18, 19)*, where they appear to be mainly localized to macrophages and endothelial cells. Our group demonstrated that both the TLR4 and MyD88 gene deficiencies in atherosclerosis-prone hypercholesterolemic mice (ApoE–/–) had a decrease in the plaque size, lipid content, and expression of pro-inflammatory cytokines and chemokines such as IL-12 and MCP-1 *(20)*. However, Mullick et al. *(21)* also reported the implication of TLR2 in the progression of the atherosclerotic lesion. It was shown that TLR2 mediated the development of the disease, but in particular the transfer of TLR2–/– bone marrow-derived cells to LDLR–/– mice revealed that endogenous TLR2 ligands may be implicated in the formation of atherosclerotic plaques. It was postulated that the activation of TLR2 in bone marrow-derived cells was not implicated in the progression of the atheroma *(21)* but just co-operated to aggravate the disease. The expression and activation of TLR2 on stromal cells, such as endothelial cells, modulated atherosclerosis, consequently promoting macrophages recruitment prone to exacerbate the pro-inflammatory milieu within the lesion. TLR2 involvement was also assessed by another group which instead used the ApoE–/– mouse model, fed on a normal chow diet, evidencing the strong implication of the immune system in a metabolic disease rather than the environmental status *(22)*. The first line of defence against invading pathogens involves not just cells with primary immune functions such as macrophages, DCs, and neutrophils but also cell types, such as endothelial cells, which seem to perform mostly surveillance. In this regard, the recent technique of the bone marrow transplantation (BMT) is an extreme and useful tool to discriminate between the role of immune cells and stromal cells as described by Mullick et al. *(21)*. The transfer of TLRs intact or deficient in bone marrow-derived cells can further elucidate the mechanisms underlying a complex pathology such as atherosclerosis.

1.2. Atherosclerosis Animal Models

Atherosclerosis is a multigenic and environmental-dependent disease. An appropriate experimental model was developed in the late 1990s in the mouse, which has aided our knowledge of the disease. However, in contrast to humans, mice are highly resistant to atherosclerosis for their lifespan. Mice also differ in the ability to carry lipids on high-density lipoprotein (HDL), while humans carry lipids on low-density lipoprotein (LDL).

Early publications investigating mouse strain-specific differences in the susceptibility to atherosclerosis revealed that C3H/HeJ mice are atherosclerosis-resistant on a high-cholesterol diet compared to C57Bl/6 mice *(23)*. This resistance is associated with unchanged lipid profiles and apolipoprotein (ApoE) composition of plasma lipoproteins, and in particular unchanged HDL levels, upon consumption of a high-cholesterol diet *(23)*. Further characterization of endothelial cells derived from C3H/HeJ mice demonstrated a lack of an inflammatory response towards minimally modified LDL (MM-LDL), supporting the hypothesis that MM-LDL indeed uses TLR4 as a receptor *(24)*. Transfer of bone marrow derived from an atherosclerosis-prone mouse strain into C3H/HeJ mice failed to reverse the phenotype of C3H/HeJ mice, supporting the important role of endothelial cells during initiation of atherosclerosis *(25)*.

One mouse strain that successfully develops a comparable human atheroma is C57Bl/6, of which the ApoE knockout (ApoE−/−) and/or the LDL receptor knockout (LDLR−/−) are widely used. ApoE is a structural glycoprotein of all lipoprotein particles and permits the binding of lipoproteins to the LDL receptor on the cell surface, modulating the lipid metabolism. Even though ApoE−/− are very commonly used as a murine model for atherosclerosis studies, it should be noted that these mice develop skin lesions characterized by eruptive xanthomas on the shoulder and back with lipids and extracellular matrix as the predominant components, and present weight loss, decreased hematocrit, and deafness. The other commonly used murine model is the LDLR−/−, a model of familial hypercholesterolemia, which lacks the receptor for the cellular influx of lipids.

Two other murine models, recently created, are ApoE and LDL receptor double knock out *(26)* and LDLR−/− $ApoB^{100/100}$ *(24)*, which develop extensive atheroma even on a normal chow diet. In contrast, all the previously mentioned animal models are fed on high-fat diet, containing 1.25% cholesterol, 15.8% fat, and no cholate (Harlan). These knockout mice are available from Jackson Laboratories.

Another less commonly used murine model supports the use of a poloxamer P407, a ubiquitous synthetic surfactant that causes massive hyperlipidemia and atherosclerosis in the rodent. The activity of this synthetic compound is based on the ability of increasing the formation of 100–200 nm pores in the liver sinusoidal endothelial cell, so that the uptake of lipoproteins is impaired and increases hypercholesterolemia *(27)*. P407 is intraperitoneally injected.

1.3. ApoE and TLR Double Knockout Mice

A link between atherosclerosis and innate immunity is based, as demonstrated by several studies, on the findings that the absence of TLR2 or TLR4 ameliorated and protected the mouse from atheroma *(20, 21)*. ApoE−/− or LDLR−/− can be backcrossed with TLRs knock out mice on a C57BL/6 background. Wright SD et al. used ApoE−/−/lps^dmice to identify the involvement of TLR4 in the atherosclerotic process. This strain was generated by backcrossing ApoE−/− with the LPS-hyporesponsive strain C57Bl/10ScN thought to involve a mutation that results in deficient TLR4 signalling. This study reported no difference in the extent of atherosclerosis between ApoE−/− and ApoE−/−/lps^d, even though there was not a direct measurement of the atherosclerotic burden. In addition, the animal model used reflected only partially a TLR4−/− genotype.

In this chapter, we aim to explain how to generate ApoE and TLR double knockout mice followed by describing how to use these mice as a tool to study atherosclerosis.

2. Materials

2.1. DNA Extraction, Polymerase Chain Reaction, and Flow Cytometry

1. Proteinase lysis buffer for tail vein: 50 mM Tris-HCl (pH 8.0), 100 mM EDTA, 0.5% SDS
2. Isopropanol 100% (Sigma) and ethanol 70% (Sigma).
3. Proteinase K (Qiagen, USA) 10 or 20 mg/ml of proteinase. Make sure the proteinase K is completely thawed, and mixed well before adding it to the sample.
4. Phenol/chloroform solution is mixed at a ratio 1:1 and stored at 4°C. Chloroform/IAA (CHCl3:IAA at 24:1) is stored in the flammable cabinet (*see* **Note 1**).
5. Master mix: 10X Taq buffer (1X final concentration; Applied Biosystem), that already contains $MgCl_2$ (*see* **Note 2**); 1.25 U/50 µl of reaction, Taq DNA polymerase (Applied Biosystem); 2 mM dNTP mix (200 nM final concentration, Qiagen); relative primers (100 nM final concentration, Operon, Qiagen), reconstituted in DNA-free water (Promega, CA, USA); 0.1–1 µg DNA template; DNA-free water as needed to reach the right volume. The amount of each component can be calculated based on the polymerase chain reaction (PCR) volume.
6. Running buffer (TBE, 10X): 108 g Tris base, 55 g boric acid, and 9.3 g Na_4EDTA (Sigma) are added to 1 L of distilled water. The pH is 8.3 and requires no adjustment. Use 1X solution as running buffer.
7. Agarose gel 1%: 1 g of agarose for each 100 ml of 1X TBE running buffer. Boil until the agarose is dissolved.
8. Ethidium bromide 1 µl of stock solution per 50 ml of gel. This quantity gives enough staining without too much background. Pour into gel tray and let cool.
9. Size standard: DNA ladder 1 kb (5 µl, Invitrogen; 1 µg/ml) added directly to the well, after dissolved with loading dye.
10. Loading dye: 10X BlueJuice™ gel loading buffer (2X final concentration, Invitrogen)

11. Red blood cell lysis buffer: 80.0 g NaCl, 11.6 g Na_2HPO_4, 2.0 g KH_2PO_4, and 2.0 g KCl are added to distilled water up to 10 L. The pH has to be 7.0.
12. CD16/32 antibody (Ab) (eBioscience): 0.5–1 μg per million of cells is added to 100 μl of staining buffer.
13. Staining buffer for FACS: phosphate-buffered saline (PBS), containing bovine serum albumin (BSA, 2%, Sigma) and sodium azide (0.1%, Sigma). It is stored at 4°C.
14. Fixing buffer for FACS: PBS with formalin at 4% (Sigma). Use gloves as formalin is toxic.

2.2. Lipoprotein Measurement, Lipid Staining, and Immunostaining

1. Anticoagulant EDTA (5 mM, Sigma) solution is prepared by dissolving the powder into PBS. It can be stored in aliquots at −20°C.
2. PBS buffer and 1% BSA (1 g in 100 ml of PBS). It is stored at 4°C.
3. Oil Red O stock solution 0.4%: 2 g Oil Red O (Sigma Aldrich) are added to 500 ml of isopropanol (60%) and stored at room temperature. Working Oil Red O solution: 120 ml Oil Red O stock solution added to 80 ml of distilled water.
4. Fast Green FCF stock: 0.1 g Fast Green (Sigma Aldrich) added to 400 ml of distilled water
5. Glycerol gel is performed by Fischer Scientific.
6. OCT compound is provided by Tissue Teck, USA.
7. Blocking solution 5%: 0.5 ml normal rabbit serum (NRS) in 10 ml of 3% BSA Solution (1.2 g of bovine serum albumin, BSA, in 40 ml PBS). Mix by gently inverting and keep on ice. Prepare this solution fresh each time.
8. Wash buffer: Make a 1:5 dilution of blocking solution in PBS, 1 ml blocking solution in 4 ml PBS. Prepare this solution fresh each time.
9. AP in ABC-AP complex: 5 ml Tris-HCl Buffer (pH 8.3), one drop levamisole solution (Vector SP).
10. One hundred and fifty microlitres of primary antibody rat anti-mouse MOMA-2 and isotype control (IgG2b). The working dilution is 1:400 in wash buffer.
11. One hundred and fifty microlitres of secondary antibody: Biotinylated rabbit anti-rat IgG monoclonal Ab (Vector; 0.5 mg, in liquid form) is prepared by using a working dilution of 1:200 in wash buffer.
12. ABC-AP reagent, alkaline phosphates substrate kit: 5 ml of PBS, one drop reagent A, one drop reagent B.

2.3. Bone Marrow Transplanta

1. Baytril (2.27% enrofloxacin, Bayer Corporation) dissolved in water at a final concentration of 25 mg/ml.
2. CD45.1 and CD45.2 antibodies (eBioscience): 0.5 μg per million cells in 100 μl total staining volume.

3. Methods

3.1. Generation of ApoE and TLR Double Knockout Mice

3.1.1 Interbreeding Micea

Interbreeding of the homozygous mutant mice is the most direct strategy to obtain the homozygous null mutant for both ApoE and TLRs. Since these genes are located on different mouse chromosomes, it is genetically possible to obtain a second generation F2 progeny, which lacks the function of two genes. Mice lacking TLRs, especially TLR4 and TLR2, MyD88, and/or ApoE are all fertile and suitable for interbreeding. Two breeding pairs can be made between ApoE−/− mice and TLR4−/− or TLR2−/− or MyD88−/−. The phenotypical normal or single mutant mice from the inbred strains need to be used as controls (**Fig. 1**). All of the progeny in the F1 generation will be double heterozygous for the ApoE allele and either TLR2, 4, or MyD88. There is

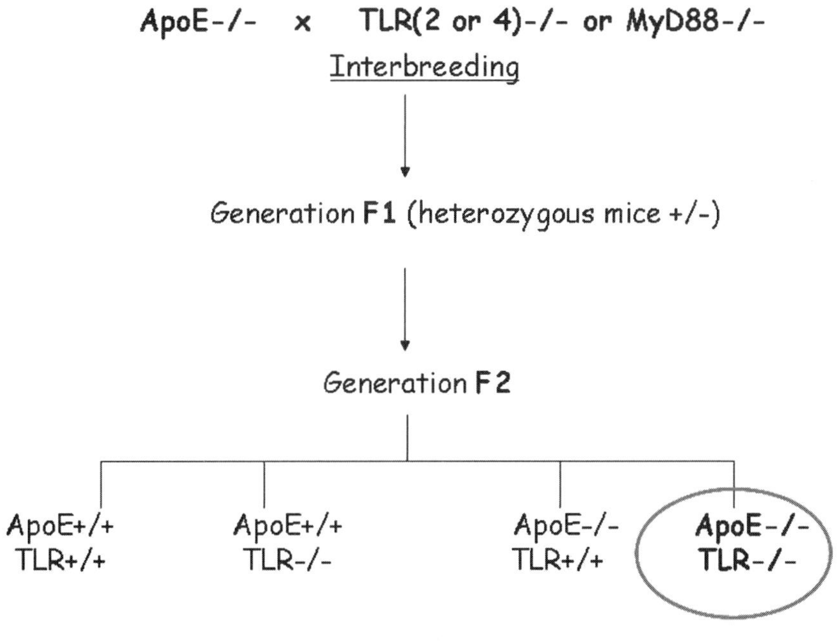

Fig. 1. ApoE knockout mice are crossed with TLR or MyD88 knockout mice. The first generation F1 generates double heterozygous for either TLR4/2 or MyD88 and ApoE alleles having a genotype TLR4 or 2+/−, or MyD88+/− ApoE+/−. The second generation F2 will generate homozygous for one target but also a lower homozygous for both targets, like ApoE−/−TLR or MyD88−/−. Polymerase chain reaction (PCR) assay can reveal the homozygosis for both targets.

no need for selection at this time. An interbreeding of the first-generation male and female has to be performed. This generation is characterized by ApoE+/+ TLR+/+, ApoE+/+ TLR−/−, ApoE−/− TLR+/+and ApoE−/− TLR−/−. Homozygous/heterozygous progeny will also be present, but these mice will only be useful as breeders. The genomic phenotype is visualized by performing a PCR assay for DNA.

3.1.2. Identification of Mutant Mice: Qualitative Polymerase Chain Reaction

1. Genomic DNA is isolated from mouse tail tips (1–1.5 cm) to determine its genotype. Mice 1–2 weeks old are used.
2. The tissue is digested in 450 μl of lysis buffer and added to a 1.5 ml microfuge tube.
3. Add 25 μl of 10 mg/ml proteinase K to the microfuge tube and incubate at 55°C overnight on a rotator (*see* **Note 3**).
4. Spin the microfuge tubes down for 3–4 min to allow the hair to form into a pellet. Pour off the supernatant into a new labelled 1.5 ml microfuge tube.
5. Add equal volume of 100% isopropanol. Gently shake the tubes until the DNA clump is seen suspended in the isopropanol and centrifuge for 1 min at the highest speed at room temperature.
6. Pour off the isopropanol without disturbing the pellet that contains DNA (*see* **Note 4**).
7. Air-dry the DNA for 1 min and then add 200 μl of sterile double-distilled water and place the tubes at 45°C water bath overnight for approximately 18–20 h. After incubation, tap tube gently to allow DNA to go into solution.
8. Add equal volume of phenol to the microfuge tubes. Shake vigorously for 10 s and then centrifuge for 10 min at the highest speed at room temperature (*see* **Note 5**).
9. Transfer the upper phase to new 1.5 ml centrifuge tubes using a pipetman. Be careful not to pick up the phenol or the interphase layer of protein by gently and carefully pulling out the supernatant along the walls of the microfuge tube.
10. Add equal volume of Chloroform (CHCl3:IAA) to the microfuge tubes and centrifuge for 10 min at the highest speed at room temperature (*see* **Note 5**).
11. Take out the upper phase, which contains the DNA, and transfer it into new microfuge tubes. Be careful not to disturb the lower layer of chloroform.
12. Store purified DNA at 4°C.
13. For genotyping, PCR is performed by using different sets of primers specific for ApoE, TLR4, 2, or MyD88. The F2 generation is genotyped. In general, PCRs (50 μl) are performed by using 0.1–1 μg of DNA template at 30–35 cycles

in the presence of 100 nM of each primer, dissolved in Master Mix. Each cycle consists of denaturation, annealing, and elongation. PCR, cycling, and gel conditions are dependent on many factors and need to be optimized as needed. The steps are as follows: 2 min of incubation at 94°C, 30–35 cycles (1 min at 94°C, 1 min at 58°C, and 2 min at 72°C), and a final extension step of 8 min at 72°C.

14. PCR products for wild-type and mutant alleles are visualized by agarose gel electrophoresis (2%) in the presence of ethidium bromide (*see* **Note 6**).
15. Pour the required volume of cooled agarose gel into the gel box, making sure that the rings are in the right position. Add the running buffer TBE.
16. Remove desired volume of PCR and add required loading dye to PCR tube (2 µl for 10–20 µl of PCR or 5 µl for 50 µl PCR). Vortex and spin down and load into well. Repeat for each sample. Load size standard (DNA ladder).
17. Set voltage. Small gels run typically at 50–70 V. Larger gels can be run at 80–100 V.
18. The presence of the corresponding ApoE or TLR band on the agarose gel, visualized by UV light, will highlight the presence or not of the target gene, revealing homozygotes or heterozygotes.

3.1.3. Alternative Method for Identification of Mutant Mice

A second method by which to determine the genetic phenotype detects the presence or absence of the genetic products from the genes of interest. Flow cytometry can be used to assess immunostaining with the antibodies that recognize surface markers that characterize and distinguish TLR4 or 2 and MyD88. Mononuclear cells can be used for flow cytometry and can be isolated by peripheral blood and stained with each antibody.

1. Draw the blood from the retro-orbital sinus using heparinized capillary tubes and centrifuge at 1000× *g* at 4°C for 5 min.
2. The pellet is then treated with a red blood lysis buffer for 2 min after which PBS is added to dissolve and dilute the buffer. The cells are centrifuged at 1000 × *g* at 4°C for 5 min. Wash the cells with staining buffer before going to **step 3**.
3. To avoid non-specific bindings by blocking Fc-mediated interactions, add CD16/32 antibody to 100 µl of staining buffer and leave on the cells for 15 min at 4°C.
4. Add the antibodies for TLRs as described by the manufacturer's instruction for 30 min at 4°C.
5. Wash the cells three times with the staining buffer by spinning at 1000 × *g* at 4°C.

6. Once the cells are stained, they are fixed in PBS with formalin at 4%. Avoid extended exposure to light as the antibody conjugates are usually light sensitive.
7. Flow cytometry analysis can be performed.

3.2. Lipoprotein Measurement, Lipid Staining, and Immunostaining

A lipoprotein assay can be performed to verify and quantify the abnormal lipid metabolism in the compound mutant mice. Blood from overnight-fasted mice can be collected from the retro-orbital plexus into capillary tubes containing the anticoagulant, EDTA (5 mM). Mouse serum is centrifuged and the resulting lipoprotein fraction (density <1.215 g/ml) can be chromatographed by fast lipoprotein chromatography (FLPC) performed by gel filtration. The same serum can be used for the assessment of different cytokines by means of ELISA.

Apolipoproteins can be analysed by SDS gel electrophoresis and Coomassie blue staining. To determine lipoproteins transport into the cells, activities of native LDL receptor and scavenger receptor can be assessed by the incubation of blood mononuclear cells, respectively, with native LDL and acetylated LDL marked with fluorescent dye or isotopes. Peripheral blood mononuclear cells can be isolated and incubated with the fluorescent ligands. Ligand binding and internalization can be performed by incubating the cells with the labelled ligands at 4°C and 37°C, respectively. The fluorescent signal can be determined by flow cytometry or also by fluorescent microscopy.

3.2.1. Morphometric Measurement of en face Mounted Aorta

After 8–12 weeks on a high-fat diet, the mice are sacrificed. The proximal aorta, including ascending aorta, aortic arch, and a portion of descending aorta can be excised and removed from the adherent connective tissue by means of an optical microscope. It can be opened longitudinally from the aortic arch to the iliac bifurcation. These processes are performed in PBS and then rinsed in a 1% BSA PBS buffer in order to allow the tissue to easily adhere onto the slide. The external part of the aorta is in direct contact with the surface of the slide and the part which is facing the operator is the lumen which will be assessed for the presence of lipid accumulation (**Fig. 2A**). The slide can be stored at –80°C or can be directly used and stained by using the Oil Red O staining technique as follows:

1. Fix the section in 4% parafolmadehyde solution for 30 min at room temperature.
2. Rinse the slide in distilled water for 2 min.
3. Remove the water from the slide by dabbing the edge on a paper towel and then immerse it in an Oil Red O working solution (*see* **Note 7**) for 20 min (*see* **Note 8**).
4. Wash with 70% isopropanol for 2 min.

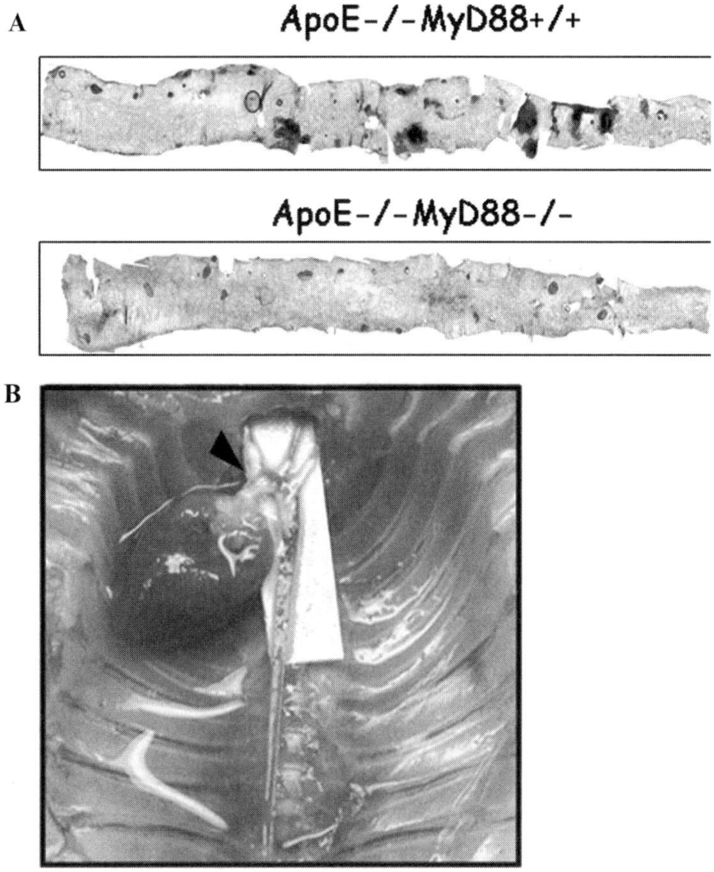

Fig. 2. (**A**) Oil Red O staining on aorta from ApoE ko versus ApoE/MyD88 ko mice. (**B**) Light photograph showing meticulous dissection of the heart, aortic arch, great vessels, and aorta in an ApoE knockout mouse. Considerable atherosclerotic plaque can be clearly seen as whitish patches (*arrow*) all along the arteries, particularly in the proximal regions.

5. Immerse the slide in a solution of hematoxylin (*see* **Note 7**) for 30 s (*see* **Note 8**).
6. Wash with water until clear, since the purple hematoxylin colour covers the whole slide.
7. Rinse 30 s in Fast Green solution.
8. Wash with distilled water until clear. Let the slide dry overnight or for 2 h on a warming plate, and then mount a cover slip by using a drop of glycerol gel to let it adhere to the slide on the top of the tissue (*see* **Note 9**). Let the gel dry overnight or on a warming plate.
9. Images of each aorta can be captured with a high resolution camera attached to a stereomicroscope and stored in digital format in a computer containing an image acquisition board,

ImagePro Plus (Media Cybernetics, Silver Spring). This software allows scoring the aortic plaques based upon the lipid accumulation, by highlighting the red stain compared to the entire background of the aortic sinus, expressed as percentage.

3.2.2. Morphometric Measurement of the Aortic Root Lesion

1. The heart is taken out of the mouse cutting along the aortic arch. The plaques should be readily visible (**Fig. 2B**). A neat cut is made at the base of the heart and the heart is then put in a Petri dish with PBS. Let the blood flow out and avoid any clogs that could interfere with the ensuing immunostaining.
2. The basal portion of the heart is embedded in OCT compound (Tissue Tek), frozen on dry ice, and then stored at −80°C until sectioning for lipid staining and immunohistochemical studies.
3. Serial 5-μm thick cryosections (every fifth section in the region from the lower portion of the ventricles to the appearance of aortic valves, every other section in the region of the aortic sinus, and every fifth section from the disappearance of the aortic valves to the aortic arch) are then collected on poly-d-lysine coated slides.

Sections are stained with Oil Red O and haematoxylin, as described (*see* **Subheading 3.2.1**), for the identification of atheromatous lesions (**Fig. 3**). Each section is reviewed under light microscope and then quantitatively evaluated by computer morphometry. The percentage of atherosclerotic plaques in the aortic sinus is calculated by using software such as ImagePro Plus.

Other stainings can be performed on the frozen aortic sinus sections, such as the identification of immune cells like macrophages by using MOMA-2 staining, calcium deposits by the alizarin red S and von Kossa techniques.

3.2.2. MOMA-2 Immunohistochemical Staining for Macrophages in Aortic Sections

1. Bring slides to room temperature for 30 min–1 h and circle sections with Pap Pen (Research products, hydrophobic slide marker) and make sure that there is no overlap into neighbouring sections and that all are well circled and sealed, otherwise, when other components are added, everything will be mixed between sections.
2. Fix slides by submerging the slides in a chamber filled with acetone for 15 min at −20°C and tap a little bit on dry tissue to remove the excess of acetone.
3. Wash with PBS three times, 10 min each. The transfer should be quick so that the sections do not dry.
4. Block the sections in 5% of blocking solution (150 μl) for 2 h in the dark (*see* **Note 10**). Use a Pasteur pipette and make

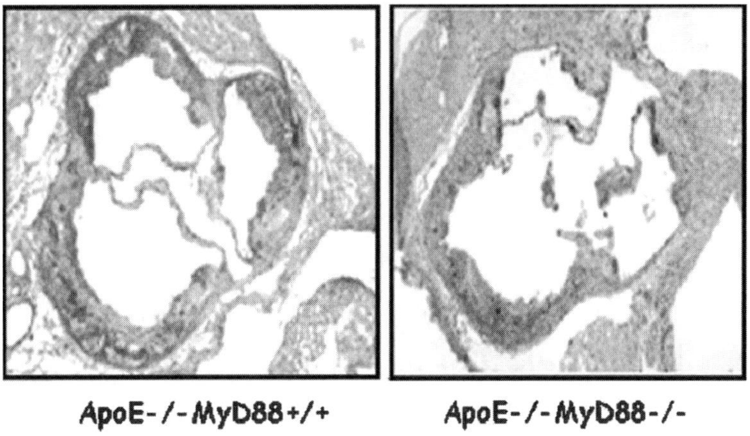

Fig. 3. Oil Red O staining shows the content of atherosclerotic lipid content in the sinus of the aorta. The above panels represent 5-μm criosection of the heart harvested from ApoE−/−MyD88+/+ and ApoE−/− MyD88−/−. ApoE−/− MyD88 +/+ mice have a serious and massive lipid stain compared to ApoE−/−MyD88−/− mice.

sure to add the solution slowly and without touching the tissue.

5. Remove blocking solution from each section, by tapping off. Dry around slides carefully; re-draw with Pap Pen if necessary.
6. Add the primary antibody (150 μl) MOMA-2 or isotype control, IgG2b to each section, without leaking or touching sections.
7. Cover and incubate at room temperature for 1 h (*see* **Note 11**).
8. Wash with PBS two times for 10 min, followed by once for 10 min in wash buffer.
9. Add 150 μl of secondary antibody prepared and incubate for 1 h at room temperature in the dark.
10. Wash with PBS for three times for 5 min.
11. Add two drops of ABC-AP reagent to each section and incubate in the dark for 1 h.
12. Wash with PBS two times for 5 min each and once for 5 min with distilled water.
13. Add Vector Red Substance, substrate for AP in ABC-AP complex and keep in the dark, checking for development until a reddish colour starts to appear (*see* **Notes 12** and **13**).
14. Counter stain with haematoxylin staining as previously described.

15. Rinse profusely in tap water, dipping slides in large container with water and rinsing, eliminating the excess.
16. Dip slides in saturated sodium bicarbonate for 1 min. This will turn counterstaining to a bluish tone, so that the contrast between red (macrophages) and blue (all else) will be better. This reagent is easier to see than red–purple contrast (Fisher).
17. Rinse with tap water again.
18. Stain in Fast Green (optional staining) for 3 min. Then rinse profusely in tap water.
19. Cover the slides with mounting glycerol medium as previously described and let dry overnight.

3.3. Bone Marrow Transplant

As previously described, the process that underlies the atherosclerotic pathogenesis is due to the collaboration of many cell types. BMT is one of the most innovative techniques by which the contribution of haematopoietic versus stromal cell effects can be distinguished. In essence, this method consists of the irradiation of a "recipient" mouse with Gamma Rad (950Y), which subsequently receives bone marrow cells (BMCs) from a "donor" mouse after a short recovery period. Recipient and donor mouse cells can be distinguished using CD45.1 and CD45.2 as cell markers depending on their backgrounds. Specifically, the transplant of TLR deficient mice into ApoE−/− mice highlights the involvement of stromal cells like endothelial cells and vascular smooth muscle cells, to participate into the atherosclerotic process. The non-stromal or hematopoietic cells, such as DCs and macrophages, would then be deficient in the TLR of interest, thereby they would not be able to induce the innate immune signalling pathway to accelerate the atheroma. If the presence of the atherosclerotic plaques is still visible, this result may represent the contribution from endothelial or vascular cells, which are still able to present the antigen, especially in the case of endothelial cells. The irradiation destroys all the cells capable of rapid proliferation and differentiation like bone marrow-derived cells.

3.3.1. Bone Marrow Chimera Method

1. Male and female mice of 8–16 weeks of age are lethally irradiated with 950 rad using a ^{137}Cs source.
2. The time of exposure depends on the type of irradiator provided. Usually, 1.02 Gy is dispensed in 1 min, so that the time of exposure is 9.5 min.
3. Irradiated mice, kept under sterile conditions, are injected intravenously after 3–6 h post-irradiation (*see* **Note 14**).
4. Meanwhile, donor mice are sacrificed and BMCs are collected from the tibias and femurs. After the removal of muscle, the bones are cut at the two ends and the cells are flushed out

the bones by means of a 25 gauge syringe using medium (RPMI 1640). BMCs are spun and red blood cells are lysed using red blood cell lysis buffer, and resuspended in sterile PBS (*see* **Note 15**).

5. The recipient mice are intravenously injected with 2–5 × 10^6 BMCs from donors via the tail vein using a 27 gauge needle in a volume of 100 μl of sterile PBS.

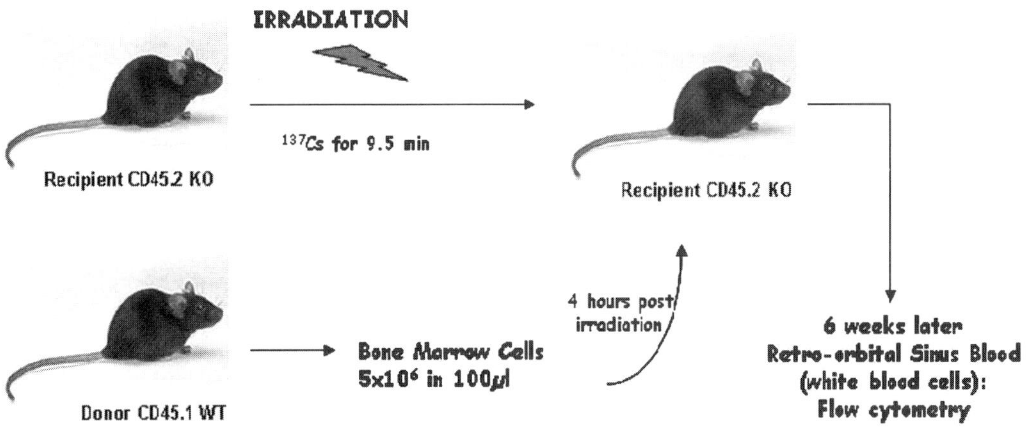

Fig. 4. Mice are lethally irradiated with 950 rad using ^{137}Cs source. Four hours post-irradiation, 5 × 10^6 bone marrow-derived cells are intravenously injected to the irradiated recipient. Six weeks later, a first confirm for the success of the technique can be performed by flow cytometric analysis on white blood cells.

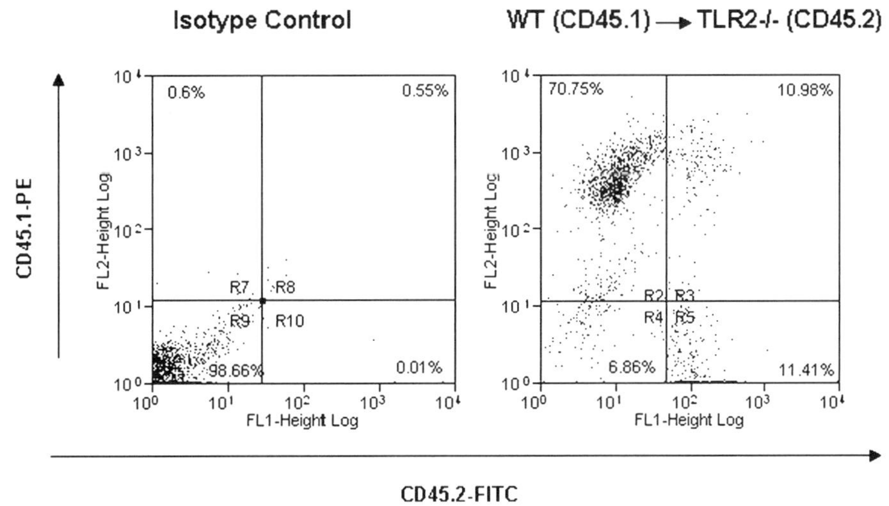

Fig. 5. White blood cells are stained with R-phycoerythrin (PE)-conjugated anti CD45.1 and fluorescein-isothiocyanate (FITC)-conjugated anti CD45.2 antibodies. The above is an example of radiation of bone marrow chimeras generated between TLR2–/– (CD45.2) mice (recipient) and C57Bl/6 CD45.1 WT mice (donor). Six weeks post-irradiation and bone marrow transfer, the circulating amount of donor-derived leukocytes was 70.75% compared to 11.41% of recipient-derived leukocytes. The left panel represents the Isotype control IgG2a.

6. Antibiotics should be administered to the mice for at least 2 weeks post-irradiation as they are more susceptible to infections. The antibiotic (enrofloxacin) could be given either orally (dissolved in the water) or subcutaneously.
7. The degree of chimerism can be assessed by measuring CD45.1 and CD45.2 expression by blood leukocytes and bronchoalveolar lavage at 8 weeks post-irradiation by means of flow cytometry, as described above. A preliminary experiment can also be performed at 6 weeks, withdrawing blood from the retro-orbital sinus, in order to assess the success of the BMT (**Figs. 4** and **5**).
8. Mice can be used at 8–12 weeks following the irradiation.

4. Notes

1. A commercial kit for DNA isolation is also available from Qiagen, which provides a protocol which can substitute the conventional solutions.
2. Mg^{2+} ions form complexes with dNTPs, primers, and DNA templates; thus too few Mg^{2+} ions result in a low yield of PCR product, and too many increase the yield of non-specific products. The recommended range of $MgCl_2$ concentration is 1–4 mM, under the standard reaction conditions specified. In our experiments, at a final dNTP concentration of 0.2 mM, $MgCl_2$ concentration ranges of 1.5 mM (in Taq buffer with KCl) and of 2.0 mM (in Taq buffer with $(NH_4)_2SO_4$) are suitable in most cases. However, it is recommended to check the amount of $MgCl_2$ present in the Taq buffer.
3. If the tail is not completely dissolved the next day, add about 15 μl of 10 mg/ml proteinase K to the microfuge tube and allow tube to rotate for a couple more hours until the tail is completely dissolved.
4. Be careful not to disturb the DNA pellet. If the DNA clump becomes detached, centrifuge for 10 s at the highest speed at room temperature.
5. Phenol or chloroform can only be temporarily stored in polypropylene Falcon conical tubes as the tubes will dissolve overtime.
6. Gels can also be stained after running. Ethidium bromide is positively charged and runs in the opposite direction to the DNA.

7. All the mentioned solutions must be filtered before use.
8. The staining time indicated must be strictly adhered to as the resulting staining could be too strong or too weak to discriminate.
9. Care must be taken to include no bubbles in the mounting media as this will affect the image quality.
10. Blocking should be performed using serum from the species of the secondary Ab. Since the secondary antibody is less specific, it may bind to certain regions of cells resembling its specificity. Blocking this with serum from the same species should reduce this effect.
11. The primary antibody step can be held overnight at 4°C, after incubating 1 h at room temperature.
12. If the section develops the colour quickly, leave for only 5–10 min. It is necessary to check constantly, under a microscope, to note if the background staining starts to develop. If so, stop developing. If it is removed too early, the colour will not be strong enough for contrast. If too late, background staining will occur. Controls should have no colour.
13. The secondary antibody can also be recognized by using 3-amino-9-ethyl-carbazole chromogen (DAKO) as a substrate. If background staining is an issue and a biotin–streptavidin sandwich is being used, commercially available biotin blocking kits can be purchased.
14. Efficacy and toxicity of this technique is affected by three parameters: dose rate, dose fractionation, and interval between irradiations. The best dose rate is 1.02 Gy/min as Salomon et al. *(30)* demonstrated that a dose rate less than 0.8 Gy/min could not effectively remove hematopoietic stem cells and furthermore, the survival rate of mice declined when the dose rate was higher than 1.3 Gy/min. As described by Cui et al. *(29)*, the time considered between the irradiation and the intravenous injection is 4 h, as it is known that tumour cell lines re-enter the mitotic cycle in vitro after 4–6 h post-lethal irradiation. A late injection could give sufficient time for the damaged cells to be repaired. The survival rate is higher. Some investigators prefer to perform this technique in two irradiation steps. Mice are irradiated with 5+5 Gy, with an interval of 1–3 h in order to effectively remove the immunocompetent cells that have just re-entered the cell cycle from the dormant stage after the first irradiation. Though, Salomon et al. *(30)* and Soderling et al. *(31)* showed that a single dose was more effective than the fractionated irradiation in terms of survival rate.

15. Some investigators also deplete T and NK1.1 T cells, representing ≈30% of the cells in the bone marrow *(32)*. The major potential source of the "graft versus host" reaction has been related to the presence of T cells in the bone marrow itself and peripheral blood which could contaminate the bone marrow during its harvest. However, our group and others have successfully obtained chimeric mice while not performing the depletion of the above mentioned cells (**Fig. 5**), since T and NK cells lifespan is also modified post-irradiation.

References

1. Munro JM, Contran RS (1988) Pathogenesis of atherosclerosis: atherogenesis and inflammation. *Lab Invest*; 58(3):249–61
2. Wautier JL (1989) Monocyte-endothelium relations. *J Mal Vasc*; 14 Suppl A:13–6
3. Yilmaz A, Lochno M, Traeg F, Cicha I, Reiss C, Stumpf C, Raaz D, Anger T, Amann K, Probst T, Ludwig J, Daniel WG, Garlichs CD (2004) Emergence of dendritic cells in rupture-prone regions of vulnerable carotid plaques. *Atherosclerosis*; 176(1):101–10
4. Kawahara I, Kitagawa N, Tsutsumi K, Nagata I, Hayashi T, Koji T (2007) The expression of vascular dendritic cells in human atherosclerotic carotid plaques. *Human Pathol*; 38(9):1378–85
5. Soilleux EJ, Morris LS, Trowsdale J, Coleman N, Boyle JJ (2002) Human atherosclerotic plaques express DC-SIGN, a novel protein found on dendritic cells and macrophages. *J Pathol*; 198(4):511–6. 5, 6
6. Kis Z, Pallinger E, Endresz V, Burian K, Jelinek I, Gonczol E, Valyi-Nagy I (2004) The interactions between human dendritic cells and microbes: possible clinical applications of dendritic cells. *Inflamm Res*; 53:413–23
7. Bobryshev YV, Cao W, Phoon MC, Tran D, Chow VT, Lord RS, Lu J (2004) Detection of *Chlamydophila pneumoniae* in dendritic cells in atherosclerotic lesions. *Atherosclerosis*; 173:185–95
8. Mendall, MA, Goggin PM, Molineaux N, Levy J, Toosy T, trachan D, Camm AJ, Northfield TC (1994) Relation of *Helicobacter pylori* infection and coronary heart disease. *Br Heart J*;, 71:437
9. Yamashiroya HM, Ghosh L, Yang R, Robertson AL, Jr. (1988). Herpesviridae in the coronary arteries and aorta of young trauma victims. *Am J Pathol*; 130:71
10. Horvath R, Cerny J, Benedik J, Jr., Hokl J, Jelinkova I, Benedik J (2000) The possible role of human cytomegalovirus (HCMV) in the origin of atherosclerosis. *J Clin Virol*; 16:17
11. Bobryshev YV (2000) Identification of HIV-1 in the aortic wall of AIDS patients. *Atherosclerosis*; 152:529
12. Hajjar DP, Pomerantz KB, Falcone DJ, Weksler BB, Grant AJ (1987) Herpes simplex virus infection in human arterial cells: implications in arteriosclerosis. *J Clin Invest*; 80:1317
13. Raza-Ahmad A, Klassen GA, Murphy DA, Sullivan JA, Kinley CE, Landymore RW, Wood JR (1995) Evidence of type 2 herpes simplex infection in human coronary arteries at the time of coronary artery bypass surgery. *Can J Cardiol*; 11:1025
14. Ishizaka N, Ishizaka Y, Takahashi E, Tooda E,t Hashimoto H, Nagai R, Yamakado M (2002) Association between hepatitis C virus seropositivity, carotid-artery plaque, and intima-media thickening. *Lancet*; 359:133
15. Vassalle C,, Masini S, Bianchi F, Zucchelli GC (2004) Evidence for association between hepatitis C virus seropositivity and coronary artery disease. *Heart*; 90:565
16. Ford PJ et al (2007) Cardiovascular and oral disease interactions: what is the evidence? *Prim Dent Care*; 14(2):59–66
17. Lehr HA, Sagban TA, Ihling C, Zahringer U, Hungerer KD, Blumrich M, Reifenberg K, Bhakdi S (2001) Immunopathogenesis of atherosclerosis: endotoxin accelerates atherosclerosis in rabbits on hypercholesterolemic diet. *Circulation*; 104:914
18. Xu XH, Shah PK, Faure E, Equils O, Thomas L, Fishbein MC, Luthringer D, Xu XP, Rajavashisth TB, Yano J, Kaul S, Arditi M (2001) Toll-like receptor-4 is expressed by macrophages in murine and human lipid-rich atherosclerosis plaques and upregulated by oxidized LDL. *Circulation*; 104:3103

19. Edfelt K, Swedenborg J, Hansson GK, Yan ZQ (2002) Expression of toll-like receptors in human atherosclerotic lesions: a possible pathway for plaque activation. *Circulation*; 105:1158
20. Michelsen KS, Wong MH, Shah PK, Zhang W, Yano J, Doherty TM, Akira S, Rajavashisth TB, Arditi M (2004) Lack of Toll like receptor 4 or myeloid differentiation factor 88 reduces atherosclerosis and alters plaque phenotype in mice deficient in apolipoprotein E. *PNAS*; 101(29):10679–84
21. Mullick AE, Tobias PS, Curtiss LK (2005) Modulation of atherosclerosis in mice by Toll-like receptor 2. *J Clin Invest*; 115:3149–56
22. Liu X, Ukai T, Yumoto H, Davey M, Goswami S, Gibson FC, Genco CA (2007) Toll-like receptor 2 plays a critical role in the progression of atherosclerosis that is independent of dietary lipids. *Atherosclerosis*; 196(1):146–54
23. Ishida BY, Blanche PJ, Nichols AV, Yashar M, Paigen B (1991) Effects of atherogenic diet consumption on lipoproteins in mouse strain C57BL/6 and C3H. *J Lipid Res*; 32:559
24. Shi W, Haberland ME, Jien ML, Shih DM, Lusis AJ (2000) Endothelial responses to oxidized lipoproteins determine genetic susceptibility to atherosclerosis in mice. *Circulation*; 102:75
25. Shi W, Wang X, Tangchitpiyanond K, Wong J, Shi Y, Lusis AJ (2002) Atherosclerosis in C3H/HeJ mice reconstituted with apolipoprotein E-null bone marrow. *Arterioscler Thromb Vasc Biol*; 22:650
26. Ishibashi S, Herz J, Maeda N, Goldstein JN, Brown MS (1994) The two-receptor model of clearance of lipoproteins: tests of hypothesis in "knockout" mice lacking the low density lipoprotein receptor, apolipoprotein E, or both proteins. *PNAS*; 91(10):4431–5
27. Cogger VC, Hilmer SN, Sullivan D, Muller M, Fraser R, Le Couteur DG (2006) Hyperlipidemia and surfactants: the liver sieve is a link Atherosclerosis.189(2):273–81.
28. Veniant MM, Sullivan MA, Kim SK, Ambriosiak P, Chu A, Wilson MD, Hellerstein MK, Rudel LL, Walzem RL, Young SG (2000) Defining the atherogenesis of large and small lipoproteins containing apolipoprotein B100. *J Clin Invest*; 106(12):1501–10
29. Cui YZ, Hisha H, Yang GX, Fan TX, Jin T, Li Q, Lian Z, Ikehara S (2002) Optimal protocol for total body irradiation for allogeneic bone marrow transplantation in mice. *Bone Marrow Transplant*; 30(12):843–9
30. Salomon O, Lapidot T, Terenzi A et al (1990) Induction of donor-type chimerism in murine recipients of bone marrow allografts by different radiation regimens currently used in treatment of leukaemia patients. *Blood*; 76:1872–8
31. Soderling CCB, Song CW, Blazar BR, Vallera DA (1985) A correlation between conditioning and engraftment in recipient MHC-mismatched T cell depleted murine bone marrow transplants. *J Immunol*; 135:941–6
32. Zeng BD, Lewis D, Sussan Dejbakhsh-Jones FL, Garcia-Ojeda M, Sibley R, Strober SB (1999) Bone marrow NK1.1– and NK1.1 + Tcells reciprocally regulate acute graft versus host disease. *J Exp Med*; 189:1073–81

Chapter 24

Generation of Parasite Antigens for Use in Toll-Like Receptor Research

Philip Smith, Niamh E. Mangan, and Padraic G. Fallon

Summary

Pathogen recognition is a central activity of the Toll-like receptor (TLR) family. Molecules from various pathogens have been widely used in TLR research as natural ligands for the receptors. TLR ligands from bacteria, viruses, and fungi are widely available from commercial companies and are increasingly being manufactured with high purity specifically for use in TLR research. Although increasingly used in TLR research, extracts from many parasites with potential TLR ligands are not generally produced commercially. Historically, parasite extracts were produced in academic laboratories for diagnostic or vaccination research, often without an emphasis on quality control. Here we describe methods for isolation of eggs from the human parasite *Schistosoma mansoni*. We also describe a protocol for generation of *S. mansoni* soluble egg antigens (SEA), which are commonly used in TLR research. This protocol has application for the isolation of extract from other parasites or pathogens as it is intended to reduce contamination that may cause spurious data in TLR research.

Key words: Toll-like receptors, Endotoxin, *Schistosoma mansoni*, Eggs, Soluble egg antigens, Quality control.

1. Introduction

Our understanding of the functions of the Toll-like receptors (TLRs) family has been facilitated by the use of pathogen-associated molecular patterns (PAMPs) from various pathogens. PAMP from different pathogens engage specific TLRs and evoke, or modulate, downstream signalling pathways. The TLR field has thus been well served by the experimental use of PAMPs from bacteria, viruses, and fungi. In contrast, parasites are as yet not

Claire E. McCoy and Luke A.J. O'Neill (eds.), *Methods in Molecular Biology, Toll-Like Receptors, vol. 517*
DOI: 10.1007/978-1-59745-541-1_24

extensively explored as a depository for potentially novel PAMPs for all aspects of investigation of innate immunity. Nevertheless, with respect to TLR family members parasite-derived molecules are potentially novel PAMPs, for example, the profilin-like molecule from the protozoan parasite *Toxoplasma gondii* is a ligand for TLR11 *(1)*, while the Hemozoin pigment from malaria (*Plasmodium falciparum*) is a potential ligand *(2)*, or carrier for plasmodium DNA *(3)*, for TLR9. The helminth (worm) parasites are a particularly interesting group of pathogens to investigate, as they have evolved modulatory mechanisms to bridge the innate to adaptive spectrums of immunity. Thus, despite helminth parasites characteristically having prolonged chronic infections, lasting years to decades, and a propensity to preferentially evoke adaptive Type 2 [T Helper (h) 2 cytokine secretion, eosinophilia, and elevated IgE] immune responses, they also are potent activators of innate responses in antigen presenting cells such as dendritic cells (DC) *(4)*. The human parasite *Schistosoma mansoni* has been widely used in experimental studies on innate and adaptive immunity, and will be used as the model helminth parasite for methods described herein.

Although *S. mansoni* infects humans in tropical countries, the mammalian-infected stage of the parasite life cycle can be maintained by infection of mice. *S. mansoni* infections of mice recapitulate, with some exceptions, many aspects of the immuno-biology of infection of humans *(5)*. In infected mice, as in humans, the worms lay parasite eggs that are trapped in the liver and also the intestines. These eggs from *S. mansoni* have been widely used in the context of the Th1–Th2 paradigm as potent inducers of in vivo Th2 responses. However, more recently with the emergence of the TLR-era, soluble egg antigens (SEA) extracted from the *S. mansoni* eggs have been shown to modulate DC through TLR-dependent and -independent interactions *(4)*. Unlike many PAMPs from bacteria (such as lipopolysaccharides, LPS), viruses, or fungi, many parasite PAMPs are not yet widely commercially available. These parasite-derived extracts or molecules are thus often prepared in academic laboratories. While scientists active in the TLR field are acutely aware of the potential for erroneous data from unknown TLR contaminants in extracts (including contaminated commercially available PAMPs such as commercial batches of LPS *(6)*), such concerns also apply to parasite extracts prepared for TLR research. Here we outline a protocol for the isolation of parasite eggs from mice and the generation of SEA that is optimized to limit endotoxin contamination. The techniques described could be adapted to isolation of antigens from different parasites, or other pathogens, for use in TLR research.

2. Materials

2.1. Preparation of Equipment/Reagents for Endotoxin Management and Screening

1. Five litres of 0.5 M sodium hydroxide (NaOH) solution in ultra-pure H_2O (*see* **Note 1**). Store in 5 L dispenser and prepare fresh a day before egg extraction.
2. Dulbecco's phosphate-buffered saline (DPBS) prepared from 10X DPBS (Sigma) with ultra-pure water (*see* **Note 2**).
3. Ovens (over 95°C) for heat-inactivation of metallic equipment.

2.2. Endotoxin Screening

1. Limulus amebocyte lysate (LAL) assay kit QCL100 (Cambrex/biowhittaker).
2. Ultra-pure LPS (Invivogen).

2.3. Infection/Mice

1. *Schistosoma mansoni* life cycle.
2. Specific pathogen-free housed mice can be BALB/c or C57BL/6 strain, or other, depending on intended application of eggs or SEA (*see* **Note 3**).
3. Irradiated fibre-free rodent pellets (SDS, Essex, UK).
4. Hydrocortisone 21-acteate (Sigma) for injection is prepared by adding 625 mg hydrocortisone 21-acteate to 24.36 ml of distilled water, with 625 μl of Cremphor EL (Sigma) added as dispersing agent. Cremphor EL is highly viscous; the use of 1 ml pipette tip with end removed will facilitate pipetting. Solution is sonicated and stored at 4°C

2.4. Egg Isolation

1. A 1.8% saline solution. Eighteen grams of cell culture-tested sodium chloride (Sigma) in 1 L of ultra-pure water.
2. Trypsin (Bovine, Merck/BDH).
3. A series of brass sieves and lids and receivers are required. Large 8 in. diameter sieves with 500, 300, 150, and 45 μm mesh-size. Smaller 4 in. diameter sieves with 106, 45, and 32 μm mesh-size, *see* **Fig. 2A**. Sieves are from Endecotts Ltd., London, UK. If available an ultrasonic bath can be used to clean sieves.
4. Laboratory orbital shaker that can be set for 37°C.
5. Standard food blender from any electrical outlet.
6. DPBS (BioWest).
7. Percoll (Sigma) prepared in 60% solution with DPBS (BioWest).
8. Disposable sterile Pasteur pipettes.
9. Glass microscope slide and 20 × 20 mm cover slips.

10. Binocular microscope.
11. Parafilm.

2.5. SEA Preparation

1. Frozen pellets of *S. mansoni* eggs.
2. Liquid nitrogen.
3. DPBS (BioWest).
4. Disposable sterile Pasteur pipettes.
5. Glass microscope slide and 20 × 20 mm cover slips.
6. Petri dishes.
7. Binocular microscope.
8. Stainless steel percussion mortar and pestle (*see* **Fig. 2B**).
9. Heavy duty lump hammer.

3. Methods

To achieve high quality egg isolation and the preparation of SEA that give reproducible immune modulating activity, the emphasis during egg extraction and SEA production is limiting the potential for contamination by endotoxin and host-derived material. In all the steps, ensure that a clean working environment is maintained, with gloves worn and frequently changed. It is important to stress that the presence of endotoxin should be viewed as merely a surrogate marker that the extract is contaminated, and indicative that other TLR ligands may also be present.

3.1. Preparation of Equipment/Reagents for Endotoxin Control

1. Use sterile disposable plastics where possible.
2. All solution used are pre-tested by LAL for endotoxin levels (*see* **Note 2**).
3. Endotoxins can be denatured either by prolonged exposure to low pH or high temperature. Glassware, such as 5 L beaker and 2 L shaking flasks or other equipment (blender), are filled with 0.5 M NaOH solution, soaked at least for 1 h or overnight, followed by extensive rinsing in ultra-pure water.
4. Metal equipment (sieves, percussion mortar) are dry-baked (>140°C) overnight. After baking, rinsed extensively in 0.5 M NaOH and then ultra-pure water.
5. For washing of glassware or metal equipment, there are detergents (PyroClean) specifically optimized for endotoxin removal.

3.2. Endotoxin Screening

1. Water and other solutions used in the methods are pre-tested for endotoxin contamination using a chromogenic LAL assay according to the manufacturer's instructions. Samples should be spiked with LPS to ensure accurate quantification.

3.3. Schistosoma mansoni Infection

1. Mice are infected with *S. mansoni* by standard methods *(7)* (*see* **Note 4**).
2. At 6 weeks post-infection, mice are injected subcutaneously with 100 μl of hydrocortisone suspension (*see* **Note 5**).
3. If eggs are to be extracted from intestinal tissue, mice should be placed on fibre-free diet 48 h before mice are culled. The modified diet reduces particulate matter from the intestinal tract and improves egg isolation (*see* **Note 6**). **Figure 1A** has the time-course of *S. mansoni* infection of mice and in vivo treatments.

3.4. Schistosoma mansoni Egg Isolation

The eggs are surrounded by a granulomatous cell infiltrate within the liver or intestines (**Fig. 1B**). The granulomas must first be removed from the tissue and then the granuloma disrupted to

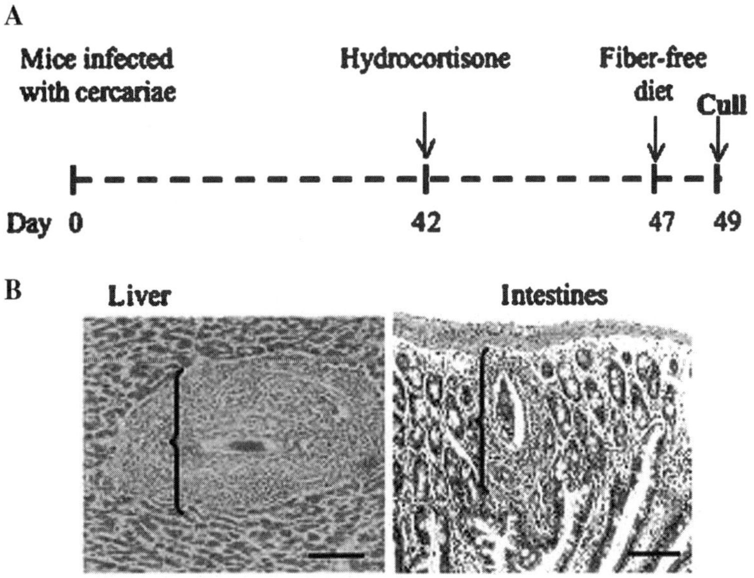

Fig. 1. (**A**) Schematic of the time-course of infection of mice with *Schistosoma mansoni* cercariae and in vivo treatment before culling for removal of livers and intestines. Hydrocortisone is injected to suppress the inflammation around the egg granuloma and thereby enhance extraction of eggs. Fibre-free diet is used to reduce intestinal lumen fibre, which can reduce the efficiency of egg extraction. (**B**) Hematoxylin and eosin-stained section of liver or small intestines from infected mice showing the egg in the centre of a granulomatous cell infiltrate (*brackets*). Note the proximity of the egg in intestines with respect to the lumen. Scale bar = 150 μm.

release the eggs. Recovery of eggs from digested granulomas involves mature eggs (approximately 140 μm × 50 μm) passing through and eventually been caught in various sieves. The following method refers to extraction of eggs from livers. The same method can be also used for extraction of eggs from intestine with modifications as stated (*see* **Note 6**). If intending to isolate eggs from both tissues, always process the livers first at every stage to prevent contamination by intestinal tissue and contents. This protocol is based on 30 infected mice, which will produce a total of ~3.5 × 10^6 eggs from the liver extraction:

1. Remove the livers from mice. Place livers in 1.8% saline (*see* **Note 7**) in a NaOH-treated 500 ml glass beaker. Wash livers 2–3 times in 1.8% saline to remove blood and tissue.
2. Add 1.5 ml of trypsin and 600 ml of pre-warmed (*see* **Note 8**) 1.8% saline. For each mouse liver, add 0.05 g of trypsin in 20 ml 1.8% saline. Transfer liver and trypsin suspension into a domestic food blender.
3. Homogenize the livers with 5–10 s bursts at lowest speed setting of blender (*see* **Note 8**). After every burst, wash down the lid and sides of blender with 1.8% saline. Continue homogenizing livers with 5–10 s bursts until there are no more large pieces of liver tissue. The egg granulomas appear as small beads that are no more than 1–2 mm in diameter and appear lighter in colour than surrounding liver debris.
4. Transfer homogenized liver suspension to 2 L flanged shaking flask. Seal flask with Parafilm. Place flask rotating in an orbital shaker at 37°C for at least 2 h (*see* **Note 8**).
5. Stack the three largest (500, 300, and 150 μm) sieves on a 5 L beaker. As the sieves are 8 in. diameter, they fit securely into the beaker top. The 500 μm sieve should be on top with the 150 μm on at the bottom. Pass the liver suspension through the sieves allowing the fall-through to collect in the beaker. Help the eggs through the sieves by washing sieves with 1.8% saline and rubbing retained tissue against the mesh with gloved fingers. Change gloves frequently. Use sterile disposable Pasteur pipette to collect a sample of tissue that is retained in sieves and place on microscope slides. Examine under a binocular microscope for presence of eggs, only move onto the next sieve when most of the eggs have been released from the tissue (*see* **Note 9**).
6. Pour the contents of the beaker, containing the released eggs, into a 500 ml centrifuge container and spin at 820 × *g* for 2 min to sediment/collect the eggs. Discard supernatants gently. Re-suspend sediment in 1.8% saline and transfer to 50 ml tubes. Spin in a bench centrifuge, allow speed to reach 500 rpm, and then stop spinning. Pour off supernatant and

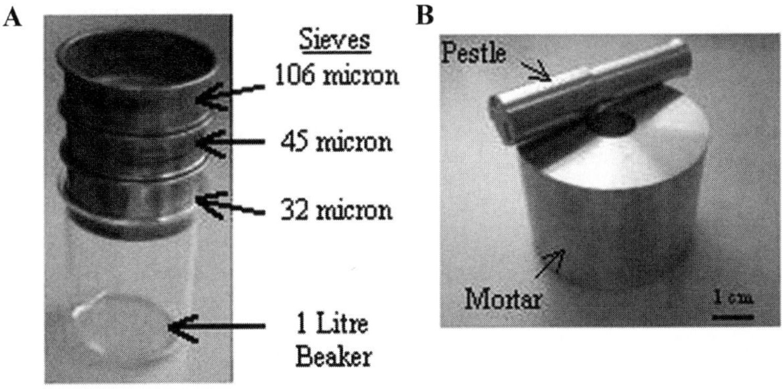

Fig. 2. (**A**) Sieves of various sizes stacked on 1-L beaker for sieving of trypsin-digested egg suspension for isolation of eggs. (**B**) Appearance of stainless steel percussion mortar used for disrupting eggs to produce soluble egg antigens (SEA).

re-suspend. Repeat this 3–6 times or until the supernatant is clear. The sediment will have a deep green colour indicating the presence of eggs.

7. Place the smaller sieves (106, 45, and 32 μm) on a 1 L beaker (**Fig. 2B**). Wash the eggs through the small sieves using 1.8% saline and agitate gently with gloved fingers to facilitate passage of eggs. Do not force residual tissue through the sieves. Most mature eggs will be caught by the 45 μm sieve. Small pieces of liver tissue, immature, damaged, and dead eggs will fall through the 45 μm sieve, to be retained in 32 μm sieves. Wash the 106 and 45 μm sieves for reuse in **step 10**.
8. Tilt the 45 μm sieve slightly and wash it unidirectionally, to allow collection of all the eggs together. Transfer eggs to 50 ml Falcon tube using a clean Pasteur pipette and wash with 1.8% saline as described in **step 6**.
9. Divide egg pellet into two 50 ml Falcon tubes. Fill each tube with 1.8% saline and allow eggs to settle by gravity for 15–20 min. Discard supernatant.
10. Add the eggs to the cleaned small sieves, stacked as in **step** 7. Wash egg gently through the sieves using 1.8% saline. Recover mature eggs from 45 μm sieves.
11. Transfer eggs to a clean 50 ml tube and spin in a bench centrifuge, allow the speed to reach 1,500 rpm, and then stop spinning. Pour off supernatant and re-suspend egg pellet in 1.8% saline. Remove 10 μl of egg suspension and transfer to a glass microscope slide place a cover slip on top and examine under a microscope. At this stage there should be no tissue visible but only eggs. If there is noticeable non-egg derived debris, repeat **step 10**.

12. Re-suspend egg pellet in DPBS. Repeat centrifugation and discard the supernatants as described in **step 11**, at least three more times (*see* **Note 7**).
13. To count the eggs, add a 50 μl aliquot from a 50 ml suspension of eggs to 1 ml DPBS. Place 5–10 μl aliquots of egg suspension on a Petri dish. Using a binocular microscope, count the number of eggs per aliquot. Determine the total number of eggs recovered.
14. Place standard 1.5 ml cryovials in a rack placed in liquid nitrogen; ensure the cryovials are also filled with liquid nitrogen. Using a sterile Pasteur pipette, the eggs are frozen by transferring droplets of egg suspension into the cryovials. Using this method, a standard 1.5 ml cryovial when full of frozen egg droplets will contain approximately 1×10^6 eggs. Frozen eggs are stored in liquid nitrogen until required (*see* **Note 11**).

3.5. Preparation of SEA

During the preparation of SEA, it is essential that the temperature is kept low during sonication. We have found that if temperature is not controlled, for example, during homogenization with a motorized glass mortar and glass pestle, key immunogenic antigens are lost from the SEA. The method described here is intended to reduce the loss of heat-labile antigens in SEA while limiting potential endotoxin contamination. As the preparation of SEA involves the use of liquid nitrogen and sonicators, ensure compliance with relevant safety legislation. Using this protocol, anticipate ~10 mg of SEA from 1×10^6 liver eggs.

1. A stainless steel percussion mortar and pestle are cooled in liquid nitrogen.
2. Approximately one million frozen eggs are placed in the mortar.
3. Use a heavy-duty hammer to strike the pestle to crush the eggs. Rotate the pestle in the mortar after each percussion. Frozen eggs are crushed by about ten percussion strikes.
4. The crushed egg powder/paste is removed to a 50 ml tube by adding liquid nitrogen to the chamber and tipping this quickly in to a waiting tube. Allowed to thaw on ice. For every million eggs homogenized add 5 ml of DPBS.
5. The crushed egg powder/paste are further disrupted by ten, 30 s pulses of sonication on ice using a probe sonicator. Each sonication pulse is separated by 30 s on ice (*see* **Note 12**).
6. To determine the efficiency of egg disruption, a drop of the egg suspension is examined under a microscope, to ensure the eggs are broken, if necessary the sonication process is repeated.

7. The disrupted eggs are then spun at 10,000 ×*g* for 60 min at 4°C. The supernatant is removed and aliquoted to 1.5 ml Eppendorf tubes and spun again for a further 15 min at 4°C.
8. After the second centrifugation, the supernatant from the 1.5 ml tubes is pooled and passed through a 0.45 μm syringe filter. The SEA is then stored in aliquots at -20°C. The protein content of SEA batches needs to be determined and batches quality controlled (*see* **Notes 14–17**).
9. Despite being spun and filtered before freezing SEA, when thawed for the first time, may produce an insoluble sediment. Spin in a micro-centrifuge for 10 min at 13,000 rpm, 4°C, to pellet the sediment and transfer SEA to a clean tube. Subsequent freeze–thaw cycles will not produce a similar sediment.

4. Notes

1. Use purest source of water available, for example, de-ionized and filtered water purification system. Water used for egg/antigen preparation should be endotoxin-free as tested by LAL assay. Always collect water direct from dispensing tap. We have noted that the use of rubber tubing attached to the tap of a water purification system is a potential source of endotoxin contamination.
2. Biological buffers or media from different commercial companies labelled as "endotoxin-free" or "low-endotoxin" can vary by tenfold in endotoxin levels. We recommend the routine screening of RPMI-1640 media, FBS, or DPBS from various suppliers using the LAL assay. We use primarily RPMI-1640 and DPBS from BioWest as they had the lowest levels of endotoxin from various suppliers tested.
3. If the eggs or SEA are to be used to stimulate cells from mice in vitro or to be injected into mice, it is essential that the mouse strain used to harbour the parasite infection should be the same as the mouse strain used experimentally. We have noted that parasites obtained from a CD-1 strain mouse activate macrophages from BALB/c mice in vitro more potently than using parasites isolated from infected BALB/c strain. This maybe due to MHC-mediated activation elicited by residual host tissue on or in the parasite. Another concern to be aware of is that the two other major schistosome species that infect man, *S. hematobium* and *S. japonicum*, are often maintained in the laboratory in hamsters or pigs, respectively.

We are not aware of influence of preparing eggs or SEA in different animal species on TLR experiments involving transfected cell lines or human cells.

4. Mice should be infected specifically for the production of eggs. Routinely we would use a minimum of 30 infected mice for egg isolation. The protocol described here is based on 30 mouse livers. The number of cercariae used to infect the mice for egg production will need to be between 250 and 400 by percutaneous infection. This is a "heavy" infection compared to conventional experimental infection, which varies usually between 25 and 75 cercariae. The optimum number of cercariae which gives maximum egg recovery with minimal mortality will vary between mouse strain, the *S. mansoni* strain used and even animal housing facilities. These factors will need to be determined by individual laboratories.
5. Immune suppression of mice with hydrocortisone at this stage of infection limits the size, integrity, cellular infiltration, and degree of collagen deposition in the granulomas surrounding the eggs within the liver. The reduced granulomatous inflammation in steroid-treated mice ensures that the extraction of eggs from tissue is more efficient.
6. Eggs extracted from intestinal tissue, or SEA prepared from intestinal eggs, should not be used in any in vivo or cell culture experiments. The levels of endotoxin contamination from eggs or SEA prepared from intestines are too high to justify use in TLR research. However, SEA or purified proteins from SEA can be used in assays, such as proteomics, where endotoxin contamination is less of a concern.
7. As schistosome eggs hatch in water, a 1.8% saline is used in egg isolation to prevent the eggs from hatching. As eggs may also hatch in DPBS, ensure final washing of eggs in DPBS is done immediately and promptly.
8. Pre-warming the 1.8% saline is optional. It does reduce the trypsin digestion time by ~30 min. The aim is to have the eggs isolated within 4–5 h of the livers being removed from mice. Some researchers store the liver, intact or homogenized, for 24–48 h at 4°C or 37°C, to allow the liver enzymes to degrade tissue to aid extraction of eggs. Due to the potential for microbial growth, we do not recommend this practice. Additionally we have observed that eggs prepared as described above are less damaged, and more viable, than when prepared using longer or more aggressive tissue homogenization or digestion protocols.
9. An adaptation can be used if a more homogeneous egg preparation is required that will consist of mainly larger

mature eggs. In **step 5**, add a fourth sieve 45 μm to catch the eggs. The collected eggs are then processed as normal. This method results in a higher quality of mature eggs for use in, for example, in vitro culture or for egg injections (*see* also **Note 11**). This adaptation will reduce the total number of eggs recovered. In all stages of egg sieving, avoid excessive physical agitation as this may force through residual connective tissue, which may have formed aggregates as a result of the preceding centrifugation steps.

10. The speed settings and duration of homogenization will vary depending on blender used. This homogenization regime used should be determined by individual laboratories. Do not over homogenize the livers as this will result in damaged eggs. Intestines require less processing, homogenize with half as many 5–10 s bursts than what is required to process the livers.
11. If eggs are to be used in vivo in the pulmonary egg granuloma model, a further step is required to ensure that eggs are of uniform size and density (correlating with maturity). The use of this additional step gives a more consistent size of peri-ova granulomas. Eggs are re-suspended in 60% Percoll in DPBS and spun at 1,000 × *g* for 3 min. The larger mature eggs pass through the Percoll and are collected in the pellet, whereas dead/immature eggs, debris, and shells are trapped in the interface. Wash egg pellet in DPBS as described above. Only liver-derived eggs should be used for in vivo studies.
12. The degree of sonication used is dependent on the machine used. The appropriate sonication regime to be adopted will need to be optimized by individual laboratories. Egg disruption from sonications is checked visually using microscopy as described. We have a sonicator probe that is designated solely for SEA preparation.
13. New batches of SEA should undergo various assays to ensure reproducible quality. These include protein estimation, LAL assay, and running on SDS-PAGE. Batches of SEA can be tested for dose-dependent stimulation of spleen cells or mesenteric lymph node cells from naïve and *S. mansoni*-infected mice. A good preparation of SEA will stimulate the production of IL-4 from spleen cells from infected mice but not cells from in naïve mice, while having little or no effect on the secretion of IFN-γ, IL-12, and TNF from spleen cells. We have found that 20–25 μg/ml of SEA in primary culture of immune cells gave maximum cell activation stimulation without any cytotoxic effects.
14. Despite rigorous attempts outlined above to limit endotoxin contamination, SEA batches will contain some endotoxin.

We have found that tolerable levels are one endotoxin unit per milligram of SEA protein (one endotoxin unit equates 100 pg *Escherichia coli* LPS). At this arbitrary level of endotoxin, SEA does not induce detectable TNF-α when incubated with spleen cells.

15. There are a number of commercial available reagents that purport to remove endotoxin contamination from solutions. It is our experience with one system (Detoxigel endotoxin gel removing gel columns, Pierce) that the columns achieve the manufacturers' recommended 99% endotoxin removal when using a new column. On secondary use, after stripping and regenerating the column, endotoxin removal is considerable less efficient. For example, endotoxin levels in commercial ovalbumin were reduced from 1.5–3 to < 0.5 endotoxin unit per milligram *(8)*. An additional concern with such columns is substantial loss of molecules, based on protein content, including potential molecules of interest. We advocate the use of endotoxin removal columns as a strategy aimed only to achieve partial removal of endotoxin from suspension used in preparation of parasite extracts, but not the extract itself.
16. The antibiotic Polymyxin B is often used to remove endotoxin from extracts, including SEA. As Polymyxin B has been shown to activate human DC *(9)*, it may interfere with the activity of parasite molecules in TLR research, and should be used with caution.
17. The methods described are based on the use of laboratory cell culture standards that involve routine screening systems for mycoplasma contamination. We have noted that murine bone-marrow derived DC generated in cell culture conditions where mycoplasma contamination was not addressed resulted in DC with an activation state (surface marker expression) markedly different than DC generated in mycoplasma-free cell culture conditions, as also reported previously using human DC *(10)*. Mycoplasma contamination influencing the activation state of DC, or other cells, is a specific concern with respect to thresholds for PAMP activation of DC and TLR interactions. There are a range of commercially available PCR- or fluorescence-based methods for mycoplasma screening that should be used.

Acknowledgements

Many of the methods described here were adapted from techniques developed in Mike Doenhoff's or David Dunne's groups. Padraic Fallon thanks both for their support and mentorship over the years. Authors are supported by Science Foundation Ireland.

References

1. Yarovinsky, F., D. Zhang, J. F. Andersen, G. L. Bannenberg, C. N. Serhan, M. S. Hayden, S. Hieny, F. S. Sutterwala, R. A. Flavell, S. Ghosh, and A. Sher (2005) TLR11 activation of dendritic cells by a protozoan profilin-like protein. *Science* 308, 1626–1629.
2. Coban, C., K. J. Ishii, T. Kawai, H. Hemmi, S. Sato, S. Uematsu, M. Yamamoto, O. Takeuchi, S. Itagaki, N. Kumar, T. Horii, and S. Akira (2005) Toll-like receptor 9 mediates innate immune activation by the malaria pigment hemozoin. *J. Exp. Med.* 201, 19–25.
3. Parroche, P., F. N. Lauw, N. Goutagny, E. Latz, B. G. Monks, A. Visintin, K. A. Halmen, M. Lamphier, M. Olivier, D. C. Bartholomeu, R. T. Gazzinelli, and D. T. Golenbock (2007) Malaria hemozoin is immunologically inert but radically enhances innate responses by presenting malaria DNA to Toll-like receptor 9. *Proc. Natl. Acad. Sci. U. S. A.* 104, 1919–1924.
4. Pearce, E. J., C. M. Kane, and J. Sun (2006) Regulation of dendritic cell function by pathogen-derived molecules plays a key role in dictating the outcome of the adaptive immune response. *Chem. Immunol. Allergy* 90, 82–90.
5. Fallon, P. G. (2000) Immunopathology of schistosomiasis: a cautionary tale of mice and men. *Immunol. Today* 21, 29–35.
6. Hirschfeld, M., Y. Ma, J. H. Weis, S. N. Vogel, and J. J. Weis (2000) Cutting edge: repurification of lipopolysaccharide eliminates signaling through both human and murine toll-like receptor 2. *J. Immunol.* 165, 618–622.
7. Lewis, F. (1998) Schistosomiasis. *Curr. Prot.* Immunol. 28, 19.11.11–18.
8. Mangan, N. E., N. van Rooijen, A. N. McKenzie, and P. G. Fallon. (2006) Helminth-modified pulmonary immune response protects mice from allergen-induced airway hyperresponsiveness. *J. Immunol.* 176, 138–147.
9. Valentinis, B., A. Bianchi, D. Zhou, A. Cipponi, F. Catalanotti, V. Russo, and C. Traversari. (2005) Direct effects of polymyxin B on human dendritic cells maturation. The role of IkappaB-alpha/NF-kappaB and ERK1/2 pathways and adhesion. *J. Biol. Chem.* 280, 14264–14271.
10. Salio, M., V. Cerundolo, and A. Lanzavecchia. (2000) Dendritic cell maturation is induced by mycoplasma infection but not by necrotic cells. *Eur. J. Immunol.* 30, 705–708.

Chapter 25

Biomarkers Measuring the Activity of Toll-Like Receptor Ligands in Clinical Development Programs

Paul Sims, Robert L. Coffman, and Edith M. Hessel

Summary

Efforts to develop therapeutic approaches based on stimulation of Toll-like receptor (TLR) pathways have increased in recent years (Nat Med 13:552–559). The effectiveness of TLR agonists is currently being tested in diseases such as cancer, asthma, allergic rhinitis, and viral infections (J Clin Invest 117: 1184–1194; Blood 105: 489–495; Proc Am Thorac Soc 4:289–294; N Engl J Med 355:1445–1455; Am J Respir Crit Care Med 174:15–20). For a successful clinical trial program, it is important to know whether the therapeutic agent under development is both pharmacologically active and activating the intended pathway in humans. A biomarker reflecting this in an accurate and sensitive manner greatly facilitates dose/regimen-finding and is a "must-have." In this chapter, we describe a polymerase chain reaction (PCR)-based method that quantifies gene expression levels indicative of TLR-stimulation in human samples. We focus on genes specifically induced in an IFN-α-dependent manner, as this pathway is activated after stimulation of both TLR-7 and TLR-9. We demonstrate that IFN-α-inducible gene expression levels can be successfully applied in a clinical trial setting as a marker of drug activity in a variety of human samples, including peripheral blood mononuclear cells (PBMC), cells derived from the airways, as well as cells from induced-sputum.

Key words: Toll-like receptors, Biomarker, Quantitative PCR, Immunostimulatory DNA sequences, TLR-9, IFN-α pathway.

1. Introduction

During the last decade, great progress has been made in our understanding of the innate immune system and how it influences adaptive immune responses. Alongside, an increasing interest developed for the translation of this knowledge into innovative therapeutic approaches *(1)*. Several efforts have focused on using

Claire E. McCoy and Luke A.J. O'Neill (eds.), *Methods in Molecular Biology, Toll-Like Receptors, vol. 517*
DOI: 10.1007/978-1-59745-541-1_25

Toll-like receptor (TLR)-stimulation to modulate a variety of diseases *(2–6)*. TLRs recognize a diverse set of molecules derived from pathogenic microorganisms and play an important role in the innate response to infection *(7)*. Four of the 10 TLRs described so far recognize nucleic acids: TLR-3 recognizes double-stranded RNA, TLR-7 and TLR-8 recognize single-stranded RNA, whereas TLR-9 recognizes bacterial and viral DNA as well as synthetic oligonucleotides containing unmethylated CpG dinucleotides (CpG ODN) *(8)*. From a therapeutic perspective TLR-7 and TLR-9 are particularly interesting; both are highly expressed on human B cells and plasmacytoid dendritic cells (pDC), and stimulation of pDC through TLR-7 or -9 leads to very high levels of IFN-α production *(8–12)*. Activation of the IFN-α pathway provides an efficient way to trigger Th1 effects and to stimulate potent dendritic cell activation, resulting in a series of immunomodulatory effects that are attractive for the manipulation of various diseases. Preclinical animal model studies have demonstrated the capability of the synthetic TLR-9 ligand CpG ODN to eradicate tumors, elicit potent and efficient antiviral responses, and prevent allergen-induced symptoms, and based on these observations several clinical programs were started. To be able to establish whether a therapeutic candidate is pharmacologically active in man and activates the intended pathway, a biomarker is highly desirable. Moreover, the ability to measure pharmacological activity greatly facilitates dose and regimen finding studies. Even though IFN-α would be a logical choice as a biomarker indicative of TLR-7 or TLR-9 stimulation, its transient kinetics makes this impractical. In this chapter, we describe a detailed method that allows quantification of IFN-α-inducible genes using a 96-well quantitative polymerase chain reaction (PCR)-based method. We include examples that document the successful application of this method and the selected set of IFN-α-inducible genes as biomarkers in several clinical trials using a variety of human samples: peripheral blood mononuclear cells (PBMC), bronchoalveolar lavage (BAL) cells, and induced-sputum cells. The IFN-α-inducible biomarker genes demonstrated stable baseline expression levels and could be used to discriminate active from placebo-treated individuals and revealed dose-dependent increases in expression levels after drug treatment in both healthy and diseased individuals.

2. Materials

2.1. Sample Preparation

2.1.1. Peripheral Blood Mononuclear Cells

1. BD Vacutainer® CPT™ Cell Preparation Tubes (BD, Franklin Lakes, NJ). Store the tubes at 18–25°C. The tubes are good for 1 year from the date of manufacture.

2. Nuclease-free phosphate-buffered salts (PBS). Dilute 10x PBS 1:10 in nuclease-free water (Ambion).
3. Trypan blue (Sigma, St. Louis, MO).
4. RLT Lysis Reagent: add 10-μL 14.3 M 2-mercaptoethanol (Sigma) to 990-μL buffer RLT (included as part of the RNeasy kit (*see* **Subheading 2.2, item 2**)). Store at 18–25°C for up to 1 month after preparation (*see* **Note 1**). Buffer RLT contains guanidine thiocyanate which is incompatible with acids and harmful to aquatic organisms. Dispose of reagent according to local regulations. Do not expose to acids or sodium hypochlorite. Use appropriate laboratory procedures.

2.1.2. Whole Blood

1. PAXgene™ Blood RNA Tube (BD). Store the tubes at ambient temperature. The tubes are good for 18 months from the date of manufacture.

2.1.3. BAL Cells

1. Cell strainers (70 μm, BD).

2.1.4. Induced-Sputum Cells

1. Dithiothreitol (DTT) Working Solution: mix one part Sputolysin (phosphate-buffered DTT solution, Calbiochem-Novabiochem, La Jolla, CA) and nine parts distilled water. DTT is a mucolytic agent and DTT Working Solution needs to be freshly prepared before use. Sputolysin can be stored at room temperature but away from light. Sputolysin has a distinct odor and if the odor has gone after adding water, start over using a fresh bottle of Sputolysin.
2. Dulbecco's PBS: addition of Dulbecco's PBS dilutes and stops the DTT reaction.
3. Cell strainers (40 μm, BD).

2.2. mRNA Preparation

2.2.1. mRNA Preparation from Lysed PBMC, BAL, and Induced-Sputum Cells

1. QIAshredder™ homogenizer columns (Qiagen, Valencia, CA).
2. RNeasy® Mini Kit (Qiagen): includes buffers RLT, RPE, and RW1; RNase-free water; spin columns; and collection tubes. Additional 2-mL collection tubes are ordered separately. Buffer RLT contains guanidine thiocyanate (see safety precautions in **Subheading 2.1.1**). Buffer RPE contains guanidine hydrochloride that is an irritant. Use appropriate laboratory procedures.
3. Seventy percent ethanol: dilute 100% ethanol (AAPER, Shelbyville, KY) to 70% in nuclease-free water. Store at –12°C to –20°C. Use ultra-pure 100% ethanol. Do not use commercially available 70% ethanol.
4. RPE Wash Reagent: add 220 mL of 100% ethanol to the RPE buffer to bring the volume to 275 mL. Store at 18–25°C for up to 1 week.

2.2.2. mRNA Preparation from Whole Blood

1. PAXgene Blood RNA Kit (Qiagen): includes buffers BR1, BR2, BR3, BR4, BR5, RDD; RNA spin columns; shredder spin columns; DNase I; proteinase K; RNase-free water; BD Hemogard™ closures; and collection tubes. Buffers BR2 and BR3 contain guanidine thiocyanate (*see* safety precautions in **Subheading 2.1.1**). DNase I (bovine) is a sensitizing agent. Proteinase K is an irritant and sensitizing agent. Use appropriate laboratory procedures.
2. DNase I Incubation Mix: add 10-μL DNase I stock solution to 70-μL buffer RDD in a 1.5-mL microcentrifuge tube. Mix gently by flicking the side of the tube, and keep on ice. Do not vortex as DNase I is sensitive to physical denaturation.

2.3. cDNA Preparation

1. Acetylated BSA (Invitrogen, Carlsbad, CA). Aliquot and store at –12°C to –20°C
2. DNase I, RNase-free 10–50 U/μL (Roche Applied Sciences, Indianapolis, IN). Store at –12°C to –20°C. DNase I is a sensitizing agent. Use appropriate laboratory procedures.
3. Superscript™ II RNaseH reverse transcriptase (Invitrogen). Includes 5x first strand buffer and DTT. Store at –12°C to –20°C.
4. Recombinant RNasin (rRNasin) 20–40 U/μL (Promega, Madison, WI). Store at –12°C to –20°C.
5. Phosphorylated $d(T)_{12-18}$ (GE Healthcare, Piscataway, NJ). Dilute to 25 A_{260} units/mL with nuclease-free water. Aliquot and freeze at –12°C to –20°C.
6. Random Hexamers (Invitrogen). Store at –12°C to –20°C.
7. dNTP (GE Healthcare, Piscataway, NJ). Mix equal amount of the dCTP, dGTP, dATP, and dTTP, aliquots, and store at –12°C to –20°C.
8. Reaction mixes: the preparation described below provides enough material for one well; therefore, increase the volumes proportionally to make enough material for the number of samples plus additional overage. All reagents should be mixed in nuclease-free polypropylene tubes on ice. Vortex and use immediately.
   (a) DNase Mix: 12-μL 5x first strand buffer, 0.4-μL rRNasin, 4-μL DNase, and 1.6-μL nuclease-free water.
   (b) Oligo d(T)/Random Hexamer Mix: 6-μL pd(T) and 6-μL random hexamer.
   (c) RT PCR Mix: 10.8-μL nuclease-free water, 12-μL 5x first strand buffer, 12-μL DTT, 6-μL dNTPS, 0.6-μL rRNasin, and 3.6-μL Superscript II reverse transcriptase (*see* **Note 2**).

2.4. Quantitative Gene Measurements

Quantitative gene measurements can be done by either using the SYBR green PCR method using the DNA-binding dye SYBR green combined with a set of primers, or with a probe PCR method using a specific probe combined with a set of primers. The SYBR green PCR method is considerably cheaper, whereas the probe PCR method has an increased specificity as both the primers and probe need to bind the amplicon.

1. Quantitect™ SYBR green PCR kit (Qiagen): includes 2x SYBR green master mix and nuclease-free water. Store kits at −20°C.
2. SYBR green Primer Mix: dilute the forward and reverse primers (Operon, Huntsville, AL) to 2 μM each in water. Aliquot and store at −20°C. See **Table 1** for list of primer sets applied.
3. SYBR green Reaction Mix: combine 12.5-μL SYBR green master mix and 2.5-μL SYBR green Primer Mix and keep on ice.
4. Quantitect Probe PCR kit (Qiagen): includes 2x Probe master mix and nuclease-free water. Store kits at −20°C.
5. Primer/probe sets for the probe PCR method are called Taqman Gene Expression Assays (Applied Biosystems, Foster City, CA). Store at −20°C. See **Table 1** for list of Taqman Gene Expression Assays used.
6. Probe Reaction Mix: mix 1.25-μL nuclease-free water, 12.5-μL 2x probe master mix, and 1.25-μL Taqman Gene Expression Assay reagent, and keep on ice.
7. For both SYBR green Reaction Mix and Probe Reaction Mix, the final volume is enough for one reaction; therefore, increase the volumes proportionally to make enough material for the number of samples plus additional overage.

Table 1

Gene Name		Primer Sequences[a]	COV[b]	Melt Temp[c]
Endogenous Controls				
HPRT	Hypoxanthine-Guanine Phosphoribosyl Transferase	TaqMan Endogenous Control (4333768) d	39	N/A[c]
UBI	Ubiquitin	Forward: CACTTGGTCCTGCGCTTGA Reverse: CAATTGGGAATGCAACAACTTTAT	35	78.5°C

(continued)

Table 1 (continued)

Gene Name		Primer Sequences	COV	Melt Temp
IFN / IFN inducible genes upregulated by CpG				
IFN-α	Interferon-alpha	Forward: CCCAGGAGGAGTTTGGCAA Reverse: TGCTGGATCATCTCATGGAGG	37	79.5°C
2', 5'-OAS	2',5'-Oligoadenylate Synthetase	Forward: AGGGAGCATGAAAACACATT-TCA Reverse: TTGCTGGTAGTTTAT-GACTAATTCCAAG	31	78.5°C
GBP-1	GTP-Binding Protein 1	Forward: TGGAACGTGTGAAAGCT-GAGTCT Reverse: CATCTGCTCATTCTTTCTTT-GCA	31	79.0°C
ISG-54	IFN-Stimulated Gene 54kDa (IFIT2)	Forward: CTGGACTGGCAATAG-CAAGCT Reverse: AGAGGGTCAATGGCGTTCTG	33	81.0°C
IP-10	Inducible Protein 10kDa (CXCL10)	TaqMan Gene Expression Assay (Hs99999049_m1)	37	N/A
Mx-B	Myoxvirus Resistance Protein B (MX2)	Forward: GAGACATCGCACTGCAGAT Reverse: GTGGTGGCAATGTC-CACGTTA	34	82.0°C
IRF-7	IFN-Regulatory Factor 7	Forward: CTGTTTCCGCGTGCCCT Reverse: GCCACAGCCCAGGCCTT	34	87.5°C
MCP-1	Monocyte Chemotactic Protein 1 (CCL2)	TaqMan Gene Expression Assay (Hs99999050_mH)	34	N/A
MCP-2	Monocyte Chemotactic Protein 2 (CCL8)	TaqMan Gene Expression Assay (Hs99999026_m1)	34	N/A
MCP-3	Monocyte Chemotactic Protein 3 (CCL7)	TaqMan Gene Expression Assay (Hs99999077_mH)	36	N/A
MCP-4	Monocyte Chemotactic Protein 4 (CCL13)	TaqMan Gene Expression Assay (Hs99999076_m1)	37	N/A

[a] Genes that have primer sequences listed were quantified using a SYBR green detection system while other genes were quantified using a hybridization probe system (TaqMan). The sequences for the TaqMan reagents are proprietary (Applied Biosystems, Foster City, CA).

[b] COV - Cut-Off Value. The Ct value above which the gene can no longer be accurately measured.

[c] Melt Temp - Melting Temperature. The Temperature at which the amplification product disassociates.

[d] Applied Biosystems catalog number.

[e] N/A - Not Applicable. Melting temperatures can not be determined for genes quantified using the hybridization probe system.

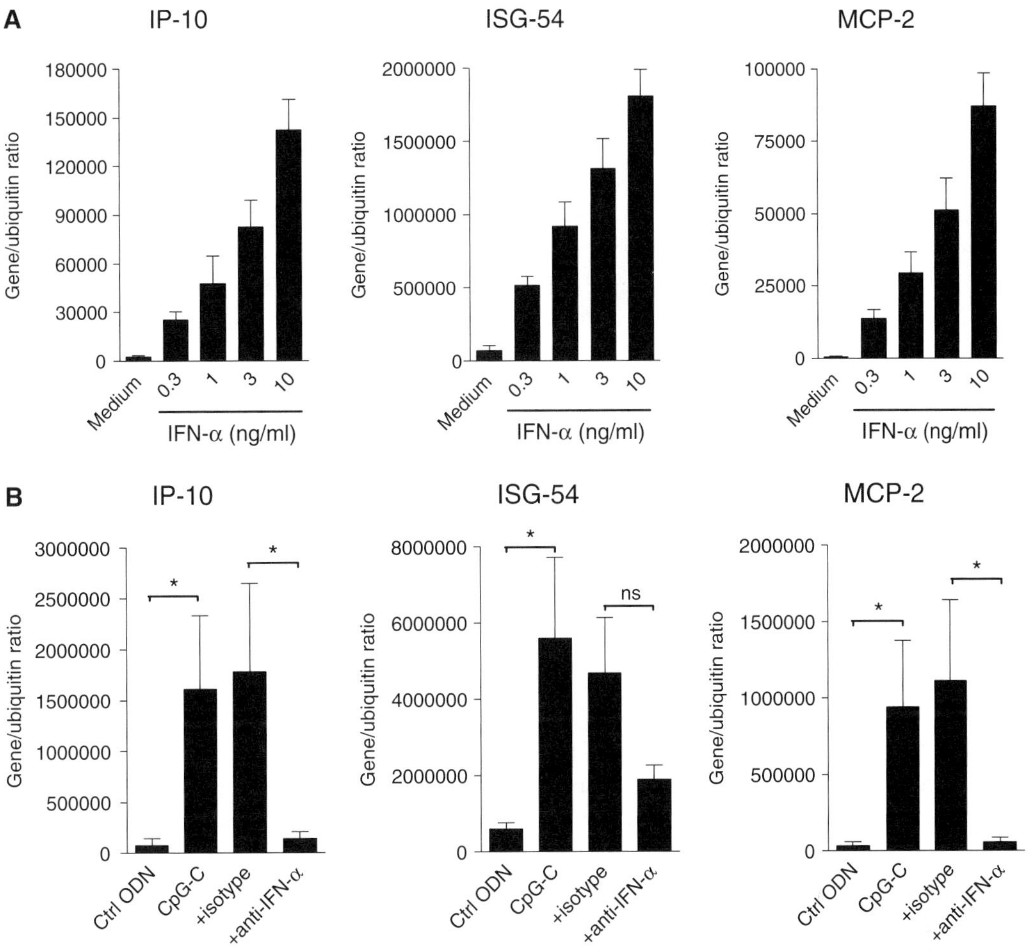

Fig. 1. Identification of genes induced by stimulation of the Toll-like receptor 9 (TLR-9) receptor pathway in human peripheral blood mononuclear cells (PBMC) in an IFN-α-dependent manner. (**a**) Human PBMC from five healthy donors were plated in 96-well plates at 2×10^6 cells/mL and incubated with medium or increasing concentrations of human IFN-α (PBL Biomedical Laboratories, Piscataway, NJ). Cells were harvested for quantitative PCR measurements after 6 h. Incubation of human PBMC with IFN-α increased the expression of IP-10, ISG-54, and MCP 2 mRNA levels in a dose-dependent fashion. (**b**) Human PBMC from five healthy donors were plated in 96-well plates at 2×10^6 cells/mL and incubated for 6 h with control ODN (1040: 5 -TGA CTG TGA ACC TTA GAG ATG A; 5 μg/mL), with CpG-C ODN (C274: 5 -TCG TCG AAC GTT CGA GAT GAT; 5 μg/mL), with C274 (5 μg/mL) plus a cocktail of isotype control antibodies (R & D Systems, Minneapolis, MN), or with C274 (5 μg/mL) plus a cocktail of anti-IFN-α, anti-IFN-β, and anti-IFNαβ-receptor antibodies (respectively 5 μg/ mL, 2 μg/mL, 5 μg/mL, all from PBL Biomedical Laboratories). CpG-C stimulation significantly increased mRNA expression levels of IP-10, ISG-54, and MCP-2, whereas coincubation with a cocktail of anti-IFN-antibodies completely prevented the induction of IP-10 and MCP-2 by CpG-C, demonstrating their dependency on IFN-α. The CpG-C-mediated induction of ISG-54 was largely reduced by incubation with anti-IFN-antibodies, however, not completely to background levels indicating a partial dependence on other factors. Data shown are mean ± SEM for five donors. *indicates a *p* value less than 0.05, *ns* indicates not significant. Data were analyzed with an ANOVA for repeated measurements followed by Bonferroni or a Dunn's multiple comparison posttest (GraphPad InStat).

3. Methods

Stimulation of the TLR-7 or TLR-9 pathway leads to rapid induction of large amounts of IFN-α which is downregulated within the first 12 h after stimulation *(13)*. The transient expression of IFN-α makes IFN-α itself impractical as biomarker. In contrast, proteins induced by IFN-α display a more prolonged kinetics and are therefore more applicable. To select a series of IFN-inducible biomarker genes, we stimulated human PBMC with various concentrations of IFN-α and evaluated which genes were induced after 6 h (*see* **Fig.1A** for example genes IP-10, ISG-54, and MCP-2). Out of a large panel of genes tested we selected a group of ten genes that fulfilled the following criteria: *(1)* induced at significant expression levels; *(2)* induced in a dose-dependent fashion (rather than on/off); and *(3)* reproducibly induced in all donors (*see* **Table 1** for list of genes). We confirmed whether these genes were induced after activation of the TLR-9 pathway by incubating human PBMC with CpG in the presence or absence of IFN-α and IFN-α receptor antibodies, and demonstrated that the genes selected were indeed induced by CpG in an IFN-α-dependent fashion (*see* **Fig.1B**).

3.1. Sample Preparation

The quality of the signal measured is greatly dependent on sample preparation. Below is a detailed description of the sample preparation for a variety of cell types that may be isolated during the course of a clinical trial. Using these methods we have been able to measure IFN-inducible gene expression levels in a sensitive and reproducible fashion (*see* also **Figs.3–6**).

3.1.1. Peripheral Blood Mononuclear Cells

1. Collect 6–8-mL blood into a 10-mL CPT tube using the standard technique for Vacutainer Blood Collection Tubes (*see* **Note 3**). Use two 4-mL CPT tubes if the centrifuge in **step 3** does not accommodate the 10-mL tubes. The CPT tube must be at room temperature (18–25°C).
2. Store the CPT tubes upright at room temperature until centrifugation. For best results, blood samples should be processed within 2 h of collection.
3. Centrifuge blood sample at room temperature in a horizontal rotor (swing-out head) for a minimum of 20 min at 1,500–1,800 RCF. The brake must be off! Use the following equation for centrifuges without an RCF setting:

$$\text{RPM} = \sqrt{\frac{(\text{RCF} \times 100{,}000)}{(1.12 \times r)}}$$

where r = radius (cm) from the centrifuge post to the tube bottom when the tube is in the horizontal position.

4. Collect the PBMC, which are the whitish layer under the plasma, using a Pasteur pipette and transfer to a 15-mL centrifuge tube. Aspirate and discard approximately half of the plasma layer without disturbing the cell layer. Carefully collect the whitish layer by placing the tip of the pipette immediately above the layer and slowly aspirate the cells. Do not have the pipette tip below the layer or aspirate the cells too quickly. If 2 CPT tubes were used, combine the cells in the same 15-mL centrifuge tube.
5. Bring the volume up to 10 mL with cold PBS and mix well by inverting the tube five times.
6. Determine the cell concentration. Transfer 50 μL of the cell suspension to a microcentrifuge tube. Add 50 μL of trypan blue and let stand for 60 s and transfer 10 μL of trypan/cells to a hemacytometer.
7. Count the cells in each of the four corners of the hemacytometer (each corner has 4 × 4 squares) and average the four cell counts. Determine the cell concentration using the following equation:

$$\text{Cells/mL} = (\text{averagecellcount}) \times (D) \times \left(10^4 \text{cells/mL}\right)$$

The dilution factor (D) is 2.

8. Make 1.5 × 10⁶ cells aliquots (*see* **Note 4**). Determine the volume of the cell suspension from **step 5** that contains 1.5 × 10⁶ cells.

$$\text{Volume} = \frac{1.5 \times 10^6 \text{cells}}{\text{Cells/mL}}$$

Pipette this calculated volume of cell suspension into fresh 15-mL centrifuge tubes.

9. Perform cell lysis with RLT Lysis Reagent (*see* **Note 1**). Bring the volume in each aliquot to 15 mL with PBS. Centrifuge at 400 RCF for 5 min at 4°C with the brake on.
10. Remove and discard the supernatant by inverting the tube. Keep the tube inverted and, using a pipette, remove any excess liquid. Do not touch the cell pellet.
11. Resuspend the cell pellet into 600-μL RLT buffer with β-mercaptoethanol (RLT Lysis Reagent). Mix carefully by pipetting up and down at least four times with a 1-mL pipette (*see* **Note 5**). Use filter tips to prevent contaminating the pipette.
12. Immediately transfer the lysed cells to a ~60°C freezer (*see* **Note 6**).

3.1.2. Whole Blood

1. Collect blood in a PAXgene Blood RNA Tube using standard phlebotomy techniques for Vacutainer blood collection tubes.
2. Let the blood tubes incubate at 18–25°C for at least 2 h.
3. The blood may be immediately processed or frozen at –20°C or <–60°C.

3.1.3. BAL Cells

1. BAL is usually collected in a so-called BAL trap. After collection, place the BAL trap immediately on ice.
2. Prepare several 50-mL tubes (depending on the total volume of BAL collected), and place a cell strainer on top of each tube.
3. Transfer the collected BAL from the trap into the 50-mL tubes with a 25-mL pipette. Do not transfer more than 40 mL of BAL to a tube.
4. Rinse the BAL trap and cell strainer with 10 mL cold PBS and add to the corresponding 50-mL tube.
5. Centrifuge tubes at 4°C for 10 min at 400 RCF with the brake off. Carefully pour off the remaining supernatant.
6. Resuspend the cell pellet in each tube with 4 mL cold PBS and pool the cell suspensions in one of the 50-mL tubes.
7. Rinse the empty 50-mL tubes sequentially with an additional 4 mL of cold PBS and transfer to the 50-mL tube already containing the pooled cells.
8. Divide the pooled cells between two 15-mL tubes and centrifuge tubes at 4°C for 10 min at 400 RCF with the brake on. Carefully pour off the remaining supernatant.
9. Resuspend each cell pellet in 0.4 mL cold PBS and combine the cell pellets in one of the tubes. Carefully resuspend cells up and down with a 1-mL pipette.
10. Rinse the empty tube with 0.4 mL cold PBS and combine with the cell suspension in the second tube.
11. Carefully measure the volume and bring the volume up to 2.0 mL with cold PBS.
12. Determine the cell concentration by transferring 10 μL of the cell suspension to a microcentrifuge tube. Add 10 μL of trypan blue and let stand for 60 s. Transfer 10 μL of trypan/cells to a hemacytometer.
13. Follow **steps 7–12** from **Subheading 3.1.1**.

3.1.4. Induced-Sputum Cells

1. Weigh a 15-mL tube and record the weight (pre-weight).
2. Transfer the collected sputum sample to a petri dish and select all portions of sputum that are free of saliva. Place the sample in the lid of the petri dish and dry the sample as much as possible by twirling it around the petri dish.

3. Transfer the sputum sample to the preweighed 15-mL tube and weigh again. Record the weight (post-weight).

$$\text{Post-weight} - \text{pre-weight} = \text{weight of cell plug}$$

4. Add a volume of freshly prepared DTT Working Solution equal to four times the weight of the cell plug, in mL. For example: if the weight of cell plug is 435 mg, use 4×435 mg = 1.740 mL DTT working solution
5. Vortex for 30 s and gently aspirate the cell plug with a Pasteur pipette (in/out, in/out).
6. Incubate the mixture on a roller mixer at room temperature for 15 min.
7. Add a volume of Dulbecco's PBS equal to four times the weight of the cell plug, in mL.
8. Vortex for 15 s and incubate the mixture on a roller mixer at room temperature for 5 min.
9. Filter the mixture through a 40 μm cell strainer into a fresh-15-mL tube.
10. Carefully measure the volume.
11. Determine the cell concentration. Transfer 10 μL of the cell suspension to a microcentrifuge tube. Add 10 μL of trypan blue and let stand for 60 s. Transfer 10 μL of trypan/cells to a hemacytometer.
12. Follow **steps 7–12** from **Subheading 3.1.1**.

3.2. mRNA Preparation

3.2.1. Preventing RNase Contamination and Cross-Contamination

1. Use clean lab coats and gloves. Change gloves whenever there is the possibility of surface contamination.
2. Use pipettes that are specifically reserved for mRNA processing. Use filter barrier tips at all times.
3. Wipe all equipment with an RNase cleaner (i.e., RNase Away (Fisher) or RNase Zap (Ambion)).
4. Avoid sneezing, coughing, or talking above the samples.

3.2.2. mRNA Preparation from Lysed PBMC, BAL, or Induced-Sputum Cells

The following procedure uses the RNeasy kits, Animal Cell I protocol; see the RNeasy Mini Handbook included with the kits for more details:

1. Thaw the lysed cell samples at room temperature until just thawed and proceed immediately to the next step.
2. Transfer no more than 700 μL of the lysed cells to the top reservoir of a QIAshredder column.
3. Centrifuge at 16,300 RCF for 2 min and retain the collection tube with the homogenized cell extract (flow-through).

4. Dilute the cell extract 1:1 by adding an equal volume of 70% ethanol to the collection tube. Mix thoroughly by pipetting up and down a minimum of three times.
5. Transfer no more than 700 μL of the diluted cell extract to the top reservoir of an RNeasy Mini Column.
6. With a fresh 2-mL collection tube in place, centrifuge at 9,300 RCF for 1 min. Discard the collection tube and flow-through. The flow-through contains RLT buffer (*see* toxicity and waste disposal warning in **Subheading 2.1.1**).
7. Repeat **steps 5** and **6** until all of the diluted cell extract has been passed through the same RNeasy column (*see* **Note** 7).
8. Transfer 700-μL Buffer RW1 to the top reservoir of the RNeasy column.
9. With a fresh 2-mL collection tube in place, centrifuge at 9,300 RCF for 1 min. Discard the collection tube and flow-through.
10. Transfer 500-μL prepared RPE Wash Reagent to the top reservoir of the RNeasy column.
11. With a fresh 2-mL collection tube in place, centrifuge at 9,300 RCF for 1 min. Discard the collection tube and flow-through.
12. Repeat **steps 10** and **11**.
13. With a fresh 2-mL collection tube in place, centrifuge at 16,300 RCF for 2 min to remove any residual ethanol from the column. Discard the collection tube and flow-through.
14. Transfer 25 μL of nuclease-free water to the middle of the membrane at the base of the reservoir of the RNeasy column. Incubate no less than 5 min at room temperature to increase mRNA yield.
15. With a fresh 1.5-mL collection tube in place, centrifuge at 9,300 RCF for 1 min to elute the mRNA. Do not discard eluate or change the collection tube.
16. Repeat **steps 14** and **15** into the same collection tube.
17. Add 40 μL of nuclease-free water to the eluate and mix by pipetting up and down a minimum of three times.
18. To create a backup sample, transfer 45 μL to a fresh 1.5-mL collection tube and freeze at <–60°C.
19. The primary mRNA sample may be immediately reverse transcribed to cDNA (*see* **Subeading 3.3**), or stored overnight at 2–8°C, or stored up to a week at < –60°C.

3.2.3. mRNA Preparation from Whole Blood (See Note 8)

The following procedure uses the PAXgene blood RNA kits; see the PAXgene Handbook included with the kits for more details:

1. Let the PAXgene tubes warm to ambient temperature (18–25°C).

2. Centrifuge the PAXgene tube for 10 min at 3,000–5,000 RCF using a swing-out rotor.
3. Carefully remove and discard the Hemogaurd tube closure. Aspirate and discard the supernatant with care as not to disturb the pellet.
4. Add 4-mL nuclease-free water and close the tube with a new Hemogaurd closure. Vortex until the pellet is dissolved.
5. Centrifuge the PAXgene tube for 10 min at 3,000–5,000 RCF using a swing-out rotor. Aspirate and discard the supernatant without disturbing the pellet. It is critical to remove as much of the water as possible.
6. Add 350-μL BR1 and vortex until the pellet is dissolved. Transfer the sample to a 1.5-mL centrifuge tube.
7. Add 300-μL BR2 and 40-μL proteinase K.
8. Vortex for 5 s and incubate for 10 min at 55°C on a shaker incubator.
9. Pipette the lysate to a PAXgene shredder column.
10. With a 2-mL collection tube in place, centrifuge for 3 min at 16,300 RCF.
11. Transfer the supernatant of the flow-through to a fresh 1.5-mL microcentrifuge tube. Do not disturb the pellet.
12. Add 350 μL of 100% ethanol and mix by vortexing.
13. Centrifuge for 1–2 s at 500–1,000 RCF to remove drops from the inside of the tube.
14. Transfer 700 μL of the sample to a PAXgene spin column.
15. With a fresh 2-mL collection tube in place, centrifuge for 1 min at 9,300 RCF.
16. Discard the flow-through and collection tube and transfer the remaining sample from **step 13** to the spin column. Repeat **step 15**.
17. Add 350-μL BR3 to the spin column.
18. With a fresh 2-mL collection tube in place, centrifuge for 1 min at 9,300 RCF.
19. Add 80 μL of the DNase I Incubation Mix directly to the spin column membrane.
20. Incubate at 20–30°C for 15 min.
21. Add 350-μL BR3 to the spin column.
22. With a fresh 2-mL collection tube in place, centrifuge for 1 min at 16,300 RCF. Discard the flow-through and collection tube.
23. Add 500-μL BR4 buffer to the spin column.

24. With a fresh 2-mL collection tube in place, centrifuge for 1 min at 16,300 RCF. Discard the flow-through and collection tube.
25. Repeat **step 24**. Replace the collection tube with a 1.5-mL centrifuge tube.
26. Pipette 40 µL of BR5 directly to the spin column membrane.
27. Centrifuge for 1 min at 9,300 RCF.
28. Repeat **steps 26** and **27** using the same 1.5-mL centrifuge tube.
29. Discard the spin column and incubate the eluate for 5 min at 65°C in a shaker incubator. Chill immediately on ice.
30. The mRNA samples may be immediately reverse transcribed to cDNA (*see* **Subheading 3.3**), or stored overnight at 2–8°C, or stored up to a week at < –60°C

3.3. cDNA Preparation

3.3.1. General Precautions During cDNA Preparation

1. Apply the precautions described in **Subheading 3.2.1** to prevent RNase contamination and cross-contamination.
2. After adding each reagent to the PCR plate, tap gently to cause all liquids to drop below the lip of the tube.

3.3.2. DNase Treatment

1. Transfer 40 µL of eluate from **Subheading 3.2.2** or **3.2.3** to the appropriate well of a 96-well PCR plate.
2. Add 2 µL of 0.5 mg/mL acetylated BSA to each well (*see* **Note 9**).
3. If the source of the mRNA is from whole blood, add 18-µL nuclease-free water and proceed directly to **Subheading 3.3.3**. The DNase step was already performed during the mRNA preparation.
4. If the source of the mRNA is not from whole blood, add 18-µL DNase Mix to each well and cover the plate with PCR caps.
5. Centrifuge the plate at 5,700 RCF for 30 s in a centrifuge that can be fitted with a plate rotor (e.g. Qiagen Model 4–15C).
6. Place the plate in a PCR instrument (e.g. iCycler) and perform the protocol listed below:

DNase Treatment	Thirty-seven degree Celsius for 20 min
Protocol	Seventy degree Celsius for 10 min
(60-µL Volume)	Four degree Celsius for 60 min[a]

[a]This step can be stopped after 5 min.

7. Centrifuge the plate at 5,700 RCF for 30 s.

3.3.3. Priming the mRNA with Oligo dT and Random Hexamers

1. Carefully remove the caps, taking care not to splash samples.
2. Add 12-µL Oligo dT/Random Hexamer Mix to each well and cover the plate with PCR caps.
3. Centrifuge the plate at 5,700 RCF for 30 s.
4. Place the plate in a PCR instrument (e.g. iCycler) and perform the protocol listed below.

Oligo dT/Random Hexamer	Seventy degree Celsius for 10 min
Protocol	Four degree Celsius for 60 min[a]
(72-µL Volume)	

[a]This step can be stopped after 5 min.

5. Centrifuge the plate at 5,700 RCF for 30 s.

3.3.4. Reverse Transcription

1. Carefully remove the caps, taking care not to splash samples.
2. Add 45 µL of a premixed cocktail of RT Mix to each well (*see* **Note 2**) and cover the plate with PCR caps.
3. Centrifuge the plate at 5,700 RCF for 30 s.
4. Place the plate in a PCR instrument (e.g. iCycler) and perform the protocol listed below.

Reverse Transcription	Forty-five degree Celsius for 50 min
Protocol	Seventy degree Celsius for 15 min
(100-µL Volume)	Four degree Celsius for 60 min[a]

[a]This step can be stopped after 5 min.

5. Centrifuge the plate at 5,700 RCF for 30 s.
6. Store the plate overnight at 2–8°C before proceeding to the next step (*see* **Note 10**).

3.4. Quantitative Gene Measurements

To be able to detect gene expression in a sensitive and reproducible fashion, the quality of the primer sets used is crucial. We carefully evaluated our primer sets for specificity by sequencing the product of the PCR reaction for each primer set (*data not shown*). In addition, we tested the linearity of the signal generated by the primer set and determined a cut-off value (COV). Per definition, the COV is reached when the sample dilution reaches the point where the gene of interest can no longer be accurately measured; therefore, the COV is inversely proportionate to the minimum copy number that can be quantified for a particular gene. The COV results for two example genes are shown in **Fig. 2** and the COV for all biomarker genes described in this chapter are shown in **Table 1**. We recommend that each lab establishes its own COVs using its reagents and equipment.

3.4.1. Determination of the COV for Each Primer Set

1. Dilute a cDNA sample which has a high level of the gene of interest twofold serially in water for 12 points.
2. Measure the gene of interest by real-time quantitative PCR (*see* **Subheading 3.4.4**).
3. Plot the Cycle Threshold or C_t values generated by the software versus the sample dilution on a linear-logarithmic scale (*see* **Fig. 2**).
4. Determine the C_t at which the signal is no longer linear. This C_t is the COV.

3.4.2. General Recommendations for Quantitative Gene Measurements

1. Use pipettes designated for PCR only, combined with filter barrier tips.
2. Prepare PCR plates and reagents on ice. Be sure not to contaminate the samples with water from the ice.
3. Multiple plates can be prepared (*see* **Subheading 3.4.4, steps 1–6**) and stored at 2–8°C for up to 1 week.

3.4.3. Measurement of and Correction for House-Keeping Gene Expression Levels

Before measuring genes of interest it is necessary to measure the level of house-keeping gene expression to determine the range of cDNA levels within a sample set. If a particular sample set displays a wide range cDNA levels, we strongly recommend adjusting the

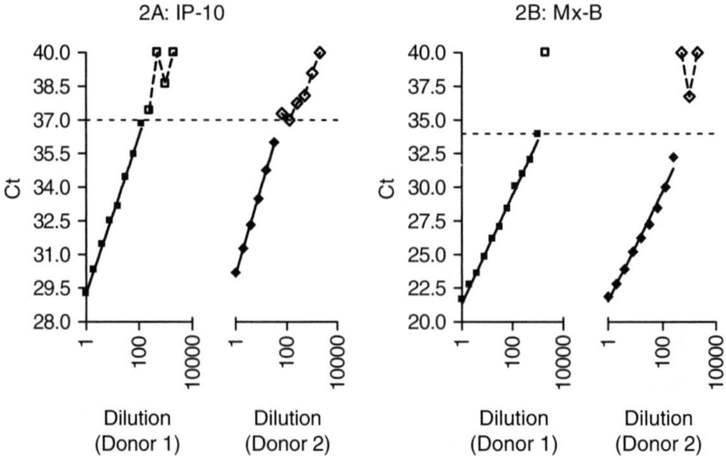

Fig. 2. Cut-off value (COV) values for IP-10 and Mx-B. cDNA transcribed from mRNA isolated from peripheral blood mononuclear cells (PBMC) of two healthy donors (■ and ♦, respectively) was diluted twofold serially in water and the levels of IP-10 and Mx-B measured. The COV values (- - - *line*) were determined to be 37 and 34 for IP-10 and Mx-B, respectively, as the C_t values above the COV were neither linear nor accurate.

dilution of cDNA for each sample, making samples within a set more comparable (*see* **Note 11**).

1. Centrifuge the cDNA samples resulting from **Subheading 3.4.4, step 6** at 5,700 RCF for 30 s.
2. Carefully remove the caps and prepare a 1:10 dilution of the cDNA by mixing 5-μL cDNA with 45-μL nuclease-free water in a fresh 96-well PCR plate.
3. Cover the source cDNA plate with PCR caps and store at 2–8°C.
4. Cover the diluted cDNA plate with PCR plate caps as well and store the plate overnight at 2–8°C before continuing to the next step (*see* **Note 10**).
5. Test each cDNA sample in triplicate.
6. Add 15-μL SYBR green Reaction Mix containing Ubiquitin primers to the required number of wells in a fresh 96-well optical PCR plate.
7. Add 10 μL of the 1:10 diluted cDNA (from **step 4**) to the appropriate wells.
8. Cover the plates with optical PCR plate caps and centrifuge at 5,700 RCF for 30 s. The plate may be stored at 2–8°C for up to 1 week.
9. Perform the PCR reaction on a real-time quantitative PCR instrument using the thermocycler conditions listed under **Subheading 3.4.4**, **step** 7.
10. Confirm that the melting temperature of the PCR products is within ±1°C from the expected temperature for ubiquitin (**Table 1**). If not, exclude this particular well from the analysis.
11. In the PCR analysis software, set the appropriate baseline cycles and threshold (*see* **Note 12**).
12. For each sample in triplicate, calculate the mean C_t value and standard deviation.
13. Calculate the average of the mean C_t for all samples that are considered part of the same experiment (*see* **Note 13**).
14. Determine the final cDNA dilution for each sample using the following equations:

$$X=[(\text{mean}\,C_t)-(\text{average of mean}\,C_t\,\text{for all samples}$$
$$\text{(rounded to the nearest whole number)}$$

If X is ≥–2 and ≤+2, the final dilution of cDNA will be 1:10 for the deep-well plates.

If X is <–2, the final dilution will be 1: ($10 * 1.8^{|X|}$).

If X is ≥+2, perform a 1:5 dilution.

If the X is larger than 3 the sample quality is questionable and the results for that sample should flagged.

15. In a 1.5-mL polypropylene block, dilute the remaining source cDNA (from **Subheading 3.3.4, step 6**) in water using the dilution values calculated in the previous step (*see* **Note 14**).
16. Cover the block with a cover and let rest overnight at 2–8°C (*see* **Note 10**).
17. To prevent contamination and multiple freeze–thaw cycles, create a working sample block by transferring an appropriate amount of diluted cDNA to a second 96-well backup block.
18. Freeze the backup block at <–60°C and continue onto the next section with the working sample block.

3.4.4. Measuring Genes of Interest by Real-Time Quantitative PCR

1. To obtain accurate results with the genes of interest, ubiquitin expression levels need to be re-measured for all samples in the working sample block.
2. For each gene, the samples are tested in duplicate; however, ubiquitin is measured in quadruplicate.
3. Add either 15 µL of the SYBR green Reaction Mix or 15 µL of the Probe Reaction Mix to the required number of wells in an optical PCR plate.
4. Tap the plate gently to cause all liquids to drop below the lip of the tube.
5. Add 10 µL of the diluted cDNA sample from the working sample block to the wells of the PCR plate (*see* **Notes 14** and **15**).
6. Cover the PCR plate with an optical PCR plate cover and centrifuge at 5,700 RCF for 30 s.
7. Perform quantitative on a real-time PCR instrument (e.g. MyIQ™ with MyIQ Optical System Software, Bio-Rad Laboratories, CA) using the following thermocycle protocols:

3.4.5. Data Analysis

1. Confirm that the melting temperature of the PCR products is within ±1°C from the expected temperature (**Table 1**). If not, exclude this particular well from the analysis.
2. In the PCR analysis software, set the appropriate baseline cycles and threshold (*see* **Note 12**).
3. The software will generate C_t values for each sample and if the C_t value for a particular sample is greater than the COV (**Table 1**), that sample for the specific gene will be assigned the C_t value equal to the COV.

SYBR Green Method	First Cycle: Taq Polymerase Activation
	95°C for 15 min
	Second Cycle: PCR Amplification (40 repeats)
	Step 1: 95°C for 15 s
	Step 2: 60°C for 1 min (Fluorescence Read)
	Third Cycle: Melt Curves (80 repeats)
	Start at 60°C
	Increase temperature by 0.5°C – Dwell Time of 10 s
Probe Method	First Cycle: Taq Polymerase Activation
	95°C for 15 min
	Second Cycle: PCR Amplification (40 repeats)
	Step 1: 95°C for 15 s
	Step 2: 60°C for 1 min (Fluorescence Read)

For example: Sample 6 has a C_t value of 38 for IP-10. Since the COV for IP-10 is 37, sample 6 is assigned a C_t value of 37 for calculation purposes.

4. For each gene, calculate the mean C_t value and standard deviation between duplicates (or quadruplicates).
5. If the standard deviation is greater than 1, then it is recommended to repeat the sample in duplicate (or quadruplicate) to determine the C_t value.
6. Calculate the gene/ubiquitin ratio using the following equation:

$$\text{genetoubiratio} = 100,000 * 1.8^{\left(C_{t\ \text{ubi}} - C_{t\ \text{gene}}\right)}$$

where

- 1.8 is the fold increase in PCR product per cycle.
- $C_{t\ \text{ubi}}$ is the ubiquitin C_t value determined for the sample.
- $C_{t\ \text{gene}}$ is the gene C_t value for the sample.
- An arbitrary multiplication factor of 100,000 is used when normalizing to ubiquitin.

3.4.6. Data Management

One of the advantages of the quantitative gene measurement technique described here is that multiple genes can be measured on one working sample block. We strongly recommend storing the data in database-form and to perform data calculations using automated formulas as much as possible.

3.5. Using Quantitative Gene Measurement as Read Out for Clinical Trials

3.5.1. Stability of Biomarker Gene Expression Levels

The gene/ubiquitin ratio is directly proportional to the fraction of total mRNA represented by the test gene and normalizes for experimental variations in RNA recovery, purification, and quality. In our hands this parameter is very robust:

1. We have confirmed that the expression of ubiquitin is linear over a broad range of RNA concentrations by processing human PBMC samples under identical conditions ranging from 5×10^3 to 10^7 cells (*data not shown*).
2. The standard deviation of the gene/ubiquitin ratio in replicate samples is generally less than 50% of the mean of the replicates.
3. We compared gene/ubiquitin ratios measured in two human PBMC samples collected 1–3 weeks apart in untreated individuals and observed that there is a strong correlation between the gene/ubiquitin ratio measured in the first and second sample (correlation coefficient $r = 0.9706$, $p < 0.0001$, *see* **Fig. 3**).

The latter observation clearly demonstrates the stability of the selected biomarker genes in human PBMC under baseline conditions.

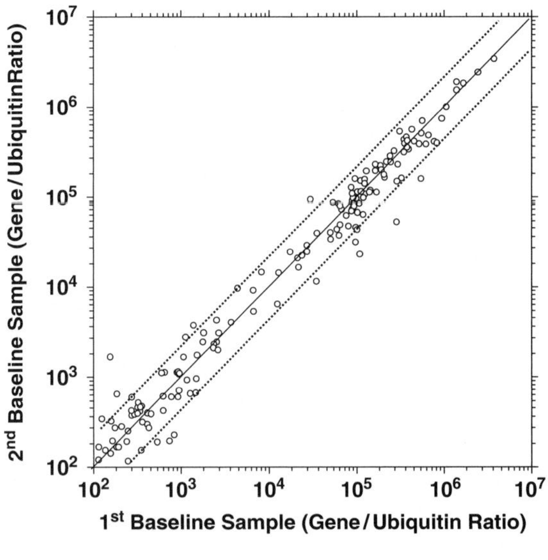

Fig. 3. Stability of IFN-inducible biomarker gene expression levels in human peripheral blood mononuclear cells (PBMC). Biomarker gene/ubiquitin ratios were measured in human PBMC derived from 19 untreated individuals from Dynavax study DV2-NHL-01. Expression levels of eight IFN-α-inducible biomarker genes (2′,5′-OAS, GBP-1, ISG-54, IP-10, IRF-1, IRF-7, MCP-1, MCP-2) and 1 IFN-γ-inducible gene (monokine induced by IFN-γ, MIG) were measured in two baseline PBMC samples taken 1–3 weeks apart. There was a strong correlation between expression levels observed in the first and second sample (correlation coefficient is 0.9706, $p < 0.0001$). Samples between the dotted lines differ <twofold. Data were analyzed with a Spearman correlation test (GraphPad Prism).

3.5.2. Measurement of Biomarker Gene Expression Levels in Clinical Trial Samples

Although a particular biomarker can appear promising under laboratory conditions, only a "real-life" test in a clinical trial setting will reveal whether it provides information that is truly useful. We have selected a series of biomarker genes indicative of TLR-9 stimulation and IFN-α induction, and included several examples of clinical studies in which we measured the expression levels of these biomarker genes (*see* **Figs. 4–6**). In all studies, patients or healthy volunteers were treated with CpG-B oligonucleotide 1018 ISS (5′-TGACTGTGAACGTTCGAGATGA-3′) and IFN-α-inducible biomarker gene expression levels were measured in human PBMC (*see* **Figs.** and **5**), in BAL cells (**Fig. 5**), or in induced-sputum cells (**Fig. 6**). Out of these data sets we conclude that baseline expression levels of our selected biomarker genes are stable and reproducible in untreated patients (**Fig. 3**) and that after in vivo treatment with CpG ODN these biomarker genes can be used to discriminate active from placebo-treated individuals (**Figs. 5** and **6**). We further conclude that different dose levels of drug treatment induce dose-dependent changes in

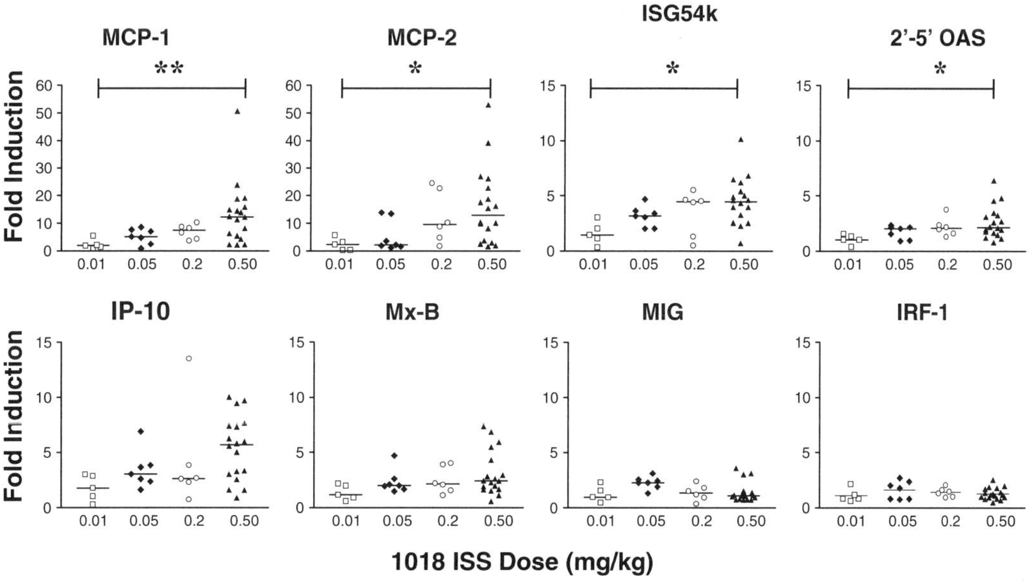

Fig. 4. IFN-α-inducible biomarker gene expression levels are elevated in peripheral blood mononuclear cells (PBMC) from 1018 ISS-treated non-Hodgkin lymphoma patients. Individuals received a subcutaneous injection with CpG-B ODN 1018 ISS as part of Dynavax clinical trial DV2-NHL-01. Blood was taken prior and 24 h after treatment and data were expressed as fold induction after 1018 ISS compared with the matching pre-1018 ISS value. Significance of the differences among groups was analyzed by the Kruskal–Wallis test, with the following values: MCP-1, $p = 0.005$; MCP-2, $p = 0.016$; ISG54, $p = 0.024$; 2 –5 OAS, $p = 0.033$; all other genes, $p > 0.05$. Significant differences between the 0.01 and 0.5 mg/kg groups by Dunn posttest are indicated by ** for $p < 0.01$ and * for $p < 0.05$ (this research was originally published in Blood 2005; 105 *(2)*: 489–495 by Friedberg et al. "Combination immunotherapy with a CpG oligonucleotide (1018 ISS) and rituximab in patients with non-Hodgkin lymphoma: increased interferon-α/β-inducible gene expression, without significant toxicity", © the American Society of Hematology).

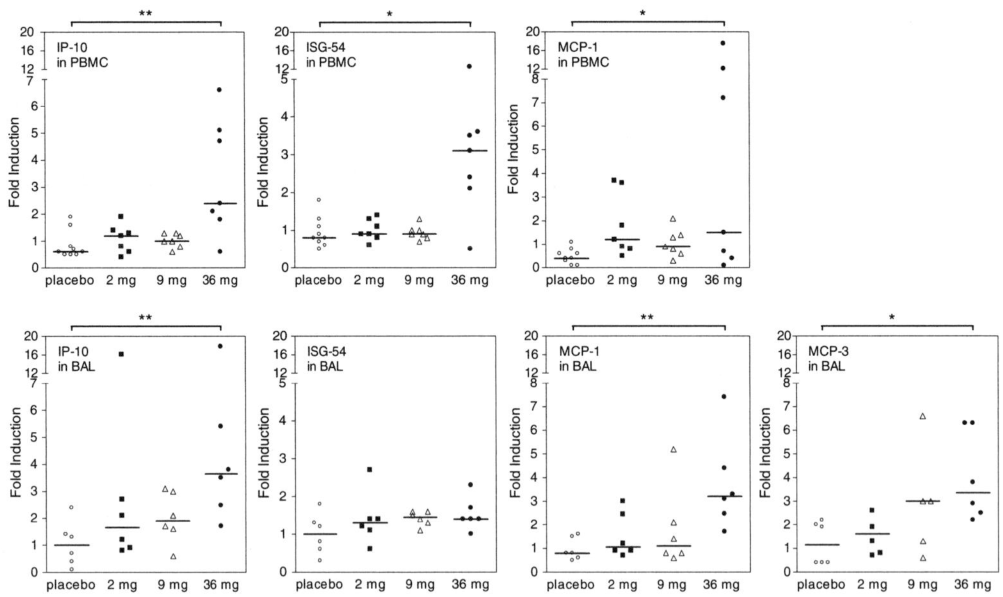

Fig. 5. Dose-dependent increase in IFN-α-inducible biomarker gene expression levels in bronchoalveolar lavage (BAL) cells and peripheral blood mononuclear cells (PBMC) from 1018 ISS-treated healthy volunteers. Healthy individuals received CpG-B ODN 1018 ISS or normal saline by inhalation as part of Dynavax clinical trial DV2-AST-01. BAL cells were taken prior and 6 h after treatment and blood prior and 24 h after treatment. Data were expressed as fold induction after 1018 ISS compared with the matching pre-1018 ISS value. In BAL cells a dose-dependent increase in IP-10, MCP-1, MCP-2, but not in ISG-54, was observed, indicating that the latter gene is not upregulated in airway cells after 1018 ISS inhalation. In contrast, in PBMC a significant increase in the expression level of IP-10, MCP1, and ISG-54 was detected (MCP-2 was not measured). Responses in cells from the airway compartment were more pronounced than in PBMC, which falls in line with the fact that these volunteers were treated via inhalation. Data were analyzed by the Kruskal–Wallis test followed by a Dunn's multiple comparison posttest (GraphPad InStat). Significant differences are indicated by ** for $p < 0.01$ and * for $p < 0.05$.

biomarker gene expression levels which can be readily detected in both healthy and diseased individuals (**Figs. 4** and **5**). And last, robust signals can be detected in cells derived from multiple compartments of the body, including PBMC, BAL cells, and induced-sputum cells (**Figs. 4–6**). In summary, this quantitative PCR method and the selected set of biomarker genes can be successfully applied in clinical development programs in which stimulation of the IFN-α pathway is a key component.

4. Notes

1. We recommend using the RLT Lysis Reagent no longer than 4 weeks. Empirically we established that the lysis capacity of this reagent decreases over time.

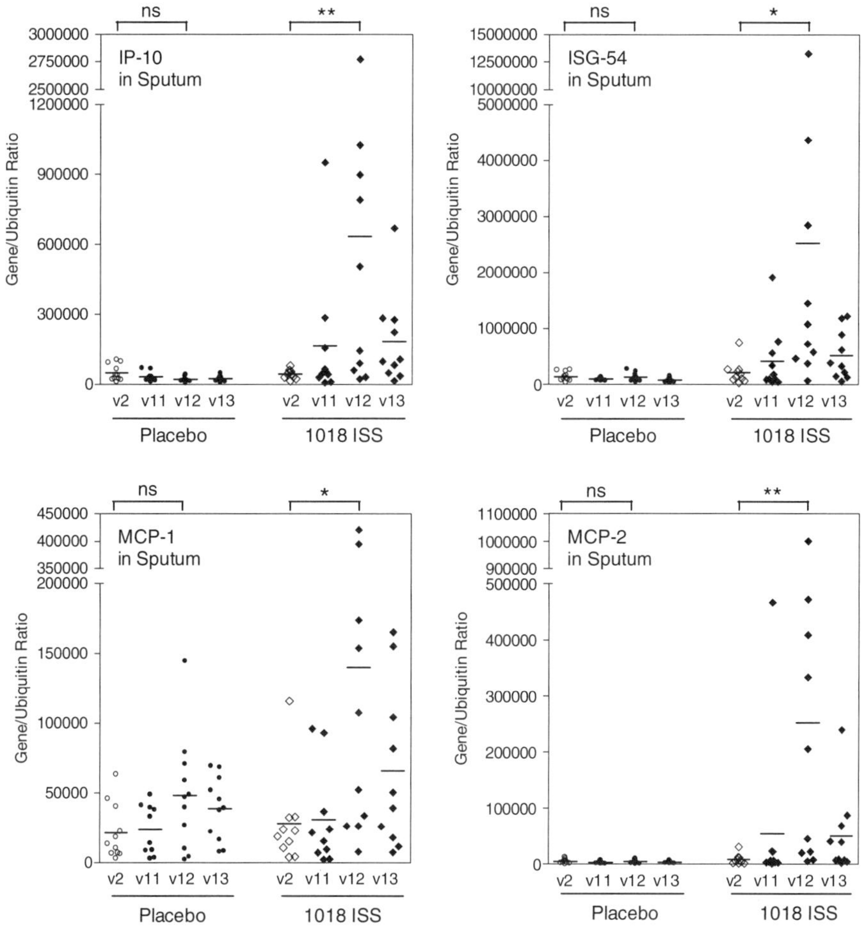

Fig. 6. Inhalation of 1018 ISS increases IFN-α-inducible biomarker gene expression levels in induced-sputum cells from mild asthmatics. Mild to moderate asthmatics received 4 weekly treatments with a dose of 36 mg CpG-B ODN 1018 ISS or normal saline by inhalation as part of Dynavax clinical trial DV2-AST-02. Induced-sputum was collected at patient enrollment (v2), before (v11), and at 30 h (v12), and at 48 h (v13) after the fourth 1018 ISS treatment. Data were expressed as gene/ubiquitin ratio. Biomarker gene expression levels for IP-10, ISG-54, MCP-1, and MCP-2 were significantly increased in induced-sputum cells 30 h after inhalation of 1018 ISS but not placebo. Comparisons between data obtained at enrollment (v2) and at 48 h (v13) revealed no significant differences. The absence of significant differences between the expression levels detectable at enrollment (v2) and at visit 11 (v11, which is 1 week after the last of 3 weekly 1018 ISS inhalations) indicated that all biomarker gene expression levels returned to baseline within a week after 1018 ISS inhalation. Data were analyzed by the Kruskal–Wallis test followed by a Dunn's multiple comparison posttest (GraphPad InStat). Significant differences are indicated by ** for $p < 0.01$ and * for $p < 0.05$.

2. When preparing the RT PCR Mix, we advise to add the Superscript II reverse transcriptase immediately before use.
3. Besides the CPT tube, other methods can be used to isolate PBMC, such as Ficoll-Paque separation, with similar results.
4. The optimal cell number is 1 to 10 million cells. If less than 1 million cells are used, the low copy genes cannot be measured. If more than 10 million cells are used, the cell lysate

becomes viscous and the efficiency of mRNA recovery is adversely affected. In addition, keeping the cell number consistent between samples facilitates later data analysis.

5. Small white clumps/threads may be visible while lysing. We recommend carefully pipetting up and down until all clumps/threads have disappeared.
6. We explored how stable the cell lysates are when frozen in RLT Lysis Reagent and established that one complete thawing cycle of the cell lysate (overnight at room temperature) resulted in good mRNA recovery and the expected gene expression levels (*data not shown*).
7. It is possible for the column to clog during **steps 5–7**. If this happens, transfer the cell extract from the top reservoir of the RNeasy Mini Column to a fresh column and proceed with both columns as described. After **step 16**, pool the two samples, skip **step 17**, and continue with **step 18**.
8. We have compared gene induction in PBMC versus whole blood within the same donor for a total of three donors, and found very comparable fold induction levels for all genes tested (two example genes shown in **Table 2**). The donors were subcutaneously injected with CpG ODN as part of a Dynavax clinical trial (DV2-ONC-01) and blood was taken prior and 24 h after injection. The whole blood method is considerably easier and worth considering if a clinical trial site is not equipped to isolate PBMC.
9. The addition of BSA increases mRNA recovery, in particular in case the amount of mRNA in a sample is low.
10. In spite of repeated efforts to shorten the protocol and to skip the overnight incubation steps (**step 6** of **Subheading 3.3.4** (after the Reverse Transcription), and **steps 4** and **16** of **Subheading 3.4.3** (after cDNA dilution)), empirically we found that each of these incubations is critical in obtaining reproducible results.
11. This is a crucial step in obtaining good results. If a sample set has a widely varying range of cDNA levels, one can correct the gene of interest levels using these ubiquitin values; however, the correction factor will vary greatly between samples and introduce a large source of error. We empirically found it is preferable for a set of samples to have ubiquitin levels within a range of "2x the SD" of the sample set as a whole.
12. To limit interrun and interoperator variation, the baseline cycles and threshold are manually set to predetermined values (e.g. for the MyIQ instruments, the baseline fluorescent reading is determined from cycles 6–15 and the threshold is set at 10).

Table 2
Comparable gene induction measured in PBMC versus whole blood derived from the same donor. Human PBMC were collected in CPT tubes (as per Subheading 3.1.1) and whole blood was collected in PAXgene tubes (as per Subheading 3.1.2) prior to and 24 h postsubcutaneous injection of CpG ODN in Dynavax study DV2-ONC-01. mRNA was isolated from CPT and PAXgene tubes as described in Subheading 3.2.2 and 3.2.3, respectively, and reversed transcribed to cDNA. Gene expression levels for IP-10 and ISG-54 were measured and the gene/ubiquitin ratio (Ratio) calculated. The fold induction (F/I) was calculated by dividing the gene/ubiquitin ratio at 24 h by the preinjection gene/ubiquitin ratio. The data indicate that whether CPT or PAXgene tubes are used for sample collection, very comparable fold induction levels are obtained

		IP-10				ISG-54			
		PBMC		Whole blood		PBMC		Whole blood	
Subject	Time	Ratio	F/I	Ratio	F/I	Ratio	F/I	Ratio	F/I
1	Pre	12	1	41	2	1,829	0	19,004	1
	24 h	14		70		790		17,788	
2	Pre	20	109	37	19	998	25	18,645	12
	24 h	2,175		687		25,310		232,785	
3	Pre	51	42	50	17	4,056	9	9,985	18
	24 h	2,153		838		34,663		183,202	

13. It increases the quality of the data if samples that are part of one experimental group are processed simultaneously as much as possible and kept on the same 96-well plate.
14. To prevent sample contamination and to increase sample recovery we recommend a 1 min centrifugation of the cDNA sample blocks before use. It is important to use an appropriately balanced counterweight as full sample blocks are at risk of cracking during centrifugation.
15. Pipetting technique is critical at this step to obtain repeatable results. Do not prewet the pipette tips. Aspirate 10 μL of sample from the from the working cDNA block. If there was any error in volume, expel the sample back into the block and discard the tips. To expel the sample into the PCR plate, lightly touch the tips to the top of the primer mix and expel. Do not mix by pipetting up and down. Discard tips between duplicates.

Acknowledgments

The authors would like to thank Mabel Chu, Kelly Beemer, Debbie Higgins, Daniela Bumbaca, Ingrid Chen, Emily Calimquim, and Nick Duran for their invaluable help during the development of this method; Franck J. Barrat, Omar Duramad, and Jennifer O. Lizcano for their contribution to the experiments shown in **Fig. 1**; and Linh Chu for help in the preparation of this manuscript.

References

1. Kanzler, H., Barrat, F. J., Hessel, E. M., and Coffman, R. L. (2007) Therapeutic targeting of innate immunity with Toll-like receptor agonists and antagonists. *Nat Med* 13, 552–9.
2. Krieg, A. M. (2007) Development of TLR9 agonists for cancer therapy. *J Clin Invest* 117, 1184–94.
3. Friedberg, J. W., Kim, H., McCauley, M., Hessel, E. M., Sims, P., Fisher, D. C., Nadler, L. M., Coffman, R. L., and Freedman, A. S. (2005) Combination immunotherapy with a CpG oligonucleotide (1018 ISS) and rituximab in patients with non-Hodgkin lymphoma: increased interferon-alpha/beta-inducible gene expression, without significant toxicity. *Blood* 105, 489–95.
4. Krieg, A. M. (2007) Antiinfective applications of toll-like receptor 9 agonists. *Proc Am Thorac Soc* 4, 289–94.
5. Creticos, P. S., Schroeder, J. T., Hamilton, R. G., Balcer-Whaley, S. L., Khattignavong, A. P., Lindblad, R., Li, H., Coffman, R., Seyfert, V., Eiden, J. J., and Broide, D. (2006) Immunotherapy with a ragweed-toll-like receptor 9 agonist vaccine for allergic rhinitis. *N Engl J Med* 355, 1445–55.
6. Gauvreau, G. M., Hessel, E. M., Boulet, L. P., Coffman, R. L., and O'Byrne, P. M. (2006) Immunostimulatory sequences regulate interferon-inducible genes but not allergic airway responses. *Am J Respir Crit Care Med* 174, 15–20.
7. Janeway, C. A., Jr., and Medzhitov, R. (2002) Innate immune recognition. *Annu Rev Immunol* 20, 197–216.
8. Akira, S., and Hemmi, H. (2003) Recognition of pathogen-associated molecular patterns by TLR family. *Immunol Lett* 85, 85–95.
9. Duramad, O., Fearon, K. L., Chan, J. H., Kanzler, H., Marshall, J. D., Coffman, R. L., and Barrat, F. J. (2003) IL-10 regulates plasmacytoid dendritic cell response to CpG-containing immunostimulatory sequences. *Blood* 102, 4487–92.
10. Heil, F., Hemmi, H., Hochrein, H., Ampenberger, F., Kirschning, C., Akira, S., Lipford, G., Wagner, H., and Bauer, S. (2004) Species-specific recognition of single-stranded RNA via toll-like receptor 7 and 8. *Science* 303, 1526–9.
11. Kadowaki, N., Ho, S., Antonenko, S., Malefyt, R. W., Kastelein, R. A., Bazan, F., and Liu, Y. J. (2001) Subsets of human dendritic cell precursors express different toll-like receptors and respond to different microbial antigens. *J Exp Med* 194, 863–9.
12. Lund, J., Sato, A., Akira, S., Medzhitov, R., and Iwasaki, A. (2003) Toll-like receptor 9-mediated recognition of Herpes simplex virus-2 by plasmacytoid dendritic cells. *J Exp Med* 198, 513–20.
13. Ito, T., Kanzler, H., Duramad, O., Cao, W., and Liu, Y. J. (2006) Specialization, kinetics, and repertoire of type 1 interferon responses by human plasmacytoid predendritic cells. *Blood* 107, 2423–31.

Index

G

H

I

J

K

L

Q

R

S

T

U

V

W

X

Z